Handbook of Research on Environmental Policies for Emergency Management and Public Safety

Augustine Nduka Eneanya
University of Lagos, Nigeria

A volume in the Advances in Wireless Technologies and Telecommunication (AWTT) Book Series

IGI Global
DISSEMINATOR OF KNOWLEDGE

Published in the United States of America by
IGI Global
Engineering Science Reference (an imprint of IGI Global)
701 E. Chocolate Avenue
Hershey PA, USA 17033
Tel: 717-533-8845
Fax: 717-533-8661
E-mail: cust@igi-global.com
Web site: http://www.igi-global.com

Library of Congress Cataloging-in-Publication Data

Names: Eneanya, Augustine Nduka, editor.
Title: Handbook of research on environmental policies for emergency management and public safety / Augustine Nduka Eneanya, editor.
Description: Hershey PA : Engineering Science Reference, [2018]
Identifiers: LCCN 2017016115| ISBN 9781522531944 (hardcover) | ISBN 9781522531951 (ebook)
Subjects: LCSH: Environmental policy--Cross-cultural studies. | Environmental management. | Emergency management. | Public safety.
Classification: LCC GE170 .H3665 2018 | DDC 363.34028/6--dc23 LC record available at https://lccn.loc.gov/2017016115

This book is published in the IGI Global book series Advances in Wireless Technologies and Telecommunication (AWTT) (ISSN: 2327-3305; eISSN: 2327-3313)

British Cataloguing in Publication Data
A Cataloguing in Publication record for this book is available from the British Library.

Advances in Wireless Technologies and Telecommunication (AWTT) Book Series

Xiaoge Xu
Xiamen University Malaysia, Malaysia

ISSN:2327-3305
EISSN:2327-3313

Mission

The wireless computing industry is constantly evolving, redesigning the ways in which individuals share information. Wireless technology and telecommunication remain one of the most important technologies in business organizations. The utilization of these technologies has enhanced business efficiency by enabling dynamic resources in all aspects of society.

The **Advances in Wireless Technologies and Telecommunication Book Series** aims to provide researchers and academic communities with quality research on the concepts and developments in the wireless technology fields. Developers, engineers, students, research strategists, and IT managers will find this series useful to gain insight into next generation wireless technologies and telecommunication.

Coverage

- Wireless Technologies
- Radio Communication
- Digital Communication
- Cellular Networks
- Network Management
- Wireless sensor networks
- Mobile Technology
- Global Telecommunications
- Virtual Network Operations
- Mobile Web Services

IGI Global is currently accepting manuscripts for publication within this series. To submit a proposal for a volume in this series, please contact our Acquisition Editors at Acquisitions@igi-global.com or visit: http://www.igi-global.com/publish/.

The Advances in Wireless Technologies and Telecommunication (AWTT) Book Series (ISSN 2327-3305) is published by IGI Global, 701 E. Chocolate Avenue, Hershey, PA 17033-1240, USA, www.igi-global.com. This series is composed of titles available for purchase individually; each title is edited to be contextually exclusive from any other title within the series. For pricing and ordering information please visit http://www.igi-global.com/book-series/advances-wireless-technologies-telecommunication/73684. Postmaster: Send all address changes to above address.

Titles in this Series

For a list of additional titles in this series, please visit: www.igi-global.com/book-series

Advanced Mobile Technologies for Secure Transaction Processing Emerging Research and Opportunities
Raghvendra Kumar (LNCT Group of Colleges, India) Preeta Sharan (The Oxford College of Engineering, India) and Aruna Devi (Surabhi Software, India)
Information Science Reference • copyright 2018 • 177pp • H/C (ISBN: 9781522527596) • US $130.00 (our price)

Examining Developments and Applications of Wearable Devices in Modern Society
Saul Emanuel Delabrida Silva (Federal University of Ouro Preto, Brazil) Ricardo Augusto Rabelo Oliveira (Federal University of Ouro Preto, Brazil) and Antonio Alfredo Ferreira Loureiro (Federal University of Minas Gerais (UFMG), Brazil)
Information Science Reference • copyright 2018 • 330pp • H/C (ISBN: 9781522532903) • US $195.00 (our price)

Graph Theoretic Approaches for Analyzing Large-Scale Social Networks
Natarajan Meghanathan (Jackson State University, USA)
Information Science Reference • copyright 2018 • 355pp • H/C (ISBN: 9781522528142) • US $225.00 (our price)

Powering the Internet of Things With 5G Networks
Vasuky Mohanan (Universiti Sains Malaysia, Malaysia) Rahmat Budiarto (Albaha University, Saudi Arabia) and Ismat Aldmour (Albaha University, Saudi Arabia)
Information Science Reference • copyright 2018 • 304pp • H/C (ISBN: 9781522527992) • US $215.00 (our price)

Routing Protocols and Architectural Solutions for Optimal Wireless Networks and Security
Dharm Singh (Namibia University of Science and Technology, Namibia)
Information Science Reference • copyright 2017 • 277pp • H/C (ISBN: 9781522523420) • US $205.00 (our price)

Sliding Mode in Intellectual Control and Communication Emerging Research and Opportunities
Vardan Mkrttchian (HHH University, Australia) and Ekaterina Aleshina (Penza State University, Russia)
Information Science Reference • copyright 2017 • 128pp • H/C (ISBN: 9781522522928) • US $140.00 (our price)

Emerging Trends and Applications of the Internet of Things
Petar Kocovic (Union – Nikola Tesla University, Serbia) Reinhold Behringer (Leeds Beckett University, UK) Muthu Ramachandran (Leeds Beckett University, UK) and Radomir Mihajlovic (New York Institute of Technology, USA)
Information Science Reference • copyright 2017 • 330pp • H/C (ISBN: 9781522524373) • US $195.00 (our price)

IGI Global
DISSEMINATOR OF KNOWLEDGE

701 East Chocolate Avenue, Hershey, PA 17033, USA
Tel: 717-533-8845 x100 • Fax: 717-533-8661
E-Mail: cust@igi-global.com • www.igi-global.com

Editorial Advisory Board

List of Reviewers

Table of Contents

Section 5
Ecosystem Management, Environmental Media, and Public Health

Detailed Table of Contents

This chapter examines both concept of global hazard and disaster and its management in the lights of its vulnerability. It categorized the different types of hazards and disasters and their components. From the research findings, it is observed that hazards and disaster are two sides of a coin. They occur at the interface between human systems and natural events in our physical environments. This chapter highlights that the major environmental changes driving hazards and vulnerability to disasters are climate change, land-use changes, and degradation of natural resources. After presenting a typology of disasters and their magnitude globally, management of disaster has transited from just being a "response and relief"-centric approach to a mitigation and preparedness approach.

Internally displaced persons (IDPs) are men, women, and children who are uprooted from their ancestral homes as victims of natural disaster or manmade occurrences for reasons often beyond their control and comprehension. The Boko-Haram insurgency in the North-East of Nigeria has caused over two million Nigerians to be internally displaced. The crisis has created management problems for the National Emergency Management Agency (NEMA) due to paucity of funds. Secondary data and human needs theory were used for the analysis. The incessant use of improvised electronic devices has created a security lacuna in the IDPs camps, which have become targets for terrorists. The concomitant is confidence deficit between the IDPs and NEMA resulting in accusations of neglect and corruption. The chapter concludes that the failure to manage IDPs is tantamount to human rights abuse and security lapse.

Countries with better disaster management plans can better signify their existence. The earthquake is one such hazardous catastrophe that leads to heavy damage. Disaster management does not avert or eliminate the threats fully; instead, it focuses on use of technology, creating plans to lessening the effect of disasters up to significant level. Depending on the magnitude and type of waves, earthquakes crush everything, from buildings to small houses, from towers to trees, and from roads to bridges. Tsunamis and landslides follow its way as well. After the earthquake comes, the settling of things back to normal involves a lot of time, money, and manpower as well. Hence, the only solution of this is leading-edge technology support. Sensor network technology is one of the proven techniques for not only varieties of early warning, prediction, and detection systems but also post-phenomenon rescue and support. This chapter will help researchers and policy makers to think fresh on disaster management with sensor network technology.

Section 4
Climate Change and Disaster Vulnerability Management

Natural hazards are potential threats to lives, properties, and the society at large. This chapter discusses psychological resilience as a strategy in overcoming the aftermath of natural hazards, which are known as natural disasters in Nigeria. Resilience consists of individual behaviours, cognitions, actions, and attitudes that enhance personal wellbeing. The ability to exhibit resilience is important in overcoming the adverse consequences of natural hazards. A review of the concept of resilience in this chapter covers an introduction to the concept of resilience and natural hazards, cases of natural hazards in Africa and Nigeria, conceptual definition and issues relating to resilience, theoretical underpinning, recommendation, and conclusion. The various factors that are critical to the development and utilisation of psychological resilience are discussed. Relevant authorities can thus be advised on the need to set up platforms on which communities can devise methods of educating individuals on methods of developing resilience in preparation for impending natural hazards.

Using macro-, exo-, meso-, and micro-level framework for assessing interdependencies in crisis management system in Poland, this chapter addresses expansion of planning and preparedness to novel hazards and threats such as precipitated mass movements of populations due to environmental changes. The Ukrainian refugee crisis served as backdrop for Poland's crisis management system testing. Benefits of planning and preparedness for mass displacements highlighted (a) an emerging all-hazard approach to crisis management, (b) proactive threat identification and assessment, (c) issuance of acts, laws, and regulations based on novel threats, (d) comprehensive multi-jurisdictional exercises, and (e) increasing importance of mutual aid agreements.

In recent years, environmental security in the coastal Niger Delta has had a growing scholarly interest from divergent perspectives seeking for broader elucidation and understanding of State policy response. These security threats notably oil spill, water and land pollution, gas flaring, acid rain, mangrove deforestation, etc are linked to the Multinational Oil Corporations (MNOCs)and oil resource exploitation. These have been perverse, resulting in ecological breakdown, vulnerability, emergency and environmental insecurity challenges since at least the 1970s when oil in the region became the main stay of Nigeria's economy. This conceptual paper builds on the political ecology framework which discusses the impact of global power asymmetry on natural resource extraction and extensive body of work in the broad field of environmental security to explore salient indicators which demonstrate the evidence of environmental insecurity threats and poor State policy response and made some recommendations.

The chapter assesses the nature and effects of climate change on sustainable development in Nigeria. It observes that climate change has a multifarious effect not only on the environment but also on the socio-economic life of the people. Therefore, if sustainable development is to be realized in the country, climate change should be properly managed through extant public policies. However, if government track records on policies and program implementations is anything to go by, the future of sustainable development is gloomy. This makes an interrogation of the interface between climate change and sustainable development germane. The chapter observes that while government has shown willingness to manage climate change for the sustainability of the environment and its people, certain limitations stand on its path. These include poor policy or program implementation, inadequate funding of climate change management, poor sensitization program on environment management, among others. However, there are opportunities that can be harvested at the state and individual levels.

Research on the impact of climate change on agriculture and food security is important, especially in the agricultural economies, not only to know the severity of impact but also the policies to be adapted to halt climate change and the technology to be used to mitigate the impact of climate change. The study was conducted in Kapiri Mposhi district of Central Province in Zambia to find out the impact of climate change on agriculture and food security. The objectives of study include to know the intensity of climate change and its impact on area under cultivation, late sowing of seed and damage of seed due to lack of water, fertilizer absorption reduction, food shortage, livestock, and productivity. The chapter also focuses on the sources of credit to the farmers.

This chapter includes an assessment of physical vulnerability of the coast, including a coastal vulnerability index composed of 9 physical variables—elevation, distance to shore, tide amplitude, significant wave weight, erosion/accretion rates, geology, geomorphology, ground cover vegetation, and anthropogenic actions—followed by a quantification of coastal recession and the data of special report on emissions scenarios (SRES) developed by the Intergovernmental Panel on Climate Change (IPCC) on the rise in average sea level. It includes an estimate of the economic value of an area of recreation based on the travel cost method. Finally, a bibliographic review is made to assess strategies and responses to the impacts of sea level rise in order to make comparisons and to develop a road map of interventions for shoreline protection. The proposed methodology was applied to a case study on the Portuguese coast corresponding to the beaches of Costa de Caparica, Almada.

Foreword

Public concern for the environment has been a catalyst for profound changes and focus on environmental policy and emergency management. Humanity faces an unprecedented challenge as our population grows, while earth and its capacity to support it does not. People around the world aspire for better lives for themselves and their children: food, shelter, a safe and healthy environment. As industries or corporations strive to produce more goods, farmers grow more crops, and human demands on forests, fields, rivers, transportation and oceans rise, it becomes inevitable for leaders in federal, state and local governments to develop environmental policies that will create, plan, respond, recover and sustain environment that supports a dignified, peaceful and equitable existence.

Climate change is one of the greatest environmental challenges in 21st Century, which calls for a global solution. Ozone depletion, global warming, hurricane, flooding, earthquake and nuclear disasters and other environmental concerns have the potential for adverse effects on a scale and of a duration previously unknown. Human activities are largely responsible for the changes in the atmospheric composition and the impacts are felt across the globe and are affecting patterns of life and living conditions of the people.

The impact of extreme natural disasters globally is equivalent to $520 billion loss in annual consumption, and forces some 26 million people into poverty each year, according to recent report from the World Bank and the Global-Facility for Disaster Reduction and Recovery (GFDRR). Under the Paris Climate Agreement of 2016, more than 190 countries are committed to cutting greenhouse-gas emissions in an effort to combat climate change. The agreement aims to keep average global temperatures from rising more than 2 degrees Celsius or 3.6 degrees Fahrenheit, above pre-industrial levels.

For effective emergency, disaster response awareness is necessary among citizens. Disasters cannot be stopped from occurring but we can be prepared for disasters by making proactive action plans to minimize loss of lives and properties. The aim of an emergency and disaster management, therefore, is to establish organizational structures and procedures for immediate response to major emergencies, in order to maintain life, improve health and support morale of the affected population until more permanent and sustainable solutions can be found.

This book is a compendium of well-researched articles on vexing issues confronting environmental policies for emergency management and public safety. It tackles environmental issues, problems, challenges and gives suggestions on how to mitigate and resolve the naughty issues besetting the quest for global environmental sustainable development. The reading public, consultants, students, scholars, public agencies, nonprofit organization and private stakeholders will find these narratives, empirical findings and discussions too attractive to resist reading them.

Samuel Iyiola Oni
University of Lagos, Nigeria

Preface

Humanity faces an unprecedented challenge as our population grows, while earth and its capacity to support the people does not. People across the globe aspire to better lives for themselves and for their children, food, shelter, a safe and healthy environment, education, jobs and other material needs and conveniences. Industries strive to produce more goods; farmers to grow more crops; and human demands on forests, fields, rivers, and oceans increase. The global challenge is to create a future in which prosperity and opportunities increase while life flourishes and pressures on oceans, earth, and atmosphere-the biosphere-diminish; to create a life sustaining earth that supports a dignified, peaceful, and equitable existence.

With shifting seasons, sea levels rising, high temperatures, changing landscapes, increased risk of drought, fire and floods, stronger storms, more heat-related illness and diseases and economic losses, the effects of climate change has become too glaring to deny. Climate change is one of the greatest global challenges of the 21st Century. Its threat is not only global but also multidimensional, invisible, unpredictable and transcends national borders. Climate change leads to changes in temperature and precipitation that are expected to reduce global crop productivity, cause changes in food production and consumption and affect global population health by changing the composition of diets and, with it, the profile of dietary and weight-related risk factors and associated mortalities.

Scientists have long shown that though the reasons for climate change are natural but most of the current changes are said to be man-made and the biggest culprit by far is burning of fossil fuels. However, disasters, both natural and man-made have left devastating impacts across the world. Recent history reminds us of several significant natural disasters that have adversely affected many communities and left them reeling for months and years. A few example include Superstorm Sandy (New York and New Jersey), several tornadoes (communities through the Midwest), Hurricane Katrina (the Gulf Coast), Hurricane Andrew (Florida), Hurricane Harvey (Houston, Texas), Hurricane IRMA (Caribbean island of Barbuda, Miami, South Carolina,

Florida, Cuba, etc.) and earthquake (Southern Mexico). These disasters left the communities involved without electricity, death of loved ones, and loss of properties.

Government at all levels are charged to prepare and react appropriately when they know natural disasters-such as hurricanes, tornadoes, earthquakes- will impact their respective boundaries. No level of preparation can mitigate the damage done to the communities after the storms dissipate. Hence, emergency planning should aim to prevent emergencies from occurring and if fails, a good action plan to mitigate the results and effects.

Disaster management (or emergency management) is the creation of plans through which communities reduce vulnerability to hazards and cope with disasters. Disaster management does not avert or eliminate the threats; instead, it focuses on creating plans to decrease the effect of disasters. Failure to create a plan could lead to damage to assets, human mortality, and lost revenue.

CHALLENGES OF EMERGENCY MANAGEMENT AND ENVIRONMENTAL CONTROL

Environmental problems and challenges are diverse. No single and sovereign explanation for a nation's continuing environmental problems will suffice. Most ecological ills are the product of several factors in different and often interacting combinations. They include the following:

1. Public managers in emergency management organizations must create and maintain an internal environment that embraces internal controls and ethical behavior to ensure that organization operates in a way that is both accountable and transparent to those it serves.

However, common challenges often identified focused on issues related to the control environment. This involves policies and procedures within emergency management organizations. They include: operational processes are not well-defined, documented and monitored, and weaknesses of internal control such as: absence of strategic plan, absence of sufficient and appropriate oversight of operating activities and resource management, inadequate documentation of the sub-recipient award process including the closeout of the grant files. Accountability about funds awarded to sub-recipients affected by disasters is weak. Emergencies do not excuse accountability.

2. Another ecological challenge is climate change. It is the greatest long-term threat facing the world in wake of droughts, flooding, storms, earthquake, hurricanes, and heat waves that have struck several countries. Climate change is one of the biggest global challenges of this century which calls for a global solution.
3. Furthermore, while practically every important environmental ill has been targeted by a major federal law, delay and difficulty in program implementation routinely impede enforcement. Implementation is frequently frustrated by decades of deficient agency staffing or budget. Enforcement of most environmental legislation also depends upon voluntary compliance by regulated interests, public and private, who frequently fail to comply. Besides, even though domestic laws and regulations may make impact on solving environmental challenges, but not with international law. International law is more than legal framework, treaty-making and institutional building. International law requires compliance and enforcement. In international regime, the question of compliance presents another challenge due to the absence of a centralized enforcement regime. Compliance goes beyond implementation. It connotes whether countries in fact adhere to the agreement's provisions and to the implementing measures that they have instituted. Some obligations are procedural; others are substantive. The implementation of substantive agreements matter, such as reducing or phasing out ozone depleting substances. Though, a country may comply with an agreement but not effective in achieving its objectives.

4. Moreover, environmental degradation, more frequent extreme weather events and resource scarcity can be expected to displace large numbers of environmental refugees within countries and across borders. Increasing desertification is expected to cause the migration of millions of people from Sub-Saharan Africa to Europe, leading to population expansion which would diminish the effectiveness of pollution controls over time.
5. Finally, recent discoveries of acid rain and global ozone depletion demonstrate that nature can surprise science. Natural systems may absorb stress for long periods without overt evidence of damage. Ozone depletion, global warming, a nuclear disaster and other modern environmental concerns have the potential for adverse effects on a scale and of a duration previously unknown. Although, not all modern environmental problems present potentially catastrophic effects, organic chemicals can pose such problems. No doubt, such recent discoveries such as acid rain, global climate warming, and upper atmosphere ozone depletion will soon become areas of policy direction. Environmental scientists are now comprehending the global scale of ecological processes and are now educating policy makers to perceive these problems in a world context.

SEARCHING FOR A SOLUTION

In response to these challenges, environmental experts have observed that every significant environmental problem is fixed in a matrix of ecological, economic, political and scientific causes and consequences that usually frustrate quick and simple solutions. This reality has down on environmentalists that technology cannot produce timely solutions to major environmental ills. Many ecological activists expected unreasonably quick and thorough environmental restoration. Much of this early optimism has been dissipated by experience with environmental regulation over the last twenty years, contemporary environmental movement is generally more realistic and cautious in its expectations about policy accomplishments.

1. The search for solution on climate change has led to series of international treaty in Montreal Protocols in 1989, amendments by delegates in Rwanda meeting and Paris

Agreement on April 22, 2016. Specifically, representatives from 195 countries gathered at the 21st Conference of Parties of the United Nations Framework Convention on Climate Change (UNFCCC) in Paris and adopted by consensus on December 12, 2015 to limit global average temperature to well below 2°C above pre-industrial levels and to pursue efforts to limit the temperature increase to 1.5°C above pre-industrial levels Data on UNFCC indicates that 61 countries have ratified the Paris Agreement accounting for 47.79% of global GHG emission. The objective is to level off Greenhouse Gas (GHG) emissions as soon as possible. This effort and others are meant to mitigate and adapt to the challenges of climate change.

2. Scientists and researchers have also advocated that geoengineering the environment would help emission control. Geoengineering is a deliberate and large-scale intervention in earth's climatic system with the aim of reducing global warming. Climate engineering has two categories of technologies-carbon dioxide removal and solar radiation management. Carbon dioxide removal addresses a cause of climate change by removing one of the greenhouse gases from the atmosphere. Solar radiation management attempts to offset effects of greenhouse gases by causing Earth to absorb less solar radiation.

3. Emergency responder organization should establish effective internal control in their policies and procedures. This would enable organizations faced with responding to emergencies affecting public safety and health document administrative actions taken using emergency procedures.
4. Finally, the growth of environmental legislation in the last two decades is evidence of the nation's increasing concern with science and technology since world war two. There have been successes in domestic environmental law and regulations in many countries of the world. The ratification and application of Paris Agreement by many countries is a step towards searching for solution.

ORGANIZATION OF THE BOOK

The book is organized into seventeen chapters. A brief description of each of the chapters are follows:

Chapter 1 reviews the issue of natural resources management. The author argues that excessive and wrong use of natural resources leads to occurrence of various human-induced disasters. Global warming, deforestation, floods, air pollution, loss of biological diversity are some of such threats that can be treated within the framework of emergency management. Minimization of human-induced disasters and prevention of such disasters can only be achieved by means of efficient and sustainable management of natural resources.

Chapter 2 assesses some aspects of natural resources exploitation and management in Nigeria. The author argues that natural resources provide the all-important basis for human existence and civilization. Natural resources management have several implications for human security and public safety. Poorly managed and severely degraded natural resources systems will interfere with other systems such as the climate to produce resource conflicts and exacerbate human insecurity and undermine public safety. Africa is highly vulnerable to natural resource degradation and its concomitant impacts, a situation that has deepened the existing poverty-environment-human security linkages. This chapter explores the dimension of natural resources and natural resource management, its implications on livelihoods and conflicts, human security and public safety and its contexts in Africa in general with particular reference to Nigeria.

Chapter 3 explains the implementation of green policy and community participation. Community participation has become important in government, policy makers and environmentalists over last few decades. The author argues that Implementation of green policy and community participation programs through representatives - organization, groups of individual enhances the benefits of polices and program and adds value to policy as well as making the policy's results and responses more effective and stronger.

Chapter 4 discusses the issues of leadership, public values and trust in emergency management. The chapters argues that managers build trust into decision making during emergencies. When managers set up the right conditions for the team to thrive, it results a better outcome. The chapters further argues that when managers share information both up and down the chain of command and make their intention clear about what winning looks like, team members are able to use their own discretion and make decisions that support the mission. The chapter concludes that an effective expression by the manager's intent must be clear and concise of what the team must do to succeed and achieve the desired end state. The manager must build trust into decision-making among the team to achieve better result in emergency situations.

Chapter 5 reviews the issue of global natural hazard and disaster vulnerability management. The chapter examined both concepts of Global hazard and disaster and its management in the lights of its vulnerability. It categorized the different types of hazards and disaster and its components. This chapter highlights that the major environmental changes driving hazards and vulnerability to disasters are climate change, land-use changes and degradation of natural resource. After presenting a typology of disasters and its magnitude globally, the authors argue that management of disaster has transited from just being "response and relief" centric approach to mitigation and preparedness approach.

Chapter 6 discusses internally displaced persons and emergency management in Nigeria. The author argues that the Boko-Haram insurgency in the North-East of Nigeria, has caused over two million Nigerians to be internally displaced. The crisis has created management problem for the National Emergency Management Agency (NEMA) due to paucity of funds. The Chapter concludes that failure to manage IDPs is tantamount to human rights abuse and security lapse.

Chapter 7 analyzes the prospect of low power sensor network technology to disaster management for sustainable future. The author argues that it focuses on use of technology, creating plans to lessening the effect of disasters up to significant level. The chapter concludes that Sensor network technology is one of the proven techniques for not only varieties of early warning, prediction and detection systems but also post phenomenon rescue and support. This chapter will help researchers and policy makers to think afresh on disaster management with sensor network technology.

Chapter 8 presents a review of psychological resilience as a response to natural hazards in Nigeria. Natural hazards are potential threats to lives, properties and the society at large. The chapter discusses psychological resilience as a strategy in overcoming the aftermath of natural hazards, which are known as natural disasters in Nigeria. Resilience consists of individual behaviors, cognitions, actions and attitudes that enhance personal wellbeing. The ability to exhibit resilience is important in overcoming the adverse consequences of natural hazards. The author further argues that various factors are critical to the development and utilization of psychological resilience.

Chapter 9 presents analysis of planning and preparing for refugee exodus in crisis, using Poland as a case study. Using macro, exo, meso, and micro-level framework for assessing interdependencies in crisis management system in Poland, the chapter addresses expansion of planning and preparedness to novel hazards and threats such as precipitated mass movements of populations due to environmental changes. The Ukrainian refugee crisis served as backdrop for Poland's crisis management system testing. The author highlights the benefits of planning and preparedness for mass displacements. They include: (a) an emerging all-hazard approach to crisis management; (b) proactive threat identification and assessment; (c) issuance of acts, laws, and regulations based on novel threats; (d) comprehensive multi-jurisdictional exercises; and (e) increasing importance of mutual aid agreements.

Chapter 10 addresses the issue of environmental security and policy response in the Niger Delta Region of Nigeria between 1990 and 2016. The author argues that security threats notably water and land pollution, gas flaring, oil spillage, acid rain, deforestation, coastal erosion and others have been perverse, resulting in vulnerability and emergency challenges since at least the 1970s when oil in the region became the main stay of Nigeria's economy.

Chapter 11 addresses the issue of climate change and sustainable development in Nigeria. The chapter assesses the nature and effects of climate change on sustainable development in Nigeria. It observed that climate change has multifarious effect not only on the environment but socio-economic life of the people. Therefore, if sustainable development is to be realized in the country climate change should be properly managed through extant public policies. The chapter argues that while government has shown willingness to manage climate change for the sustainability of the environment and its people, certain limitations stand on its path. These include: poor policy or program implementation, inadequate funding of climate change management, poor sensitization program on environment management, among others. The chapter concludes that there are still opportunities that can be harvested at the state and individual levels.

Chapter 12 analyzes the impact of climate change on agriculture and food security. The author argues that climate change reduces rain days, delays sowing of seed, reduces fertilizer absorption to the plant, area under cultivation and production per hectare. Climate change had negative impact on food security and livestock. The study also revealed that the sources of finance to the farmers were limited. The author suggests that the solutions to this problem includes: development of new varieties of crops which mature faster and are heat resistant ; provision of rural credit facility; investment on research for climate resistant crops varieties, water harvesting and irrigation schemes; provision of alternative employment opportunities in non-farming activities; allocation of adequate funds to meteorological department to increase its capacity in climate data collection, storage, analysis, forecasting and dissemination to the farmers in their local languages.

Chapter 13 presents analysis of impact of sea level rise on coastal regions and strategic responses. The author explains that the methodology includes the assessment of physical vulnerability of the coast, including a coastal vulnerability index composed by 9 physical variables – elevation, distance to shore, tide amplitude, significant wave weight, erosion/accretion rates, geology, geomorphology, ground cover vegetation and anthropogenic actions - followed by a quantification of coastal recession, and the data of Special Report on Emissions Scenarios (SRES) developed by the Intergovernmental Panel on Climate Change (IPCC), on the rise in average sea level. It includes an estimate of the economic value of an area of recreation, based on the travel cost method. Finally, a bibliographic review will be made to assess strategies and responses to the impacts of sea level rise in order to make comparisons and to develop a road map of interventions for shoreline protection. The proposed methodology was applied to a case study in the Portuguese coast, corresponding to the beaches of Costa de Caparica, Almada.

Chapter 14 reviews the issue of land use policy and urban sprawl in Nigeria. Land use policy is very central to the development of urban life and the emergence of cities. The author argues that in many developed capitalist societies, both the planning and expansion of the cities are usually anchored on sustainable urban land policies such that the growth of urban sprawl is effectively controlled. In most developing countries, land use policies are not only disparate, they are usually not connected to the growth of cities because policy makers are after the money they could make from private investors. This chapter argues that though, the coming of the Land Use Act, 1978 ended the era of disparate land law regimes, it has over the years, sealed the control of urban lands to State Governors, a development, which has given massive corruption and arbitrariness in the allocation and utilization of urban lands. This approach to land administration has also hindered effective and sustainable urban and regional planning in many Nigerian cities. This chapter suggests the review of the 1978 Land Use Act and effective utilization of modern technologies in the monitoring of urban sprawls.

Chapter 15 assesses the impact of transportation on climate change, using North-East Region in India as case study. The author argues that India's North-East Region has greater demand for road and personalized modes of transport powered by fossil fuels. And due to emissions there has been evidence of climate change. It has been found that diesel cars cause greater emissions (per kilometer travelled) as compared to petrol cars therefore, the use of diesel should be discouraged. The chapter suggests that the emissions in case of public transport passenger-km are lesser than other modes of transport. However, in the North-East region, there is negligible share of public transport due to poor infrastructure and service facilities. Therefore, improvements should be made with regard to public transport system so that considerable number of passengers shifts to public transport modes. Further it is inferred that use of alternate vehicle or fuel technologies like hybrid electric vehicles, biofuel, biodiesel, hydrogen fuel need to be initiated to mitigate the climate change.

Chapter 16 reviews issues, trends and challenges and the politics of environmental pollution in Nigeria. The author argues that In Nigeria, the oil boom in the early 1970s led to a major boost in industrialization and the subsequent environmental pollution, and the attendant undercurrent political implications. In the light of the above, the chapter concludes that the environmental problems in Nigeria and the inherent political dynamics propelled and shaped the call for a cleaner environment.

Chapter 17 concludes and reviews the issue of integrating ecosystem management and environmental media (air, water and land) for development of public policy on public health and safety. The author argues that over the past three decades, the relationship between ecology and public policy has changed because of the increasing role of scientific uncertainty in environmental policy-making. Previously, policymakers use of scientific data to support their decision-making disciplinary boundaries are less useful because uncertain environmental policy problems span the natural sciences, engineering, economics, politics and ethics. The aim of the chapter is to demonstrate the linkages between the environmental policy from a holistic perspective and its interaction with air, water, land and human on public health and safety.

Augustine Nduka Eneanya
University of Lagos, Nigeria
30th September 2017

Acknowledgment

The publication of this book reflects the culmination of a great deal of effort by a large number of people. I express our profound gratitude to contributing authors for their efforts in the process of writing, reviewing, rewriting, editing, and proofreading. This has been a cooperative, and collaborative effort. My special gratitude to experts appointed to help in the peer-review process, namely: Drs Vide Adedayo, Nna Uluocha of Department of Geography, University of Lagos, Oluwole, Solomon Tolutope of Centre for Information Technology and System, also University of Lagos and other specialist reviewers and authors, involved in the peer-review process. I thank the authors and reviewers.

I am also grateful to the Dean, Faculty of Social Sciences, University of Lagos, Yaba, Lagos, Nigeria, for writing the foreword for this book. I want to thank Ms Jan Travers, Director of Intellectual Property and Contracts for her cooperation and quick response to issues that needed urgent attention. My special gratitude to Marianne Caesar, the Development Editor and members of the Editorial Advisory Board for their assistance and guidance during the book development process. I am indebted to them for their time and efforts.

Finally, I thank my wife and children for their encouragement, patience and understanding during the development, reviewing, proofreading and editing processes. I hope that readers will gain as much enjoyment in consuming this product as we have had in creating it.

Augustine Nduka Eneanya
University of Lagos, Nigeria

Section 1
Natural Resources Management

Chapter 1
Natural Resources Management

Funda Varnaci Uzun
Aksaray University, Turkey

ABSTRACT

As a result of the rapidly growing population in the last century, the pressure of people on natural resources has considerably increased. Excessive and wrong use of natural resources leads to occurrence of various human-induced disasters. Global warming, deforestation, floods, air pollution, loss of biological diversity are some of such threats that can be treated within the framework of emergency management. Minimization of human-induced disasters and prevention of such disasters can only be achieved by means of efficient and sustainable management of natural resources. In this chapter, the emphasis will be put on the definition of natural resource management that plans the sustainability of economic activities governing the relationship between humans and the use of nature, such as land use plan, water management, biological diversity and agriculture, mining, tourism, fishing, and forestry, and its importance within the context of emergency management will be discussed.

INTRODUCTION

As a result of unconscious and uncontrolled use of rapidly developing industry and technology and human-induced causes such as population growth and urbanization, the pressure on natural resources has increased and the balance naturally existing in the world has started to change in a negative direction. Contamination of resources, desertification, climate changes, endangered species, habitat destruction, erosion, flood, overflow, avalanche, and landslide are accelerated by the activities of humans and rapidly destroying the natural life; namely the biological diversity, of which human is a part. Nearly half of the forest lands in the world have already been destructed and a large part of this destruction has occurred in the last 30 years. Moreover, each year 6 million hectares of forest lands are destroyed. The fact that nearly half of all the biodiversity exists in forests is a clear indication of how this destruction can affect the whole ecosystem. Such natural disasters becoming too severe with the contribution of human-induced activities should be addressed within the emergency management (Kervankıran&Eryılmaz, 2015).

DOI: 10.4018/978-1-5225-3194-4.ch001

All these negative conditions have become a subject of interest for the concerned circles and the need to take scientific, cultural, educational, visual,and ecological precautions to protect natural resources arose. The primary objective of developing such conservation plans is the prevention of the deterioration and destruction of natural resources and making it possible for future generations to utilize them(Kervankıran&Eryılmaz, 2015). Minimization and prevention of the negative impacts of human-induced disasters can only be achieved through good and sustainable management of natural resources.

In this section of the book, the main focus will be on the definition of natural resources management regulating the relationships between humans and their exploitation of natural resources and planning the sustainability of land use, water management, biological diversity and agriculture, mining, tourism, fishery and forestry and its importance within the context of emergency management.

Moreover, in this section, protected areas one of the most important instruments employed by the traditional use of natural resources and today's resources management and created to ensure the sustainability of natural resources and to enhance environmental quality will also be discussed.

Finally, this section will look at the roles of stakeholders in natural resources management and the importance of participatory planning.

BACKGROUND

There are many studies conducted within the context of natural resources management and emergency management. When some studies on the topic of natural resource management are examined, it is seen that they have focused on the different aspects of the issue; for example, Behnken (2007) conducted a thesis study to investigate the effect of the management decisions made for the Cache River natural resources area on the use of resources by community and long-term management plans. Mountjoy (2014) evaluated the effect of the community-based natural resources management in Illinois on the success of ecosystem management. He worked with 10 different indicators to monitor the changes seen in the ecological unity as a result of the community-based natural resources management. Sanderson (2004) aimed to develop a management plan integrating community and ecological initiatives to ensure the sustainability of the coastal resources in the Olango Island. Jampolsky (2014) wrote in his thesis against the dominant framework by examining Ute Mountain Ute resource management, through the development of the federally funded Integrated Resources Management Plan and Cultural Resources Management Plan, as a proxy for understanding Native property as an enmeshment of law, space, and power. Berg (2015) investigated the effect of two management plans adopting a cooperative approach in the ACE Basin National Estuarine Research Reserve on natural resources. In the thesis study conducted by Sehlke (2016), the main focus is on the benefits of the integrated water resources management to the management of water resources in America. In this thesis, the basic principles of integrated water resources management and water and water-related regulations in America are addressed. In his thesis, Peters (2016), answered these problems: How ecological theory can inform management strategies for achieving specific conservation objectives? How information used to inform natural resource management decisions can be generated through partnerships between volunteers and small conservation organizations? How scientific information is communicated to and shared among natural resource managers in Federal Land Management agencies?

In the doctoral study by Mann (2011) conducted within the context of emergency management, the extent of human resources departments' participation in the process of emergency planning at the local level was explored. In the study, a short history of emergency management since 11th September is given and its importance in human resources management and emergency planning is discussed. Chisholm (2015) studied on the organization of protection and disaster management systems. Mardis (2015) focused on local and state emergency managers particularly those who had a direct role in the formulation of response strategy and had more than five years of emergency management experience. The purpose of the study was to contribute to the expansion of the knowledge in the field of emergency management. Samuel (2014) explored the current state in the field of danger reduction within the context of emergency management. He also aimed to expand the knowledge and theory on the issue of danger reduction. Williams (2016) investigated whether the community preferences theory would be sufficient for the future emergency management in the Virginia Islands. In the study by Coveleski (2014), it was aimed to determine the university students' perception of emergency, preparation for natural disasters and intervention plans.

MAIN FOCUS OF THE CHAPTER

What Is Natural Resources Management?

All the resources naturally found or coming into being in the nature without any influence of human wisdom, effort, science and technique are defined as natural resources (Doğanay&Altaş, 2013). There are various criteria developed to classify natural resources. Natural resources are classified by their origin of emergence as organic and inorganic resources; by their place of occurrence as underground and terrestrial resources; by their role in nutrition as those that can be food and those that cannot be food and by being processable or not as economic resources and uneconomic resources (Doğanay&Altaş, 2013). According to another classification, they are classified as renewable resources such forest, water, wild life, soil and nonrenewable resources such as coal, petrol and mineral ores. According to this classification, natural resources are constituted by many resources each of which has its own unique renewability, dynamic and activity (Yılmaz, 2002). The conceptual model of natural resource values deals with four interrelated systems: (1) Human and wild life populations, the natural environment system of biosphere elements such as natural resources and ecosystems; (2) the social system of human attitudes, values, behaviors and technologies; (3) the economic system focusing on human attitudes, institutions and behaviors related to land, labor and capital and (4) the political system of policies, laws, courts and public institutions.

Exploitation of natural resources started with the appearance of man on earth. Yet, in periods when the number of human beings was small and natural resources were in abundance, the natural balance could be maintained without deteriorating (Anderson, 2013). This balance started to be destroyed as a result of increasing human population, expanding and diversifying human needs and the advance of land cultivation equipments. Since their existence, human beings have not felt satisfied with what is offered by the nature and attempted to create new living spaces for themselves and thus established a dominance on all the other species and at the same time started to rapidly destroy the nature and their own living opportunities (Küçükali, 2005). Humans' unrestricted and irresponsible behaviors demonstrated while using natural resources cause many problems to arise in today's world (İlke et al., 2011). In the last century, pressure of humans on natural resources has increased a great deal as a result of rapidly growing population and many human-induced natural disasters have been experienced as a result of overt and erroneous exploitation of natural resources. Global warming, deforestation, flood, air pollution and decreasing biodiversity are some of them. In addition to these, one of the main problems faced by our world and humanity is the pollution of nature and environment. While human beings have been struggling to survive, they have not accepted nature and natural resources as well as other living entities as they are; thus, they have been able to manage to change nature and themselves to improve their living conditions but while realizing these changes, they have consciously or unconsciously caused serious pollutions detrimental to natural resources, nature and themselves. Unrestricted, relentless and unconscious exploitation of natural resources by humans has been causing rapid deterioration of the balance between nature and life. Environmental pollution defined as the pollution of air, water or foods in such a way as to bring about actual or potential harms to human health and welfare has accelerated in the last century due to increasing industrialization (Yılmaz et al., 2005:16).

The data belonging to 1990s show that 14,800 million hectare land area was deteriorated to some extent (13% of all the cultivated areas in the world) and that 84% of this deteriorated land area was subjected to erosion caused by wind and water (Wild, 2003). According to global evaluations made by FAO for 2005 and 2010, the total forest area in the world is estimated to be 4 billon hectare (30% of all the land area). In the period of 1990-2010, the total amount of deforestation is 13 million hectare per annum; yet, in the same period, the annual global forest loss was 8.3 million hectare. The reasons for this loss are conversion of forest areas into agricultural lands and lack of efforts invested on forest planting, landscape restoration and natural extension of forest areas. All these problems make it necessary to adopt sustainable strategies in agricultural and rural development (Pretty, 1998; McLuskey, 2001).

Rapid industrialization experienced in the 19th century and the widespread belief that natural resources were infinite led to the danger for natural resources not to be able to renew themselves and their depletion in the 20th century (Arat, 1995). It is an erroneous conception that environmental destruction is only related to the ecology of plants and living organisms. The environment also suffers from the unrestricted consumption of raw materials (Çağlar, 1998). As in the past, one of the most important factors destroying the ecological balance and negatively affecting the sustainable management of natural resources is the unconscious consumption of natural resources and generation of so many wastes that they can cause environmental problems (Palabıyık, 1998).

Problems caused by crowding cities and disorganized urbanization such as pollution of water, air and soil as well as destruction of the historical city texture are compounded by destruction of forests and desertification and soil erosion create adverse effects on healthy human habitats (Önen, 2002). Parallel to the increasing population, production efforts invested to meet the increasing needs of people and industrialization resulted in allocation of more resources taken from different parts of nature for these production efforts, increasing time and variety of processes through which the resources have to go and their distribution to larger areas and as a result, accelerating consumption of natural resources bring about some environmental problems (Türköz, 1997). Due to the increasing pressure on natural resources, it has become an obligation to manage these resources sustainably to prevent their extinction and to hand them down to future generations (Yılmaz, 2002). There is a great deal of research aimed at investigating abiotic and biotic factors of the environment, flora-fauna characteristics and food-sources to determine the needs of people changing as a result of population growth. This research argues that management strategies should be developed for the survival of habitats, resources and species and for the protection of endangered species.

The term "management" is used to define a comprehensive process of duties and actions to be carried out to achieve pre-determined goals and objectives. Management of public areas can be a complicated system starting with brain-storming sessions and requiring creative planning, organization and leadership to accomplish the goals and objectives. This process is the management process in which decisions are made about various issues including human resources, finance, time and equipment. The concept of "management" in natural resources management refers to the functioning of ecosystem and thus making and implementing decisions to change its reactions to external factors (Yılmaz, 2002).

Natural resources management means managing natural resources such as soil, water, plants and animals in such a way as to enable present and future generations to utilize these resources (Rohlmann, 1992). Natural resources management also aims to manage human-nature interaction. It also aims the sustainability of agriculture through land use planning, water management and protection of biological diversity and the sustainability of economic activities relying on natural resources such as mining, tourism, fishery and forestry (Anderson, 2013). Protection of natural resources refers to the sensible use of the resources of the world by humans. Natural resources management includes the management of valuable natural resources such as wood, fish, land, minerals, forests, wild life and water reservoirs (Rim-Rukeh et al., 2013).

Natural resources management addresses the issues related to the use of planning techniques developed to separate strict protected reserves, sustainable, protected and usable areas, the selection of the content and type of planning and planning process (Küçükali, 2005: 5). However, the issues of natural resources management are quite complex including ecological cycles, hydrological cycles, climate, animals, plants and geography. All these issues are dynamic and interrelated. Changes affecting one of these issues also affect others in the long run. Natural resources management; together with natural systems, also manages the interests of various stakeholders, rules, political views, geographical borders and economic situations. It is very difficult to manage all these elements in such a way as to satisfy everybody simultaneously and it is inevitable to encounter various conflicts. Growing population, expanding economies and climate change give rise to great difficulties in managing natural resources to protect and improve resources. Food security and underestimation of environmental management and poverty are the main factors contributing to the complexity of natural resources management. In the management of the complexity regarding the use of natural resources, agriculture and rural development at global, national and local levels, there is a need for sustainable strategies (Karamidehkordi, 2012). Some global and national agreements propose approaches to sustainable natural resources and land management. Examples are the UN Convention to Combat Desertification, United Nations Framework Convention for Climate Change (UNFCCC), Kyoto Protocol for reducing GHG carbon emission (through soil and forest conservation), international and national water resources conservation and Millennium Development Goals (Dumanski, 2006; FAO, 2006; United Nations, 2006).

In the management of natural resources, there is a new approach called "Integrated Natural Resources Management" (Rohlmann, 1992). This approach integrates many interrelated goals such as poverty reduction, food security and environmental sustainability. Thus, the integrated natural resources management intends to contribute to the solution of the complicated problems affecting natural resources in the ecosystem. The integrated management of natural resources defines a successful strategy to accomplish multiple goals through various ways. What is the most suitable and effective way of achieving these goals dependent on the environmental system but including the mixture of environmental, economic, social and political issues? In fact, definition of all the important issues means developing a general view of the whole system to select characteristics that can contribute to the achievement or prevention of the goals of management (Lewis et al., 2014). This approach supports the formerly used integrated natural resources management, which attempted to combine the management of sectors separately addressed in terms of resource exploitation. With the achievement of this integration, it is believed that plans can be made to maximize benefits without negatively affecting natural resources. In addition to the obtained benefits, if management plan is successfully implemented, all the parties will be positively affected from the change of management. There is no single universally accepted definition of the concept of the integrated natural resources management. Different scientists or organizations have offered different definitions of the concept. Some of these definitions are given below (Yılmaz, 2002):

- The integrated natural resources management is a concept understood and emerging as a broad-based management responsible for the sustainability of land, water, forest and biological resources (including genes, too) required for the prevention of the destruction to the potential efficiency and the sustainability of agriculture (Hagmann et al., 2002),
- The integrated natural resources management is a conscious process of including different aspects of natural resource use (e.g. biophysical, socio-politic or economic factors) into a sustainable management system aiming to achieve the particular production goals of rural people and other use-related goals (e.g. food security, profitability, risk reduction) as well as more comprehensive social goals (e.g. poverty reduction, wellbeing of future generations, environmental conversation) (Yılmaz, 2002),
- The integrated natural resources management is an approach intending to determine land use applications that can protect natural resources at local and global levels, provide the services of the ecosystem and enhance production levels as well (Izac&Sanchez, 2001),
- The integrated natural resources management is an approach integrating the research exploring the different issues of natural resources into the adaptation management and innovative processes by considering the interest groups to carry out interventions and to generate effects for the improvement of sources of income, ecosystem, agricultural efficiency and environmental services (Thomas, 2002),
- It is the management of two or more natural resources (e.g. water, land, wood, pasture, fish, wild life and forest) in the same area and time period. In this approach, planning, management goals, strategies and policies are decided in cooperation with all the resource users (Dunster&Dunster, 1996).

The management-oriented objectives of the research on the integrated natural resources management are expressed as follows (Noordwijk et al., 2001):

- Setting of more precise, realistic and/or comprehensive goals and more effective performance criteria defining the state of accomplishing these goals,
- Development of models in order to understand the relationships between outcomes and results and inputs and how multi-dimensional reasons lead to the similar effects. This is important in terms of new technologies expanding the choices of rural people, - Use of models helping how desired outcomes can be attained with the least input and labor, one goal of multi-directional management,
- Improving the applications related to the management plan and scenarios,
- A good analysis of the existing state,
- Determination of how factors outside the control of managers affect the system,
- Understanding how the system reacts against the changes occurring as a result of ecological, economic, social and political interactions.

On the other hand, natural resources management should be based on a continuous exchange of communication and thoughts between the concerned parties. This encompasses adaptation management as well as political processes related to the conflicts experienced between the concerned groups. As a result, all the managerial activities are experimental in a sense and managers are included in all the research (Sayer&Campbell, 2001).

The concept of ecosystem is the heart of natural resources management. When the protection of species for them to survive or the management of natural resources is concerned, the focus should be on the ecosystem of natural resources (Botkin & Keller, 1995: 99). Another management approach aiming to enable all living entities to survive in a healthy and balanced environment with the existing resources is the environment management approach. This approach emphasizes that natural resources on earth are finite and they are being irreversibly destroyed; thus, what is aimed should be to maintain economic development without disturbing the ecological balance. That is, economic development is addressed together with the conservation of natural resources (Baki & Cengiz, 2002). In Turkey with the increasing awareness of environmental problems, the concept of environmental management, which is a widely used term in European countries, was introduced to our agenda meaning that "It is necessary to establish a feasible system of communication, planning, coordination and supervision with the participation of both the private and public organizations to enable all living things to survive in a healthy and balanced environment and to protect, make use of and develop natural resources and to found an organization to operate this system" (Keleş & Hamamcı, 2002). As can be understood from this definition, the purpose of environmental management is to seek an answer to the question of how and how efficiently the protection of natural resources such as air, water and soil and ecological balance by using a given management structure will be achieved (Sandal, 2001). The concept of environmental management built on the idea that there should be a balance in the use of environment and natural resources complies with the concept of sustainable development claiming that economic development cannot be addressed independent of the protection of nature and defined as development through which people can meet their current needs without hindering future generations from meeting their own needs.

The Importance of Natural Resources Management Within the Context of Emergency Management

Zdziarski (2006), defined a disaster as "an event, which is often sudden or unexpected, that disrupts the normal operations of the institution or its educational mission and threatens the well-being of personnel, property, financial resources, and/or reputation of the institution". Disasters are the products of the relationship between natural dangers and human activities. Population growth, urbanization, poverty and urban sprawl are among the prominent human factors aggravating the effects of natural disasters (Hutton, 2012).

The effect of a natural hazard on the objects or people of a particular area exemplify the complex interrelationships and emerging domino effects. The UN (1991) and the UNDP (2004) define a conceptual superstructure as follows: Risk = Hazard × Vulnerability

Thus, risk results from a future interplay of a hazard and the various components defining vulnerability. Risk as result of the interaction of the hazard and the vulnerability embedded in the various phases of the disaster or emergency management (Taubenbock et al., 2008). Furthermore, vulnerability likewise shows relevance to resistance (e.g., safer engineering) and resilience (e.g., the ability of social, political, economic, and cultural systems to rebound after a disaster) perspectives as well. Risk and vulnerability are often used interchangeably in the research literature. Vulnerability is also discussed frequently in those circles formulating homeland security policy – especially as it relates to epidemiology and bioterrorism. The concept of vulnerability may incorporate many of the views of all other disaster and emergency management paradigms to date (McEntire, 2004).

Then, what is emergency management? In addition to the research focusing on identifying the focal point of this issue, there is also some research conducted in an attempt to determine its scope. The research has revealed that the concept of "emergency management" has at least three important associated aspects (McEntire, 2004):

First, academicians are not interested in emergencies but disasters. Second, focusing on the term "emergency" prevents the activeness of the field and the applications of those having the prime responsibility. Third, emergency management might mean that we can completely control our ability to overcome adverse events called disasters. Thus, the concept of emergency management both accommodates an erroneous name and a contrast. However, no better definition has been found yet (McEntire, 2004).

Emergency management contains a multitude of definitions and a dynamic developmental history. Its various definitions can be broad and encompassing, and it has been response-based throughout history, responding to specific events rather than prescribing solutions (Myer et al., 2010). Emergency management is essentially the "discipline of dealing with risk and risk avoidance" (Haddow et al., 2007). Emergency management activities are concentrated on three main objectives. These objectives are the protection of life, property and environment. Planning of emergency management programs is initiated with the detection of emergency problems. For instance, in case of an earthquake; what are the developments taking place in the primary impact area of the earthquake fault? What can be the extent of damage on the basis of the expected magnitude of the earthquake, characteristics of the land and other geological data? What are the facilities that should be restored or moved? What are the facilities located in high-hazard areas (bridges, highways, overpasses, dangerous substance storage warehouses, hospitals etc.). Reduction of adverse impacts can be achieved through the implementation of regulations limiting construction in earthquake and flood zones. Another way of reducing the negative impacts is the use of fire-resistant materials in areas susceptible to fire (Johnson, 2000).

Preparation stage in an emergency management includes activities necessary for actual emergencies. When an emergency occurs and ends, then rescue efforts should be started. The rescues stage should be planned as short-run and long-run. The short-run rescue aims to bring vital services and systems back. This might include the provision of food, water and shelter for people who have lost their houses in case of a hurricane or a big forest fire. The long-run rescue plan means restoring all the services or making them better. The long-run rescue (construction of houses, water and wastewater management systems, roads, hospitals, bridges, schools etc.) might last for several years (Johnson, 2000).

One of the most effective ways of reducing the losses caused by natural disasters is the risk-reduction applications that should be conducted prior to the occurrence of a disaster (Genç, 2007). One of the concepts coming to the fore in relation to the issue of minimizing harms is risk management. In the broadest terms, risk management can be defined as "A series of precautions taken to inform people about the dangers that can come from different sources and to increase the control on these sources". Risk management in natural disasters refers to the development of precautions to be taken to evaluate the risk level, to eliminate the potential risk factors and to improve the existing conditions in a residential area so that this residential area could be rendered less vulnerable to natural disasters. For example, while making settlement decisions in an earthquake region, factors such as seismic sensitivity of the region, geological structure of the land, gradient of the landscape and underground water level should be taken into consideration and by creating buffer zones, the impact of a possible earthquake is attempted to be reduced (Genç, 2007).

During the stage of determination of the potential risks prior to the occurrence of a natural disaster, management of natural resources comes into play. For example, in regions where land use planning is not done well and urbanization does not proceed properly, it is possible to argue that more harm is given to city centers by earthquakes. Throughout the history, settlements in Anatolia have been centered between geological masses and around fertile wetlands and plateaus without taking the proper use of natural resources into account. Such settlements have been followed by the construction of industrial and transportation centers in their close surroundings. Besides this, poor management of natural resources has aggravated potential destruction to be incurred by disasters such as floods, overflows and earthquakes (Haddow et al., 2007).

The main challenge within the field of disaster reduction is to change people's perception so that they can recognize this notion of disasters as the outcome of a development process whereby societies have implicitly generated vulnerabilities and risks which become evident during the disaster (Villagran de Leon, 2006). But, assessing vulnerability and risk to natural hazards such as earthquakes can be regarded as an ill-structured problem (i.e., a problem for which there is no unique, identifiable, objectively optimal solution) (Rashed&Weeks, 2003).

The Use and Management of Natural Resources From Past to Present

Natural resources management is not only affected from natural processes and the dispersion of species but also from the land use by humans. Past and present anthropogenic land use experiences affect organizational strategies of natural resources management. This is related to who possesses and manages the land and follows the natural processes (Behnken, 2013).

In the last 50,000 years, human communities constructing small and large settlements first selected natural caves located close to seas, lakes and rivers as their shelters and then they started to spread towards inlands. However, the close connection of primitive settlements with water sources is apparent. The first natural disasters witnessed by human communities must have been floods and tsunamis. Hence, it can be maintained that they must have recognized that the first natural resource that should be managed is water sources. Aside from water sources, the first natural resources becoming important for humans were agricultural lands and wild animals to be domesticated (Appiah-Opoku, 2007; Rim-Rukeh et al., 2013).

In old times, local people developed various resources management applications surviving up to now in Africa, Asia, South America and other parts of the world (Appiah-Opoku, 2007). The use of traditional methods helping people organize their interactions with their natural environments is one of the natural resources management approaches. The role of traditional beliefs in the conservation of bio-diversity is as old as human history (Berkes et al., 2000; Turner et al., 2000; Shastri et al., 2002; Rim-Rukeh et al., 2013). Chacon (2012) and Krech (2005); though argue that traditional beliefs and taboos do not guarantee the sustainability of natural resources, the traditional African religious and cultural experiences observed in many African communities are environment friendly and sustainable. Thus, they ensure the sustainability and conservation of natural resources. The role of traditional beliefs in the conservation of natural resources has led to emergence of various areas including sacred sites. For example, in India, the forests determined as the sacred sites are protected by the people without making any production. In such forests, bio-diversity is quite rich and many species such as rare weeds and medicinal herbs are hosted.

More than two thousand years ago, Plato wrote that the forest in the Attica hill was depleted, which led to loss of fertile water sources and ground cover. He even likened this situation to "the skeleton of a body exhausted by a disease". In the same age, the first forest was declared to be under protection in India. Again in India, the Emperor Ashoka put first hunting laws into effect to protect some bird and fish species (Kurdoğlu, 2007). From the early periods of history, Mount Ida has been considered to be a holly site. The allocation of holiness to the site dates back to the early periods of history and though the powers ruling the region have continuously changed, it has always remained as a holly place. Due to its holiness and natural resources it offered to people living in the area for centuries, these people had always naturally protected the area up to its declaration as a national park (Arı & Soykan, 2006).

As a result of rapid population growth, industrialization and urbanization, the pressures on and threats to natural resources have become one of the most important problems of our day. There are many national and international attempts to give the status of "protected area" to natural resources found in many countries to prevent the excessive and erroneous use of natural resources, to use and manage them on the basis of scientific principles and to carry out an effective conservation (Kuvan, 1991). After the United Nations Conference on Environment and Development (UNCED) held in Rio de Janeiro in 1992, many countries resorted to new principles for the integrated management of lands, waters and forests. Though the names of the programs change from country to country, all of them state similar objectives (Akten&Akten, 2014). Protected areas that can make important contributions to economic and social development through the conservation of natural and cultural resource values are considered to be an indispensable instrument of conservation and resource use policies (Hepcan, 1997). The primary function of conservation is to ensure the sustainability of bio-diversity. While habitat loss and deterioration play an important role in decreasing bio-diversity, introduction of pollutants and alien species into the system accelerates the deterioration. Moreover, global warming might result in change and deterioration of bio-diversity from different respects. Such changes also lead to changes in the habitats of some natural unisons (Bridgewater, 1993).

It seems to be necessary to make integrated and participatory "management plans" in order to be able to make rational use of the existing natural, cultural and visual resource values of protected natural areas and to hand them in a sustainable manner down to future generations (Gül & Özgüner, 2005). Conservation and sustainable management of biological diversity and natural resources have continuously increasing their importance in the international agenda. Conservation and maintenance of genetic diversity together with the diversity of species and ecosystem is of a vital importance for future generations. In this regard, the most powerful tool at hand is the effectively managed protected areas (Thomas, 2006).

Establishment of protected areas led to the development of the science of ecology, natural resources planning and management. The protected area classification of IUCN in 1994 indicates the conviction that conservation of bio-diversity and natural resources is the main function of protected areas. Many protected areas serve environmental functions such as protection of water resources and the reduction of the adverse impacts of floods and climate change (Eagles, 2002).

According to IUCN's definition, protected areas "*are landscapes or seascapes allocated for the conservation of natural and cultural and particularly biological diversity and managed through laws or other effective regulations* (IUCN, 1994). On the other hand, according to the up-dated definition of the protected area proposed by IUCN in 2008; the protected area is an area managed by laws or other effective methods within its clearly defined geographical borders for the long-term conservation of the nature and its associated ecosystem services and cultural values.

According to the definition of IUCN, protected areas in fact are managed to serve many different purposes. For the better understanding of the purposes of protected areas, IUCN developed a protected area system consisting of six categories. According to this, the protected areas are as follows:

1. **Strict Nature Reserve/Wilderness Area:** Protected area managed mainly for science or wilderness protection
    a. **Strict Nature Reserve:** Protected area managed mainly for science.
    b. **Wilderness Area:** Protected area managed mainly for wilderness protection.
2. **National Park:** Protected area managed mainly for ecosystem protection and recreation.
3. **Natural Monument:** Protected area managed mainly for conservation of specific natural features.
4. **Habitat/Species Management Area:** Protected area managed mainly for conservation through management intervention.
5. **Protected Landscape/Seascape:** Protected area managed mainly for landscape/seascape conservation and recreation.
6. **Managed Resource Protected Area:** Protected area managed mainly for the sustainable use of natural ecosystems.

Among the management objectives of protected areas is the conservation of wild life, species and bio-diversity, natural characteristics and the sustainable use of ecosystems.

The Role of the Stakeholders in Natural Resources Management

Natural resources management planning refers to a process of integrating sociological, ecological and legal information into a natural resources management plan. A natural resources management plan is designed through a process based on the cooperation of many sectors and disciplines desired for the determination and prioritization of the issues of resources management, the promotion of consensus and the generation of support for the resources management program (Sanderson, 2004). The goal of management planning is to develop a management plan including effective strategies and actions to address. In the preparation of a natural resources management plan, participation of local society and its administrators is required. Collaborative management requires the dissemination of power among administrators and local stakeholders and cooperation (Sanderson, 2004).

Borrini-Feyerabend (1996) observed that stakeholders are aware of their own interests, have certain capacities for participating in resources management and are willing to have a say in decisions. However, usually not all the stakeholders are interested in protecting resources and social welfare or deserve an equal role in decision-making process. In addition to this, many managements do not want to take the responsibility for participating in all the decisions rather they prefer to have a say in some important moments. Thus, there might be importance differences between the areas of interest, knowledge, motivation and power of the stakeholders involved in cooperative management.

The most important problems encountered in the participatory management processes concern the questions such as "Who should participate?", "Which roles should be taken by whom?", "How can effective participation be achieved?", "Which group should have stronger influence on the decisions made?" (Conroy&Peterson, 2013).

Such decisions should be made through participation and over time when the process is advanced and more information is obtained, these decisions should be revised. Moreover, early determination of who the stakeholders will be, what their roles within the cooperative management will be and how they will contribute to the process is important (Güneş, 2011).

Of the potential stakeholders to be involved in natural resources management, users should be the primary associate group as they represent the most complex group and they are the group to be affected the most from the management decisions to be made. Due to the diversity of interests, associate groups are rarely homogenous. Therefore, the concept of interest group can be a useful tool to help identify the sub-groups. The most apparent sub-groups are income source-based sub-groups. For example, animal owners, foresters and hunters represent diverse interest groups related to the particular use of forests (Conroy&Peterson, 2013).

Governments involved in associate groups might include the government having domination on a specific natural resource and other external governments having various regional or international interests concerning this resource. Regions where large rivers extend beyond the national borders is an example of a classic case where external governments have some interests in the management of natural resources. Governments are important stakeholders because everybody interested in resources management should be in contact with the government at one point (Conroy & Peterson, 2013).

Law-makers including politicians approach to natural resources in terms of how they should be managed in line with the wider development targets of the country and how they can contribute to their local political agenda. Senior government authorities, local administrators and other members of the government strive for adhering to laws and can exert great influence on what can be done by the government in some particular areas. As these people can exercise influence on laws and how they are implemented, they are in the position of the most powerful stakeholders. Personnel of the government working in the field are another sub-associate group because they can create direct links between the conditions determined by the government and needs and priorities of local stakeholders. In general, governments have different interests, information and power than other associate groups and they have different roles in the formation and promotion of cooperative management. Moreover, due to structural and other institutional differences, there is a great diversity among governments. Therefore, it is quite difficult to be specific in terms of governments' interests and participating approaches (Conroy & Peterson, 2013).

Development agencies are another associate group providing funds and other services for national development programs. This group includes international funding organizations, consultants hired to revise and evaluate development programs and civil personnel involved in projects supported by international benefactors. Benefactors are in the position of fund managers; thus, they have to abide by the designated rules. Consultants; on the other hand, can exert great influence on how development programs and projects are shaped and conducted because they provide information and direction for governments and benefactors about what should be done and how programs should be supported. International funding projects should consider the regulations of the government, the state of benefactors and the design of the project and should adjust them to local conditions. As a result, the influence of this associate group is stronger than that of the other associate groups in cooperative management (Conroy & Peterson, 2013).

Private non-governmental organizations include few international, national or local organizations slightly interested in natural resources management. Such private organizations provide funders, consultants or the project group for the support of the development on their own. In other words, they can be associations and unions promoting the positions shared, providing services for their members and coordinating certain types of common activities. These organizations involve various types of organizations ranging from local agrarian unions to international community services groups. Other private stakeholders are quite extensive and complex. These involve many different kinds of organizations that cannot be included in the above-mentioned associate groups. For example, private businesses and entrepreneurs not directly making use of natural resources but their production activities depend on these natural resources constitute an important sub-group of stakeholders. These stakeholders have reasonable interests and commercial influences on resources management decisions (Conroy & Peterson, 2013).

The most important step of the associate participation in natural resources management is the participation of the local community. Participation of the community in natural resources management is usually seen to be necessary for success (Mendis-Millard & Reed, 2007). Non-governmental organizations such as World Bank, Conservation International, The Nature Conservancy (TNC), The Ford Foundation, The MacArthur Foundation, and USAID support this new community-based management approach (Agrawal & Gibson 1999), and academicians and administrators enforce concerned parties to support this movement. The opinions of the community previously ignored are today regarded as a component of a successful ecosystem management (Folke et al. 2005). One of the important factors in the conservation of natural resources and enhancing the efficiency of natural resources management is the existence of a sensitive community highly cognizant of and organized on the environmental issues (Sayer & Campbell, 2003)

It has been discussed in the literature whether community-based approaches would make natural resources management more integrated. These discussions are bidirectional. While one group argue that making management systems more integrated will serve both the interests of the local community and western scientific world, another group maintain that use of natural resources not only by local people but also many countries makes the management applications unreliable from the perspective of local people (Welp & Stoll-Kleemann, 2006).

In general, though the community-based approach relies on the traditional practices and information systems to develop and regulate the principles of resources management, it promotes the participation of community members, resource users and local organizations in the management process (Armitage, 2005). Thought there are slight differences among initiatives, Gray et al. (2001) identify four main principles of the community-based ecosystem management:

1. Stakeholders see the ecosystem services as a critical part of the community;
2. Decisions concerning the resources should be made through cooperative processes together with the participation of the people knowing the field and those to be affected from these decisions.
3. Equality is required in the distribution of the benefits of an ecosystem (ecosystem services, resource use and processes) for the sustainability of local capacity.
4. Citizens and their communities are accepted as the main components of an ecosystem (Mountjoy, 2014).

The community-based resources management is a process of decentralization aiming to assign the responsibility for making decisions on resources and their control to well-established organizations. The main motivation of local communities for participating in conservation efforts is to protect their rights on resources. The community-based natural resources management strategy defines the participation of communities as an instrument to promote the sustainable use of resources by freeing rural communities to utilize the resources they rely on (Nhantumbo et al., 2003). As a conclusion, participation of stakeholders in the planning process has become the basis of natural resources management. The aim of natural resources management can be redefined as setting common goals for various groups of stakeholders (Cain, 2001).

SOLUTIONS AND RECOMMENDATIONS

With the emergence of man, the use of natural resources for his aims started. Increasing consumption of natural resources over the course of the century has led to a gradual deterioration of the balance between man and nature. The necessity of managing these natural resources has emerged in order to ensure this balance and transfer of natural resources to future generations. The management of natural resources is a complex and potentially contentious undertaking that encompasses issues from the environmental, economic, social, and political realms. However, population growth and increased urbanization, combined with climate change-driven sea-level rise and increased rainfall, have brought increased attention and urgency to environmental issues in coastal communities. In order to meet these complex challenges, it has become increasingly clear that natural resource managers must draw on a wide range of resources, and engage all relevant stakeholders, in order to create legitimate and lasting outcomes. Collaborative or participatory processes offer a potential solution to the challenge of inclusion, providing managers with a range of tools to engage, educate, and learn from the communities affected by their decisions (Berg, 2015). One of the most important things to be considered in the planning and management of natural resources is to make all stakeholders benefitting from natural resources participate in the management process. To ensure the protection of natural resources, local people must be involved in the management. It is not possible to ensure the sustainability of natural resources when local people are not included in management, and when management and protection are centralized.

Due to excessive consumption of human beings, harms caused by natural disasters are becoming greater. For this reason, the concept of emergency management, which includes the implementation of risk reduction measures prior to the disaster, is becoming more and more popular to reduce the damages caused by natural disasters. In order to increase the effectiveness of emergency management, it is necessary to identify the areas under the risk and properly manage the resources in these areas.

FUTURE RESEARCH DIRECTIONS

This chapter draws a conceptual framework for natural resource management. Therefore, it adopts a more theoretical approach to the definition of natural resources management and utilization and protection of natural resources. Further research may include applied field surveys to be conducted in various areas where natural resources are at risk and should be managed by emergency management.

CONCLUSION

In this chapter, it is aimed to define the natural resource management and to explain its importance within the context of emergency management. Natural resource management refers to the planning and management of naturally existing resources such as land, water, plants and animals in such a way as to enable current and future generations to benefit from them.

Emergency management; on the other hand, focuses not on preventing or eliminating incidents, but on reducing their effects. Incidences dealt with by emergency management may include terrorism, fire, dangerous substances, earthquakes, floods, tsunamis, storms, hurricanes, insurrection, famine, drought or major accidents. It is important to correctly assess the state of existing natural resources in order to minimize the damage that can be experienced within the context of emergency management. At the same time, the unconscious consumption of natural resources causes harms of natural disasters to be greater. Natural resources must also be properly planned and managed in order for emergency management to be implemented more effectively.

REFERENCES

Agrawal, A., & Gibson, C. C. (1999). Enchantment and Disenchantment: The Role of Community in Natural Resource Conservation. *World Development, 27*(4), 629–649. doi:10.1016/S0305-750X(98)00161-2

Akten, S., & Akten, M. (2014). Doğal Kaynak Yönetimi İçin Sürdürülebilir İndikatörler ve İlkeler, II. Ulusal Akdeniz Orman Ve Çevre Sempozyumu, 22-24 Ekim 2014, Isparta.

Anderson, D. A. (2013). *Environmental Economics and Natural Resource Management.* Routledge.

Appiah-Opoku, S. (2007). Indigenous beliefs and environmental stewardship: A rural Ghana experience. *Journal of Cultural Geography, 22*(2), 79–88. doi:10.1080/08873630709478212

Arat, Z. (1995). Çevre Yönetimi ve Planlama, *Yeni Türkiye Dergisi,* 5.

Arı, Y. & Soykan, A. (2006). Kazdaği Milli Parki'nda Kültürel Ekoloji ve Doğa Koruma. *Türk Coğrafya Dergisi,* (44), 11-32.

Armitage, D. (2005). Adaptive Capacity and Community-Based Natural Resource Management. *Environmental Management, 35*(6), 703–715. doi:10.1007/s00267-004-0076-z PMID:15940398

Baki, B. & Cengiz, E., (2002). Toplam Kalite Çevre Yönetimi. *Uludağ Üniversitesi, İktisadi ve İdari Bilimler Fakültesi Dergisi, 21*(1).

Behnken, J. A. (2013). *Natural Resource Management Knows No Bounds: A Case Study Of The Cache River Joint Venture Partnership* (Unpublished master dissertation). Southern Illinois University, Carbondale, IL.

Berg, C. (2015). *Evaluating Collaborative Natural Resource Management in The Ace Basin National Estuarine Research Reserve* (Unpublished master dissertation). The University of Charleston.

Berkes, F., Colding, J., & Folke, C. (2000). Rediscovery of traditional ecological knowledge as adaptive management. *Ecological Applications, 10*(5), 1251–1262. doi:10.1890/1051-0761(2000)010[1251:ROTEKA]2.0.CO;2

Borrini-Feyerabend, G. (1996). *Collaborative management of protected areas: Tailoring the approach to the context.* Gland.

Botkin, D., & Keller, E. (1995). *Environmental Science.* John Wiley & Sons.

Bridgewater, P. B. (1993). Landscape Ecology, Geographic Informations Systems and Nature Conservation. In R. H. Young, D. R. Gren, & S. Cousins (Eds.), Landscape Ecology and GIS (pp. 23–36). Academic Press.

Çağlar, İ. (1998). Çevre Yönetimi. *Çorum Çevre Dergisi*, 5.

Cain, J. (2001). *Planning improvements in natural resources management; guidelines for using Bayesian networks to manage development projects*. Wallingford, UK: Institute of Hydrology.

Chacon, R. (2012). Conservation or Resource Maximization? Analysing Subsistence Hunting Among the Achuar (Shiwiar) of Ecuador. In R. Chacon & R. Mandoza (Eds.), *The Ethics of Anthropology and Amerindian Research: Reporting on Environmental Degradation and Warfare* (pp. 311–360). New York: Springer. doi:10.1007/978-1-4614-1065-2_13

Chisholm, V. (2015). *Preventative Conservation and Disaster Management Planning in Cultural Institutions* (Unpublished master dissertation). The State University of New Jersey.

Conroy, M. J., & Peterson, J. T. (2013). *Decision Making in Natural Resources Management: A Structured Adaptive Approach*. Wiley-Blackwell. doi:10.1002/9781118506196

Doğanay, H., & Altaş, N. T. (2013). *Doğal Kaynaklar*. Ankara: Pegem Akademi.

Dumanski, J. (2006). *Soil Conservation and The International Environmental Conventions*. Paper presented at the 14th International Soil Conservation Organization Conference: Water Management and Soil Conservation in Semi-Arid Environments, Marrakech, Morocco.

Dunster, J., & Dunster, K. (1996). Dictionary of Natural Resource Management, the Comprehensive, Single-Source Guide to Natural Resource Management Terms. CAB International (International for Agriculture and Bioscience).

Eagles, P. F. J., McCool, S. F., & Haynes, C. D. (2002). Sustainable Tourism in Protected Areas. World Commission on Protected Areas, Spain: Best Practice Protected Area Guidelines Series:8.

Erkal, T., & Değerliyurt, M. (2009). Türkiye'de Afet Yönetimi. *Doğu Coğrafya Dergisi*, *22*, 147–164.

FAO. (2006). *Global Forest Resources Assessment 2005: Progress Towards Sustainable Forest Management*. Rome: Forestry Department, Food and Agriculture Organization of the United Nations.

Folke, C., Hahn, T., Olsson, P., & Norberg, J. (2005). Adaptive governance of social-ecological systems. *Annual Review of Environment and Resources*, *30*(1), 441–473. doi:10.1146/annurev.energy.30.050504.144511

Genç, F. N. (2007). Türkiye'de Doğal Afetler ve Doğal Afetlerde Risk Yönetimi. *Stratejik Araştırmalar Dergisi*, *9*(5), 201–226.

Gray, R. H., Javad, M., Power, D. M., & Sinclair, C. D. (2001). Social and environmental disclosure and corporate characteristics: A research note and extension. *Journal of Business Finance & Accounting*, *28*(3/4), 327–356. doi:10.1111/1468-5957.00376

Gül, A., & Özgüner, H. (2005). *Ülkemizdeki Korunan Doğal Alanlarda Yönetim Zonlarının Oluşturulması. Korunan Doğal Alanlar Sempozyumu*. Isparta: SDÜ Orman Fakültesi Poster Bildiri Kitabı.

Güneş, G. (2011). Korunan Alanların Yönetiminde Yeni Bir Yaklaşım: Katılımcı Yönetim Planları. *Ekonomi Bilimleri Dergisi*, *3*(1), 47–57.

Haddow, G., Bullock, J., & Coppola, D. P. (2007). *Introduction to Emergency management*. Elsevier.

Hagmann, J., Chuma, E., Muriwira, K., Connoly, M., & Ficarelli, P. (2002). Success Factors in Integrated Natural Resource Management R & D: Lessons from Practice. *Conservation Ecology*, *5*(2), art29. doi:10.5751/ES-00298-050229

Hepcan, Ş. (1997). *Milli Parklarda Yönetim Zonlarının Belirlenmesi Amacıyla Manisa Spil Dağı Milli Parkı Örneğinde Bir Yöntem Araştırması* (Unpublished doctoral dissertation). Agean University, İzmir, Turkey.

Hutton, D. (2012). Lessons Unlearnt: The (Human) Nature of Disaster Management. In Emergency Management (pp. 1-10). In Tech.

İlke, E.F., Uzun, O., Çetinkaya, G., Erduran, F., & Açıksöz, S. (2011). Doğal Kaynaklarin Kullaniminda Peyzaj Yönetimi, Koruma ve Planlamasi. *I. Konya Kent Sempozyumu*, 319-332.

IUCN. (1994). *Guidelines for Protected Area Management Categories*. Gland.

Izac, A. M. N., & Sanchez, P. A. (2001). Towards a Natural Resource Management Paradigm for International Agriculture: The Example of Agroforestry Research. *Agricultural Systems*, *69*(1-2), 5–25. doi:10.1016/S0308-521X(01)00015-4

Jampolsky, J. A. (2014). *Integrated Resource Management Planning with The Ute Mountain Ute Tribe: Property, Place, and Governable Space* (Unpublished doctoral dissertation). University of Colorado.

Johnson, R. (2000). *GIS Technology for Disasters and Emergency Management*. An ESRI White Paper.

Karamidehkordi, E. (2012). Sustainable Natural Resource Management, a Global Challenge of This Century. In Sustainable Natural Resources Management (pp. 105-114). In Tech.

Keleş, R., & Hamamci, C. (2002). *Çevrebilim*. Ankara: İmge Kitabevi.

Kervankıran, İ., & Eryılmaz, A. G. (2015). Milli Parkların Sürdürülebilir Kullanımı ve Yönetim Planı Önerisi: Isparta İli Örneği. *SDU Faculty of Arts and Sciences Journal of Social Sciences*, *34*, 173–190.

Krech, S. (2005). Reflections on Conservation, Sustainability, and Environmentalism in Indigenous North America. *American Anthropologist*, *107*(1), 78–86.

Küçükali, U. F. (2005). *Havza Tanımı ve Doğal Kaynak Yönetimi* (Unpublished master dissertation). Yıldız Teknik Üniversitesi, İstanbul, Turkey.

Kurdoğlu, O. (2007). Dünyada Doğayi Koruma Hareketinin Tarihsel Gelişimi ve Güncel Boyutu. *Artvin Çoruh Üniversitesi Orman Fakültesi Dergisi*, *8*(1), 59–76.

Kuvan, Y. (1991). Avrupa topluluğu tarafından yapılan korunan alan sınıflandırması ve ülkemizdeki korunan alanların bir değerlendirilmesi. *Istanbul Üniversitesi Orman Fakültesi Dergisi*, *41*(1-2), 133–137.

Lennon, J. (2006). Cultural Heritage Management. In M. Lackwood, G. L. Worboys, & A. Kothari (Eds.), *Managing Protecting Areas: A Global Guide* (pp. 448–473). London: Earthscan.

Lewis, F., Johnston, R., & Hamdard, M. (2014). *Tools and Approaches for Integrated Natural Resource Management*. European Commission / FP7 Collaborative project.

Mann, S. C. (2011). *Creating Resilient, Sustainable Local Governments: Merging Human Resource Management and Emergency Management* (Unpublished doctoral dissertation). Mississippi State University.

Mardis, R. M. (2015). *Crisis, Strategy, And Response: Investigating The State of Strategic Leadership in Emergency Management* (Unpublished doctoral dissertation). Capella University.

McEntire, D. A. (2004). *The Status of Emergency Management Theory: Issues, Barriers, and Recommendations for Improved Scholarship*. Paper Presented at the *FEMA Higher Education Conference*, Emmitsburg, MD.

McLuskey, D. (2001). *Extension practice in creating a sustainable agriculture*. Australasia-Pacific Extension Network International Conference 3-5 October 2001, Toowoomba, Australia.

Mendis-Millard, S., & Reed, M. G. (2007). Understanding community capacity using adaptive and reflexive research practices: Lessons from two Canadian biosphere reserves. *Society & Natural Resources*, *20*(6), 543–559. doi:10.1080/08941920601171915

Mountjoy, N. J. (2014). *Community-Based Natural Resource Management: Group Capacity, Resource Management Planning and Assessing Success* (Unpublished doctoral dissertation). Southern Illinois University.

Myer, R. A., James, R. K., & Moulton, P. (2010). *Or a Tornado or Earthquake Drill. This is Not a Fire Drill: Disaster Intervention and Prevention on College Campuses*. Hoboken, NJ: John Wiley& Sons. doi:10.1002/9781118269909

Nhantumbo, I., Norfolk, S., & Pereira, J. (2003). *Community Based Natural Resources Management in Mozambique: A Theoretical or Practical Strategy for Local Sustainable Development? The Case Study of Derre Forest Reserve*. Sustainable Livelihoods in Southern Africa Research Paper 10, Institute of Development Studies.

Noordwick, M., Tomich, T. P., & Verbist, B. (2001). Negotiation Support Models for Integrated Natural Resource Management in Tropical Forest Margins. *Conservation Ecology*, *5*(2).

Önen, Y. (2002). *Kentsel Mekan, Çevre, Çoğulculuk ve İnsan Haklarına Genel Bir Yaklaşım, İnsan, Çevre, Kent*. İstanbul: Demokrasi Kitaplığı.

Palabiyik, H. (1998). Çevre Sorunu Olarak Kentsel Katı Atıklar (Çöpler) ve Entegre Katı Atık Yönetimi. *Türk İdare Dergisi*, 420.

Peters, C. B. (2016). *Theory and Information for Natural Resource Management* (Unpublished doctoral dissertation). University of California.

Pretty, J. (1998). Supportive policies and practices for scaling up sustainable agriculture. In N. G. Röling & M. A. E. Wagemakers (Eds.), *Facilitating sustainable agriculture: participatory learning and adaptive management in times of environmental uncertainty*. Cambridge, UK: Cambridge University Press.

Rashed, T., & Weeks, J. (2003). Assessing vulnerability to earthquake hazards through spatial multicriteria analysis of urban areas. *International Journal of Geographical Information Science*, *17*(6), 547–576. doi:10.1080/1365881031000114071

Rim-Rukeh, A., Irerhievwie, G., & Agbozu, I. E. (2013). Traditional beliefs and conservation of natural resources: Evidences from selected communities in Delta State, Nigeria. *International Journal of Biodeversity and Conservation*, *5*(7), 426–432.

Rohlmann, M. (1992). *Integrated Natural Resources Management: A Question of Property Institutions?* (Unpublished master thesis). York University, Canada.

Samuel, C. (2014). *The Current Status of Hazard Mitigation in Local Emergency Management: An Examination of Roles, Challenges, and Success Indicators* (Unpublished doctoral dissertation). University of North Texas.

Sandal, A. (2001). Sekizinci Beş Yıllık Plana Çevre Yönetimi Açısından Bakış. *Çevre ve İnsan Dergisi*, 48.

Sanderson, N. (2004). *Application of Collaborative Resources Management to Developing a Coastal Resources Management Plan for Olango Island, Philippines* (Unpublished master dissertation). University of Calgary.

Sayer, J. A., & Campbell, B. (2001). Research to Integrate Productivity Enhancement, Environmental Protection, and Human Development. *Conservation Ecology*, *5*(2).

Sayer, J. A., & Campbell, B. M. (2003). Research to Integrate Productivity Enhancement, Environmental Protection and Human Development. In J. A. Sayer & B. M. Campbell (Eds.), *Integrated Natural Resources Management, Linking Productivity, the Environment and Development* (pp. 1–14). Cromwell Press. doi:10.1079/9780851997315.0001

Sehlke, G. (2016). *A Conceptual Framework for Developing and Implementing Integrated Water Resources Management (Iwrm) Programs In The United States* (Unpublished doctoral dissertation). University of Idaho.

Shastri, C. M., Bhat, D. M., Nagaraja, B. C., Murali, K. S., & Ravindranath, N. H. (2002). Tree species diversity in a village ecosystem in Uttara Kannada district in Western Ghats, Karnataka. *Scientific Correspondence*, *82*, 1080–1084.

Taubenbock, H., Post, J., Roth, A., Zosseder, K., Strunz, G., & Dech, S. (2008). A conceptual vulnerability and risk framework as outline to identify capabilities of remote sensing. *Natural Hazards and Earth System Sciences*, *8*(3), 409–420. doi:10.5194/nhess-8-409-2008

Thomas, L. (2006). *Türkiye Korunan Alan Yönetiminde IUCN Kategori Sistemi. Çevre ve Orman Bakanlığı.* Doğa Koruma ve Milli Parklar Genel Müdürlüğü. Biyolojik Çeşitlilik ve Doğal Kaynak Yönetimi Projesi.

Thomas, R. (2002). *What is Integrated Natural Resource Management. Integrated Natural Resource Management: Putting Practice into Action.* Paper presented at the meeting of Fourth INRM Task Force, Aleppo, Syria.

Türköz, O. İ. (1997). Çevre Yönetimi ve Metotları. *Standart Ekonomik ve Teknik Dergi*, 421.

Turner, N. J., Ignace, M. B., & Ignace, R. (2000). Traditional Ecological Knowledge and Wisdom of Aboriginal Peoples in British Columbia. *Ecological Applications, 10*(5), 1275–1287. doi:10.1890/1051-0761(2000)010[1275:TEKAWO]2.0.CO;2

United Nations. (1991). Mitigating Natural Disasters: Phenomena, Effects, and Options: a Manual for Policy Makers and Planners. New York: UN.

United Nations. (2006). *The Millennium Development Goals Report.* Available online: http://unstats.un.org/unsd/mdg/Resources/Static/Products/Progress2006/MDGReport2006 .pdf

United Nations Development Programme (UNDP). (2004). Reducing Disaster Risk – A challenge for development – A Global Report. New York: UNDP.

Villagran de Leon, J. C. (2006). Vulnerability – A conceptual and methodological review. Studies Of the University: Research Council Education, 4.

Welp, M., & Stoll-Kleemann, S. (2006). Integrative Theory of Reflexive Dialogues. In S. Stoll-Kleemann & M. Welp (Eds.), *Stakeholder Dialogues in Natural Resources Management* (pp. 43–78). New York: Springer. doi:10.1007/978-3-540-36917-2_3

Wild, A. (2003). *Soils, Land and Food: Managing the Land During the Twenty -First Century.* Cambridge, UK: Cambridge University Press. doi:10.1017/CBO9780511815577

Williams, M. A. (2016). *A Whole Community Approach to Emergency Management for the United States Virgin Islands* (Unpublished doctoral dissertation). Walden University.

Yavaş, H. (2001). Doğal Afet Yönetimi ve Yerel Gündem 21 Çalışmaları Kapsamında İzmir'de Deprem Riski. *Dokuz Eylül Üniversitesi Sosyal Bilimler Enstitüsü Dergisi, 3*(3), 118–138.

Yılmaz, A., Bozkurt, Y., & Taşkın, E. (2005). Doğal Kaynakların Korunmasında Çevre Yönetiminin Etkinliği. *Dumlupınar Üniversitesi Sosyal Bilimler Dergisi, 13*(1), 15–30.

Yılmaz, E. (2002). Integrated Natural Resource Management and Its Workshop. *Eastern Mediterranean Forestry Research Institute Journal of DOA, 8,* 1-30.

Zdziarski, E. L. (2006). *Crisis in the context of higher education. Crisis management: Responding from the heart.* Academic Press.

KEY TERMS AND DEFINITIONS

Emergency Management: Emergency management is the organization and management of resources and responsibilities for dealing with all humanitarian aspects of emergencies—preparedness, response, and recovery—in order to reduce the harmful effects of disasters.

IUCN: The international union for conservation of nature (IUCN) is a membership union uniquely composed of both government and civil society organisations. It provides public, private, and non-governmental organisations with the knowledge and tools that enable human progress, economic development and nature conservation to take place together.

Natural Resources: Natural resources are resources that exist without actions of humankind. This includes all valued characteristics such as magnetic, gravitational, and electrical properties and forces. On earth it includes: sunlight, atmosphere, water, land (includes all minerals) along with all vegetation and animal life that naturally subsists upon or within the heretofore identified characteristics and substances.

Natural Resources Management: Natural resource management refers to the management of natural resources such as land, water, soil, plants and animals, with a particular focus on how management affects the quality of life for both present and future generations.

Protected Areas: Protected areas or conservation areas are locations which receive protection because of their recognized natural, ecological or cultural values. There are several kinds of protected areas, which vary by level of protection depending on the enabling laws of each country or the regulations of the international organisations involved.

Stakeholders: An accountant, group, organization, member, or system that affects or can be affected by an organization's actions.

Chapter 2
Some Aspects of Natural Resource Exploitation and Management in Nigeria

Mayowa Fasona
University of Lagos, Nigeria

Olatunde J. Ogunkunle
University of Lagos, Nigeria

ABSTRACT

Natural resources provide the all-important basis for human existence and civilization. Natural resources management has several implications for human security and public safety. A quality environment supports virile natural resources, which provide the necessary goods and services to satisfy life quality needs and mitigate famine and food shortages and their concomitant effects, including food riots and public disorder. Poorly managed and severely degraded natural resources systems will interfere with other systems, such as the climate, to produce resource conflicts and exacerbate human insecurity and undermine public safety. Africa is highly vulnerable to natural resource degradation and its concomitant impacts, a situation that has deepened the existing poverty-environment-human security linkages. This chapter explores the dimension of natural resources and natural resource management, its implications on livelihoods and conflicts, human security and public safety, and its contexts in Africa in general with particular reference to Nigeria.

INTRODUCTION

In no other continent is the strong nexus between the natural resources, environment, livelihoods, poverty, food security, human security and public safety better exemplified than Africa. Livelihoods in the vast rural interiors of Africa are tied to resources of nature. More than ever before, harsh political, economic and social conditions and low level of technology have considerably diminished people's resilience and make them vulnerable to impacts of global environmental change. Although indigenous societies across Africa are known for their resilience and have survived changes for thousands of years based on

DOI: 10.4018/978-1-5225-3194-4.ch002

traditional coping strategies, increasing population and diminishing natural resources - especially water and agricultural and grazing lands – has become additional stressor that have stretched the traditional resilience to the hilt. Accelerated land and water degradation, increasing aridity and competition for finite resources are impacting the social fabric and increasing resource conflicts. This further drives the flywheel of poverty and human insecurity and public safety concerns across the continent.

Land is Africa's most important resources. Africa is a massive continent in size and population. With about 30.3million square kilometers (22% of the world's land area) and over one billion people across 58 sovereign states. Africa remains a land of tremendous potentials in material (natural) resources. Half of the world's usable arable lands are in Africa (AfDB, 2016). About 70% of Africa's land is communally owned or under customary tenure (Peters, 2006). Africa's land resource is increasingly affected by the competition of mutually exclusive uses. The sustainability of natural goods and services depend to a large extent on the natural resources governance arrangement (ownership and access regime). Land resources in most rural Africa are regarded as common pool resources where exclusion is difficult, and their use is rivalrous. Poor natural resource governance and management drives conflicts and creates the tragedy of the commons with attendant composite environmental, social and public safety challenges.

Nigeria is Africa's most populated country with about 183.6 million in 2015 (representing 15.4% of Africa) and an annual growth rate of 2.8% (CBN, 2015). It is Africa's 14th largest country in terms of land mass with 923,768 km^2 (consisting of 909,890 km^2 of land area and 13,879 km^2 of water area). The average population density is 200.1 persons/km^2 and land per capital is 0.507ha in 2015. Nigeria has about 71.2 million hectares of cultivable land (Adeniyi, 2002) but with only 34 million hectares actually under cultivation. This makes the country richly endowed in land resources which are still highly underutilized like in most other countries of Africa. Nigeria spans over 10 degrees of latitude which presents a wide range of climatic, vegetation and soil conditions that provide great potentials for agricultural production. This rich renewable natural resource base is also complemented by rich mineral resources including energy resources, metallic minerals and non-metallic minerals. Despite these huge potential, Nigeria is plagued by natural resource management issues including poor resource governance, unsustainable exploitation, land and environmental degradation, unsustainable resource management, and resource conflicts. Thus, Nigeria is also plagued by the 'rich but poor' and 'resource curse' syndrome that pervades the African landscape.

The objective of this chapter is to explore the dimensions of natural resources issues including livelihood, resource governance and management, and resource conflicts in Africa generally and Nigeria in particular. Section two sets the context for Africa and Nigeria in terms of natural resources and the new resource exploitation drive, section three deals with the main issues including contextualizing natural resources and natural resource management and rural livelihoods, as well as resource conflicts and its connection with public safety in Africa and Nigeria. The last sections highlight some solutions and recommendations, future research directions and conclusions.

BACKGROUND

Several African countries have large tracts of cultivable land, something that is at a premium in most developed countries. Thus in a new deal being offered African countries, in recent years, millions of hectares of land are being leased on a long-term basis by nations as well as private corporations outside Africa (Mbow, 2010). Africa has become the new frontier and a beacon for land grabbing by the developed countries in deals that makes the local population poorer and dispossessed. Like a new era of slavery and resource exploitation, food and energy crops grown on the leased lands are shipped to feed people, animals and industries outside Africa, while Africa remains poor. This is a mystery orchestrated by poor land and other resources governance across Africa. As an extension of the mystery of the capital, African countries have the land resources, but lack the capital and technology to become master of the land and improve productivity and output. The lack of productivity of African agriculture exacts a high human and economic cost and high rates of poverty with over 50% of people live on less than US$1.25 a day and more than 232m people under-nourished (AfDB, 2016). In another era of plunder, the land resources of Africa is becoming another source of wealth and (food) security outside Africa for pittance while countries of Africa remain poor and food insecure and food stamp, and food aid and food import remain high with very devastating human security and public safety effect.

Livelihood in the vast rural interiors of Africa is tied to land resources. Poor resource governance and management and its concomitant effects including poor land management, tenure insecurity, land degradation, and resource conflicts, among others, further drive African societies into poverty. Apart from the globalization of the economies, many countries of Africa, with support from development partners, have recognised that the ability of successful economies to thrive on regular and predictable income from land resources, deliver security of tenure, and attract development is made possible through land reform and land resource management and administration. Variants of land reform including land restitution and redistribution, land consolidation, and most pervasively, changes in the tenure of the land by abolishing complex traditional and customary rights in favour of simple and streamlined mechanisms of land transfer are now being considered or instituted across several African countries.

However, land resources is only an aspect of natural resources. For example, due to the spatial variation in climatic regime and vegetation and soil types, Nigeria is endowed with land resources that can support the growth of different types of crop including: cereals, roots and tubers, legumes, cash/tree crops and fruits and vegetables. This rich agricultural endowment is complemented by rich mineral resources including energy resources (oil and gas, coal and lignite, bituminous sand, hydropower, biomass and solar), metallic minerals (iron ores, columbite, manganese, molybdenum, tungsten, titanium, tin, lead, zinc, etc.) and non-metallic minerals such as gravel, sand, various kinds of clays (benthonic clays, limestone and gypsum, marble, feldspar etc). Thus, natural resource management entails the framework for comprehensive planning and management of both renewable and non-renewable resources in order to safeguard nature and also get maximum societal benefits from these resources. Natural resources are the basic bedrock for societal development. Natural resources can also orchestrate issues of concern that severely undermine public security and safety where the resource governance and management is poor. Issues around unsustainable resource exploitation such as conflicts between farmers and pastoralists over grazing lands and water points have severely undermined public security and safety in virtually every parts of the Nigeria as at today with no solution yet in sight. Poor and unsustainable resource exploration resulting in land degradation and water pollution in the Niger delta area have severely compromised rural livelihoods and led to agitation for resource control and a better deal. Neighbouring communities are known to have gone to war over land and/or water boundary in order to claim ownership of particular resources including agricultural lands. Poor resource management can also have secondary and even tertiary effects including increasing corruption through 'bourgeoisie' intermediaries between communities and resource exploiters and destruction of social bond and cohesion in families, people and communities. These are all serious issues of concern to public safety in resource exploitation and management.

NATURAL RESOURCE MANAGEMENT ISSUES IN AFRICA AND NIGERIA

Natural Resources

Defining what constitutes a resource is not straightforward because it connotes different thing to people in different environmental, social, economic, political, cultural and technological milieu. A resource is not only perceptive, but also contextual and subjective. Adeniyi et al. (1992) and Mitchell (1989) agreed that a resource connotes any part of the environment that has or is known to possess a potential of being developed to satisfy human needs. Natural resources (a resource of nature) is a subclass resource and it refers to anything produced naturally that is needed by a group of organisms (Arms, 1994) and man. Natural resources include air, fresh water, the ocean, soil, land, food, forest, woodland, energy (wind, radiation), and genetic resources (biodiversity). They form the basic biological life support systems on the earth (Arms, 1994, EOLSS, 1996, Oyebande, 1998).

Natural resources, according to Arms (1994), can be irreplaceable or replaceable. Irreplaceable resources cannot be substituted, which means once they are gone, they are forever gone. Examples include species of organisms that go into extinction. Sun's energy is irreplaceable because life could not exist without it. Replaceable natural resources can be replaced or have close substitute or can substitute for others. The replaceable resources group can be further classified into renewable and non-renewable groups. Non-renewable resources can be used up completely or depleted into such a degree that it becomes too expensive to obtain them. Examples include fossil fuels and minerals. Renewable resources can theoretically last forever because they are either produced continuously, or come from sources that can essentially never be exhausted. Examples include solar energy, freshwater, air, soil, trees, crops and other living organisms. The non-renewable resources constitute the basic life (biological) support systems on earth (Arms, 1994, EOLSS, 1996). The renewable natural resources of the earth are the greatest heritage of mankind, the real wealth of the nation, and the only sound basis on which to build hope for future prosperity (UNEP, 1990). In essence, the renewable natural resources encompass all the basic life support systems on which the long term sustainability of the earth rests. The encyclopedia of life support system (EOLSS, 1996) attempts to delineate the constituents of renewable natural resources as follow:

- **Air:** (As a sink and its meteorological aspects)
- **Fresh Water:** (For agriculture and food, mineral, energy, fisheries)
- **Ocean and Seas:** (For fisheries, minerals, plant, sink, energy)
- **Land:** (For forests, grassland, deserts, agric lands, minerals, suburban, urban, soil, sink)
- **Genetic Resources:** (Biodiversity)
- **Solar Resources:** (Radiation, wind, ultraviolet radiation).

It is important to note that all these resources are both source and sink for the different type of energy that gives abundant life on the earth. When the normal equilibrium of any of these is altered it results into what is termed pollution or degradation which poses serious threat to human health, sometimes in ways that go unnoticed for decades and sometimes dramatically (UNEP, 1990).

Another common classification of natural resources according to de Zeeuw (2000) include: non-renewable and non-recyclable resources such as fossil fuels; non-renewable but recyclable resources such as minerals; quickly renewable resources such as fish; slowly renewable resources such as forests; environmental resources such as air, water and soil; and flow resources, such as solar and wind energy.

Non-renewable resources will when extracted, be depleted in the long run. Renewable resources may however also be depleted if the extraction rate exceeds the renewal rate. To sustain a renewable resource, the consumption rate should therefore be maintained within the natural system's capacity to regenerate.

Natural Resource Management

Natural resources management (NRM) refers to the sustainable utilization of major natural resources, such as land, water, air, minerals, forests, fisheries, and wild flora and fauna. Campbell et al. (2001) defines NRM as a conscious process of incorporating the multiple aspects of natural resource use (bio-physical, socio-political or economic) into a system of sustainable management to meet the production goals of direct users (including food security, profitability, risk aversion) as well as the goals of the wider community which include poverty alleviation, welfare of future generations, and environmental conservation. The European Commission (2002) posits that in a sense most natural resources are renewable and are only distinguished by the length of time it takes for them to be reproduced. For example, crude oil and bitumen are produced from sedimentation over long years (typically thousands or millions of years). Thus, their next production may have to wait that long too. Therefore, two effects of interest in considering natural resource management are: the depletion of resources which concerns non-renewable resources, where use will reduce the total stock; and over exploitation of renewable resources and degradation of resources where the waste or side-effects of consumption and production processes may degrade natural resources, resulting in a reduction in the level of physical and environmental services. Thus, defining the targets or guidelines for sustainable use of resource as well as levels of resource use that are consistent with a desired improvement in overall human welfare are key issues in sustainable natural resource management.

According to Barlett (1997) in his seventeen laws of sustainability, the larger the population of a society, and /or the larger its rates of consumption of resources, the more difficult it will be to transform the society to the condition of sustainability. The size of a population that can be sustained (i.e. the carrying capacity) and the average standard of living of the population is inversely related to one another. That suggests that sustainability requires that the size of the population be less than or equal to the carrying capacity of available natural resources for the desired standard of living. The growth in the rate of consumption of a non-renewable resource such as a fossil fuel causes a dramatic decrease in the life-expectancy of the resource. The time of expiration of non-renewable resources can be postponed, possibly for a very long time, by technological improvements in the efficiency with which the resources are recovered and used, using the resources in accord with a program of "sustained availability", and recycling and the use of substitute resources. These are the thrust of modern day resource reduce, reuse, and recycle philosophy which aims to extend the life of non-renewable resources. Sustainable development emphasizes futurity. This recognizes that humanity need to use available renewable natural resources without degrading the capacity of the natural systems to continue to produce these resources; and by implication in order not to jeopardize the survival chances of future generation. One of the ways of doing this is to reduce humanity's ecological footprint on the global ecosystems.

Ecological footprint (EF) shows the extent of human demand on global ecosystems. It measures the area of biologically productive land and water needed to provide ecological resources and services – food, fibre, and timber, land on which to build, and land to absorb carbon dioxide (CO_2) released from burning fossil fuels. According to Dearden and Mitchel (2012) the EF had exceeded the earth's biocapacity since the late 1980s. As at 2007, humanity's ecological footprint had grown to 50% more than biocapacity. It is becoming clear that the earth's regenerative capacity can no longer keep up with demand. This means people are turning resources into waste faster than nature can turn waste back into resources. Humanity is no longer living off nature's interest but now drawing down its capital. The growing pressure on ecosystems is causing habitat destruction or degradation and permanent loss of productivity, threatening both biodiversity and human well-being. This suggest that our rate of consumption threatens future viability of the system and by feedback loop-mechanism, our survival on earth. This is a grave situation with serious public safety and human security concerns.

Natural Resources and Livelihood

Nature is the basis of human existence and civilizations. It supplies the provisioning services which are products obtained from the ecosystems including food, water fuel wood, fibers, biochemical and genetic resources. It also provides regulating services which benefits are obtained by regulating ecosystems processes including climate, disease, water and pollination. In addition, nature provides cultural services which are non-material benefits obtained from the ecosystems. These include spiritual, religious, educational, aesthetics, recreational and ecotourism, sense of place and cultural heritage. Finally, nature provides supporting services including soil formation, nutrient cycling and primary production which are necessary for continued provision of all the other services (MEA, 2005). All these natural provisions support our aspirations and constituents of well-being on the earth including security and safety, basic materials for good life, health and good social relations.

According to FAO (2004), the rural poor in developing countries remain the most directly dependent on natural resources for their food and livelihood security. Subsistence farmers, fishers, hunters and gatherers, and agricultural wage workers depend on the availability of usable land, water, and plant and animal species for their livelihoods. Thus, the agricultural livelihoods of poor rural women and men depend on the condition of natural resources, particularly livelihoods of people living on fragile lands (World Bank, 2005). Livelihood in the words of Carney (1998) comprises the capabilities, assets (including both material and social resources) and activities required for a means of living. A livelihood is sustainable when it can cope with and recover from stresses and shocks and maintain or enhance its capabilities and assets both now and in the future, without undermining the natural resource base. Livelihood in its simplest sense is a means of gaining a living (Chambers and Conway, 1991). It is understood as the ensemble or opportunity set of capabilities, assets, and activities (IPCC, 2014). WCED (1987) defines livelihoods as adequate stock and flows of food and cash to meet basic need; while Concern (2005) refers to livelihoods as the adequate and sustainable access to and control over resources, both material and social, to enable households to achieve their rights without undermining the natural resource base. A livelihood thus comprises the resources (including skills, technologies and organizations) and activities required to make a living and have a good quality of life (Pasteur, 2011).

Central to the definition of livelihoods are livelihood assets, i.e., the means of production available to a given individual or group that can be used to generate material resources sufficient enough to reduce poverty. The greater and more varied the asset base, the more sustainable and secure the livelihood (IUCN, SEI, IISD, 2003, Concern, 2005). The emergence of livelihoods approaches has led to new understandings on how poverty, and the ability to move out of it, reflects the (lack of) capabilities and assets available to the poor (IUCN, SEI, IISD, 2003). The DFID (1999) framework for sustainable livelihood views livelihoods as systems and proposes ways for understanding: the assets people draw upon; the strategies they develop to make a living; the context within which a livelihood is developed;

and those factors that make a livelihood more or less vulnerable to hazards, shocks and stresses. Assets may be tangible (such as food stores and cash savings, as well as trees, land, livestock, tools, and other resources) or intangible (such as claims for food, work, and assistance as well as access to materials, information, education, health services and employment opportunities) (IRP 2010).

Livelihoods are dynamic and people adapt and change their livelihoods with internal and external stressors. Ultimately, successful livelihoods transform assets into income, to improve living conditions, a prerequisite for poverty alleviation. In addition to the direct and physical benefits, adequate and decent livelihoods can improve capabilities by providing conditions and opportunities for widening choices, promoting self-respect, reinforcing cultural and moral values and in other ways improving the quality of living and experience (Chamber and Conway, 1991). The poor are more heavily dependent on ecosystem services and therefore most severely affected by deteriorating environmental conditions and factors limiting natural resource access. In other words, for the livelihoods to be sustainable, the natural resources must be sustained (Rennie and Singh, 1996). Thus a strong nexus exists between natural resource management and public order, safety and security. In other words, natural resource degradation is a threat to livelihoods, as well as public safety and security.

Land, a scarce resource to fulfilling livelihood aspirations, is increasingly affected by the competition of mutually exclusive uses including urban growth, food security, mitigating and adapting to climate change, protecting biodiversity and initiating economic growth, protecting people from natural disasters, and preventing land conflicts. Land use planning can help to find a balance among these competing and sometimes contradictory uses. According to Wehrmann (2011), land use planning is a systematic and iterative procedure carried out in order to create an enabling environment for sustainable development of land resources which meets people's needs and demands. It assesses the physical, socio-economic, institutional and legal potentials and constraints with respect to an optimal and sustainable use of land resources, and empowers people to make decisions about how to allocate those resources. It is also a culmination of all activities and decisions concerned with guiding the allocation and use of land in patterns that enable improvements in peoples' way of living (Chigbu and Kalashya, 2015). Land use planning has become a central prerequisite for any development that aims at social, ecological and economic sustainability. The objective of land use planning is to create the preconditions required to achieve a type of land use that is environmentally sustainable, socially just and desirable and economically sound and fulfills people's livelihood. It thereby activates social processes of decision making and consensus building concerning the utilization and protection of private, communal or public areas.

Land use planning, according to Wehrmann (2011) presents a development approach that contributes to the prevention of land use conflicts, the adaptation of land uses to physical and ecological conditions, the lasting protection of land as a natural resource, the lasting productive use of land and a balanced use that fulfils all social, ecological and economic requirements. It creates the preconditions required to achieve a type of land use that is environmentally sustainable, socially just and desirable and economically sound. It thereby activates social processes of decision making and consensus building concerning the utilization and protection of private, communal or public areas. At the core of land use planning is the joint balancing of competing land uses by all stakeholders and the joint identification of those uses for which the highest consensus can be achieved – ideally for the purpose of sustainability. The use of adequate, locally adapted information systems providing information on, for instance, land availability, existing land rights and land uses can create transparency at national and local levels.

A natural resource-insecure nation with little resources to import ecological capital is an insecure nation. Incidence of droughts leading to food shortages, starvation and hunger, malnutrition, food riots and food stamps and humanitarian crises are great public order and public safety concerns. The nexus between people, development and environment and the natural resources it supports is unassailable in all climes. In most less and least developed countries, weak resource governance ensures that gains from natural resource exploitation are never effectively translated into physical and human development. This disconnection which pervades most resource rich countries seems to be the bane of Africa's development.

Natural Resource Management in Africa

The huge size of Africa partly explains the diversity of resources and environments that characterize the continent. Africa's ecology can generally be classified into the humid (swamp and tropical forest), moist sub-humid (savanna woodland or guinea savanna), dry sub-humid (Sudan savanna), semi-arid (Sahel), arid and hyper arid zones. This divisions runs north and south of the equator except around the high Africa to the east. Ecosystems, agriculture and food production in Africa are highly interconnected. Africa is a primary producer and its economy depends significantly on the natural resource capita. Poor management of the ecosystems and natural resource capita reflects in the economic, social and cultural characteristics of individuals, households and social groups (UNDP-UNEP, 2009). Agriculture which accounts for about 17% of the gross domestic product (GDP) is the main source of livelihood in Sub-Saharan Africa (Muller, 2009). But one of the major concentration of population at risk of hunger is located in tropical savannas of Africa (Falkenmark and Rockstrom, 2008). The natural ecosystems of Africa are facing increasing pressure from global change drivers including land-use change, land degradation and fragmentation, pollution of the natural systems and overexploitation of natural resources. Tropical dry lands of Africa are especially characterized by highly variable and unpredictable rainfall and high ecological vulnerability to droughts (Falkenmark and Rockstrom, 2008). Tropical Africa countries especially in western, eastern and southern Africa that are located in drought prone arid and semi-arid regions may be worst-hit and experienced some of the worst public safety, human security and humanitarian crisis in Africa.

Forests are a significant sink of global carbon (Pielke *et al.*, 2007) and vegetative cover plays a significant role in human and environmental health. About 6.7million km^2 (23%) and 3.5million km^2 (12%) of Africa presently consists of forest and wooded land respectively (FAO, 2011). But the annual rate of deforestation is about 0.53% between 1990 and 2010. The annual rate of deforestation was as high as 1.01%, 1.11% and 1.10% respectively in the dry countries of Botswana, Ethiopia and Somalia; and 2.72% in Uganda, 4% in Nigeria, and 5.75% in Togo. Only few small countries including Rwanda, Gambia and Reunion recorded marginal growth in forest. The distribution of the forests according to purpose suggests that 30% of the total forested land area is for production, 14% for conservation of biodiversity, 17% for multiple uses and 3% for the protection of soil and water (this is lower than the global average of 8%). The carbon storage in living forest biomass has also generally been decreasing across the countries of Africa. The Congo Basin forests represent the most important terrestrial ecosystem block in Africa and about 30million people depend directly on this forest block for livelihood (Sonwa *et al.*, 2011).

The challenge of the natural resource-energy nexus is particularly graphic. Africa has one of the lowest per capita energy production and consumption rates and some of the least energy-efficient economies. The share of the global total primary energy supply (TPES) from Africa was 5.3% in 2008 in which combine renewables (mostly biomass from forests and woodlands) and waste accounted for 48% (OECD/IEA, 2010). The contribution of the combined renewables and waste to total primary energy supply is about 10.4% for South Africa, 24% for Botswana, 41.7% for Senegal, 77% Kenya, and 81.3% for Nigeria

(OECD/IEA 2010). This put pressure on the ecosystems and natural resource stock. Across the sub-Saharan Africa wood remains the major source of fuel especially to majority of the rural communities. Expansion of agricultural land, overharvesting of industrial wood, fuel-wood and pole-wood extraction, charcoal production for rural as well as urban uses and overgrazing remain the key drivers of deforestation (UNEP, 2002, Geist and Lambin, 2001). The deforestation scenario has substantially been aided by population growth, and in some countries such as Nigeria, the energy deficit and the traditional shifting cultivation and slash and burn agriculture and savanna burning to encourage new forage for herds.

Ecosystems influence climate through multiple pathways, primarily by changing the energy, water, and greenhouse-gas balance of the atmosphere (Chapin *et al.*, 2008). In the Sahel and Savannas where agriculture is predominantly rainfed, a decline in precipitation occasioned by change in land-use and land-cover may have considerable ecological and economic consequences (Hoffmann & Jackson, 2000, Omotosho & Abiodun, 2007). The recurring droughts (and concomitant food shortages) across the Sahel (from Mauritania through Niger, Chad, Sudan to the Horn of Africa) are clear testimony to the public safety effects of this. Sub-Saharan Africa has been projected to get drier in future scenario. Current temperatures are predicted to increase in the order of 1.4°C to 5.8°C by 2100 (Solomon, et al., 2007). Erratic distribution of rainfall in space and time has already been noted in the West Africa region (Abiodun *et al.*, 2008, Afiesimama *et al.*, 2006, Nicholson, 2000). The water footprint which is highly correlated with change in terrestrial ecosystems may become critical in defining future trajectory of agricultural lands in rainfed areas of Africa and by implication, the dividing line between areas of human security and insecurity.

Land resources remain the primordial resources of Africa; but land in most African countries are dead capital because so far they lack the representational process to create capital from it. African countries have not been able to institute sustainable land governance. The high rate of poverty, food insecurity, hunger, malnutrition, land degradation, and widespread environmental deterioration across Africa can be substantially trace to the poor land governance regimes. Without a reform that institute efficient land administration and management, African countries continue to be trapped by the poverty cycle created by inability to draw out land from its dormant shelf to the global land market to contribute to economic growth and development. According to Williamson, et al. (2010) wealthy and successful economies thrive on regular, predictable, and institutionalized access to land. They provide reliable and trusted institutions to manage land and to deliver security of tenure, equity in land distribution, sensible and attractive development, and fair land taxation. The beginning of this is what Hernando de Soto (2000) described as a representational process that make assets (land resources) to lead an invisible, parallel life alongside their material existence and thus be used as collateral for credit.

Natural Resource Management in Nigeria

Nigeria's economy is substantially tied to both renewable and non-renewable natural resources. Oil and gas exploration is the major non-renewable natural resource of Nigeria and the backbone of Nigeria's economy. According to the CBN (2015), the oil revenue (gross) stood at N3,830.1 billion, (or 4.0% of Nigeria's GDP representing 55.4% of total federally-collected revenue). Crude oil export accounted for 78.3% of the total export while gas export constituted 14.2%. This means the oil sector constitutes 92.5% of Nigeria's total export in 2015. Agriculture is a key economic mainstay of the majority of households and a significant sector in Nigeria's economy contributing significantly to the GDP. Agriculture and land resources contribute about 40% of Nigeria's GDP and remains the most important sector of Nigeria's economy in terms of employment, engaging 60-65% of the labour force, predominantly smallholders in the vast rural country sides. Some of the key issues of natural resource management in Nigeria include desertification and loss of productive arable lands in the Sahel and Sudan savanna, land degradation including erosion, flooding and salinization, deforestation and forest degradation in the forest zones, desiccation of mangroves through oil spill and oil exploration activities, unsustainable logging and tree felling, poor range management, poor and unsustainable agricultural practices including poor land practices and agrochemicals usage, bush burning and savanna fires, pollution of water bodies, etc.

Natural resource management issues are scattered across different ministries, departments and agencies (MDAs) of government including environment, agriculture, water resources, energy, solid minerals, works and housing, etc. The national policy on environment aims to ensure sustainable development based on proper management of the environmental/natural resources based on the fundamental re-thinking and a clearer appreciation of the interdependent linkages among developmental processes, environmental factors as well as human and natural resources. The policy aims to provide a rational, practicable, coherent and comprehensive approach to the pursuit of economic and social development in a way that minimizes contradictions and duplications, while enhancing inter and intra sectoral co-operation and effectiveness at all levels. This is being pursued through various strategies including biological diversity management, natural resources conservation, land use and soil conservation, forestry, and wildlife and protected natural areas. The objective of the national forest policy is to achieve sustainable forest management that would ensure sustainable increase in the economic, social and environmental benefits from forests and trees for the present and future generation including the poor and the vulnerable groups.

The Nigerian natural resource governance documents and arrangements have been well discussed in Adedayo, et al. (2013) and Fasona et al. (2014). Nigeria has some of the best policies, programmes and projects that address the connection between natural resources, environment, climate change, rural livelihoods and food security. The documents addressed the cross-cutting issues in natural resources-use and human livelihood and human security including: supporting national income accounting systems that include the degradation and depletion of natural resource stocks; promoting farming systems based on natural adaptations across ecological zones and encouraging the production of high yield crop varieties requiring minimum agro-chemicals input; strengthening forest protection programmes to ensure adequate vegetation cover in critical areas and to discourage developments likely to cause harmful changes; combining desirable features of traditional approach (of natural resource management) with modern scientific methods of conservation; increasing support for non-governmental organizations and community tree planting programmes; encouraging viable afforestation and reforestation programmes using tested drought resistant and/or (local) economic tree species and conserving indigenous tree species that are endangered; cultivating fast growing tree species needed to accelerate the regeneration of forests; reducing the percentage contribution of fuel wood consumption in the domestic, agricultural and industrial sectors; encouraging the use of alternative sources of energy e.g. coal briquettes, efficient wood stoves, solar energy, wind energy, biogas; encouraging the establishment of private and community woodlots for supply of fuel wood on sustainable basis; promoting ecosystems compatible agro-forestry practices for food security; developing an appropriate integrated land use plan that emphasizes the establishment of grazing reserves to reduce wanton destruction of vegetation by humans and animals; and integrating human development issues including income generation, increased local control of resources, local institution strengthening, capacity-building and greater involvement of community based non-governmental organizations and lower tiers of government as delivery mechanisms.

However, evidences on the ground suggest a very wide gap between the policy and programme intentions and reality. Policy intentions are not matched with actions or actionable projects. None of those cross-cutting issues has been substantially addressed across the nation. Land degradation and unsustainable land practices remains a challenge. Unsustainable logging and tree felling for fire wood and charcoal continues and unsustainable grazing has become a formidable source of human insecurity and a concern for public safety across the nation. The UNU-IHDP and UNEP Inclusive Wealth Report (IWR) assessment for 20 countries for the period from 1990 to 2008 reports that Nigeria is one of the five countries (with Colombia, Russia, Saudi Arabia, and Venezuela) that had negative growth rates for the Inclusive Wealth Index (IWI) which suggest an unsustainable track where most of the GDP growth has come at the expense of the natural capital base (UNU-IHDP and UNEP, 2012). In spite of the contribution of natural resources to the nation's economic development, degradation and depletion of natural resource stocks is yet to be acknowledged as a draw-down on the nation's capital base.

Natural Resource Conflicts in Nigeria

One of the impacts of poor resource governance and unsustainable management of natural resources with serious implications for public safety and human security is natural resource conflicts. This has become a common place in Nigeria. Odusote (2015) posits that conflict arises from contradictions that occur in social interactions. This may include actions of one or more members of a group that are unacceptable to and thus resisted by one or more groups or member. Social and economic deprivations, natural resource degradation and lack of sustainable livelihood are strong recipe for conflicts. Natural resource conflicts can result from different factors including increasing population and diminishing resources, poor resource governance arrangements and lack of clear demarcation of ownership rights, livelihood fragility, social exclusion, climate change and its concomitant impacts, as well as political and ethnic depravations and dislocations, among others.

Land is a scarce resource increasingly affected by the competition of mutually exclusive uses including urban growth, food security, mitigating and adapting to climate change, protecting biodiversity and initiating economic growth, protecting people from natural disasters, and preventing land conflicts (Wehrmann, 2011). The sustainability of natural goods and services especially with regards to a common depend to a large extent on the ownership and access to common pool resources. Grafton et al. (2004) posits that a property right exist over an asset whenever a recognizable entity is able to exclude, at least partially, others from either using it or enjoying a flow of benefits of its use. Property rights can be individual right (e.g. individual land holdings) and can also be shared between individuals and groups (e.g. family and communal land holdings). Who holds property rights over natural assets and the nature of these rights has very important implications in terms of environmental sustainability and outcomes. Public goods, unlike private goods, are inherently non rival in use i.e. their use is mutually in-exclusive. Land and pasture, like many other environmental assets in the Nigerian savanna, are neither pure public goods nor private goods, but common pool resources where exclusion is difficult (but not totally impossible if the resource governance framework are well implemented), and their use is rivalrous. Although there is no exclusivity, one person's use of the resource reduces the ability of the other to either use or enjoy it. This is one of the roots of violent resource conflicts in the savanna. Due to poor land management and lack of land administration procedures, property rights regimes are completely absent. The state governor is the custodian of all lands in Nigeria by virtue of the Landuse Act of 1978 and individuals are vested only with "usufructuary rights", in the rural areas. Individuals and communal lands must be granted a statutory or customary rights of occupancy certificate by the governor before the 'usufructuary rights' on the land becomes legal. The vast rural lands does not have this due to the very cumbersome land registration process required. Thus, every actor on the land tries to maximize their own benefits at the expense of others. The result is a tragedy of the common where individual users consider only their private costs and not the costs their actions impose on other resource users. Thus, what obtains is Pareto efficiency where it is not possible to make someone better off without making someone else worse off. Violent conflicts and human insecurity is the natural outcome of such exclusivity. This is one of the drivers of conflicts between the Fulani pastoralists and the farmers across the nation (Grafton et al. 2004, Fasona, et al. 2016).

Again due to lack of registration and poor boundary demarcation and documentation procedures, rural communities are always at logger-head over the ownership of lands and who has the right to use a natural resource such as water. This has been worsened by increasing population and diminishing productivity of arable lands. In the Niger Delta area of Nigeria, while the land-use Act of 1978 cemented state ownership of the land, the Petroleum Act of 1969 and other mineral Acts, through the principle of 'eminent domain', placed all minerals and its exploration in the hand of the Federal Government. Thus, all the lands where minerals and other natural resources are located have been parcelled into Oil Prospecting Leases (OPL) (which becomes Oil Mining Leases (OML) once oil has been found on the land). Oil exploration activities on productive lands coupled its externalities including oil spills and ecosystems disruption drive the local population to the margin of survival. This leads to resource agitations and conflicts. Thus, the lack of access to productive natural resources by a group of people to fulfil their livelihood aspirations and advance their desires to be able to live the 'good life' will always lead to conflicts and human insecurity and erode public safety because survival is the first law of nature and a hungry people do not care about sustainability.

SOLUTIONS AND RECOMMENDATIONS

Natural resources is the backbone of the African and Nigerian economies. Natural resource and environmental degradation and unsustainable resource degradation remain some of the greatest challenges in Africa and Nigeria. This is exacerbating poverty, hunger, and resource conflicts with attendant social dislocations. Unsustainable resource exploitation in particular is leading to community deprivation and pollution of local community assets such as water, air and land which also aggravates health and poverty challenges. The resource management and governance arrangement is weak. Africa countries including Nigeria, have vast lands which are dormant because they lack the representational process to draw capital out of them. Thus, the most important resources to the rural peasant life remains hugely unexplored for rural development.

Africa and Nigeria requires innovative solutions to be able surmount some of these challenges. Some of these include:

- Comprehensive natural resource management plans that incorporate all resource aspects including land, water, forest resources, agriculture, fisheries, rangeland etc is required. These plans are very vital to knowing the past state of natural resources and guide present and future exploitation of natural resource base. This resource inventory and utilization plans are vital to safeguard the resources against unsustainable exploitation. This will also assist in proper land-use planning and allocation and reallocation of resources in space and time among competing user-groups.

- Resource governance remains a key issue in Africa and Nigeria. For example, Nigeria has excellent policies on natural resources including forest, rangeland, agriculture and ecosystems, etc. But the policy intentions are never translated into actions. This explains why natural resource and environment degradation is persistent and resource conflicts is pervasive all around the country. Implementation of policies on natural resource management is therefore an important steps toward resolving conflicts on natural resources.
- There is also the need to recognize the multidimensional impacts and influences of natural resource exploitation and degradation for human security including food shortages and hunger, conflicts, as well as disaster vulnerability. Well maintained natural resource systems improve people's resilient and reduces their vulnerability to natural disaster; but when they are degraded, they can exacerbate the impacts of natural disasters.
- Land reform and land administration and management is critically important in Nigeria to draw land resources out of its dormant shelf to make it contribute to capital formation. A spatially enabled land administration driven by Geoinformation technology and systems is prerequisite for efficient land management that will further unlock national wealth creation especially in the vast rural interiors. This will shift the focus from poverty reduction to wealth creation among the teeming rural population.
- Recognizing the intrinsic and intricate relationship between natural resources and rural livelihood is paramount to national security. Natural resource exploitation should be guarded by all means possible to prevent them from impairing rural livelihoods and well-being. Resource exploitation processes, at all times, should be guided by the basic principles of sustainable development including the futurity, polluter pays, precautionary, as well as do no harm principles.

FUTURE RESEARCH DIRECTIONS

Resource inventory is a first step to natural resource management. In Nigeria and across Africa, only little information is available on how much of a resource is available and what is the status. These can only be known by the use of geographic (or geospatial/geoinformation) information systems. Geospatial information are a class of geographic information that are obtained from remote sensing and global positioning systems platforms and managed within a geographic information systems (GIS) framework. Geographic information refers to information which can be related to specific locations on the earth (DOE, 1987). This means information about a place, area or locations of natural resources and their characteristics. Geographic information is handled by GIS which has been described as the biggest step forward in handling geographic data since the invention of the map (DOE, 1987). The innovations in the field of geospatial technology and mapping have contributed immensely to decision making and knowledge-based economy (Kufoniyi, 2009). The need for remote sensing technology for resource assessment in developing countries such as Nigeria is a cost saving method to obtain the first generation of information on some of their natural resources and raw data to update existing information. Remote sensing is the most important tool for rapid natural resource inventorying, development and management. It provides means through which natural resource data and information on hitherto difficult and unmapped locations can be acquired.

Nigeria and other countries of Africa need to improve on collection, processing, integration and use of geospatial data and information to guide resource exploitation and natural resource management. More studies are also needed on the assessment of impacts on resource exploitation on the natural environment and on rural livelihoods. A better understanding of the nexus between natural resource, rural livelihoods, vulnerability, and human security and safety are also important. Experience have shown that the difference in severity of disaster between the developed and developing countries is not really in the number and magnitude of occurrence of events, but in the level of disaster management - disaster preparedness (risk reduction), mitigation, and emergency responses. Nigeria and Africa needs to be able to translate the land resources into tradable assets in formal market. This is critical to operating in today's global market place. More researches are required in the area of land administration and management to transform the huge land resources of Africa into tradable assets.

CONCLUSION

Natural resources are the basic platform of every civilization on earth. Degradation and draw down of the earth's natural resource base has significant implication for human security and public safety and long term sustainability of our planet. Nigeria and other countries of Africa are highly vulnerable to the concomitant effects of loss of natural resources with significant impacts on the economy, environment and climate change, food security, human security and public safety. Natural resource conflicts are already a commonplace in Nigeria. They result from different factors including increasing population competing on diminishing resources, poor natural resource governance arrangements and lack of clear demarcation of ownership rights, livelihood fragility, social exclusion, climate change and its concomitant impacts, as well as political and ethnic depravations and dislocations, among others. Resource conflicts have tremendous impacts on social cohesion, human security, public safety and economic development as natural resource capital is drawn down. There is the need to bring back the spirit of community and responsibility by inspiring respect for the land and our shared resources; reduce our demographic momentum and balance population with resource carrying capacity; restore climatic stability and protect natural resources from excessive degradation and pollution; curb deforestation and excessive ecosystems fragmentation and stabilize soils and restore degraded lands and enforce environmental and natural resources regulations as well as land management laws. Land reforms that institute new regimes of easy, simple and quick titling and registration of lands should be expedited to improve ownership rights on land and natural resources across the Nigeria to guarantee sustainable access to livelihoods among social groups to stem the tide of natural resource conflicts and concern for human security and public safety.

REFERENCES

Abiodun, B. J., Pal, J. S., Afiesimama, E. A., Gutowski, W. J., & Adedoyin, A. (2008). Simulation of West African monsoon using RegCM3 Part II: Impacts of deforestation and desertification. *Theoretical and Applied Climatology*, *93*(3-4), 245–261. doi:10.1007/s00704-007-0333-1

Adedayo, V., Fasona, M., Olorunfemi, F., Elias, P., & Oloukoi, G. (2013). An Evaluation of Policy on Natural Resource Management and Its Impacts on Rural Livelihoods in Nigeria. The Nigerian. *The Journal of Business and Social Studies*, *7*(1), 49–71.

Adeniyi, P. O. (2002). Guest Statement on "Change, Turbulence and Uncertainty in Nigeria". In Resource and Environmental Management (2nd ed.). Prentice Hall.

Adeniyi, P. O., Omojola, A. S., & Soneye, A. S. O. (1992) Mapping Landuse – Landcover Resources of Wamako Local Government Area, Sokoto State, Nigeria With SPOT Satellite Imaging System, *Proceedings of 35th Annual Conference of N.G.A.*

Afiesimama, E. A., Pal, J. S., Abiodun, B. J., Gutowski, W. J. Jr, & Adedoyin, A. (2006). Simulation of West African monsoon using the RegCM3.Part I: Model validation and interannual variability. *Theoretical and Applied Climatology*, *86*(1-4), 23–37. doi:10.1007/s00704-005-0202-8

African Natural Resource Center-African Development Bank (ANRC-AfDB). (2016). *Review of land tenure policy, institutional and administrative systems of Botswana – Case study. African Development Bank.* Abidjan: Cote d'Ivore.

African Union Commission, Economic Commission of Africa, African Development Bank (AUC-ECA-AfDB) Consortium. (2011). Land policy in Africa: West Africa Regional Assessment. Addis Ababa: ECA Publications and Conference Management Section.

Arms, K. (1994). *Environmental Science* (2nd ed.). Saunders College Publishing.

Bartlett, A. A. (1997). *Environmental Sustainability*. Paper presented at the Annual Meeting of the American Association of Physics Teachers Meeting, Denver, CO. Retrieved from http://www.albartlett.org/articles/art1997aug16.html

Campbell, B., Sayer, J. A., Frost, P., Vermeulen, S., Ruiz-Pérez, M., Cunningham, A., & Prabhu, R. (2001). Assessing the performance of natural resource systems. *Conservation Ecology*, *5*(2), 22. Retrieved from http://www.consecol.org/vol5/iss2/art22/

Carney, D. (1998). *Sustainable Rural Livelihoods. What contribution can we make?* London: Department for International Development.

CBN. (2015). *Central Bank of Nigeria Draft Annual Report 2015.* Available at https://www.cbn.gov.ng/Documents/cbnannualreports.asp

Chambers, R., & Conway, G. (1991). *Sustainable Rural Livelihoods: Practical Concepts for the 21st Century*. IDS discussion paper 296.

Chapin, F. S. III, Randerson, J. T., McGuire, D. A., Foley, J. A., & Field, C. B. (2008). Changing feedbacks in the climate–biosphere system. *Frontiers in Ecology and the Environment*, *6*(6), 313–320. doi:10.1890/080005

Chigbu, U. E., & Kalashyan, V. (2015). Land-Use Planning and Public Administration in Bavaria, Germany: Towards A Public Administration Approach to Land-Use Planning. *Geomatics, Landmanagement and Landscape*, (1): 2015.

Concern. (2005). *Approaches to Disaster Risk Reduction*. Retrieved from http://earthmind.org/files/risk/Concern_ApproachestoDRR-paper-final.pdf

De Soto, H. (2000). *The Mystery of the Capital: Why capitalism Triumph in the West and Fails Everywhere Else*. London: Black Swan Books.

Dearden, P., & Mitchel, B. (2012). *Environmental Change and Challenge* (4th ed.). Oxford University Press.

Department of Environment. (1987). *Handling Geographic Information- Report of the Committee of Enquiry chaired by Lord Chorley*. London: HMSO.

DFID (Department for International Development). (1999). *Sustainable Livelihoods Guidance Sheets*. Retrieved from http://www.ennonline.net/dfidsustainableliving

EOLSS. (1996). *Encyclopedia of Life Support System, 1996 Workshop Reports*. EOLSS Publishers Company Ltd.

European Commission. (2002). *Analysis of Selected Concepts on Resource Management. A Study to Support the Development of a Thematic Community Strategy on the Sustainable Use of Resources*. Retrieved from http://ec.europa.eu/environment/archives/natres/pdf/cowlstudy.pdf

Falkenmark, M., & Rockstrom, J. (2008). Building resilience to drought in desertification-prone savannas in sub-Saharan Africa: The water perspective. *Natural Resources Forum, 32*(2), 93–102. doi:10.1111/j.1477-8947.2008.00177.x

FAO. (2004). *The State of Food and Agriculture*. Rome: FAO.

FAO. (2011). *Global Forest Resource Assessment (FRA 2010)*. Retrieved from FAO Yearbook-resources – Landuse: http://www.fao.org/economic/ess/ess-publications/ess-yearbook/ess-yearbook2010/yearbook2010-reources/en/

Fasona, M., Fabusoro, E., Sodiya, C., Adedayo, V., Olorunfemi, F., Elias, P., & Oloukoi, G. et al. (2016). Some Dimensions of Farmers'-Pastoralists' Conflicts in the Nigerian Savanna. Journal of Global Initiatives –Policy, Pedagogy. *Perspectives, 10*(2), 87–108.

Fasona, M., Oloukoi, G., Olorunfemi, F., Elias, P., & Adedayo, V. (2014). *Natural Resource Management and Livelihoods in the Nigerian Savanna*. Ibadan: Penthouse Publication.

Geist, H. J., & Lambin, E. F. (2001). *What Drives Tropical Deforestation? A meta-analysis of proximate and underlying causes of deforestation based on subnational case study evidence. LUCC Report Series No. 4*. CIACO Louvain-la-Neuve.

Grafton, R. Q., Adadmowicz, W., Dupont, D., Nelson, H., Hill, R. J., & Renzetti, S. (2004). *Economics of the environment and natural resources*. Malden, MA: Blackwell Publishing. doi:10.1002/9780470755464

Hoffmann, W. A., & Jackson, R. B. (2000). Vegetation–Climate Feedbacks in the Conversion of Tropical Savannah to Grassland. *Journal of Climate, 13*(9), 1593–1602. doi:10.1175/1520-0442(2000)013<1593:VCFITC>2.0.CO;2

International Institute for Sustainable Development (IISD), International Union for Conservation of Nature and Natural Resources (IUCN) and Stockholm Environment Institute (SEI). (2003). *Livelihoods and Climate Change.* A Conceptual Framework Paper Prepared by the Task Force on Climate Change, Vulnerable Communities and Adaptation. Retrieved from https://www.iisd.org/pdf/2003/natres_livelihoods_cc.pdf

IPCC. (2014). *Climate Change 2014: Livelihoods and Poverty. The Working Group II Contribution to the Intergovernmental Panel on Climate Change Fifth Assessment Report.* Cambridge, UK: Cambridge University Press.

IRP. (2010). *Guidance Note On Recovery: Livelihood.* International Recovery Platform, Kobe, Japan. Retrieved from http://www.recoveryplatform.org/

Kufoniyi, O. (2009). *Enabling environment for geospatial information technology uptake in Africa.* UNECA Committee on Development Information, Science and Technology, Addis Ababa, Ethiopia.

Mbow, C. (2010). *Africa's risky gamble.* IGBP.

Millennium Ecosystem Assessment (MEA). (2005). *Millennium Ecosystem Assessment.* Washington, DC: Island Press. Available at www.millenniumassessment.org/en/index.aspx

Mitchell, B. (1989). *Geography and Resource Analysis.* Longman Group.

Nicholson, S. (2000). Land surface processes and Sahel climate. *Reviews of Geophysics, 38*(1), 117–339. doi:10.1029/1999RG900014

Odusote, A. (2016). Nigeria: The Matrix between Fragility of Livelihoods and Conflicts. Journal of Global Initiatives –Policy, Pedagogy. *Perspectives, 10*(2), 9–30.

OECD/IEA. (2010). *IEA Energy Statistics.* Available online: http://www.iea.org/statist/index.htm

Omotosho, J. B., & Abiodun, J. (2007). A numerical study of moisture build-up and rainfall over West Africa. *Meteorological Applications, 14*(3), 209–225. doi:10.1002/met.11

Oyebande, L. (1998). *The Water Cycle: Courier of Health and Wealth, How Sustainable is it? Inaugural Lecture Series.* University of Lagos Press.

Pasteur, K. (2011). *From Vulnerability to Resilience (V2R).* Warwickshire, UK: Practical Action. doi:10.3362/9781780440583

Pielke, R. A. Sr, Adegoke, J., Beltran-Przekurat, A., Hiemstra, C. A., Lin, J., Nair, U. S., & Nobis, E. et al. (2007). An overview of regional land-use and land-cover impacts on rainfall. *Tellus, 59B*(3), 587–601. doi:10.1111/j.1600-0889.2007.00251.x

Rennie, J. K., & Singh, N. (1996). *Participatory Research for Sustainable Livelihoods.* Winnipeg: IISD, Small study for Directorate B Environmental Instruments Directorate General Environment European Commission (B4-3040/2000/258081/MAR/B2). Retrieved from http://ec.europa.eu/environment/enveco/waste/pdf/zeeuw.pdf

Solomon, S., Qin, D., Manning, M., Alley, R. B., Berntsen, T., Bindoff, N. L., & Wratt, D. et al. (2007). Technical Summary. In S. Solomon, D. Quin, M. Manning, Z. Chen, M. Marquis, K. B. Averyt, & H. L. Miller et al. (Eds.), *Climate Change 2007: The Physical Science Basis. Contribution of Working Group I to the Fourth Assessment Report of the Intergovernmental Panel on Climate Change*. Cambridge, UK: Cambridge University Press.

Sonwa, D. J., Walker, S., Nasi, R., & Kaanninen, M. (2011). Potential synergies of the main current forestry efforts and climate change Mitigation in Central Africa. *Sustainability Science*, *6*(1), 59–67. doi:10.1007/s11625-010-0119-8

Stohlgren, T. J., Chase, T. N., Pielke Snr, R. A., Kittel, T. G. F., & Baron, J. S. (1998). Evidence that local land use practices influence regional climate, vegetation, and stream flow patterns in adjacent natural areas. *Global Change Biology*, *4*(5), 495–504. doi:10.1046/j.1365-2486.1998.t01-1-00182.x

UNDP-UNEP. (2009). *Mainstreaming Poverty-Environment Linkages into Development Planning: A Handbook for Practitioners*. Available online at www.unpei.org

UNEP. (2002). *Global Environmental Outlook 3 – Past, present and future perspectives. Synthesis Report. UNEP, Earthscan United Nations Environment Programme*. Cambridge, UK: UNEPWCMC.

UNU-IHDP & UNEP. (2012). *Inclusive Wealth Report 2012.Measuring progress toward Sustainability*. Cambridge, UK: Cambridge University Press.

WCED. (1987). *Global Policies for Sustainable Agriculture, a Report of the Advisory Panel on Food Security, Agriculture, Forestry and Environment to the World Commission on Environment and Development*. London: Zed Books.

Wehrmann, B. (2011). *Land use planning: concept, tools and applications*. Deutsche Gesellschaft für Internationale Zusammenarbeit (GIZ) GmbH, Division Agriculture, Fisheries and Food Sector Project Land Policy and Land Management, Eschborn/Germany.

World Bank. (2005). *Agriculture Investment Sourcebook*. Washington, DC: World Bank.

Zeeuw, A. de (2000). *Resource Management: Do we need public policy?* Retrieved from http://ec.europa.eu/environment/enveco/waste/pdf/zeeuw.pdf

KEY TERMS AND DEFINITIONS

Natural Resources (a Resource of Nature): A subclass resource and it refers to anything produced naturally that is needed by a group of organisms and man. Natural resources include air, fresh water, the ocean, soil, land, food, forest, woodland, energy (wind, radiation), and genetic resources (biodiversity).

Natural Resources Management (NRM): The sustainable utilization of major natural resources, such as land, water, air, minerals, forests, fisheries, and wild flora and fauna.

Resource: A resource connotes any part of the environment that has or is known to possess a potential of being developed to satisfy human needs.

Chapter 3
Environmental Policies for Emergency Management and Public Safety:
Implementing Green Policy and Community Participation

Amidu Owolabi Ayeni
University of Lagos, Nigeria

ABSTRACT

Policy refers to the commitment of people or organization to the laws, regulations, and other green mechanisms concerning environmental issues. Community participation has become important in government, policy makers, and environmentalists over last few decades, and as a result, it is now an established principle as it is widely used not only in academic literature but in policy-making documents, international discussions, as well as in local debates when considering issues dealing with decision-making to achieve sustainable development. Implementation of green policy and community participation programs through representatives—organization, groups of individuals—enhances the benefits of polices and program and adds value to policy as well as making the policy's results and responses more effective and stronger.

INTRODUCTION

Management of any programme and success requires the participation of various concern stakeholders. The stakeholders may include government, environmentalists, and scientists as well as communities and amongst others. In green policy, the role of communities and their participation requires the understanding of the environment, problems and appropriate mechanism to address the problems. Although, it should be noted that the wide gaps and weak relationships between government, policy/decision makers and grass roots have results to rejection of communities while in most they are not carried along in the planning and crucial national and/or state decisions. In addition, the relationships of various human facets - existing

DOI: 10.4018/978-1-5225-3194-4.ch003

politics, socio-economic and cultural environment also form the lacuna. As a result, local communities have forced to be hostile in reaction and commitment to issues requires their efforts and knowledge.

In response to the above, this chapter considers implementation of green policy and community participation as an important agenda for achieving sustainable development particularly in the developing nations. The paper examines importance of local communities if actively involved in decision-making to facilitate the achievement of sustainable development. It argued that involvement of community in decision- making produces many important benefits. Finally, its underscores conditions under which green policy and community participation produce the greatest gains in effective citizen governance to guide policy makers towards sustainable development.

BACKGROUND

Over the years now, the need to understand community participation and community-based environmental management initiatives have been addressed different studies (Richardson, 2003, Roe, Nelson, Sandbrook, 2009; van Breugel, 2013; Wasilwa, 2015). Community participation is seen as a proven approach to address issues that have been long utilized in application environmental policy and decisions for sustainable development (Bass, Dalal-Clayton and Pretty 1995; Muhanna, 2006; Wasilwa, 2015). Any policy, decision and programs that fail to consider community participation in any environmental issues is attempting impossible solution Irvin and Stansbury, 1996; Ferreira, 1996; Nour, 2011). As a result of this statement, developing countries' poor communities often argued that environment is an expensive luxury due to lack of community opinion in policy and decisions (Nath, 2001; McCandless and Karbo, 2011; van Breugel, 2013; Happer and Philo, 2013).

Indigenous communities and/or traditional communities around the world have over the centuries developed a close unique connection with the environment – lands and waters around them for their livelihoods (Campese, et al., 2009). These connections have established various bound of participation in various communities. Community Participation is basically concerned with the involvement of individuals, groups and communities in the affairs of managing the environment they live in (Ogunleye-Adetona and Oladeinde, 2013; Aloni et al., 2015; Vanclay et al. 2015). Globally today, Community Participation in environmental issues is being underscored because of the realization of the necessity to involve the stakeholders in the conception of a lasting solution to environmental and human problems (Campese, et al., 2009; Aloni et al., 2015, Ayeni et al., 2016). Human activities intended to improve the socio-economic well-being of man have continue to contribute to the degradation of the environment, therefore, to address, the problems sustainably, major stakeholders in the environment must necessarily be involved (Costanza, et al., 2009; Aloni et al., 2015).

In alliance to the goal of Sustainable Development to create and maintain prosperous social, economic, and ecological systems, today, the participation of communities in managing their environment is obviously a global interest to incorporates communities in decision making that can sustainably influence environmental development (Bass et al., 1995; Aloni et al., 2015; Vanclay et al. 2015).

However, most developing country including Nigeria have long established laws and formal government structures to address serious environmental problems (Bell, 2002) which in turn affects sustainable development of such nation. Regulations and policies are the most common approach to environmental problem (Adelegan, 2005). Meanwhile experience from Nigeria environmental policies and implementation has showed that community has significant roles to play in the interaction of man with his environment. Environmental law provides a cohesive and comprehensive framework for a city or local government to identify significant aspects and manage its environment - both opportunities and risks - and to document, evaluate and communicate its environmental plans and programmes to local stakeholders (UNEP-IETC 2004).

It is this brief introduction that this chapter examines Green policy and community participation implementation for sustainable development. It also underscores how community participation can aid the policy through individual participation at different tiers of government vis-à-vis local, state and federal via ways communities meet their present needs without compromising the ability of future generation to meet their own needs.

GREEN POLICY

Green policy refers to the laws, regulations, ethics and other policy mechanisms established for the purpose of environmental issues (Harrison and Coussens 2007; OECD, 2008, Bernstein and Sweeney, 2012). It is also the totality of the government efforts to protect, conserve and promote environmental issues for sustainable development. The issues may be seen as air, water, land pollution, waste and toxic management, ecosystem (plant and animal) management, maintenance of biodiversity, wildlife, endangered species and most especially protection of natural resources (Scherr and McNeely, 2008). Green policy focuses on problems arising from human impact on the environment, which has negative impact in human values. A green policy is a documented strategy of the organization in terms of its actions relating to the environment which include need for and approach to compliance with regulation (McCormick, 2011).

However, policies can be instituted to overcome harmful effects on biophysical environment, natural resources, and ensure changes in the environment do not affect humans negatively (Daily and Ehrlich, 1992; de Sherbinin et al., 2007). Environmental issues that are addressed by environmental policy are beyond the control of one person rather a collective effort of all concern stakeholders. For example, a factory produces waste pollution which may be dumped into a river, ultimately contaminating water (Meiners and Yandle, 1998). The members of the community will have to find means to clean the water before drinking which does not concern the factory. It should be noted that the tragedy of the commons is the problem – meaning that; communities will assume the river belongs to nobody while each of them will now have an incentive to utilize their common resources as much as possible (Meiners and Yandle, 1998; Roach, Harris and Codur, 2015). This issue will perhaps, continues and beyond control if government and concerns stakeholders fail to intervene.

The Concept of Green Policy

For example, in 1988, the Federal Government of Nigeria established "the Federal Environmental Protection Agency (FEPA)" now Federal Ministry of Environment with its major objective to protect, restore and preserve the Nigeria ecosystem. FEPA was to establish Environmental guidelines and abatement and control of all forms of pollution as well as empowered to initiates specific programs of environmental protection and monitoring stations or network to locate sources of a danger associated with pollution (Adelegan, 2004). These form the basis for environmental law and regulations. On a lighter note, environmental law/policy is synonymous with green policy.

Environmental policy is "any action deliberately taken to manage human activities with a view to prevent, reduce, or mitigate harmful effects on nature and natural resources, and ensuring that man-made changes to the environment that do not have harmful effects on humans or the environment" (McCormick, 2001; Ries, Jenkins, and Wise, 2009). Environmental law comprises of two major aspects vis-a-vis environment and policy. Environment is the physical ecosystem as well as social dimension which look into man's quality of life, health, resources management and biodiversity while policy is an action or principle adopted or proposed by a government, party, business and even individual. Thus, environmental policy focuses on problems arising from human impact on the environment which in turn unto human society by having mostly negative impact on human values such good health and green environment (Rushefsky, Mark 2002). Any regulation concerning toxic substances including pesticides and many industrial wastes are part of the focus of green policy. This policy can be deliberately taken to direction and oversee human activities within the environment and thereby prevent harmful or dangerous effects on the biophysical environment and that of natural resources as well as to make that the changing environment will not have harmful effects on humans. In order for green policy to be implemented and be made real there are different instruments employed by those in charge of the policy. These tools are used by government to implement their environmental policies. Examples of such instrument use by government include constituted law. For example, 1999 constitution of the Federal Republic of Nigeria recognizes the importance of improving and protecting the environment and makes provision for it (Ijaiya and Joseph, 2014). The provision is made in the following sections of constitution;

- Section 12 establishes, though impliedly, that international treaties (including environmental treaties) ratified by the National Assembly should be implemented as law in Nigeria;
- Section 20 makes it an objective of the Nigerian State to improve and protect the air, land, water, forest and wildlife of Nigeria, and
- Section 33 and 34 which guarantee fundamental human rights to life and human dignity respectively, have also being argued to be linked to the need for a healthy and safe environment to give these rights effect (Environmental Law Research Institute, 2011):
    - Economic incentive and market based instruments such as taxes and tax exemptions tradable permits and fee can be very effective to encourage compliance with environmental policy (Eccleston and March 2010).
    - Environmental Impact Assessment Act of 1992 (EIA Act)
    - Harmful wastes (special criminals' provisions etc) Act of 1999 (harmful waste Act)
    - World against indiscipline(WAI)

All these laws and policy are set to manage the environment in such a way that it will be useful for next generation.

It is a fact that Nigeria recognizes various challenges facing the country in moving its economy to a more green - environmentally friendly, and sustainable development (Federal Ministry of Environment, 2016). The economic, social and environment pillars of sustainable development have been adequately embedded in many parts of the Nigeria 1999 Constitution. It is stated in the Constitution that the Federal Republic of Nigeria is *"a State based on the principles of democracy and social justice"*. The Constitution also promises to all Nigerian citizens' justice encompassing the social, economic, political, equality of status opportunity and the dignity of the individual. Emphatically on the environment and sustainable development, Article 20 (sub-section 2), of the Nigeria Constitution states that *"the State shall protect and improve the environment and safeguard the water, air and land, forest and wild life of Nigeria".*

As a result, Nigeria is one of the nations committed to a national environmental policy that will ensure sustainable development based on proper management of the environment. The goal of the Nigeria Policy on the Environment (Federal Ministry of Environment, 2016) is to achieve sustainable development, and, in particular to: secure a quality of environment adequate for good health and well-being; conserve and use the environment and natural resources for the benefit of present and future generations; restore, maintain and enhance the ecosystems and ecological processes essential for the functioning of the biosphere to preserve biological diversity and the principle of optimum sustainable yield in the use of living natural resources and ecosystems; and raise public awareness and promote understanding of the essential linkages between the environment, resources and development, and encourage individual and community participation in environmental improvement efforts.

For example, one of the basic thrust of agricultural Policy is rational utilization of agricultural resources, improved protection of agricultural land resources from drought, desert encroachment, soil erosion and flood, and the general preservation of the environment for the sustainability of agricultural production while that of energy policy is to guarantee adequate, reliable and sustainable supply of energy at appropriate costs and in an environmentally friendly manner, to the various sectors of the economy, for national development (Wehrmann, 2011; Adedayo et al., 2013; FMARD, 2016). The National Water Policy of 2004 (Federal Republic of Nigeria, 2004) reiterated that water quality and quantity are interdependent and shall be managed in an integrated manner, which is consistent with broader environmental management approaches. Water quality management options shall include the use of economic incentives and penalties to reduce pollution; and the possibility of irretrievable environmental degradation as a result of pollution shall be prevented.

In addition, Agencies in Nigeria have been created to focus attention to some specific environmental problems and/or issues. These agencies include but not limited to the *National Oil Spill Detection and Response Agency* (NOSDRA) and *National Environmental Standards and Regulations Enforcement Agency* (NESREA), which were created respectively in 2006 and 2007 (Federal Ministry of Environment, 2016). NOSDRA has the mandate to implement the national oil spill contingency plan. NESREA has responsibility to enforce all environmental laws, guidelines, policies, standards and regulations in Nigeria, as well as enforce compliance with the provisions of all international agreements, protocols, conventions and treaties on the environment to which Nigeria is a signatory.

COMMUNITY

According to Dalby and Mackenzie, (1997) and Fortmann and Roe, (1993) Communities are typically defined on the basis of geographical establishments with respect to, and occupying a particular geographical space. Communities can be defined by characteristics that the members share, such as culture, language, tradition, law, geography, class, and race. In the opinion of (Anyanwu, (1999), community is a group of people reside in the same geographical area, a work place, even an idea or a theme/issue, or on the basis of gender/age. The group may be a village, a family, tribe but the people must have common basic conditions of life. According to Shaeffer (1998) some communities are homogeneous while others are heterogeneous; and some united while others conflictive. Some communities are governed and managed by leaders chosen democratically who act relatively autonomously from other levels of government, and some are governed by leaders imposed from above and represent central authorities. The shape and size of a community varies, and hence definitions of communities have also varied. Regardless of the definition, the community must be bonded by certain principles, such as, a shared territory, beliefs, bonds of fellowship, common culture, set standards and administration.

The community has a unique administrative leadership responsible for making laws and ensuring that such rules and regulations are followed-to maintains sustainability. Those who abide and bring glory to the community are rewarded. Community could arise in different perspectives. for instance, in 1996, Bray identifies three types of communities vis-à-vis geographic community, ethnic, racial, and religious, and shared family or educational. According to Bray (1996) geographic community relatively defined according to its member's place of residence, such as a village or district. The ethnic, racial, and religious communities based its membership on ethnic, racial, or religious identification, and commonly cuts across membership based on geographic location. The family or educational communities based on shared concerns of family or educational and include parents' associations and similar bodies that are based on families, shared concern for the welfare of students.

In addition, community is recognized based on particular attribute or function or characteristics (Zenter, 1964). In Zenter's opinion, community is a group structure, whether formally or informally organized; in which members play roles which are integrated around goals associated with the problems from collective occupation and utilization of habitational space. Secondly, community's members have some degree of collective identification with the occupied space and lastly, community has a degree of local autonomy and responsibility.

Participation

Participation is a development strategy that has been tested and adopted as a useful tool that would assist a people not only to identify priority needs, but also to partake in planning, implementation, and evaluation of community projects that are expected to improve the lives of such communities. Participation also involves positively engaging the people and their leaders in making contributions on projects that would be relevant to the development of their communities. In a nut-shell, participation is the involvement of a community in development project (Raymond, 1974).

Community Participation

Community participation also carries different interpretations and inclinations although the interpretations and inclinations cannot be divorced from the broader aim of encouraging the active participation of local people in the process as a whole (Oakley, 1989). Community participation either through groups of individuals or organizations optimizes the benefits of policies and programs, add value to them and results in a stronger and more effective response. It is the effective environmental promotion and practice which places people at the center of all activities through elected representative with identifying and pursuing the common goals (Laird, 1993). The involvement of community ranges from activities defined by outsiders to the management as well as ownership of activities developed primarily by community members themselves (Aubel and Samba, 1996). Environmental promotion needs to be carried out by the people and with the people, rather than on the people or to the people (Prizzia, 2005). This requires us to engage with the communities in ways that allow people to have ownership and individual involvement with immediate environment.

Community participation has also involved community engagement or community action with regards to four common levels of community engagement. The first basic level is to provide information to individual or community about decision making, policy and activities underway in such a way it will be understandable by all the citizens. The second level is to consult with individual or community and get feedback on the proposed activities. The third level is working with the community explore issues. It may be done by organizing a program that involve community member the fourth level which is the highest is a process whereby committees involve a special program will be in partnership with the community in decision making (Cooper and Vargas, 2004). In the context of protecting the environment for future use there is also need to recognize differences in perspective to achieve a balance of increased participation, greater expertise and efficiency in the decision-making process (Prizzia, 2005).

More so, because most environmental decision are concern with establishing right and responsibilities over the use of common natural resources (water, land, air and biodiversity) environmental law must require substantial rights of community participation to provide checks and balances on administrative government and to improve the quality of decision. Community participation can also enable advocacy on behalf of interests not normally represented (Spyke, 1999). Community involvement in protecting the environment of their impacted community and related environmental issues is generally supported for its potential and ability to provide low sources of information to the government agencies increased acceptance of and confidence in government decisions, empowered community members on issues that affect them and advancement of democratic ideals (Fiorino, 1990, 2000, Heinman, 1997, Shepard and Bowwels, 1997, Prizzia, 2005).

Participation can take different forms and can vis-a-vis passive, contractual, consultative, and collaborative, community self-mobilization and participation among colleagues. Passive participation is practiced when a group of people receive information about something that will occur or has already occurred (Bass, Dalal-Clayton and Pretty, 1995 and Cornwall, 2008). The idea for community participation comes from outside the community. The participatory interaction occurs in only one direction from those who have made decisions towards those who must listen. Contractual participation is when community's participation takes place through a formal arrangement. Consultative participation refers to a situation when the initiative to participate comes from outside basing on the desires and needs of the people. The agent from outside defines the problems and solutions. In collaborative participation, an agent and community participate in diagnosing the problem till implementation, monitoring and evaluating the initiatives. Participation among colleagues emphasizes on activities that can increase the abilities of local people.

The involvement of these local people in environmental issues have become a valuable resource in resource management and environmental security. It has been observed that there an increasing realization in the observations and assessments of indigenous groups in providing valuable local information and local verification for global models (Kirsty, 2010; Ayeni et al., 2015 & 2016).

In Africa, local communities had well-developed traditional indigenous knowledge systems for environmental management and coping strategies, making them more resilient to environmental change (Ayeni et al., 2014 & 2016). The local knowledge has continued to receive a high degree of acceptability amongst the majority of populations in which it has been preserved and are easily identify with this knowledge and it facilitates their understanding of certain modern scientific concepts for environmental management including disaster prevention, preparedness, response and mitigation (Ayeni and Olorunfemi, 2014; Wangu et al., 2015). It is a fact that environmental resources vis-à-vis land, water, animals and plants have their place within the sanctity of nature and are not just production factors with economic significance (Dyson et al., 2003; Nehal et al., 2004). Certain places particularly in Africa are known and valued for their special spiritual significance and are such used as rituals and sacrifices abodes with specific custodians (Abah et al., 2015; Gandile et al., 2017). The focus of Indigenous Knowledge is to establish the role of cultural practices through which communities consume, sustain and conserve natural resources for livelihood and maintain equilibrium (Kalafatic, 2007; Ayeni and Olorunfemi, 2014; Gandile et al., 2017). It also encourages the local people to account through transformation knowledge the local ecology, environment and sustainability and evolve strategies which could be environmentally sustainable and based on local knowledge and practices (Wehrmann, 2011). For example, there are various high biodiversity locations well conserved and protected by the African communities as sacred forest, shrines, mountains, grooves and rivers.

Factors Influencing Community Participation

Community participation depends on many factors vis-a-vis individual attitude, leaders 'qualities, decision-making, transparency, rules, availability of resources and personal attributes.

- **Attitude:** A persistent tendency to feel and behave in a particular way toward some object (Luthans, 2005). Attitude is used in describing people and explaining their behavior. It is on the hand refers to as an evaluative statement either favorable or unfavorable concerning objects, people or events (Luthans, 2005).
- **Leadership:** A process of influencing subordinates so that they cooperate enthusiastically in the achievement of group goals and / or objectives (Chandan, 2003; Jain, 2005 Igbaekemen, 2014). Leadership is the ability to influence through communication, the activities of others, individually or as a group toward the accomplishment of worthwhile meaningful and challenging goals Leadership can be formal or informal (Chandan, 2003). Formal leadership occurs when a person is appointed or elected as an authority figure while informal leadership emerges when a person uses interpersonal influence in a group without designated authority or power. Leaders should be effective and efficient and as well endowed with good leadership qualities. Success or failure of community depends on the quality of leadership (Jain and Saakshi, 2005).
- **Decisions:** Can range from those of a vital, once for all nature to those of a routine and relatively trivial in nature (Cole, 2004). There are three principal decision areas: strategic, operating and administrative ((Koontz and Weihrich, 1998; Cole, 2004). Strategic decisions are the basic long-term decisions which settle the organizations relationship with its environment. Operating decisions are the short-term decisions which settle issues such as output level, pricing and inventory levels. Okali and Farrington (1994) uses the level of involvement in decision-making as a basis for classifying different types of participation and critical decision-making powers need to be transferred to communities. Decision-making is very important all levels of participation and therefore, need to be transferred to communities (Paul and Demarest, 1984; Botes and van Rensburg 2000).
- **Transparency:** An effective way to encourage community participation as it can potentially change power relations between communities and development organizations and between interests within communities (Ghaus-Pasha, 2005; Wasilwa, 2015). Merely transferring funds to committees is not adequate to introduce community control, as communities need to be protected from the abuses of committees hastily assembled to present them. Agrawal et al (2006) further say

that accountability of the actors can be increased if information on the roles and obligations of the government is made available by the media.

- **Resources:** Include physical assets, such as plant, equipment, and location, human assets, in terms of the number of employees and their skills, and organizational assets, such as culture and reputation (Hunger and Wheelen, 2007). human resources are conceived in terms of experience, knowledge and understanding that managers bring to the context of the organization. Organizational resources include structure and its systems for planning, coordinating and controlling as well as informal aspects such as the nature of internal and external relationship.

Chandan (2003) defines rules as a specific policy statement about conduct of certain affairs. Jones and George (2001) further narrate that rules are formal written instructions that specify actions to be taken under different circumstances to achieve specific goals. They govern specific questions, issues and circumstances (Griffin and Ebert, 1991). Some rules are positive requirements that certain actions be taken while others are prohibitions against unethical, illegal or undesirable actions (Anthony and Guvindarajan, 2004). Rules are substitutes for managerial direction and they may be used to specify the goals of the workers, indicate the best method for performing tasks, show which aspects of task accomplishment are most important and outline how individuals will be rewarded (Thomson, 1998; Clark, 2003).

It is widely accepted that households' participation depends on the characteristics of the household. Education has been reported to influence farmers' participation and stimulates social participation (Lise, 2001). Education is a significant instrument to stimulate local participation in a variety of development management initiatives. This means that younger people are more willing to participate than older ones. A study done by Batwel (2008) showed that there was relationship between age of the people and their participation in communal projects whereby younger people participated more than older ones.

Community participation is very importance in achieving green policy goals because community is the grass root of every nation (Wasilwa, 2015). The best person to interact with in any environment is those present in the particular community (Ayeni et al, 2016). some of the importance of community participation include the following.

1. People organize best around problems they consider most important
2. Local people tend to make better economic decisions and judgments in the context of their own environment and circumstances
3. Voluntary provision of labor, time, money and materials to a project is a necessary condition for breaking patterns of dependency and passivity
4. The local control over the amount, quality and benefits of development activities helps make the process self-sustaining (Wasilwa, 2015).

By utilizing economic tools, early theorists offered that policies to protect the environment could also promote innovation and turn a profit. According to Pigou in 1920, the presence of incidental, uncharged services act as a barrier to achieving equilibrium in the market (Emas, 2015). Pigou noted that the divergence between marginal private costs and benefits and marginal social costs and benefits create what we now call "externalities" (Pigou, 1920). These externalities are conceived as transaction spillovers, or costs and benefits unaccounted for in the given price of a good or service. In order to correct the market failure, Pigou proposed a tax on those activities that produce negative externalities at a rate equal to those external costs. By levying this charge, called a Pigouvian tax, the market price will more accurately reflect the comprehensive costs and benefits of the activity.

Community Participation in Green Policy for Sustainable Development

Sustainable development is a process of meeting human development goals while sustaining the ability of natural systems to continue to provide the natural resources and ecosystem services upon which the economy and society depends (Mensah and Castro 2004; Keengwe, Mbae and Onchwari, 2016). As the concept developed, it has shifted to focus more on economic development, social development and environmental protection for future generations. managing the environment of using natural resources sustainably and of disposing and keeping the wastes generated to a minimum has become an increasing priority over the last few decades (Kibert, et al., 2012; Chukwuma et al., 2012). The unbridled industrial growth of the last century and this present has created impacts, mostly negative, on the environment that is only now being realized (Kibert, et al., 2012).

Community Participation in Environmental Management is basically concerned with the involvement of individuals, groups and communities in the affairs of managing the environment they live in. Community Participation in Environmental matters is today being emphasized globally because of the realization of the necessity to involve the stakeholders in the conception of a lasting solution to their problems (Musa and Ifatimehin, 2012). Various human activities aimed at improving the socio-economic well-being of man have contributed enormously to the degradation of the environment. Any attempt to redress or arrest the problems must necessarily involve the major stakeholders, if it must be sustainable and stand the test of time.

The involvement of communities in managing the various activities in their environment today is appreciably a global order in incorporating host communities in decision making in any developmental activities that may affect their environment. The goal of Sustainable Development is to create and maintain prosperous social, economic, and ecological systems. These systems are intimately linked: humanity depends on services of ecosystems for its wealth and security. Moreover, humans can transform ecosystems into more or less desirable conditions. Despite the efforts of many governments to implement strategies as well as international cooperation to support national governments, there are continuing concerns over global economic and environmental developments in many countries (Voumik, 2014; Mukuna and Shisanya, 2015).

The participation and role of the ‘community’ in these processes has become an important component of all environmental programmes and projects, both in terms of decision making processes, and of creating an enabling environment for the community to have a say over aspects that affect their lives (Wasilwa, 2015). In general, in all development programmes of government, local level participation has occupied an important position in project implementation (Ulleberg, 2009; Hora, 2014; Wasilwa, 2015). Meanwhile participation of community in environmental law makes it easy for both community and the government to have equal benefits for policy makers, public servants and civil society (Bass, Dalal-Clayton and Pretty 1995). To highlight few: Participation helps government to make more appropriate decisions based on the local needs of people. The more informed people are, or the more participation, the better they understand about effect of greenhouse gas and we can sustain the environmental budget and resource limitations in managing environmental programmes through sustainable use of natural resources and of disposing and keeping the generated wastes judiciously (Asnani, 2006).

SOLUTIONS AND RECOMMENDATIONS

The interaction and consideration of application of green policy by community for sustainable development is below average mainly because most people at the local level are really aware on how to relate with the environment and the lay down policy to guide them. It has even reach the level of most community having traditional believe of what might happen in a particular setting. However, evidence that has been accumulating in diverse regions all over the world suggests that natural and social systems behave in nonlinear ways, exhibit marked thresholds in their dynamics, and that social-ecological systems act as strongly coupled, complex and evolving integrated systems (Berkes, and Folke, 1998; Gunderson, and Holling, 2002)

Individuals from the community may be giving opportunity to engage in community work e.g. monitoring the exploitation of the environment, thus ensuring that locally important factors are assessed (OECD, 2013). The recognition of the need for active community participation in sustaining environmental management will make the environment conducive in the near future for coming generation.

Developing trust with communities in which programs are being implemented can be used to ensure that solutions will be workable, accepted by, appropriate to the needs of the local communities, and can be effectively implemented. Community participation builds community education where those involved will be informed about green policy with programmes and projects for the community are incorporated (Bass, Dalal-Clayton and Pretty 1995; Wasilwa, 2015. A clear understanding of the necessity of community participation will increase the wiliness of stakeholders to contribute their time and efforts. Developing processes that are responsive and effective for incorporating input from the community into green policy, program will result in a wider range of perspective being heard. Community participation in green policy further increases accountability as it is in the community's interest to ensure that work is being carried out responsibly and effectively (Meier, B. M., C. Pardue and L. London, 2012).

This therefore calls for people's skills and knowledge to participate in planning, implementing and managing green resources via the local governments and the citizens' commitment and readiness to accept equal responsibilities and activities in managing their environments (Kusek and Rist, 2004; DESA 2008). The honest inclusion of a community's representatives as "partners" in decision-making makes for successful community participation.

There are five important factors that illustrate the importance of community participation and involvement in green policy (Chirenje, Giliba and Musamba, 2013; Wasilwa, 2015; Kilewo and Frumence, 2015), and include:

1. Choices and preferences on quality of life and lifestyle - are made at the community, household and individual levels;
2. It is important to maintain subsidiarity of decision-making, since local, daily decisions need to be taken at the local and community levels;
3. Community participation calls for clear commitment and involvement of all members of a community in order to ensure success of various joint activities;
4. Community participation pools resources and diverse skills and working strategies from within the community, creating pride and ownership of a programme or project, and
5. Community participation will ensure that checking and corrective action through monitoring/ evaluation can be done by and for the community itself.

FUTURE RESEARCH DIRECTIONS

Future research should focus on green policy reforms, institutional challenges and sustainable development: The nexus and the implication in a climate change scenario.

CONCLUSION

In managing environmental, communities as one of major stakeholders should seriously considered through effective policy that recognizes their knowledge Implementing Green policy and community participation is very important in sustainable development. Green policy makes it easy although hard to comply with how man can relate positively with environment rather than negative reaction. This is because man remains the major contributors of various impacts during the interaction with the environment as man cannot be in total isolation with the factors that can tamper with the environment. Nevertheless, involving the community participation in various environmental issues e.g. planning and management has been suggested as a remedy to environmental problems. Positive impacts can be ensured through involvement of community knowledge and skills as well as consensus-driven visions and goals of all stakeholders for effective and efficient coordination and monitoring man's immediate environments.

Neglecting our rural communities as well as their traditional values and knowledge in environmental management could have a very damaging backwash effort. The rural communities should be considered alongside their urban counterparts. From all indications, knowledge of local institutions and cultural practice is a very crucial success factor in environmental management. Tradition is considered the bedrock of society on which any kind of development is built. For this reason, there is need for implementing green policy - community-based programme which will focus on environmental resources management. Communities should be allowed to participate in managing natural resources by recognizing and accepting communities' legal, institutional and attitudinal rights (Luning, 1995). These form part of effective tools of environmental management and protection from the perceptive and knowledge of cultural values, norms and practices of the communities.

The government and non-governmental environmental agencies should undertake or sponsor research studies and documentary on indigenous knowledge, values and practices of environmental management across the country. Also, the private organizations, corporate multinational enterprises, environmentalists, the academia and even the press should equally be actively involved in the effort to fathom the largely neglected traditional ways of dealing with environmental problems to forge a harmonious and lasting relationship between man and environment. The ultimate purpose of this is to conserve our environment (natural or otherwise) not by scientific knowledge but also involving the indigenous peoples and/ or local communities.

In conclusion, values of rural communities' participation, perception and knowledge need to be adequately discovered, identified, duly recognized, properly documented, jealously adopted and adapted into government environmental action policies. This will effectively bridge the gap in the environmental policies' formulation and implementation.

REFERENCES

Abah, J., Mashebe, P., & Denuga, D. D. (2015). Prospect of Integrating African Indigenous Knowledge Systems into the Teaching of Sciences in Africa. *American Journal of Educational Research*, *3*(6), 668–673.

Adedayo, V., Fasona, M., Olorunfemi, F., Elias, P., & Oloukoi, G. (2013). An Evaluation of Policy on Natural Resource Management and Its Impacts on Rural Livelihoods in Nigeria. *The Nigerian Journal of Business and Social Studies*, *7*(1), 49–71.

Adelegan, J. A. (2005). The history of environmental policy and pollution of water sources in Nigeria (1960-2004). The Way Forward, 3.

Agrawal, A. G., Yadama, R., & Bata, C. A. (2006). *Decentralization and Environmental Conservation: Gender effects from participation in joint forest management*. Washington, DC: US Agency for International Development Bureau for Africa.

Aloni, C., Daminabo, I., Alexander, B. C., & Bakpo, M. T. (2015). The Importance of Stakeholders Involvement in Environmental Impact Assessment. *Resources and Environment*, *5*(5), 146–151.

Anthony, R. N., & Govindarajan, V. (2004). *Management Control Systems*. McGraw- Hill Inc.

Anyanwu, C. N. (1999). *Introduction to Community Development*. Ibadan: Gabesther Educational Publishers.

Asnani, P. U. (2006). *Solid Waste Management*. Retrieved from http://www.iitk.ac.in/3inetwork/html/reports/IIR2006/Solid_Waste.pdf

Aubel, J., & Samba, K. (1996). Community participation lessons on sustainability for community health projects. *Journal of International Health Development*, *17*(1), 54–63.

Ayeni, Kapangaziwiri, Soneye, & Engelbrecht. (2015). Assessing the impact of global changes on the surface water resources of Southwestern Nigeria. *Hydrological Sciences Journal*, *60*(11-12), 1956–1971.

Ayeni, A., Soneye, A., & Badru, F. (2014). Adaptation to Water Stress in Nigeria Derived Savanna Area: The Indigenous Knowledge and Socio-Cultural Nexus of Management and Humanitarian Services. *Journal of Management Policies and Practices*, *15*(3), 78–87.

Ayeni, A. O., Cho, M. A., Mathieu, R., & Adegoke, J. O. (2016). The local experts' perception of environmental change and its impacts on surface water in Southwestern Nigeria. *Environmental Development*, *17*, 33–47. doi:10.1016/j.envdev.2015.09.007

Ayeni, A. O., & Olorunfemi, F. B. (2014). Reflections on Environmental Security, Indigenous Knowledge and the Implications for Sustainable Development in Nigeria. *Journal of Research in National Development*, *12*(1), 46–57.

Bass, S., Dalal-Clayton, B., & Pretty, J. (1995). Participation in Strategies for Sustainable Development. *Environmental Planning Issues*, *7*, 118. Retrieved from http://pubs.iied.org/pdfs/7754IIED.pdf

Batwel, A. (2008). *Factors affecting community participation in primary education development projects in Makete District of Tanzania* (MA dissertation). Sokoine University of Agriculture, Morogoro, Tanzania.

Berkes, F., & Folke, C. (Eds.). (1998). *Linking Social and Ecological Systems. Management Practices and Social Mechanisms for Building Resilience*. Cambridge, UK: Cambridge University Press.

Bernstein, A. B., & Sweeney, M. H. (2012). Public Health Surveillance Data: Legal, Policy, Ethical, Regulatory, and Practical Issues. *Morbidity and Mortality Weekly Report*, *61*(03), 30–34. PMID:22832995

Botes, L., & van Rensburg, D. (2000). Community participation in development: Nine plagues and twelve commandments. *Community Development Journal: An International Forum*, *35*(1), 41–58. doi:10.1093/cdj/35.1.41

Bray, M. (1996). *Decentralization of Education: Community financing*. Washington, DC: World Bank. doi:10.1596/0-8213-3724-6

Campese, J., Sunderland, T., Greiber, T., & Oviedo, G. (2009). *Rights-based approaches Exploring issues and opportunities for conservation*. IUCN-CIFOR. Retrieved from www.cifor.org/publications/pdf_files/Books/BSunderland0901.pdf

Chandan, J. S. (2003). *Management Theory and Practice*. New Delhi: Vikas Ltd.

Chirenje, L. I. R. A. G., & Musamba, E. B. (2013). Local communities' participation in decision-making processes through planning and budgeting in African countries. *Chinese Journal of Population Resources and Environment*, *11*(1), 10–16. doi:10.1080/10042857.2013.777198

Chukwuma, Obinna, Richard, Ephraim, & Ikechukwu. (2012). *Biotechnological Tools for Environmental Sustainability: Prospects and Challenges for Environments in Nigeria—A Standard Biotechnology*. Research International. doi:10.1155/2012/450802

Clark, R. E. (2003). Fostering the work motivation of individuals and teams. *Performance Improvement*, *42*(3), 21–29. doi:10.1002/pfi.4930420305

Cornwall, A. (2008). Unpacking 'Participation': Models, meanings and practices. *Community Development Journal: An International Forum*, *43*(3), 269–283. doi:10.1093/cdj/bsn010

Costanza, R., Hart, M., Posner, S., & Talberth, J. (2009). *Beyond GDP: The Need for New Measures of Progress*. Retrieved from https://www.bu.edu/pardee/files/documents/PP-004-GDP.pdf

Daily, G. C., & Ehrlich, P. R. (1992). Population, Sustainability, and Earth's Carrying Capacity: A framework for estimating population sizes and lifestyles that could be sustained without undermining future generations. *Bioscience*. Retrieved from http://dieoff.org/page112.htm

de Sherbinin, A., Carr, D., Cassels, S., & Jiang, L. (2007). *Population and Environment Annual Review of Environment and Resources*. Academic Press. doi:10.1146/annurev.energy.32.041306.100243 PMID:20011237

DESA. (2008). *Achieving Sustainable Development and Promoting Development Cooperation*. United Nations. Retrieved from http://www.un.org/en/ecosoc/docs/pdfs/fina_08-45773.pdf

Dyson, M., Bergkamp, G., & Scanlon, J. (Eds.), *(n.d.). Flow. The Essentials of Environmental Flows.* IUCN.

Eccleston, C. (2010). *Global Environmental Policy: Principles concepts and practice.* CRC Press Inc.

Emas, R. (2015). Successes and Shortcomings in the Implementation of National Sustainable Development Strategies: From the Greening of Governance to the Governance of Greening. Retrieved from http://digitalcommons.fiu.edu/cgi/viewcontent.cgi?article=3229&context=etd

Eradiauwa, U. A. (1999). *The role of Traditional Institution in Nation-building in a democratic society.* Academic Press.

Federal Ministry of Environment. (2016). *National Policy on The Environment (Revised).* Retrieved from http://environment.gov.ng/assets/downloads/revisednationalpolicyontheenvironmentfinaldraft.pdf

Federal Republic of Nigeria. (2004). *National Water Policy.* Federal Republic of Nigeria July 2004.

FEPA. (1991). National Environmental Protection (effluent limitation) Regulation (S.1.8). FEPA, FGPE, Lagos Nigeria.

Ferreira, B. R. (1996). The Use and Effectiveness of Community Policing in A Democracy. In *Policing in Central and Eastern Europe: Comparing Firsthand Knowledge with Experience from the West, College of Police and Security Studies.* Retrieved from https://www.ncjrs.gov/policing/use139.htm

Fiorino, D.J. (1990). Citizen participation and Environmental risk: A survey of institutional mechanisms. *Science, Technology and Human Value, 15*(2), 226-243.

FMARD. (2016). *The Agriculture Promotion Policy (2016–2020): Building on the Successes of the ATA, Closing Key Gaps. Federal Ministry of Agriculture & Rural Development (FMARD).* Policy and Strategy Document.

Gandile, Tessema, & Nake. (2017). Biodiversity conservation using the indigenous knowledge system: The priority agenda in the case of Zeyse, Zergula and Ganta communities in Gamo Gofa Zone (Southern Ethiopia). *International Journal of Biodeversity and Conservation, 9*(6), 167–182.

Ghaus-Pasha, A. (2005). *Role of Civil Society Organizations in Governance.* 6th Global Forum on Reinventing Government Towards Participatory and Transparent Governance, Seoul, South Korea.

Grafton, R. Q., Adadmowicz, W., Dupont, D., Nelson, H., Hill, R. J., & Renzetti, S. (2004). *Economics of the environment and natural resources.* Malden, MA: Blackwell Publishing.

Gunderson, L. H., & Holling, C. S. (Eds.). (2002). *Panarchy: Understanding Transformation in Human and Natural Systems.* Washington, DC: Island Press.

Happer, C., & Philo, G. (2013). The Role of the Media in the Construction of Public Belief and Social Change. *Journal of Social and Political Psychology, 1*(1), 321–336. doi:10.5964/jspp.v1i1.96

Harrison, M., & Coussens, C. (2007). *Global Environmental Health in the 21st Century: From Governmental Regulation to Corporate Social Responsibility.* Roundtable on Environmental Health Sciences, Research, and Medicine. Retrieved from https://www.ncbi.nlm.nih.gov/books/NBK53989/pdf/Bookshelf_NBK53989.pdf

Heinman, M. (1997). Science by the people; Grass root environmental monitoring and the debate over scientific expertise. *Journal of Planning Education and Research*, *16*(4), 291–299. doi:10.1177/0739456X9701600405

Hora, E. A. (2014). Factors that affect Women Participation in Leadership and Decision-Making Position, *Asian Journal of Humanity. Art and Literature*, *1*(2), 97–118.

Igbaekemen, G. O. (2014). IMPACT of Leadership Style on Organisation Performance: A Strategic Literature. *Review Public Policy and Administration Research*, *4*(9), 126–135.

Ijaiya, H., & Joseph, O. T. (2014). Rethinking Environmental Law Enforcement in Nigeria. *Beijing Law Review*, *5*(4), 306–321. doi:10.4236/blr.2014.54029

Irvin, R. A., & Stansbury, J. (1996). *Citizen Participation in Decision-Making: Is it Worth the Effort? 'CBEP (Community-Based Environmental Protection) is designed to maximize the use of scarce resources, encourage local support, and consider the economic well-being of communities'.* Environmental Protection Agency. Retrieved from https://smartnet.niua.org/sites/default/files/resources/citizen%20participation%20is%20it%20worth%20the%20effort.pdf

Ivancevich, M. J., Donnelly, J. H., & Gibson, L. J. (2003). *Management: Principles and Foundations.* AITBS Publishers.

Jain, N. K. (2005). *Organisational Behaviour* (Vol. 1). Atlantic Publishers.

Kahle, L. R., & Gurel-Atay, E. (Eds.). (2014). *Communicating Sustainability for the Green Economy.* New York: M.E. Sharpe.

Kalafatic, C. (2007). *Indigenous peoples' sustainable livelihoods.* FAO Livelihood Support Programme. Retrieved from ftp://ftp.fao.org/docrep/fao/010/aj033e/aj033e02.pdf

Keengwe, J., Mbae, J. C., & Onchwari, G. (2016). Handbook of Research on Global Issues in Next-Generation Teacher Education. Information science Reference.

Kibert, C. J., Thiele, L., Peterson, A., & Monroe, M. (2012). *The Ethics of Sustainability.* Retrieved from http://rio20.net/wp-content/uploads/2012/01/Ethics-of-Sustainability-Textbook.pdf

Kilewo, E. G., & Frumence, G. (2015). Factors that hinder community participation in developing and implementing comprehensive council health plans in Manyoni District, Tanzania. *Global Health Action*, *8*, 26461. Retrieved from https://www.ncbi.nlm.nih.gov/pmc/articles/PMC4452651/pdf/GHA-8-26461.pdf

Kusek, J. Z., & Rist, R. C. (2004). *Ten Steps to a Results-Based Monitoring and Evaluation System.* The International Bank for Reconstruction and Development / The World Bank. doi:10.1596/0-8213-5823-5

Laird & Frank, N. (1993). Participatiory analysis, democracy, and technological decision making. *Science, Technology and Human Values, 18*(3), 341-361.

McCandless, E., & Karbo, T. (2011). *Peace, Conflict, and Development in Africa.* University for Peace. Retrieved from http://www.upeace.org/pdf%5CREADER_webpages.pdf

McCormick J. (2001). *Environmental policy in the European Union.* The European Series. Palgrave.

Meier, B. M., Pardue, C., & London, L. (2012). Implementing community participation through legislative reform: A study of the policy framework for community participation in the Western Cape province of South Africa, International. *Health and Human Rights*, *12*(15), 1–14.

Meiners, R., & Yandle, B. (1998). *The common law: How it protects the environment*. PERC. Retrieved from http://www.perc.org/articles/common-law-0 Retrieved 12/3/2017

Mensah, A. M., & Castro, L. C. (2004). *Sustainable Resource Use & Sustainable Development: A Contradiction?!* Zentrum fur Entwicklungsfourschung (ZEF), Center for Development Research University of Bonn. Retrieved from http://www.zef.de/fileadmin/downloads/forum/docprog/Termpapers/2004_3b_Mensah_Castro.pdf

Muhanna, E. (2006). Sustainable Tourism Development and Environmental Management for Developing Countries. *Problems and Perspectives in Management*, *4*(2), 14–30.

Mukuna, T. E., & Shisanya, C. A. (2015). *Milestones in Green Transition and Climate Compatible Development in Eastern and Southern Africa*. Organisation for Social Science Research in Eastern and Southern Africa (OSSREA). Retrieved from http://www.ossrea.net/publications/images/introduction-ccd.pdf

Musa, S. D. (2005). Sustainable Grassroots Development: A major challenge of Globalization. In The Humanities and Globalization: the Africa perspective. Aboki Publishers.

Nath, V. (2001). *White Collar Invasion: Developed country policies leading to environmental degradation in South, Inlaks Fellow*. London School of Economics. Retrieved from www.vikasnath.org

Nehal, A. F., Majila, B. S., & Kala, C. P. (2004). Indigenous Knowledge Systems and Sustainable Management of Natural Resources in a High-Altitude Society in Kumaun Himalaya, India. *Journal of Human Ecology (Delhi, India)*, *16*(1), 33–42.

Nour, A. M. (2011). Challenges and Advantages of Community Participation as an Approach for Sustainable Urban Development in Egypt. *Journal of Sustainable Development*, *4*(1), 79–91.

Oakley, P. (1989). *Community Involvement in Health Development: An examination of critical issues*. Geneva, Switzerland: World Health Organization.

Oakley, P. (1991). *Projects with People: The Practice of Participation in Rural Development*. Geneva: International Labour Office.

OECD. (2001). *Policies to Enhance Sustainable Development*. Organization for Economic Co-Operation and Development (OECD). Retrieved from https://www.oecd.org/greengrowth/1869800.pdf

OECD. (2007). *Instrument Mixes for environmental policy*. Retrieved from http://www.oecd.org/department

OECD. (2008). *Promoting Sustainable Consumption Good Practices in OECD Countries*. OECD Publications. Retrieved from https://www.oecd.org/greengrowth/40317373.pdf

OECD. (2013). *Local Strategies for Youth Employment: Learning from Practice*. Retrieved from http://www.oecd.org/employment/leed/Local%20Strategies%20for%20Youth%20Employment%20FINAL%20FINAL.pdf

Office of Technology Assessment. (1995). *U.S. Congress, Office of Technology Assessment, Environmental Policy Tools: A User's Guide. U.S. Congress, Office of Technology Assessment, OTA-ENV-634.* Washington, DC: U.S. Government Printing Office, September.

Ogunleye-Adetona, C. I., & Oladeinde, C. (2013). The role of community self-help projects in rural development of Kwara state, Nigeria. *International Journal of Development and Sustainability*, *2*(1), 28–45.

Omran, A., & Read, A. D. (2008). "Waste Not, Want Not" A Study of Household Attitude toward Recycling of Solid Wastes. *Environmental Engineering and Management Journal*, *7*, 1–8.

Paul, B. D., & Demarest, W. J. (1984). Citizen participation over planned: The case of health project in the Guatemala community of San Pedro Laguna. *Social Science & Medicine*, *19*(3), 185–192. doi:10.1016/0277-9536(84)90209-0 PMID:6484608

Paul, S. (1987). *Community Participation in Development Projects: The World Bank Experience.* World Bank Discussion papers. Retrieved from http://www.ircwash.org/sites/default/files/Paul-1987-Community.pdf

Pigou, A. (1920). *The Economics of welfare*. London: Macmillan and Company.

Prizzia, R. (2005). Community involvement in protecting the environment: The role of Restoration Advisory (RABS). *The Innovation Journal*, *10*(1).

Raymond, P. (1974). Community Ideology. In *An essay in applied social philosophy*. London: Routledge and Kegan Paul Ltd.

Richardson, D. (2003). *Community-based solid waste management systems in Hanoi, Vietnam* (MSc Thesis).

Ries, C. P., Jenkins, J., & Wise, O. (2009). *Improving the Energy Performance of Buildings: Learning from the European Union and Australia. Technical Report*. Rand Corporation.

Roach, B., Harris, J. M., & Codur, A. (2015). *Microeconomics and the Environment*. Medford, MA: Global Development and Environment Institute, Tufts University. Retrieved from http://www.ase.tufts.edu/gdae/education_materials/modules/Microeconomics_and_the_Environment.pdf

Roe, D., Nelson, F., & Sandbrook, C. (Eds.). (2009). *Community management of natural resources in Africa: Impacts, experiences and future directions, Natural Resource Issues No. 18*. London, UK: International Institute for Environment and Development.

Rushefsky, M. E. (2002). Public policy in the united states at the down of the twenty-first century network. M. E. Sharpe, Inc.

Scherr & McNeely. (2008). McNeely Biodiversity conservation and agricultural sustainability: towards a new paradigm of 'ecoagriculture' landscapes. *Phil. Transactions of the Royal Society. B*, *363*, 477–494

Shaeffer, S. (1998). *Partnerships and participation in basic education: A series of training modules and case study for educational planners and managers.* Paris: UNESCO, International Institute of Educational Planning.

Shepard, A., & And Bowler, C. (1997). Beyond the requirement improving public participation in EIA. *Journal of Environmental Planning and Management*, *40*(6), 725–738. doi:10.1080/09640569711877

Spyke, N. P. (1999). Public Participation in Environmental Decision making in the new millennium structuring new spheres of public influence. *Boston College Environmental Affairs Law Review. Boston College. Law School*, *26*(2), 263–313.

The UNEP - DTIE International Environmental Technology Centre. (2016). *Participation: Enhancing Local Programmes*. Author.

Thomson, T. M. (1998). *Management by Objectives* (2nd ed.). Jossey-Bass/Pfeiffer. Retrieved from http://home.snu.edu/~jsmith/library/body/v20.pdf

Ulleberg, I. (2009). *The role and impact of NGOs in capacity development: From replacing the state to reinvigorating education.* International Institute for Educational Planning. Retrieved from http://unesdoc.unesco.org/images/0018/001869/186980e.pdf

van Breugel, L. (2013). *Community-based tourism: Local participation and perceived impacts - A comparative study between two communities in Thailand Research Master Social and Cultural Science.* Faculty of Social Sciences Radboud University Nijmegen.

Vanclay, Esteves, Aucamp, & Franks. (2015). *Social Impact Assessment: Guidance for Assessing and Managing the Social Impacts of Projects*. International Association for Impact Assessment. Retrieved from https://www.iaia.org/uploads/pdf/SIA_Guidance_Document_IAIA.pdf

Voumik, L. C. (2014). A green Economy in the Context of Sustainable Development and Poverty Eradication: What are the Implications for Bangladesh? *Journal of Economics and Sustainable Development*, *5*(3), 119–131.

Wangu, G., Nyariki, D., & Sakwa, M. (2015). The Role of Indigenous Knowledge in Socio-Economic Development. *International Journal of Science and Research*, *4*(5), 32–37.

Wasilwa, C. (2015). *Effect of Community Participation on Sustainability of Community Based Devt. Projects in Kenya.* Retrieved from https://www.linkedin.com/pulse/effect-community-participation-sustainability-based-calebwasilwa Retrieved 5/3/2017

Wehrmann, B. (2011). *Land use planning: concept, tools and applications. Deutsche Gesellschaft für Internationale Zusammenarbeit (GIZ).* GmbH, Division Agriculture, Fisheries and Food Sector Project Land Policy and Land Management.

Zenter, H. (1964). The State and the Community. *Sociology and Social Research*, *48*, 414.

KEY TERMS AND DEFINITIONS

Community Participation: In environmental management, is basically concerned with the involvement of individuals, groups, and communities in the affairs of managing the environment they live in.

Green Policy: Any regulation concerning toxic substances including pesticides and many industrial wastes.

Sustainable Development: Means to create and maintain prosperous social, economic, and ecological systems.

Section 2
Leadership, Natural Hazard, and Emergency Management

Chapter 4
Leadership, Public Values, and Trust in Emergency Management

Augustine Nduka Eneanya
University of Lagos, Nigeria

ABSTRACT

In today's environment, emergency managers get things done through team leadership. Good leaders know how to follow others when the situation calls for it. Being the right kind of leader is critical to getting committed and engage followers. The chapter argues that managers build trust into decision making during emergencies; when managers set up the right conditions for the team to thrive, it results a better outcome. The chapter further argues that when managers share information both up and down the chain of command and make their intention clear about what winning looks like, team members are able to use their own discretion and make decisions that support the mission. The chapter concludes that an effective expression of the manager's intent must be clear and concise of what the team must do to succeed and achieve the desired end state. The manager must build trust in decision making among the team to achieve better results.

INTRODUCTION

One of the most universal cravings of our time is the thirst for compelling and effective leadership. The crisis for leadership today is the mediocrity of men and women in Public Agencies and institutions involved in emergency management and public safety. Emergency management at the state and federal levels pose significant challenges to the ablest of public managers. Emergency managers need to align the values of the Agencies, institutions, private organizations, non-governmental organizations and citizens to their own values, build teams and trust in their assignments. Hence, those taking on leadership of these organizations must appreciate the difficulties in managing risks because emergency management involves responding to and managing risks within the environment.

DOI: 10.4018/978-1-5225-3194-4.ch004

In today's environment, emergency managers get things done through team leadership. To make it happen, team members must trust each other and their leaders. Managers build trust into decision-making and when emergency managers set up the right conditions for a team to thrive, it sometimes results in a harder job with a better outcome. To do so, they share information both up and down the chain of command and make their intentions clear. In the emergency management world, unanticipated situations and opportunities arise, requiring the team to exercise its own initiative. When managers are clear about public values of what winning looks like, they are able to use their own discretion and make decisions that support the vision and mission of Public Agencies and institutions.

BACKGROUND

For two millennia at least, the concept of leadership has been in literature long before modern study of leadership grapple with the "vexing problem of the rulers versus the ruled" (Burns, 1979, p 2). Plato analyzed not only philosopher–kings, but the influences on the rulers upbringing. Thinkers were busy examining the concept of leadership in moral teaching. Long before modern biography, Plutarch was writing about the living of a host of Roman and Greek Rulers and orators, arguing that philosophers "ought to converse especially with men in power" (Burns, 1979, p. 2). Arish literature on leadership and followers flourished in the classical and middle ages. Later – for reasons we would examine in this chapter, the study of rulership and leadership ran into serious intellectual difficulties. Leadership as a concept has dissolved into small and discrete meanings. Immense reservoir of data, analysis and theories have developed but no central concept of leadership has yet emerged, in part because scholars have worked in separate disciplines and sub-disciplines in pursuit of different and often unrelated questions and problems.

Hence, it is time that the study of leadership be lifted out of the anecdotal and the eulogistic and placed squarely in the structure and processes of human development, public safety and political action (Burns, 1979). This is because leaders have a critical effect on society in the present and future. They can determine on the success or failure of a society, country and community.

The need for emergency or disaster management came to fore in Nigeria as a result of the impact of extreme natural disasters globally . There have been reports that natural disasters forced 26 million people into poverty, cause $520 billion loss annually (World Bank (2016); Global Facility for Disaster Reduction and Recovery (GFDRR) (2016) .

In Nigeria, the impact of flooding and Boko Haram insurgency activities that displaced men, women and children motivated the Federal government to establish National Emergency Management Agency (NEMA). NEMA formulates policy on all activities relating to disaster management functions. It coordinates the activities of other stakeholders in disaster management. Through the National Emergency Management Agency (NEMA), a national disaster management framework was established. The framework objectives are to establish functional disaster management institutions at all levels of governance to prepare for, prevent, mitigate, respond to and recover from disaster events in Nigeria. The framework also develops capacity of relevant institutions and stakeholders for effective and efficient disaster management in Nigeria. NEMA facilitates the establishment of enabling legislation and monitor the activities of State Emergency Management Agency (SEMA) and Local Emergency Management Authority (LEMA).

The purpose of this chapter is to demonstrate that leadership is nothing if not linked to collective purpose and trust of the led. Therefore, the chapter focuses on collaborative leadership which includes leading in a shared-power world where citizens must have broad access, trust and engagement in emergency management and public safety activities; and where more public organisations, agencies, nonprofit organisations and private organisations must be included in polices and solutions on environmental hazards.

Against this background, the chapter is divided into five sections. Section one examines introduction, background and analytical framework. Section two examines the methodology, research questions and explanation of public sector activities in emergency management and public safety. Section three discusses the relationsip between leadership, organizational culture and public values in emergency management. Section four examines the challenges of leadership teamwork, trust, networking and collaborative activities in multi-organizations. While, the final section concludes the chapter with discussion on solutions/recommendations on leadership question and future research directions in emergency management and public safety.

CONCEPTUAL CLARIFICATION

The term "leadership" can be expressed in a broad sense to refer to all people who occupy positions of authority through which they can influence the behaviors and actions of other people. Leadership involves a relationship between people in which influence and power are unevenly distributed on legitimate basis. A leader cannot function in isolation; it requires someone to follow. Leadership is thus inseparable from followers' needs and goals. According to Burns (1979, p. 19), "the essence of the leader-follower relation is the interaction of persons with different levels of motivations and of power potential, including skills in pursuit of a common or at least joint purpose". Leadership can also be viewed as the "exercise of authority, whether formal or informal in directing and coordinating the work of others" (Shafritz et.al. 2011, p. 372).

Regardless of sector, all leaders have to deal with people, processes and resources. With people, the leadership challenge is to recruit, motivate and inspire the highest productivity and ingenuity from the workforce talent. With processes, we find the right process to yield the outcome you want. That takes time. If a particular process does not achieve the objective desired, it is incumbent upon leadership to modify the process or change it. Finally, leadership in any sector requires mastery of resource management, especially financial resources. An executive has to know how resources derived, the motivations of resource providers and how the process works to maximize resources.

However, there is the need to define leadership in public sectors. In defining the broad boundaries of public leadership, Morose, Buss, and Kinghom, (2007) note that there are at least three types of leadersip in public sector. There is political leadership, which involves legislators, elected executives in the policy process such as presidents, governors, and mayors, and the various stakeholders in the political process. There is organizational leadership, also known as bureaucratic or administrative leadership, primarily aimed at those leading and managing employees, programs, and organizations for the public good. Third, there is collaborative leadership, which focuses on leading in a shared – power world where citizens must have broad acces s and engagement, where more broadly distributed. Each of these types is important.

The concept of emergency management or disaster management can be defined as the creation of plans through which communities reduce vulnerability to hazards and cope with disaster. Disaster management does not avert or eliminate disasters (such as earthquakes,hurricue threats; instead; it focuses on creating plans to decrease the effect of disasters. Failure to create a plan could lead to damage to assets, human mortality and lost revenue. Events covered by disaster management include acts of terrorism, industrial sabotage, fire, natural disasters (such as earthquakes, hurricanes, flooding, etc,), public disorder, industrial accidents, and communication failures (Drabek, 1991).

The primary focus of this chapter, however, is collaborative leadership affecting emergency management and public safety. Though it is difficult to disentangle the types because of the enormous overlap, at the same time, it is important to provide a narrower focus of collaborative and networking leadership for heuristic purposes.

THEORETICAL UNDERPINNINGS

Leadership has been "researched and appeared more than any other concepts in Social Science literature" (Eneanya, 2009, p. 91). In the context of this chapter, literature to be reviewed would focus on theories that can contribute key concepts we need for the analysis. They include: Trait theory, moral leadership theory, transactional leadership theory, situational (contingency) theory, transformational leadership theory, integral leadership theory, leadership in crisis management paradigm network and collaborative leadership theory.

Trait Theory

This approach focuses entirely on the physical and personality characteristics of a leader. This theoretical approach assumes that some people have unique leadership characteristics and qualities that enable them to assume responsibilities not everyone can execute. Therefore, leaders are "born" leaders. However, there were many criticisms that trailed trait theory.

First, trait theory has largely fallen out of favour because reality never matched the theory. Second, it is no longer fashionable to contend that people will be effective leaders because they possess certain traits, without considering other variables that influence leadership effectiveness. Another criticism on trait theory, is its lack of ability to identify which traits make an effective leader. Even among the traits that have been most commonly cited – intelligence, energy, achievement, dependability socio-economic status, etc; there is a lack of consensus across studies (Shafritz, et al. 2011,p. 375). The most obvious proof that leadership involves more than possessing certain traits is the simple fact that a leader may be effective in one setting and ineffective in another. It all depends on situation. Situation is now viewed as an enormous influence in determining the qualities, characteristics and skills needed in a leader (Shafritz, et al. 2011,p.375).

Moral Leadership Theory

Burns (1979) defines moral leadership in three perspectives. First, moral Leadership means that leaders and the led have a relationship not only of power but of mutual needs, aspirations and values. Second, that in responding to leaders, followers have adequate knowledge of alternative leaders and programs and the capacity to choose among those alternatives. Third, that leaders take responsibility for their commitments, that is, if they promise certain kinds of economic, social and political change when they assume leadership positions, they would bring about that change (P.4).

Moral leadership, therefore, is not mere preaching or uttering of pieties or insistence of social conformity. Moral leadership emerged from fundamental wants, needs, aspirations and values of followers. It is "a kind of leadership that can produce social change that will satisfy followers and authentic needs" (Burns, 1979, P.4).

Transactional Leadership Theory

Such leadership occurs when "one person takes the initiative in making contact with others for the purpose of an exchange of values" (Burns, 1979:19). The transactional approach to leadership uses the path-goal concepts as its framework. In this context, the leader identifies what followers want or prefer and helps them achieve level of performance that results in rewards that satisfy them. In using the transactional style, the leader relies on contingent reward and on management by exception. That is, followers believe that accomplishing objectives will result in their receiving desired rewards; and by using management by exception, the leader would not be involved unless objectives are being achieved.

It has now become the standard practice to view leadership as a relationship, an interaction between individuals. The interaction is called "a transaction, so that, the term transactional leadership has become the umbrella label encompassing many theories of the leadership" (Shafritz, et.al. 2011, p.235).

Transformational Leadership Theory

In contrast, transformational leadership is about radical change. Sometimes, the radical change calls for co-optation, the inclusion of new, potentially. Second, that, in responding to leaders, followers have adequate knowledge of alternative leaders and programs and the capacity to choose among those alternatives. Third, that, leaders take responsibility for their commitment. That is, if they provide certain kinds of economic, social and political change when they assume leadership positions, they would bring about that change. Moral leadership, therefore, is not more preaching or uttering of pieties' or insistence of social conformity. Moral leadership emerges from fundamental wants, needs, aspirations and values of followers. It is a kind of leadership that can produce "social change that will satisfy followers' authentic needs" (Burns, 1979:4).

Situational or Contingency Theory

The search for the "best" set of traits or behaviour has failed to discover an effective leadership mix and style for all situations. Situational leadership theories evolved and suggest that leadership effectiveness depends on the fit between personality, task, power, attitudes and perceptions (Fleishman, 1973). A number of situation–oriented leadership studies have been researched and publicized. Researchers, however, recognize that the leadership behaviour needed to enhance performance depends largely on the situation – what is effective leadership in one situation may be disorganized incompetence in another.

This is because, its situation (or context) is different and the choice of a style needs to be contingent on the situation. The contingency or situational approach, therefore, assert that different leadership styles will differ in their effects in different situations. Thus, the situation (not traits or styles themselves), determines whether a leadership style or a particular leader will be effective. The basic foundation of situational (contingency) theory is that an effective leader must be flexible enough to adapt the differences among subordinates and situations.

Integral Leadership Theory

Integral leadership tends to focus on leadership as a community process, democratizing and decentralizing leadership as much as possible. McCrimmon (2007, p. 1) asserts that "leadership needs to be reframed from a digital and postmodern age. The world is losing its stable and hierarchical character. Life is now more dynamic, chaotic; final authorities have vanished" Edwards, (2009) in his essay "seeing integral leadership through three important lenses: developmental, ecological, and governance" which incorporates the focus on followers, the environment, and community. Integral leadership themes are common in the popular literature because of the concern for corporate social responsibility among leaders and private organizations, as well as the public administration literature because of its focus on serving the community and doing good.

Integral leadership focuses on cross – boundary problem – solving that elevates the community (Crosby and Bryson, 2005; Crosby and Kiedrowski 2008) in the tradition established by Burns (1978). In line with Burn's distinction, it emphasizes "transforming" leadership (raising the consciousness of followers to solve problems through enlightenment as much as self-interest) rather than merely transformational (that is, change-oriented) leadership, which tends to be more executive and top-down oriented.

Fairholm (2004) frames leadership more broadly from a public values perspective. He emphasizes "five leadership perspectives (ranging from leadership as equivalent to scientific management, to leadership being a whole – soul or spiritual endeavour) held by public managers and discusses their implications for public administration". In doing so, he provides a classical apology for administrative leadership.

Leadership in Crisis Management Paradigm

All organizations can and do have crises from time to time, but their frequency and severity are much affected by the quality of leadership (Tichy and Bennis 2007; Boin and Hart 2003). Good leaders have contingency plans (mitigation) to prevent many crises altogether, prepare for a variety of plausible events, response quickly and effectively when crises occur, and are able to move the affected community and responding organizations back to normal after the event in a reasonable timeframe. For example, emergency response agencies have a special challenge in dealing with catastrophic events that are very large or unusual, or simply catch agencies off guard (Farazmand, 2001). This can amount to leadership failure as experienced in New Orleans with Katrina/Rita crisis (Kapricu and Van Wart (2006, 2008).

Network and Collaborative Leadership Theory

The inter-connectedness of problems, the regionalization and globalization of solutions, and the decrease in government resource, have emphasized the need to move increasingly from government to government and from hierarchy to networks (Maak and Pless, 2006). This requires that leaders have a new worldview, different competences, and additional tools.

Focus has, therefore, shifted to collaborative leadership, collaborative public management, a concept that describes the process of facilitating and operating in multi-organizational arrangements to solve problems that cannot be solved or easily solved by single organizations. The term collaborative means to co-labor, to cooperate to achieve common goals, working across boundaries in multi-sector relationships, cooperation is based on the value of reciprocity" (O' Leary, Gerard, and Bingham, 2006, 7). Don Kettle (2006, p.10) explains why the contemporary imperative to collaborate is more important than historic boundaries. He notes, "working effectively at these boundaries requires new strategies of collaboration and new skills for public managers. Failure to develop these strategies – or an instinct to approach boundaries primarily as political symbolism – worsens the performance of the administrative system". Thompson & Perry (2006) dissect collaboration into five variable dimensions that leaders have to understand and master for maximum effectiveness: governance, administration, organizational autonomy, mutuality, and norms of trust and reciprocity. However, some researchers point out when collaboration is less than ideal (McGuire 2006) and discuss the aim of collaboration (Bevir, 2006).

GAP IN LITERATURE

The major gap identified by literature reviews show that collaborative leadership in emergency management in public sector has not received much studies. It has now blossomed from being marginalized to being a recognized area of research interest. Public leadership theory has formal aspects concentrated on postmodern theory and integrated, collaborative, collective and net work theory. However, as organizations have flattered in the corporate and agency worlds, team and shared leadership, collaborative leadership, and integrate frameworks have matured, and much work needs to be done in emergency management and public safety.

Hence, this chapter articulates this conceptual framework of analysis, borrowing concepts from integrated, public value, collaborative, collective and network theories to construct this model for explanation, as illustrated below:

METHODICAL APPROACH

From the conceptual tramework in Figure 1, key concepts were identified, analyzed and explained in the context of the pattern of relationship between leadership, public values and trust in emergency management and public safety. The conceptual framework for analysis and explanation is collaborative leadership in public organization. The method of research design is qualitative research and research method is secondary data analysis using conceptual framework for explanation.

RESEARCH QUESTIONS

The explanatory perspectives are however driven by the following three research questions:

1. To what extent have leadership activities satisfied public value and trust in emergency management within the political environment?
2. What are the challenges of leadership in managing organizational culture and values to enhance policy decisions in emergency and public safety management?
3. To what extent can teamwork leadership, trust, networking and collaborative activities with multi-organizations enhance the effectiveness and efficiency of emergency and public safety managers in political environment?

Figure 1. Conceptual model for coordinating disaster responses and internal and external forces constraining emergency manager in emergency management and public safety

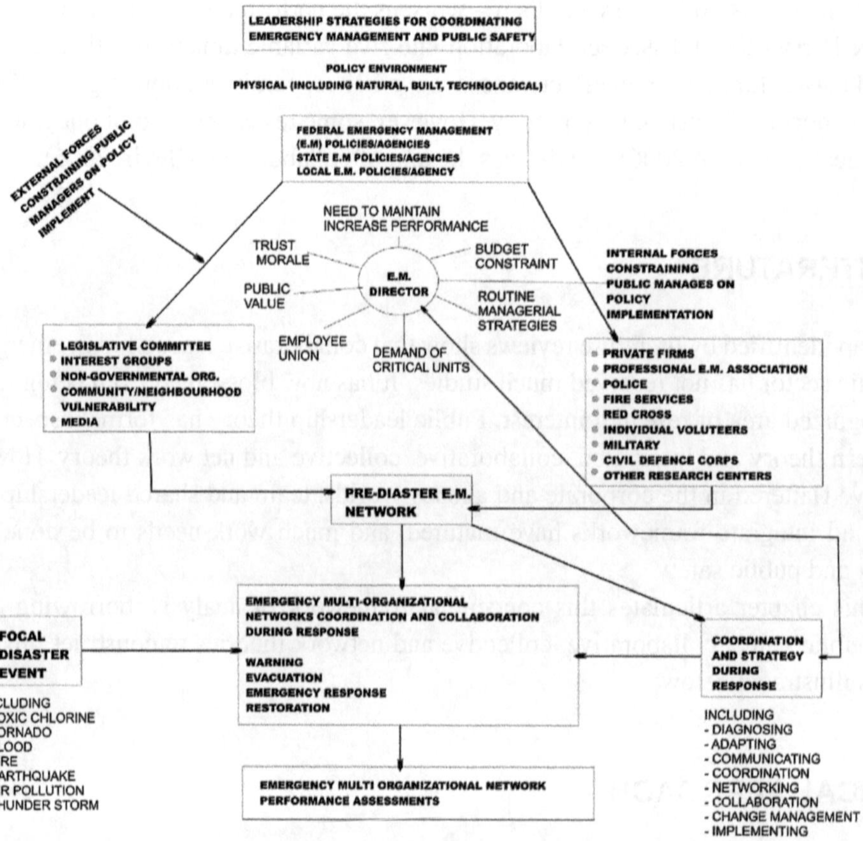

ANALYTICAL AND EXPLANATORY FRAMEWORK

Analysis and explanation of these key concepts in figure 1 above demonstrates the role of leadership in shaping public values and building trust in emergency management and public safety as demonstrated by answers to research questions.

Research Question 1: To what extent have leadership activities satisfied public value and trust in emergency management within the political environment?

Public Sector Leadership, Public Value, and Trust in Emergency Management and Public Safety Within Political Environment Issues

Policy making cannot adequately be studied apart from the environment or context in which it occurs. According to system theory, demands for policy actions stem from problems and conflicts in the environment and are transmitted to the political system by groups, officials and others.

At the same time, the environment both limits and directs what policy –makers can effectively do. The environment broadly viewed, includes geographic characteristics such as climate, natural resources, and topography; demographic variables such as population size, age distribution, racial composition, and spatial location; political culture, social structure, or the class system; and the economic system. Other nations become an important part of the environment when climate change, migration, foreign policy and defense policies are involved.

Leaders articulate common objectives and values to which the network commits and around which it can coalesce. Control must be replaced by empowerment through self-initiation, with the network members being given the freedom to find the most appropriate route to achieve project goals. This is illustrated in the conceptual model in figure 1 above, showing leadership process, public values and trust in emergency management and public safety in Nigeria

The network will be held accountable for delivering its objectives. Leadership's responsibility is to ensure that systems and structures are in place that enable the members of the network to collaborate, learn, share knowledge, and execute their responsibilities. The network's output is the generation of capabilities (Saint Onge & Armstrong 2004, p.191).

Thus, in cross-organizational endeavors, such as emergency management and public safety, "leaders are key "Champions", who sustain organizational commitment, and "alliance managers", who enable people to work together efficiently and role model trust and collaboration" (Holbeche, 2005, 179). Leaders, then, are responsible for "generation of capacities, promoting the flow of knowledge within the organization and with the organization and other entities, synchronization of the key collaborating organizations, examining mutual capabilities, and calibrating internal organizing structures to external needs" (Agranoff, 2011,p. 274).

To perform leadership role in external mobilization during emergency management and public safety has an internal and external task perspectives. Internally, it is important that collaborative managers receive the proper leadership signals and support, and this is a primary function of the collaborative "champion". Without such top-level support, the internal, natural, bureaucratic resistances and turf-protection behaviors (of which there are many as illustrated in fig. 1. can flourish). This champion role also includes support by agency executives, such as department heads. Empowerment is also critical. Leaders must go beyond lip service; they must facilitate an internal atmosphere that generates program level participation, sharing of information, joint learning, and participative decision-making. In many ways, for this model of empowerment of staff to work outside the agency is a contemporary extension of more traditional executive leadership functions.

Externally, targets involve convincing or working with critically important organizations outside government – the regulated, competing business, alternative service deliveries, and advocacy groups – to participate in a collaborative endeavor. This normally requires initial enlistment of a series of top nonprofit organizations' executives, private industries and public safety agencies (police, and fire service) and advocacy groups, after which technical and managerial people can be brought into the collaborative enterprise for the detailed negotiations. These tasks are more non-traditional, convincing sets of external parties to enter into new working relationships with government (Agranoff 2003; Bryson and Crosby, 1992).

Public Value Issues

Public values could refer to individually held values that focus on the welfare and just treatment of others. Such public values are not only endorsed collectively and ratified by government policy; they pertain to the welfare of someone other than oneself and also concerned about relationships in the society (Mansbridge 1990; Wilson 1974). The values are not only expressed by individuals but it creates latent possibility in the polity that can be roused to different kinds of individual, social, political life and public policy. Public values could be expressed as those particular values that are articulated and embraced by political actors through democratic process or deliberations.

In liberal democracies, individuals, and collectives do develop, hold and act to produce "public values" through voluntary, civic action as well as through public policy decisions that require them to make contributions to public purposes. Individuals can have public values that they express on their own as volunteers, charitable donors and philanthropists. They are arbiters of values. Collectives can pursue the material well-being of their members, such as professional associations, labour unions or consumer cooperatives, and Red Cross Organizations. Some of these collectives are formed to animate voluntary contributions of time and money to contribute to the welfare of others or enable each of them to act in accord with what they deem as moral duties to others. Still others are formed as policy advocacy or political organizations, such as child trafficking, domestic violence, National Rifle Association, and so on, that seek to influence public policy choices.

The role of voluntary sector and civil society in defining and producing public values in liberal democratic societies and often succeed in having its conception encoded in legislation and other forms of public policy. Bozeman (2007) focuses on the policy at societal level. According to him, a society's public values are those providing normative consensus about:

1. The rights, benefits, and prerogatives to which citizens should (and should not) be entitled;
2. The obligations of citizens to society; the state, and one another; and
3. The principles on which governments and policies should be based(P.17).

Although public values in a democracy are typically contested, a relative consensus is discernible from constitutions, legislative mandates, policies' literature reviews, opinion polls, and other formal and information sources (Beck Jorgensen & Bozeman, 2007).

Whereas Bozeman focuses on the policy at societal level, Moore (1995, 52-55) focuses on "Public Managers on government performance perspectives". Moore believes that citizens want from their governments some combination of the following that together encompass public values:

(1) high-performing service-oriented public bureaucracies; (2) public organizations that are efficient and effective in achieving desired social outcomes; and (3) public organizations that operate justly and fairly and lead to just and fair conditions in the society at large. While Moore's definition of public value is vaguer than Bozeman's, it highlights reasonably specific public values: efficiency, effectiveness, socially and politically sanctioned desired outcomes, procedural justice, and substantive justice. Like Bozeman, Moore's definition of public value can encompass input, process, outputs, and outcome measures.(PP.52-55)

Stoker (2006) moving beyond Moore's primary focus on public manager at the top of a public bureaucracy delivering services or obligations, proposed "public value management" as a new paradigm better suited to network governance than traditional public administration or the New Public Management. Stoker (2006) focuses on network inter-organizational and cross-sector relations and governance, though, he vaguely considers how leader and managers in specific instances would achieve efficiency, accountability, and equity along with broader domestic values (Williams and Shearer, 2011, O'Flynn, 2007). Stoker (2006) also fails to explain how leaders and managers should cope with a democracy having problems with low voter turnout, divided government, competing organized interests, and competing conceptions of what public value might be in any situation (Davis and West 2009, Jacobs, 2014).

Critics of public values argue that it has been used as a rhetorical strategy to protect and advance the interests of bureaucrats and their organizations (Roberts 1995). Though, the criticism may have their merits in particular cases, creating public values by government has "its benefits: (a) focus on public values enables one to bring together debates about values, institutions, systems, processes and people; (b) it enables one to link insights from different analytical perspectives, including public policy, policy analysis, management, economics and Political Science"(Smith, 2004, pp.68-69) .

Bozeman (2007) policy-level and societal public values have also been used in variety of ways. The approach incorporates a broad range of value considerations in policy decision-making processes by helping: (1) identifying public values; (2) assess whether or not public value failures have occurred; (3) map relationships among values; and (4) graphically represent relationship between public failures and market failures (Welch, Rimes, and Bozeman, 2014). The approach has been used primarily in the science and technology field (Bozeman and Sarewitz 2011), but "increasingly on other fields" (Bozeman and Moulton 2011, p. 317).

However, Peters & Pierre (1998) observe that in a market – based democracy, government is not the owner of all the processes and institutions having public value potentials or obligations. Hence, Moore's (1995) focus on cross-sector collaboration, integrative leadership and network governance is relevant to the benefits of public value. To him, public managers must frequently collaborate with nonprofits businesses, the media and citizens to accomplish public purpose on emergency management and public safety. This cross-sector collaboration, integrative leadership and network governance is adapted in the context of public values in this chapter.

Trust Issues

Trust has continued to be an important topic in the mainstream and in the public sector. Trust is invariably high-lighted by popular writers such as Covey (2009), and the main literature has shown the power of positive organizational behaviors as emerging from select traits, such as hope, optimism, resiliency and other variables (Luthans, 2007), and has provided careful analysis of the sub-elements of trust (Burke et al. 2007).

Extensive research has been done about trust in government. This includes Newell, Reeher and Ronayel's edited volume (2008) on building trust through value-based leadership, self-awareness, coaching, using teams effectively, providing high performance (good value), collaborating, and using good networking skills. Empirical research has shown that although citizen participation and involvement can positively affect trust, agency or government performance is the stronger factor ultimately, (Vigoda – Gadot and Mizrahi, 2008; Wang and Van Wart, 2007). In the final analysis, the manager who builds trust and responsibility among team members and set the conditions for its execution achieves result in emergency management and public safety.

Analysis

Thus, the description of leadership, public value and trust in Nigeria context is not the case. The leadership and management of funds for Internally Displaced persons in Maiduguri, North East of Nigeria is nothing to write home about. Cases of corruption and diverting of funds meant for management of IDPs camps have made copious literature in daily news reading in Nigeria. Coupled with this is poor funding for handling displaced persons in IDPs camps. The under-supply of funds because of poor budgeting has left the camps dependent on international donor agencies. This has resulted in deficient infrastructure, good shelter, food and medicine. Leadership commitment on these basic issues has been a thing of concern. Besides, public safety agencies- the police and military personnel were alleged to be fighting over bags of rice allocated for displaced persons. Corrupt officialks from government handlinling IDPs camps were often accused of diverting needed funds and relief materials for their personal use or the relief materials were sold to traders for profit. So, corruption of officials are common issues of concern, leading to distrust by displaced persons in IDPs camp and civil society in general. Collaborative groups such as nonprofit organizations and Red Cross personnel were left out of managing food items allocated to Internally Displaced persons. Corruption in the distribution of relief materials supplied was rife. The Secretary to Federal Government of Nigeria was also accused of awarding contract to a private company in which he has interests and currently investigated after his suspension. Besides, the Air force personnel could not differentiate Internally Displaced Persons camp as one of the camps was bombed in error, leading to the death of more than 95 persons. Boko Haram terrorists and armed robbers often attached displaced persons in the camp. On 11th September, 2015, seven people were killed in Malkohi IDP Camp near Yola, Adamawa State. On 10th February, 2016, 80 displaced persons were killed in Dikwa IDP Camp in Bornu State. So, there is high level of insecurity of internally displaced persons. All these have led to public outrage and public distrust on the issues of leadership, public value and trust in emergency management situations in Nigeria.

CHALLENGES OF LEADERSHIP IN EMERGENCY MANAGEMENT AND PUBLIC SAFETY ACTIVITIES

Lack of Collaboration Between Federal, State and Local Government

Emergency management at the federal, State and local levels poses significant challenge to the ablest of public managers. Emergency management involves responding to and mitigating risks and those taking on leadership of organizations at these levels must know how to manage risks. Public managers in emergency management organizations must create and maintain an internal environment that embraces internal controls and constrains managers on policy implementation. As illustrated in figure 1 above, these include, public values, trust, labor union employees' demands, and budget constraints. Ethical behaviors must be observed to ensure that the organization operates in a way that is both transparent and accountable to those it serves.

There were also other critical demands and challenges from legislative committees, bureaucracy, private firms, professional associations, interest groups, nonprofit organizations, community media, fire service department, police, Red Cross, individual volunteers, military and Advocacy groups over policy decisions and procedures within each organizations.

In Nigeria, for example, collaborative activities between the Federal, State and Local governments in coordinating activities in Iternally Displaced Persons (IDPs) Camps revealed that there is a big gap of unmet expectations. The National Emergency Management Agency (NEMA) in Nigeria could not coordinate and collaborate with State governments, Local governments involved, nonprofit organizations, safety agencies such as: police, military, redcross and civil defence corps. Hence, it was easy for the military and police to hijack distribution of food aids/relief materials and divert them for their own benefits or profit. Public managers coordination and collaboration are in disarray and questionable. There are reported cases of corruption, rape and sex for money by military/police personnel assigned to IDPs Camps. There are also reported cases of cholera and other health challenges. So establishing an effective control environment is an important responsibility of senior management personnel, particularly in organizations faced with responding to emergencies affects public safety and health.

In disaster and other incidents, all levels of government are supposed to be involved. Many disasters require fire department, country, municipal, community, state and federal government departments, personnel and resources. In these multi-organizational responses, the local National Emergency Management Agency (NEMA) is "one of the best resources with which to connect locally to understand the process of progressing disaster response through levels of government". (Randatt, Winter, 2015 – 2016, p. 22). This does not exist in Nigeria IDPs Camps' situations.

Leadership, Organizational Culture and Public Values Nexus in Emergency Management and Public Safety

Every society has a culture that differentiates its members' values and lifestyles from those of other societies. Culture is part of the environment and shapes social action but does not fully determine it. It gives form and direction to human action. The political culture is the determinant of any community's attitudes toward the quality and vigor of its governing institutions (Sharitz, 2011: p. 68). Our focus in this context is on values, belief and attitudes on what public managers and Administrators should try to do; how they should operate, and the relationship between the citizen and government in emergency Management and Public safety.

The quality of bureaucratic operations is measured by levels of citizen satisfaction and efficiency. The quality or style of operations is determined by the values and resources to be applied to any given problems. The values help to establish the obligations of citizenship and establish the parameters of activities in which an official may participate. In other words, the political culture ensures that the decisional process is filtered through its values system before administrative action is taken.

Inside the cultural environment is the organization culture – that is, culture which exists within an organization. It is made up of intangible things, such as: values, beliefs, assumptions and perceptions. It is the "pattern of these beliefs and attitudes that determines members' behaviors in and around the organization, persists over extended periods of time, and pervades all elements of the organization" (Shafritz et al. 2011, p. 70). It is transmitted to new members through cultural process (socialization), maintained and transmitted through a network of interaction patterns. It is "enforced and reinforced by group norms and organization's system of rewards and control" (p. 70).

Organization culture provides a framework for shared understanding of events, defines behavioral expectations, and serves as a source of and focus for members' commitment and as an organization "control system". It provides meaning, direction and mobilization for administration. However, while a strong organizational culture can control organizational behavior; it can also block an organization from making those changes needed to adapt to a changing environments. There is a challenge in this regard because there is absence of organization culture and value of efficiency and effectiveness in emergency management in National Emergency and Management Agency(NEMA) in Nigeria

SOLUTIONS AND RECOMMENDATIONS

Research Question 3: To what extent can team leadership, public value, trust, networking and collaborative activities with multi-organizations enhance the effectiveness and efficiency of emergency and public safety managers in political environment?

Team Leadership and Trust

Leadership entails providing direction for a team- providing vivid pictures of what the team is try to achieve. This involves managing performance of team members in the form of series of tasks relating to the internal dynamics of the group. These tasks include (Eneanya, 2015: p.476):

- Reaching agreement on and communicating performance expectations and standards
- Keeping performance on track
- Attending to the needs of team members
- Resolving all forces (or competition of vested interest, petty rivalry, personality classes) which might threaten to split the team members and vice versa, to enhance cohesiveness and networking.

Good leaders know how to follow others when the situation calls for it. Leaders should listen to the team ideas and reserve giving direct orders for situations when they must do so. Letting the team members decide can give everyone greater sense of control over their jobs, the outcome and ultimately the process itself.

In today's environment, emergency managers get things done through team leadership. To make things happen, team members must trust each other and their leaders. Leaders build trust into decision-making. When public managers set up the right conditions for a team to thrive, it sometimes results in a leader's job with a better outcome, to do so. It is important to share information to team members vertically or horizontally in the organizational chain of command. The Public manager must make the intention clear and when they need to use their own initiatives to make decisions that support the mission. In emergency management world, the public managers' intent must show the purpose, key tasks end state, appropriate guidance and priorities. The purpose must be clear and concise of what the team members must do to succeed and achieve the desired end state for the environment in which it operates. In a value-based organization where individuals can be trusted to do the right thing, it is better for the team to understand the mission and intent. This would make public managers better equipped and empowered to execute assigned tasks effectively.

Networking and Collaboration With Multi-Organizations

A related conductive role for the public manager involves the need for public agencies to re-invent themselves by encouraging the creation and flow of information across public agencies and outside of government. Clearly, the existing deep involvement of public managers as alliance managers in generating partnerships, networks, challenge grants, venture capital pools, and contracts – for – service programs, as well as electronic information networks, underscores this role. Partners in emergency management and public 'safety do not automatically come together, but they can be leveraged. Public managers are in pivotal positions to perform, such convening actions by looking for an including potential stakeholders that can transform human capital into live systems of problem identification and solution (Bryson and Crosby, 1992).

It follows that public agencies need to establish conducive support through electronic communication. Because of the collaborative nature of their tasks, they use such tools as; e-mail, twitters, face-book, teleconferencing, web-based geographic information system, decision-support software in addition to face to face meetings. Thus, at the centre of such relationships, the Public Agency, for example, Federal Emergency Management Agency (FEMA) in United States or Nigeria's National Emergency Management Agency (NEMA) needs to foster information and communication technology links along with more direct community building engagement practice.

Thus, networking and collaboration during disaster require multi-organizational responses. Getting public, private and nonprofit agencies to work together before, during and after disasters is likely to reduce liabilities and raise capabilities. For instance, encouraging developers and land owners to support safe development will reduce liabilities, while increased contact and cooperation among disaster response organizations will build capabilities.

In USA for example, citizens depend upon their local responders in the initial hours following a crisis. In order to serve others, our responders train in their area of specialty and also prepare contingency plans to shelter their families out of harm's way. Responders either rely on a pre-formulated plan and or may need external support in the first hours of a large–scale crisis to ensure safety and shelter of family members and /or other dependents. Again, firefighters serve in a volunteer capacity. Whether career or volunteer, all firefighters are essential to the local response capacity. These volunteer firefighters may not have the resources or support mechanisms in place to readily formalize a contingency plan to care for family members and optimize the likelihood of their availability to respond. So, there is need to send firefighter to specialized training, and neighborhood financial support.

In developing community preparedness plans for a large-scale crisis, leaders should consider establishing neighborhood policy and planned effort for first responders by ensuring there is a viable plan for the shelter and safety of their families. This could include neighborhood volunteer and responder family buddy system, enabling and affording the local responder the ability to remain committed to crisis response rather than focused on concerns for family emergency shelter and safety.

There is need to develop a neighborhood contingency policy and nurture a culture of support for families of local responders. Doing so would ensure responder's availability by minimizing a potential conflict between duty to respond and family in the first hours of a larger scale crisis. A neighborhood contingency plan to care for its first responders' families in the first hours of a crisis could help reinforce the local response capacity and thereby sustain community resiliency (Wyatt, 2016).

Furthermore, working with local Emergency Agency to determine what is needed allows first responders to initially collect information, like the infrastructure and number of structures damaged in addition to what their initial reconnaissance mission finds. Developing a resource guide for neighboring communities is a great way to not only identify available resources, but also connect the community's department heads with the personnel needed to deploy the resources where they are needed most. The guide can "provide the basis for an informal hazard and vulnerability analysis and create a gap analysis comparing the resources required with those available" (Randall, 2016, P. 22).

Special operations teams can also be used to build comprehensive plans and garner logistics to support operations. Firefighters and police officers receive the bulk of their training in operations, would produce a solid "planning section chief and logistics chief and enabling emergency service organizations to focus on the tasks at hand while the team develops the plans" (Randall, 2016, P. 23).

Furthermore, utilizing task forces and established teams can minimize overhead management, as many teams have an internal management structure and logistical support. This "frees up limited overhead management personnel to manage adhoc task forces and strike teams" (p. 23).

Finally, collaborating and networking with other response and government agencies are one of the best actions an emergency services leader can take in disaster planning or response activities. Simple informal conversations can build friendships and set expectations prior to an incident. This would provide an understand that the local department would need to call employee to contact firefighters and a police officer. It is, therefore, advisable to ensure proper planning to prevent poor performance. A way out is to investigate resources and relationships, write formal plans and exercise with players likely to attend unexpected disaster. This is the only way a department can prepare for the next "big" disaster event.

Reducing Unethical Leadership Performance Assessment in Emergency Management and Public Safety

Performance assessment of emergency management and public safety has a direct impact on the way organizations function and provides a multitude of services. The importance of an effective and accurate performance assessment takes on a significant degree of transparency when it involves the inherent safety of the public.

Performance assessment in Public safety does the following:-

- Directly connect the employees' performance with the organization's strategic goals
- Allow the organization to communicate these goals with employees
- Influence pay and promotion
- Identify the strengths of the employee or organization
- Create opportunities to develop and improve an employee or organizations overall performance.

There are different ways of reducing inflation or deflation of data in performance assessment of emergency management and public safety. To avoid unethical performance assessment, emergency and public safety managers should:

- Set outputs that recognize the cost and consequences of the methods used to achieve them.
- Recognize when legitimate outcomes have been achieved
- List the dimensions of value in the mission statement and develop a way to measure them, if they are being accomplished.
- Implement a method for recording client's satisfaction

- Reduce reliance on bottom line objectives in assessing overall performance of the agency and
- Decrease quota driven forms of law enforcement.

These strategies would help articulate clear, complete and compelling idea of the public value the agency exists to produce. They will also assist in developing a set of measures to record the agency's performance in producing public value, while inviting external accountability for creating and defining value. More importantly, the opportunity to partner in these ways with the community will increase agency transparency. Finally, it will assist in the elimination of tunnel vision and broadens the scope of effective policing, decreasing the pressure to produce. In this regard, law enforcement agencies must take a definitive approach in aligning performance assessment outcomes for the public. The more consistently this occurs, the closer we will be in decreasing the practice of performing unethical performance assessment in public safety.

Leadershiship as Agent of Change in Emergency and Public Safety Management

Leaders are agents of change whose acts affect other people more than other people's acts affect them. Real change means transformation to a marked degree in the attitudes, norms, institutions and behaviors that structure our daily lives. Such changes embrace not only new cultural patterns and institutional arrangement and new psychological dispositions but changes in material conditions, in the explicit, felt existence, information and communication technology, the flesh and fabric of peoples' lives. The leadership process is defined as carrying through from the decision-making stages to the point of concrete changes in people's lives, attitudes, behaviors, and institutions.

Real change means a continuing interaction of attitudes, behavior, and institutions, monitored by alterations in individual and collective hierarchies of values. Leadership brings about real change that leaders intend; though, leaders may seem to cause the most titanic of changes. The test is purpose and intent drawn from values and goals of leaders, high and low resulting in policy decision and real intended change.

In leading, there are three general skills or competencies or strategies that public managers need and should have in emergency management and public safety. They include:

1. **Diagnosing:** Being able to understand the situation you are trying to influence. Diagnosing is a cognitive or cerebral competency. It is understanding what the situation is now and knowing what you can reasonably expect it to be in the future. The discrepancy between the two is the problem to be solved. This is what the other competencies are aimed at changing.
2. **Adapting:** Being able to adapt your behavior and the other resources you have available to meet the contingencies of the situation. Adapting is a behavioral competence. It involves adapting your behaviors and other resources in a way that helps to close the gap between the current situation and what you want to achieve.
3. **Communicating:** Being able to communicate in a way that people can easily understand and accept. Communicating is a process competency. Even if you are able to understand the situation, even if you are able to adopt behavior and resources to meet the situation, you need to communicate effectively. If you cannot communicate in a way people can understand and accept, the whole process will not have the impact you will like it to have.

These leadership strategies would be of immense importance for leaders in emergency management and public safety.

FUTURE RESEARCH DIRECTIONS

There is the need for future research direction to use selected case studies of five other countries across the globe to assess the performance of Emergency Management Agencies or Environment Protection Agencies in terms of collaborating with multi-organizations in planning, response activities and management in vulnerable situations such as: earthquake, flooding or hurricane and other natural hazards. The results of such investigation would lend credence to any generalization about paucity of leadership in emergency management and deficiency of trust of citizenry.

CONCLUSION

The chapter has shown that the quality of leadership determines the response activities of emergency agencies, espectially when dealing with catastrophic event. For effective performance, there is the need to network and collaborate with multi-organizations. Getting public, private and nonprofit agencies to work together before, during and after disasters is a challenge without good leaders. But when good leaders are in place, they can make contingency plans (mitigation) for a variety of plausible events, respond quickly and effectively when crises occur, and are able to move the affected community and responding organizations back to normalcy after the event in a reasonable timeframe.

In today's environment, emergency managers get things done through team leadership. They get right things done without constant prodding or course correction. To make it happen, team members must trust each other and their leader. Managers build trust into decision-making. When emergency managers set up the right conditions for a team to thrive, it results with a better outcome. To do so, they share information both up and down the chain of command and make their intentions clear. In emergency management world, unanticipated situations and opportunities arise, requiring the team to exercise its initiatives. When managers are clear about what winning looks like, they are able to use their own discretion and make decisions that support the mission.

Leadership is a process of morality to the degree that leaders engage with followers on the basis of shared values and goals. Only followers can "define their own true needs and they can do so only when they are exposed to competing alternatives and values of their leaders" (Burns, 1979: 20). Leadership is so sensitive to the motives of potential followers that the roles of leaders and followers become virtually inter-dependent. As moralists, emergency managers maintain a strong sense of values, have respect for doing things right and hold the moral high ground-all so they and their teams know it is important to do the right thing, not simply do things right in emergency management situations.

Thus, response of governmental agencies is one of the best actions an emergency services leader can take in disaster planning or response activities to all environmental hazards. Proper planning would prevent poor performance. In leading, there are three general skills or competencies or strategies that public managers need in emergency management and public safety. They are: diagnosing, adapting and communicating. Diagnosing is understanding what the situation is now and knowing what you can reasonably expect it to be in the future. Adapting involves adapting your behaviors and other resources in a way that helps to close the gap between the current situation and what you want to achieve. While, communicating is being able to communicate in a way that people can easily understand and accept. These leadership skills would help bring about real change in emergency management and public safety. Real change in this context means transformation to a marked degree in the attitudes, norms, institutions and behaviors that structure our daily lives. Such changes embrace not only new cultural patterns and institutional arrangement and new psychological dispositions but changes in material conditions, in the explicit, felt existence, information and communication technology, the flesh and fabric of peoples' lives.

REFERENCES

Agranoff, R. (2011). Collaborative Public Agencies in the Network Era. In *In the State of Public Administration: Issues, challenges and opportunities*. New York: M.E. Sharpe.

Beck, J. T., & Bozeman, B. (2007). Public values: An Inventory. *Administration & Society*, *39*(3), 354–381. doi:10.1177/0095399707300703

Bevir, M. (2006). Democratic governance: Systems and radical perspective. *Public Administration Review*, *66*(3), 426–436. doi:10.1111/j.1540-6210.2006.00599.x

Boin, A., & Hart, P. (2003). Public Leadership in times of crisis: Mission impossible? *Public Administration Review*, *63*(5), 544–553. doi:10.1111/1540-6210.00318

Bozeman, B. (2007). *Public Values and Public interests: Counter-balancing Economics individualism*. Washington, DC: Georgetown University Press.

Bozeman, B., & Moulton, S. (2011). Integrative Publiciness: A Framework for Public Management Strategy and Performance. *Journal of Public Administration Research and Theory, 21*(3), i363-i380.

Bryson, J., & Crosby, B. (1992). *Leadership for the Common Good*. San Francisco: Jossey – Bass.

Burke, C. S., Sims, A. E., Lazzara, E. H., & Salas, E. (2007). Trust in Leadership: A multi-level review and integration. *The Leadership Quarterly*, *18*(6), 606–632. doi:10.1016/j.leaqua.2007.09.006

Burns, J. M. (1978). *Leadership*. New York: Harper & Row Publishers Inc.

Covey, S. M. R. (2009). Building Trust: How the best leaders do it. *Leadership Excellence*, *21*(1), 15.

Crosby, B. C., & Bryson, J. M. (2005). *Leadership for the Common Good: Tackling Public*. Jossey-Bass.

Crosby, B. C., & Kiedrowski, J. (2008). Integrative Leadership: observation from a University of Minnesota Seminar Series. *Integral Leadership Review, 8*(3), 1-14. Retrieved from www.integralleadershipreview.com/archives/2008-06/2008-06-article-crosby.html

Davis, P., & West, K. (2009). What Do Public Values mean for Public Action, Putting Public values in their plural place. *American Review of Public Administration*, *39*(6), 602–618. doi:10.1177/0275074008328499

Drabek, T. (1991). *Emergency Management: Principles and Practice for Local Government*. Washington, DC: International City Management Association.

Edwards, M. (2009). Seeing Integral leadership through three important lenses: Developmental, ecological and governance. *Integral Leadership Review*, *9*(1), 1–13.

Eneanya, A. N. (2015). *Theory and Practice of Public Administration*. Lagos: University of Lagos Press.

Fairholm, M. R. (2004). Perspectives on the practice of Leadership. *Public Administration Review*, *64*(5), 577–590. doi:10.1111/j.1540-6210.2004.00405.x

Farazmand, A. (2001). Crisis and Emergency Management. In A. Farazmand (Ed.), *Handbook of Crisis and Emergency Management*. New York: Marcel Dekker.

Fleishman, E. A. (1973). Twenty – years of consideration and structure. In E. A. Fleishman & J. C. Hunt (Eds.), *Current Development in the study of Leadership*. Carbondale, IL: Southern Illinois University Press.

Jacobs, L. (2014). The Contested Politics of Public Value. *Public Administration Review*, *74*(4), 480–494. doi:10.1111/puar.12170

Kaiser, R. B., Hogan, R., & Craig, S. B. (2008). Leadership and the fate of organizations. *The American Psychologist*, *63*(2), 96–110. doi:10.1037/0003-066X.63.2.96 PMID:18284278

Kapucu, N., & Van Wart, M. (2006). The evolving role of the public sector in managing catastrophic disasters: Lessons learned. *Administration & Society*, *38*(3), 1–30. doi:10.1177/0095399706289718

Kapucu, N., & Van Wart, M. (2008). Making matter worse: Anatomy of Leadership failures in catastrophic events. *Administration & Society*, *80*(7), 711–740. doi:10.1177/0095399708323143

Kettl, D. (2006). Managing boundaries in American administration: The collaboration imperative. *Public Administration Review*, *66*(s1), 10–19. doi:10.1111/j.1540-6210.2006.00662.x

Luthans, F., & Youssef, C. M. (2007). Emerging Positive organizational behavior. *Journal of Management*, *33*(3), 321–349. doi:10.1177/0149206307300814

Maak, T., & Pless, W. M. (Eds.). (2006). *Responsible Leadership: A Rational Approach*. London: Routledge.

Mansbridge, J. (Ed.). (1990). *Beyond Self interest*. Chicago: University of Chicago Press.

McCrimmon, M. (2007). Reframing Leadership for a modern age. *Integral Leadership Review*, *7*(2), 1-6.

McGuire, M. (2006). Collaborative Public Management: Assessing what we know and how we know it. *Public Administration Review*, *66*(s1), 33–43. doi:10.1111/j.1540-6210.2006.00664.x

Moore, M. H. (1995). Creating Public value: Strategic Management in Government. Cambridge, MA: Harvard University Press.

Morse, R. S., Buss, T. F., & Kinghorn, C. M. (Eds.). (2007). *Transforming Public Leadership for the 21st Century*. Armonk, NY: M.E. Sharpe.

Newell, T., Reeher, G., & Ronayne, P. (Eds.). (2008). Relationships that make Government work. Washington, DC: CQ Press.

O'Flynn, J. (2007). From New Public Management to Public value: A paradigmatic change and managerial implications. *Australian Journal of Public Administration, 66*(3), 353–366. doi:10.1111/j.1467-8500.2007.00545.x

O'Leary, R. C., General, R. C., & Bingham, L. B. (2006). Intrdocution to the symposium on collaborative public management. *Public Administration Review, 66*(s1), 6–9. doi:10.1111/j.1540-6210.2006.00661.x

O'Shea, P. G., Foti, R. J., Havenstein, N. M. A., & Bycio, P. (2009). Are the best leaders both transformational and transactional? A pattern analysis. *Leadership, 5*(2), 237–259. doi:10.1177/1742715009102937

Petery, B. G., & Pierre, J. (1998). Government without Governance?:Rethinking. *Public Administration Research and Theory, 8*(2), 223–243. doi:10.1093/oxfordjournals.jpart.a024379

Randall, H. (2015-2016). All hazards Response starts at the local level. *PA Times Winter, 2*(1), 22-3.

Roberts, A. (1995). Civic Discovery as a Rhetorical Strategy. *Journal of Policy Analysis and Management Policy Analysis and Management, 14*(2), 291–307. doi:10.2307/3325154

Schriesheim, C. A., Castro, S. L., Zhou, X., & Dechurch, L. A. (2006). An investigation of path-goal and transformational leadership theory predictions at the individual level of analysis. *The Leadership Quarterly, 17*(1), 21–38. doi:10.1016/j.leaqua.2005.10.008

Shafritz, J. M., Russell, E. W., & Borick, C. P. (2011). *Introducing Public Administration* (7th ed.). New York: Longman.

Smith, R. H. (2004). Focusing on Public Value: Something New and Something Old. *Australian Journal of Public Administration, 63*(4), 68–79. doi:10.1111/j.1467-8500.2004.00403.x

Stoker, G. (2006). Public Value Management: A New Narrative for Networked Governance. *American Review of Public Administration, 36*(1), 51–57. doi:10.1177/0275074005282583

Thompson, A. M., & Perry, J. L. (2006). Collaboration Processes: Inside the black box. *Public Administration Review, 66*(s1), 20–32. doi:10.1111/j.1540-6210.2006.00663.x

Tichy, N. M., & Bennis, W. G. (2007). *Judgment: How Winning Leaders make Great calls*. New York: Penguin.

Trottier, T. C., Van Wart, M., & Wang, X. (2008). Examining the nature and significance of leadership in government organizations. *Public Administration Review, 68*(2), 319–333. doi:10.1111/j.1540-6210.2007.00865.x

Vecchio, R. P., Justin, J. E., & Peace, C. L. (2008). The utility of transactional and transformational leadership for predicting performance and satisfaction with a path-goal theory framework. *Journal of Occupational and Organizational Psychology, 81*(1), 71–82. doi:10.1348/096317907X202482

Vigoda-Gadot, E., & Mizrahi, S. (2008). Public Sector Management and the democratic ethos: A 5 – year study of key relationship in Israel. *Journal of Public Administration: Research and Theory, 18*(1), 79–108.

Wang, X., & Van Wart, M. (2007). When Public participation in administration leads to trust: An Empirical Assessment of manager's perceptions. *Public Administration Review, 67*(2), 265–278. doi:10.1111/j.1540-6210.2007.00712.x

Wart, M. V. (2011). Changing Dynamics of Administrative leadership. In *In the State of Public Administration: Issues, Challenges and Opportunities*. New York: M. E. Sharpe.

Welch, J., Rimes, H., & Bozeman, B. (2014). Public Value Mapping. In Valuing Public value. Minneapolis, MN: University of Minnesota.

Williams, L., & Shearer, H. (2011). Appraising Public Value: Past, Present and Future. *Public Administration Review, 89*(4), 1367–1384. doi:10.1111/j.1467-9299.2011.01942.x

Wilson, J. O. (1974). *Political Organizations*. Princeton, NJ: Princeton University Press.

Wyatt, R.J. (2016). *Encouraging a Neighborhood contingency for America's First Responders*. Retrieved in http//patimes.org/encouraging.neighborhood_contingencypolicy-america

KEY TERMS AND DEFINITIONS

Emergency Management: Emergency management involves responding to and mitigating risks and those taking on leadership of organizations at these levels must know how to manage risks.

Leadership: Leadership is often thought of as one and the same thing as management. We feel, however, that there is an important distinction between the two concepts. In essence, leadership is a broader concept than management.

Management: Management is thought of as a special kind of leadership in which the achievement of organizational goals is paramount. The key difference between the two concepts, therefore, lies in the word organization. Leadership occurs anytime one attempt to influence the behavior of an individual or group, regardless of the reason. It may be for one's goals or for those of others, and they may or may not be congruent with organizational goals.

Organizational Leadersip: There is organizational leadership, also known as bureaucratic or administrative leadership, primarily aimed at those leading and managing employees, programs, and organizations for the public good.

Political Leadership: There is political leadership, which involves legislators, elected executives in the policy process such as presidents, governors, and mayors, and the various stakeholders in the political process.

Public Values: Public values could refer to individually held values that focus on the welfare and just treatment of others. Such public values are not only endorsed collectively and ratified by government policy; they pertain to the welfare of someone other than oneself and also concerned about relationships in the society.

Chapter 5
Global Natural Hazard and Disaster Vulnerability Management

Nkemdilim Maureen Ekpeni
University of Lagos, Nigeria

Amidu Owolabi Ayeni
University of Lagos, Nigeria

ABSTRACT

This chapter examines both concept of global hazard and disaster and its management in the lights of its vulnerability. It categorized the different types of hazards and disasters and their components. From the research findings, it is observed that hazards and disaster are two sides of a coin. They occur at the interface between human systems and natural events in our physical environments. This chapter highlights that the major environmental changes driving hazards and vulnerability to disasters are climate change, land-use changes, and degradation of natural resources. After presenting a typology of disasters and their magnitude globally, management of disaster has transited from just being a "response and relief"-centric approach to a mitigation and preparedness approach.

INTRODUCTION

From previous researches, it is on record that past actions that led to the destruction of properties, displacement of population across various states of the country(Nigeria) and loss of lives ranged from ethnic differences, land-use conflicts, communal violence to mention but a few. In recent times the negative effects in the country ranging from loss of lives to displacement of persons are due to climate change irregularities such as abnormal rise in precipitation, rising sea levels, unprecedented desertification, biodiversity loss and in addition to these climate changes, large parts of the country are being subjected to over population resulting to over-exploitation of natural resources such as air, water, soil pollution, hazards, disasters amongst others, the challenges are numerous (YandeMwape, 2009). These disasters to

DOI: 10.4018/978-1-5225-3194-4.ch005

a very large extent are responsible for the loss of lives and properties in Nigeria and therefore pushed a very large number of people into abject poverty. Disasters and Hazards are becoming a major developmental and economic issue of urgent concern to the government, various related agencies, organizations, stakeholders and the people, hence the essence for this paper. For the purpose of this study, we shall be concentrating on flooding as the major natural hazard/disaster because, of all the environmental hazards, flooding is the most common which on a regular basis claims over 20, 000 lives yearly and has tremendous negative impacts on about 75million people world-wide (Smith, 1996). Urbanization means more development and by implication, it means the area is more equipped in terms of infrastructures, good level of commercial activities, manufacturing ports in the country, great businesses coupled with just moderate rainfall have contributed immensely to flooding by simply replacing vegetation and soil with parking lots and buildings, highways, of which leads to quick water runoff by exposing the earth surface. Urbanization therefore, results in high concentration of residential, industrial, commercial, educational and military facilities which automatically causes growth in population. Thereby making housing more expensive and limited to the low income earners, consequently making the poor/low-income earners relocate to floodplains areas where they can afford, and these areas are highly susceptible to flooding most especially the urban poor who seem not to have anywhere else to stay except along the coast and flood plain areas. These areas experience increased likelihood of flood events and consequently, storm surges frequency may be increased by sea-level rise, therefore, resulting in high concentration of residential, industrial, commercial, educational and military facilities which automatically causes growth in population. In general, about 20% of the country's population face the risk of one form of flooding or another because flooding does not discriminate, it just act because it usually occurs when a normally dry land is suddenly covered up with water from either an overflowing river or from heavy rains without signals/warnings, worse hit especially are those communities which does not prepare for its occurrences (Kesiena, 2011). This disaster has been witnessed in different parts of the country. These are examples of past flood attacks in different states in Nigeria with its corresponding levels of impacts and effects, such as: destruction of bridges, roads, houses, farmlands, infrastructures, polluted water resources, increased risks of diseases and even loss of lives especially in the Northern and Western states of the country (Etuonovbe, 2011):

1. In 2012, Nigeria was one of the most challenging in recent history due to unprecedented flood that ravaged several states in the country, Benue, Adamawa, Tarabaand Plateau states were the worse hit. According to the National Emergency Management Agency (NEMA), 30 out of 36 states were affected by the floods. It affected well over 7million people, displaced 2.3million people, killed over 363 people, destroyed over 597,476 buildings, damages and losses caused by this floods that was termed the worst in 40years was estimated at N2.6 trillion. This to a large extent showed how unprecedented it was in saving the lives of its people and a drastic measure is thereby anticipated. While we cannot do away with natural hazards/disasters, we can help to eliminate those caused by human factors, minimize those we individuals exacerbate and try to reduce our vulnerability levels.
2. It would be recalled that there was heavy amount of rainfall on July 10, 2010 in Lagos that led to over 140,000 people being displaced in Gaza province and several lives were lost as a result of flooding in the region. Various results in Nigeria, showed that most of the wetlands had been severely disorganized as a result of human factors which most times lead to negative effects such as ;Malnutrition, Starvation, Displacement etc.(Clark et al., 1984).

3. 2016, was a National Tragedy again. According to the statistics presented by NEMA, no fewer than 53 people died with over 100, 420 people displaced from 11states in the country from Floods. The affected states includes: Kano, Katsina, Sokoto, Kebbi, Zamfara, Kaduna, Jigawa, Adamawa, Yobe, Gombe and Bauchi.
4. Floods kills 8 in Kogi state and over 2million people were displaced in Kogi and Edo states following the release of water from dams and the overflowing of the banks of River Niger in the areas (NEMA).
5. September, 2010 experienced flooding in Kagara close to Goronyo town caused by the opening of the dam gate, which led to about 49 deaths, about 50 villages were submerged, the destruction of farmlands, thousands of houses, about 130,000 people displaced. Other areas affected by the disaster include Isa, Kebbe, Sokoto-North, Sokoto-South, Rabah, Binji, Goro nyo, Silame, Shagari, Binji and Kware local government councils.
6. In Kwara state, over 15communities were affected with farm plantations worth millions of naira destroyed, 500 hectare farms of cassava, rice, guinea corn, maize crops including livestocks such as ; cattle, sheep and goats submerged by flood in Shonga district, Edu local Government Area. This was caused by the release of water from kainji dam which also flooded all banks along river Niger areas.
7. Also in Bayelsa, recently flood displaced over 5000 people in Sagbama and Kolokuma/Opokuma local government areas. The flood was caused as a result of overflow of river Nun.
8. In Niger state, 1987, 1991and 1994, flooding caused by overflow from Benue and Sokoto basins, destroyed properties and farmlands worth millions of naira.

Though internationally, countries are prone to Tsunamis, Earthquakes, Avalanches, Hurricanes and so on which is not peculiar to Nigeria as a country? On the other hand, here in Nigeria, floods are the main leading cause of natural disasters causing loss of lives and properties (Askew, 1999). Flooding is the most universally experienced hazard, this is because there are many streams and rivers across humanized landscapes that periodically overflows, which is caused mainly because about 70% of the earth surface is occupied by water. Nigeria experiences two main seasons: dry and rainy seasons. During the rainy season, seasonal flash floods occur which are most times destructive especially in the rural communities. These flood disasters are usually traced to human activities and increased effects on those who live in flood plain areas in the country (ActionAid, 2006). Flood is a natural event that is caused by the presence of too much water in an unwanted area resulting from overflowing its banks. There has been rapid growth in number of people killed or severely affected by flood disaster worldwide (UN-Water, 2011). A large proportion of people in low-lying coastal zones with other areas that are at risk of flooding and extreme weather conditions are affected by economic damages. Flood is seen to have caused about half of disasters with about 84% disaster deaths worldwide (UN-Water, 2011).

The common and recurrent phenomenon of flooding in Nigeria occurs on a regular perennial basis in some parts of the nation. The following geographical areas are more badly hit from the hazard than others in Nigeria. This is according to NEST, 1991:

1. Low-lying areas in the southern parts of the nation where annual rainfall is very heavy.
2. The Niger Delta zone
3. The floodplains of the larger rivers of the Niger, Benue, Taraba, Sokoto, Hadeja, Cross River, Imo, Anambra, Ogun, Kaduna etc.

4. Flat low-lying areas around and to the south of Lake Chad which may be flooded during, and for a few weeks after the rain. Because of its disposition as flat, low-lying swampy area of alluvial deposition across which the tributaries of the Niger meander, the Niger Delta is by far the largest single area subject to annual flooding in Nigeria. Niger Delta is a natural reservoir for the Niger-Benue catchment area with its inefficient outlet to the sea. Its rivers are bordered by levees, which are overtopped and extensive areas submerged when these rivers are in flood. The zone is subject to two types of flood:
   a. The rainy seasons floods which are the result of rainfall within Nigeria and especially rainfall in the delta area itself which ranges from 2000mm to 4000mm per year and is concentrated in few months.
   b. The river flood, almost all of which are caused by watercoming from as far as the Fouta Djallon Highlands in Guinea. The water is primarily the run off from theprevious year's rainfall on these highlands slowly making its way down to Nigeria. Almost all parts of the NigerDelta, with the exception of the northern parts, suffer from flooding at one time of the year or the other from which Town, villagesand even agricultural lands in the Sagbama and Yenagoa areas in Balyesa state are usually affected. Also, often affectedare parts of the riverine areas of Edo, Delta, Cross River and Akwa Ibom states.

This study is a further contribution to flood studies, and it has the overall objective of analyzing the general effects of flood menace in Nigeria and subsequently recommending remedial and management strategies to curbing it.

Floods as natural disaster/hazard have particularly become frequent and likely to occur in the future due to climate change. A rise in the amount of greenhouse gases in the atmosphere will most likely cause a rise in the amount of temperature over most land surfaces and hence varying climate change will be observed regionally (Pinter, 2005). Hence the need to understudy the past and present events to have a foresight on how to mitigate flooding effects in the nearest future in Nigeria.

The processes that occur naturally includes; physical (heat, noise, vibration), chemical (gas leaks or acid leaks and biological (outbreak of flu or nuclear radiation). These changes modify the landscape. Also, some processes occur internally, such as earthquakes, volcanic eruptions etc. These changes are driven deep into the earth while other processes which act close or even on the earth surface includes landslides, flooding, coastal erosion, violent storms and wildfires (Laven, 1996).

In general, the scope of this study generated the following objectives for disaster management:

1. To involve in activities that may be of assistance that would help to water down the effects caused by disaster especially in the low income communities.
2. Reduce or avoid losses from hazards.
3. To create awareness amongst the people about hazard and disasters and their corresponding effects on the people and to help prepare them in advance when faced with such situations.
4. To make the roles of hazard and disaster management agencies and NGOs explanatory enough, so that they know how to handle such situations
5. Ensure prompt assistance to victims
6. To go as far as taking courses which will enhance and facilitate in the preparation of rural development schemes and ensure they are carried out effectively and also publish details of the previous development policies.

7. Keeping of information- data base which would help to predict future occurrences, help in risk assessment and vulnerability analysis.
8. To engage in self-help program such as spreading literacy among the minorities, economically deprived communities /.
9. To establish and sustain educational and training institutions for educating the communities on topics related to both hazards and disasters.
10. Achieve rapid and effective recovery.

BACKGROUND TO THE STUDY

Natural Hazard

Natural hazards are natural phenomenon which has the potential to cause harm to the people or property and its events can be the same that we can learn from. With natural hazards, there is a threat to life, health, environment or property. Hazards have related causes for example, a landslide can be provoked by heavy rainfall whereby its severity can be determined, for example, deforestation. In other words, one hazard is usually triggered by another hazard i.e. an earthquake may trigger the destruction of buildings and infrastructures such as dams, which will further result in other hazards such as floods (Cruz et al., 2006). Its events can be grouped under 2 broad categories:

1. Geophysical which can be subdivided into geologic, meteorological and climatic factors such as earthquakes, drought, flood, and coastal erosion to mention but a few.
2. Biological hazards can be seen as severe array of diseases and infestations outbreaks (Blodgett, 2006).

Table 1. Groups and types of hazards

Hazard Group	Hazard Types	Examples
Purely geologic	Geological	Earthquake, volcanic eruption, tsunami, landslides, subsidence
Geo/ Anthropogenic	• Metrological • Oceanographic • Hydrologic • Biological	Cyclones, lightning, & fire, drought, avalanche, hail storm, cool spell, sea storm, storm surge, flood, flood flash, epidemics, crop blight, insect infestation
Purely anthropogenic (technological)	• Explosion • Release of toxic materials • Severe contamination • Structural collapse • Construction or manufacturer accident	
Purely anthropogenic (Conflict)	• Crowd-related • Terrorist activity • Political conflict	• Riot, crowd crush • Bombing, shooting, hijacking • International and civil war, revolution and 'coup d'etat'

Source: Smith, 2000.

Forms of Flooding in Nigeria

1. **Coastal Flooding:** It occurs in the low-lying belt of mangrove and fresh water swamps along the coast which is sometimes exacerbated by wind-induced storms deposited along the coast and it is usually associated with Tropical Cyclones known as Typhoons and Hurricanes.
2. **River Flooding:** It occurs in the flood plains of the larger rivers caused by precipitation over large catchment areas. It encompasses many independent river basins with tributaries that may drain large geographic areas.
3. **Flash Floods:** They usually occur within 6hours of the beginning of rainfall which are associated with rivers in the inland areas where sudden heavy rains can change them into destructive torrents within a short period with towering cumulus clouds, severe thunderstorms and tropical cyclones. Other causes of Flash floods include dam failures or sudden break up of other river obstructions.
4. **Urban Flooding:** This is dependent on flash flood, river flooding and coastal floods. It occurs in towns located on low lying or flat terrain especially where so little or no provision has been made for surface drainage, or where existing drainage has been blocked with municipal waste, refuse and eroded soil sediments. Urban flooding is a regular occurrence in every rainy session in some areas such as; Lagos, Maiduguri, Aba, Warri, Benin and Ibadan.

The manufacturing industries, schools, colleges and research institutes, markets and commercial centers, hospitals and maternity homes, recreation centers, hotels,restaurants and bars, chemicals and food processing plants, households and others release and dump solid (degradable and non-degradable) and liquid wastes into the immediate and distance communitiesindiscriminately without thinking of the consequences on the environment. As a result, the drainagechannels, canals and erosion passages are blocked inadvertently and or deliberately. Waste generation and management are challenges to both governments and urban dwellers, (Mabogunje, 1987, Oyebande, 2005 and Aderogba, 2010).

Natural Disaster

Natural Disasters on the other hand are also natural phenomenon but are sudden and calamitous,resulting from the combination of hazard, vulnerability and inability of a community to reduce the potential chances of risk from unexpected events that cause heavy damage, destruction of property from hazard impact on the population that is vulnerable which results to human casualty. It can result to a serious widespread of human, material or environmental loses which most times surpasses the capability of the affected community to handle left alone to manage with its resources. No two disaster events are the same, hence pre-plans may not work. The magnitude of its damage is very high and it is mostly immensurable and varies with either its geographical location, earth surface type, climatic factors or degrees of vulnerability. Disasters cn be classified under 2 categories: 1. Natural which includes; flood, drought, earthquake, thunderstorm, cold waves, cyclones, heat waves amongst others.

1. Anthropogenic which include; fire outbreak, road accident, food poisoning, building collapse, plane crash, deforestation, chemical pollution and war to mention but a few.

Figure 1. The causes of floods and intensifying conditions
Source: Pickering and Owen, 1995.

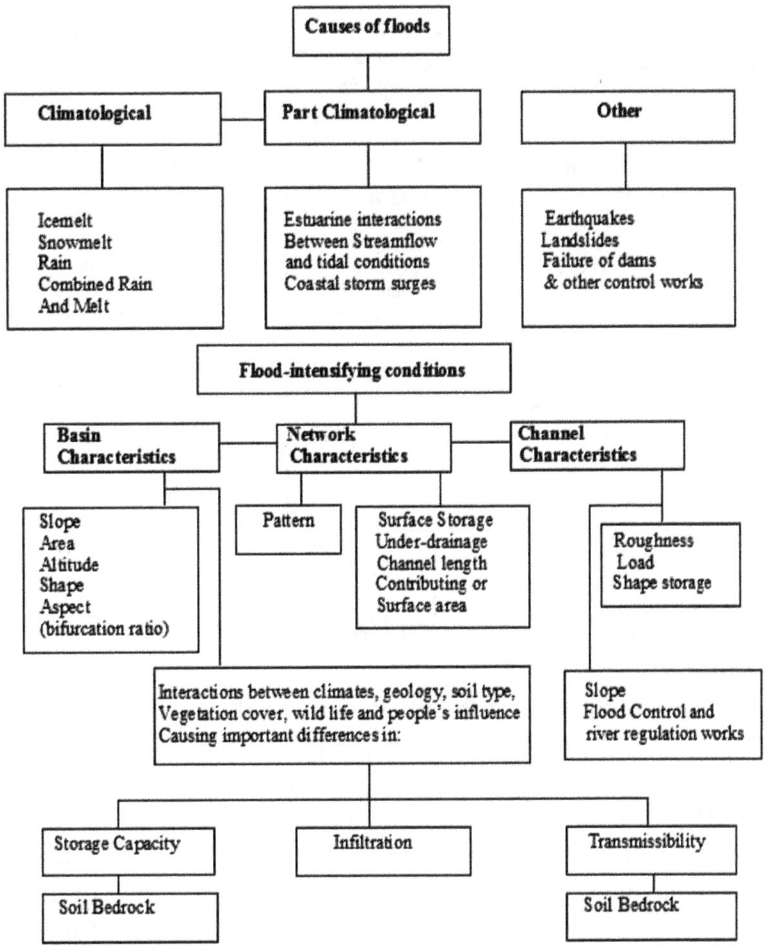

Any form of hazard/disaster such as earthquake or flood with high vulnerability (inadequate access to resources, lack of awareness, and large number of sick and old people) would lead to disaster causing loss of property and life. For example, if an earthquake happens in an uninhabited desert, it won't be seen as a disaster without a direct effect on the people no matter the intensity of the disaster. In essence, disaster happens only when hazards and vulnerability come together (Singh, 2004).

Disasters and Hazards are very similar, they are like two sides of a coin. One will require more than just social and physical science to explain its mode of operations. Hazards occur at the interface between natural events and human systems, hence the various ways in which humans adapt and relate to hazards can help to modify natural occurrences. Again the major changes that occur environmentally which pushes hazards and vulnerability to disasters are climate change, land-use changes and degradation of natural resource (Gupta et al., 2011).

Figure 2. Major changes that occur environmentally which push hazards and vulnerability to disasters
Source: Gupta, 2011.

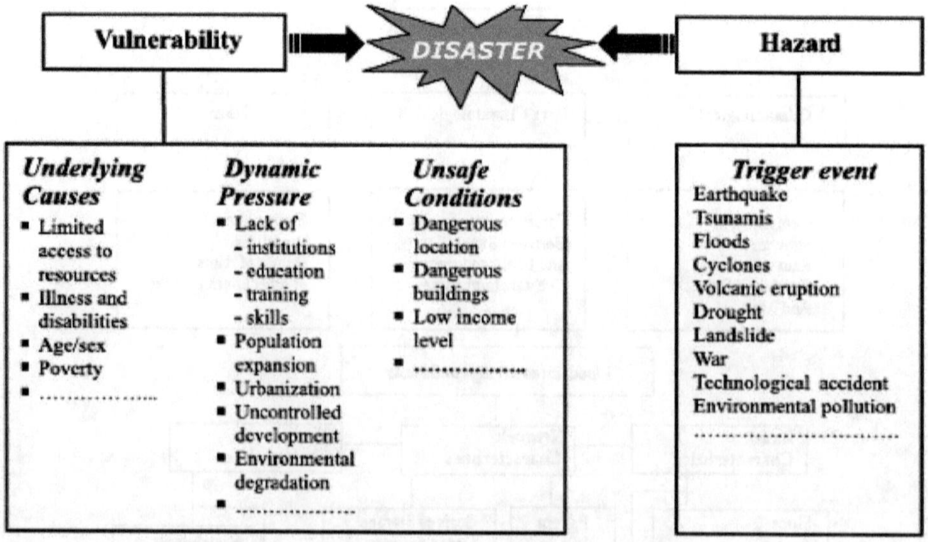

The Concept of Vulnerability to Natural Hazards and Disasters

The term Vulnerability can be looked at from different perspectives. It can be seen as the quality of being easily hurt or attacked that is, being exposed to injury or appearing as if you are. Vulnerability to natural hazard or disaster events can also be seen as the potential of being harmed by such hazards and disasters which influences the people and make them have the ability to adjust, cope, anticipate, resist or even recover from natural hazard (Garatwa & Bollin, 2001)..Vulnerability is related to poverty. Commonly, the poorer communities naturally have lesser resources at their disposal and hence lesser opportunities of reducing vulnerability (Chambers, 1989). Vulnerability is also dependent on the cultural content (Chapman, 1994). Following different cultures, risk is considered an underlying cause of disaster (White, 1974).

WISNER (2001) shows that the term 'vulnerability' is often used with studies involving hazard and disaster in different ways and was able to identify the following groups:

- Structural, engineering vulnerability,
- Lifeline infrastructural vulnerability,
- Communications system vulnerability,
- macroeconomic vulnerability,
- Regional economic vulnerability,
- Commercial vulnerability (including insurance exposure),
- Social vulnerability.

Vulnerability changes in its characteristics with time which makes it dynamic. These changes can happen either slowly due to development or it may change rapidly as a result of hazard or disaster thereby rendering the community homeless and helpless and even become refugees.The introduction of the topic on population's capacity to 'absorb' and 'recover' was presented by Susman et al., (1983). Due to different community's vulnerability;"The degree to which different classes in society are at risk, both in terms of the probability of its occurrence of an extreme physical event and also the degree to which the community absorbs the effects of extreme physical events and further helps different classes to recover" vary.

Chambers, (1989) and Bohle,(2001), states that the *internal* dimension of vulnerability can be seen as the defencelessness and the insecurity of the people, the capacity to anticipate, cope with, resist, and recover from the impacts of a hazard while the *external* dimension involves exposure to risks and shocks. The most suitable definition is the working definition that combines the various thoughts of IFRC (1999), Blaikieet al. (1994), Garatwa & Bollin (2001) and Mani (2002). It goes thus:

A vulnerable human being is:

1. *Capable of being physically (emotionally or spiritually) wounded;*
2. *Open to attack or damage (physical, emotional, or spiritual); and*
3. *Lacking in defence or support mechanisms (at the levels of state/government; community; household; and individual).*

Within the context of food insecurity, SEN (1981) initiated a reconsideration of classic drought and flood-driven famines by claiming that starvation is based on a lack of access to food rather than food availability. Social, Political and Economic rather than natural circumstances are also part of the causes of limited access to food supply. Although practitioners, experts and researchers agreed on the need to further emphasize the socio-economic situation, but their views on the concept of vulnerability and the underlying definitions strongly differed which was dependent on the approach adopted. Alwang et al. (2001) tried to address this by comparing the understanding of vulnerability from the different scientific disciplines of which their differences ranged from the standpoint of poverty dynamics, food security, disaster management, sociology, anthropology, ecology based environment, to mention but a few. But then, a list that is of common interest was drawn out:

- The poor are more vulnerable.
- Households are vulnerable to a variety of negative outcomes that can be measured in different ways.
- Vulnerability is caused by multiple sources of risk.

According to Chambers, (1989), he points out the problem of assuming these two terms 'vulnerable' and 'vulnerability' are substitutes for 'poor' and 'poverty' and states that "vulnerability is not the same as poverty. Poverty, looking at it from the angle of low income, can be reduced by borrowing and investing; but such debt usually makes households more vulnerable". In general, poor people have fewer assets and opportunities to withstand and respond to harmful impacts. By considering four main natural hazards, the UNDP calculates that those countries classified under 'high human development' represent15% of the population exposed, but only 1.8% of deaths (UNDP 2004b). Consequently, the statistics of the IFRC (2001) for the years 1990-2001 reveal that the number on the average of people killed by natural disaster was 353 for 'medium and low' developed countries and 22for 'highly' developed countries (Garatwa &

Bollin, 2001). This suggests that underdevelopment and poverty play an important role for vulnerability assessment without leaving out other factors.

Due to the various negative outcomes and sources of risk necessitates the need to specify causes and effects of disasters in order to describe vulnerability. The World Food Progamme (WFP 2004, p 2) states that in order "to be useful, vulnerability has to be defined in terms of what it is that a population is considered to be vulnerable to and specify". In the context of disaster management this means specifying exogenous events and shocks and in the context of food security, the research tends to focus on the consequences rather than the causes of a disaster (in this case: famines). Chambers, (1989), initiated the outcome-oriented approach which was further developed by Downing, (1991), who states, that "vulnerability is better referred to as a consequence rather than a cause, this is because vulnerability in reference to a cause insinuates a negative consequence without completing the reference". This approach is now broadly accepted in the food security literature (Blaikie et al. 1994; Dilley & Boudreau 2001).

Risk Equation

This concept was visualized by Crichton, (1999) using the 'risk triangle' with emphasis on the dependency of risk (R) on the following components: hazard (H), exposure (E) and vulnerability (V).

This concept in its original version or modifiedslightly has been widely accepted and applied for research on natural disasters (for example Peduzzi et al. 2002; Granger 2003; Dilley et al. 2005). It can be expressed thus:

$$R_{ah} = H_{ah} \times E_{a} \times V_{ah} \quad (1)$$

a = area affected;
h = type of hazard.

Subscript 'h' relates to the type of hazard (determined in its severity and its temporal extent) and subscript 'a' is the geographical region affected by hazard 'h'. Exposure is the number of people located in area 'a'. The resulting risk refers to the potential lives lost regarding hazard 'h' in area 'a'. Vulner-

Figure 3. The "risk triangle"
Source: After Crichton 1999.

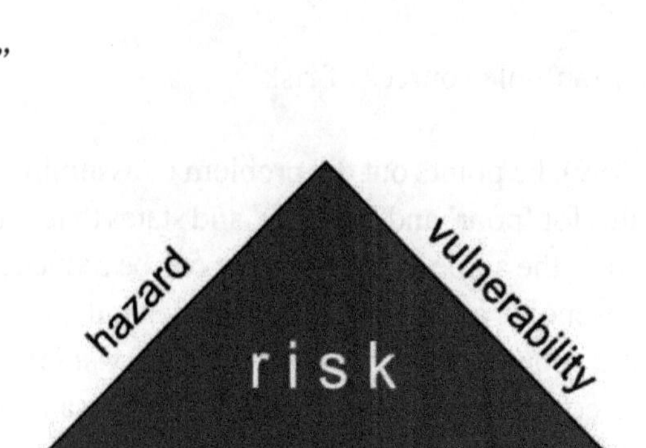

ability is people's ability to cope with hazard 'h' in area 'a'. Since the degree of vulnerability of the people living in the affected area may vary, the vulnerability in Equation 1 has to represent the average vulnerability of a single individual within area 'a'. The risk equals 0 if one of the three components of hazard, exposure or vulnerability is 0.

In the case of earthquakes there is no risk if (1) there is no possibility that an earthquake would occur and / or (2) that the area affected is scarcely populated and / or (3)that the population in question is not vulnerable (all houses are built to a high level of earthquake security).

It should be noted that:

1. Vulnerability changes with the severity and type of hazard (the houses built might be earthquake resistant but vulnerable to floods).
2. That 'people' is the only *element at risk* considered. Other possible elements could be physical assets such as built-up areas, transport lines or similar types of infrastructure.
3. Knowledge of the spatial distribution of hazardous events and the elements at risk are needed for Risk determination.

Physical parameters and demographic datasets can be used to determine 'Hazard'' and 'Exposure'. The concept of vulnerability is more complex and more difficult to describe. It is necessary to rely on approximating methods such as proxy indicators when attempting to quantitatively estimate a population's vulnerability.

TYPES OF VULNERABILITY

Vulnerability can be categorized under the following headings;

1. **Physical Vulnerability:** It deals with questions of 'who' and 'what' may have been destroyed during the process of natural hazard/disaster such as floods. This is solely based on the physical conditions of the people affected and elements at risk. This includes; buildings and infrastructures, together with considering the technicality of the structures and buildings to verify if they can be able to resist the forces acting upon them during the events.
2. **Socio-Economic Vulnerability:** This handles the extent to which a population is affected by hazards/disaster. The intensity/impact of hazard/disaster would always be determined by the socio-economic conditions of the people involved.
3. **Capacity:** It is the combination of the resources, strengths and means which obtains in households and communities and which enables them to cope, prepare, withstand, prevent, mitigate and even quickly recover from a disaster.

In our society, poverty is normally seen as related to low income and poor availability to goods, resources and wealth. Looking at the bigger picture, not being poor may not necessarily mean not being vulnerable and vice versa, after all, an household may have access to relatively high income and may still be seen as being more vulnerable mainly because of the economic development of its community when compared with an household that has relatively lower income but fortunate enough to be located in a better community, development wise. In essence, the state of development is mostly used as an in-

Figure 4. Vulnerability in relation to the level of economic development and mitigation gap
Source: Alexander, 2000.
A – Poorest societies
B – Countries with slightly increased economies
C – Industrialized society with high economic development
D – Very high developed society
E – Reduction of vulnerability achieved with technology
F – 'Mitigation gap'

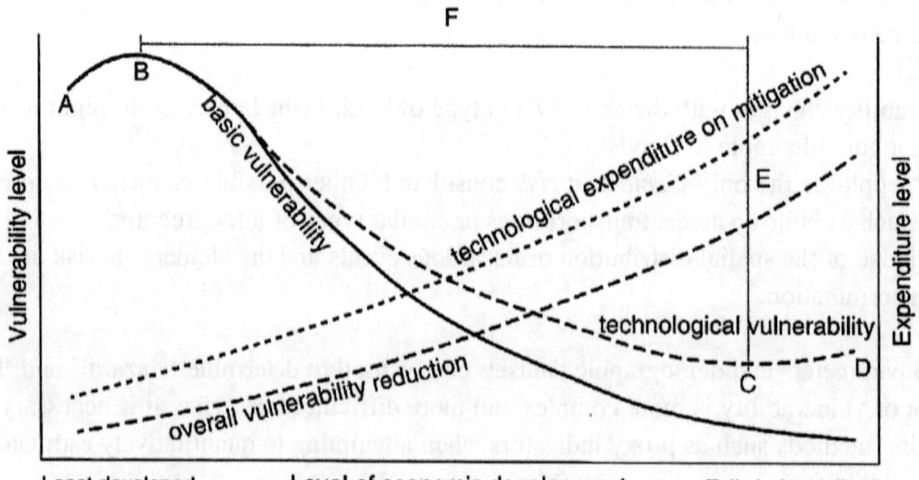

dicator to determine vulnerability in correlation with the economic development standard of a country (Alexander, 2000). Vulnerability as a concept can be differentiated by different levels and magnitudes (IFRC, 1999b)

Who Are Most Vulnerable to Natural Hazards and Disasters?

Natural Disasters occur most times across the world. It doesn't spare both the developed and developing countries. Although, some communities are obviously more vulnerable and equally more susceptible to its impacts than others. Past studies revealed that the majority of lives and property affected and lost to natural disasters are from the developing countries which shows that poverty and vulnerability are also

Table 2. Description of hazard's characteristics

Magnitude	Only those occurrences that exceed some common level of Magnitude are extreme
Duration	The length of time over which a hazardous event persists
Areal Extent	The space covered by the hazardous event
Speed of Onset	The length of time between the first appearance of an event and its Peak
Spatial Dispersion	The pattern of distribution over the space in which its impacts can Occur
Temporal Spacing	The sequencing of events, ranging along a continuum from random to periodic.

Source: Gravley, 2001.

two sides of a coin (Oluduro, 1988). Which goes to show that the major disparity between the developed and developing countries vulnerability level is 'poverty'. In other words, the wealthier a country is, the better access the country has to resources that will help manage vulnerability and develop coping strategies when such hazards take place. The earth due to its nature is prone to disasters and so we cannot completely rule out the continuous effects of these natural hazards and disasters with severe impacts in poor countries which further results in these countries even becoming poorer following the aftermath effects. Besides, untold hardship is experienced, especially by the most vulnerable groups (women, elderly and school children) whenever there is flood disaster (Oluduro, 1988; Durotoye, 1999; Folorunsho and Awosika,).These days, what we now see in developed countries is that disasters are becoming less deadly but more costly due to their advancement in science and technology. Though, the destruction and damages to properties are costing so much more but fewer people are dying from disaster outbreaks which is also due to the fact that their mitigation capabilities are better. Such as:

- Forecasting Disasters
- Building stronger structures that can withstand strong physical impacts
- More resources in circulation to run to the masses aid.

This increases the people's resilience to hazards but with the continuous growth in population, the society willlose more of value that is exposed to hazards and disasters thereby skyrocketing the amount spent on addressing the issue of damage.

The urban poor use their location as the basis for organizing livelihood activities (for example; street hawking, crafts, crimes, casual labour, prostitution and so on) as long as they can be very close to economic opportunities (Hardoy & Satterthwaite, 1989; Fernandes & Varley, 1998).

Vulnerability is a dynamic characteristic of a population that changes continuously and possibly significantly when this population is exposed to hazardous events. A pre-event and post-event vulnerability can be identified. In the pre-event phase, vulnerability tends to be determined by physical assets and is based on availability of financial resources before the disaster. Socio-cultural aspects such as the level of fatalism amongst the population may also play an important role, for example in implementing mitigation measures. In the post-event phase, vulnerability is mostly determined by the institutional setting, organisational infrastructure and strength of social networks. We cannot rule out the importance of quality of health services or, as mentioned earlier, access to financial resources when a disaster occurs.

Some authors have divided the concept of vulnerability in a similar way not related to temporal change. UNISDR, (2003) and Jigyasu, (2002), differentiated the subordinate components of a physical (or primary) vulnerability and a 'socio-economic' (or secondary) vulnerability. Some other authors (e.g. Alwang et al. 2001; Wisner et al. 2004, Plate, 2006) based their concepts on the dynamic and temporal aspects of vulnerability. However, most authors focus on the post-event phase and designate 'coping capacity', 'resilience' and 'ability to recover' as the most important characteristics of vulnerability. As a description of pre-event vulnerability, the term frequently used is 'susceptibility'. Which is predominantly explained as 'exposure' to hazards (WHO 1999). The terms 'susceptibility' for vulnerability in the pre-event phase and 'resilience' for vulnerability in the post-event phase are proposed. Resilience however, can then be further subdivided into an immediate aftermath 'coping capacity' and a longer term post-event 'recovery' (Figure 5). Susceptibility would be predominantly determined by physical features, while resilience by socio-economic characteristics.

NATURAL HAZARD AND DISASTER MANAGEMENT

This is the combination of all the activities, measures, preparedness and mitigation programmes which can be looked at before, during and after a disaster/hazard with the sole aim of avoiding disaster, reduce its impacts and recover quickly from its losses. Three important stages of activities are addressed within the disaster risk management according to ADPC, (2003) . They include;

Preparation/Pre-Event

It is the precautionary stage which involves activities such as awareness campaigns, strengthening existing weak structures and buildings, formulation of viable emergency policies, upholding of warning signs and systems, and training of personnel. This is the preparedness stage which involves measures that are put in place before an occurrence. It allows the government, communities and individuals embrace measures to assist them in responding quickly to disaster events to enable the people adjust to them effectively. The major aim here is to minimize the extent of destruction and damages of property, disorganization of critical services and reduction in loss of lives (WHO, 2002).

Impact/Disaster Occurrence

These are initiatives taken to ensure that the provision and the needs of the victims are met to a reasonable extent and suffering minimized. It is an emergency response stage to provide assistance immediately to maintain and preserve lives, improve health and boost the victims emotional stability. The focus here is meeting the basic needs of the victims until an alternative or permanent and sustainable solutions are

Figure 5. Terms in relation to a sudden hazard impact
Source: WHO, 1999.

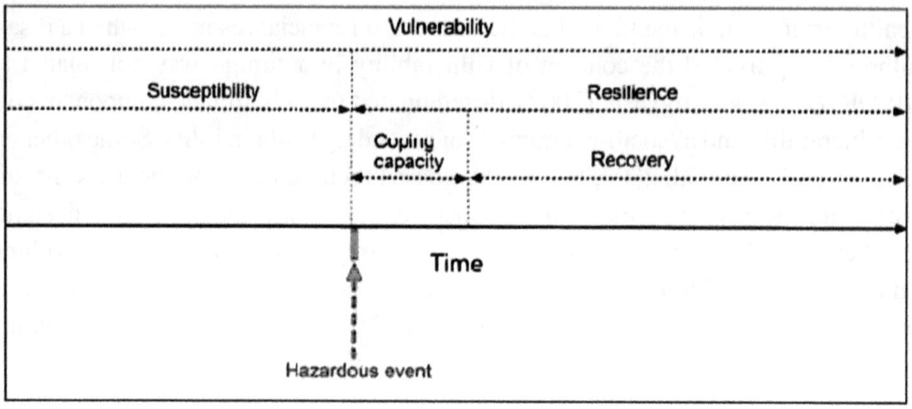

proffered. At this stage, the activities the rescue operatives are involved in are called emergency response activities which includes; search and rescue, planning (emergency access/ relief, evacuation routes), training (emergency teams, practice drills), supply (providing emergency response equipment). This is the stage of humanitarian organizations.

Response/Post-Event

This involves activities taken in response to a disaster with the aim of achieving quick and rapid recovery and rehabilitation of the affected people or communities as soon as the disaster strikes to restore services to their regular order, long-term reconstruction of health, utility and communication facilities. In addition, efforts to reduce future risk factors should be included in the plan. For example; temporary housing, grants, medical care, health and safety education, counselling programs, reconstruction to mention but a few. After the emergency has been controlled, the victims would now be made to under several activities which would help restore their lives and the infrastructures that they support themselves with. Here restoration process continues until all systems are brought to normal. At this stage, the activities the rescue operatives are involved in are called response and recovery activities.

From Figure 6, it is seen that the disaster management cycle is divided into a pre-event/ preparation stage, the main stage in temporal relation to the main impact and post-event/response stage. This illustration covers both the activities usually associated with(short-term) disaster preparedness and emergency actions as well as the long-term processes of recovery, development and mitigation of future risk.

Figure 6. Disaster management cycle
Source: ADPC, 2003.

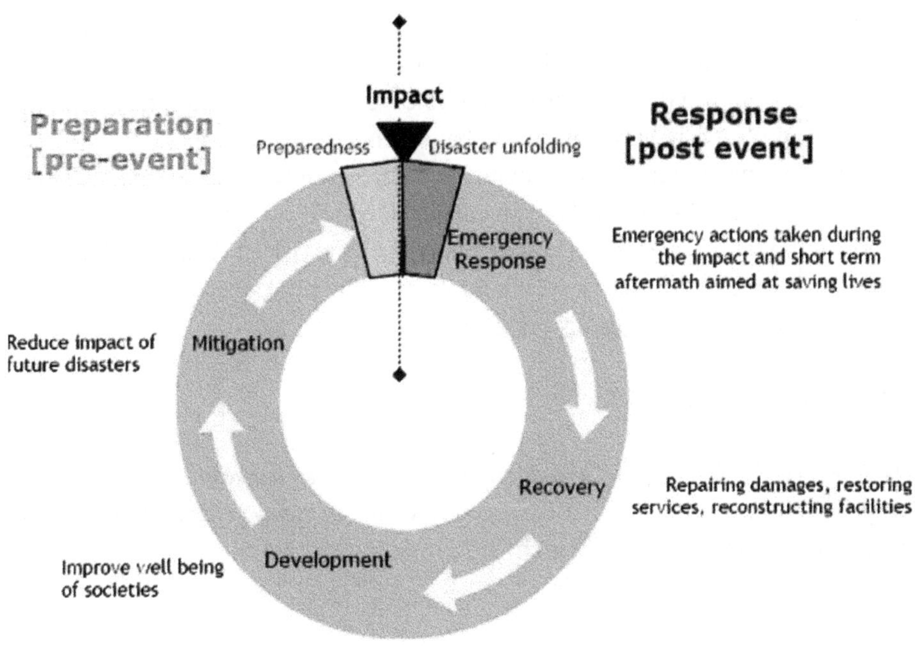

POSSIBLE MEASURES FOR REDUCING DISASTER/HAZARD

Risks

1. **Mapping of Flood Prone Areas:** Historical records give the indication of the flood inundation areas, the periods of occurrence and event, It goes further to give the extent of the coverage. This makes mapping of the flood prone areas a very important step that should be taken in reducing the risk of an area. In the coastal areas the tide levels and the land characteristics will determine the submergence areas. Flood hazard mapping will give the proper indication of water flow during floods thereby making it possible for warnings to be issued by looking into the earlier marked heights of the water levels in case of potential threat.
2. **Land-Use Control:** The number of casualties is usually related to the population in the area at risk. In areas where people already have built their settlements, measures should be taken to relocate them to better sites so as to reduce vulnerability. Major development should not be permitted in the areas which are subjected to high flooding. Important facilities like hospitals, schools should be built in safe areas away from floodpains. In urban areas, water holding areas can be created like ponds, lakes or low-lying areas. This will reduce the danger on life and property when waters inundate the floodplains and the coastal areas.
3. **Flood Shelters:** This is one of the ways of how communities can protect themselves from the worst of the floods. Banks of earth are raised by up to 5 metres and cover an area of several kilometres. The people dig a huge pond in the middle and use this earth to raise the ground. Whenever the floods come, people can bring their livestock, possessions – even their homes – to safety. The pond in the middle becomes an important source of food, as it is used to farm fish.
4. **Construction of Engineered Structures:** Better and stronger structures engineered should be constructed in floodplains areas built specially to withstand flood forces and seepage. The buildings should be constructed in elevated areas. Where necessary build on stilts or platform.
5. **Flood Control:** This can be done by decreasing the amount of runoff with the help of reforestation (to increase absorption could be a mitigation strategy in certain areas), protection of vegetation, clearing of debris from streams and other water holding areas, conservation of ponds and lakes etc. Ways of Flood Diversion include levees, embankments, dams and channel improvement. Dams can be used to store water which can be controlled to release water at a manageable rate. The disadvantage of this is that, failure of dams in earthquakes and operation of releasing the water can cause floods in the lower areas. Flood Proofing reduces the risk of damage. Measures include use of sand bags to keep flood water away, blocking or sealing of doors and windows of houses etc. The aim is to reduce flood damage in a community.

Preventing Disaster in Nigeria

In the late 60s and early 1970s, disaster management function was the remit of the offices of the Head of State, state governors and the National Emergency Relief Agency, NERA which came into being following the devastating drought of 1972 and 1973.

Then, the Federal Government set up an inter-ministerial body to find ways to reduce natural disaster risks, in line with the 1990 United Nations International Decade for Natural Disaster reduction. Three years later, the Federal Government expanded the scope of risk reduction to encompass all types of disaster by passing Decree 119 and subsequently making NERA, an independent body under the office of the President. After then more organizations were formed to assist other environmental bodies for efficiency such as NEMA.

National Emergency Management Agency (NEMA), grew out of the Environmental Policy Development process known as The Consultative National Environmental Policy process (CONNEPP). This process involved public participation extensively. NEMA, came into effect on 29th January, 1999 to satisfy the following purposes in the country:

- To provide for co-operative environmental governance by establishing principles for decision making on matters affecting the environment.
- To provide for institutions that will promote co-operative governance and procedures for co-coordinating environmental functions exercised by organs of the state.
- To provide for certain aspects of the administration and enforcement of other environmental management laws.
- To save lives and provide necessary assistance at disaster site. It also involves resettlement of affected persons where necessary.

Events of the past necessitated the enactment of NEMA as a body. For example in August 2011, when Ibadan, the Oyo State capital witnessed one of the worst flood disasters in its history that resulted in the death of many residents and the destruction of property worth more than N20 billion. Even the University of Ibadan suffered damages worth more than over N10 billion, which included the washing away of its fish farm valued at N300 million, flooding of the zoological garden, and damage to the Teaching and Research Farm and books worth N2 billion.

It is sad to note that while the disaster lasted, there was no significant presence of disaster management officials from any of the government agencies to rescue those trapped in the flood because there wasn't enough government agencies to handle such situations and even those organizations that were already in existence had too much rescue missions at hand, thereby making them ineffective which happened to be the case of NEMA. Turning up to offer relief materials to the victims of the Ibadan flood was slow.

It is interesting to note the recent proactive measures announced by National Emergency Management Agency (NEMA) to prevent occurrence of disaster instead of reactionary measures to such crises. Recently the Director General of National Emergency Management Agency (NEMA), AVM Mohammed Audu-Bida (Rtd.), called on all disaster management related agencies in the country to be proactive in their approach to curtail disasters in view of the fact that the cost implication of reactionary measures is enormous compared to preventive strategies put in place. The damage caused by disaster is mostly immensurable and varies with the geographical location, climate and the type of earth surface or degrees of vulnerability. This influences the mental, socioeconomic, political and cultural state of affected areas and the people. Audu-Bida (Rtd), further said that disasters derail and delay socio-economic progress and put millions of people into dire poverty or make the poor even poorer. According to him the world is witnessing a rapidly increasing impact of disasters caused by the combination of natural hazards and vulnerability, which constantly threatens people's lives and livelihoods.

Despite the fact that there is an adage which says prevention is better than cure, so many people still fall victims due to their ignorance and negligence. Not until recently after some disasters, especially plane crashes that made the public become aware of an agency responsible for disaster management in Nigeria. Awareness of these agencies related to handling disasters are very poor. Past events have shown that disasters on several occasions completely disrupt the normal day to day life, negatively influences the emergency system while normal needs and process like food shelter, health, etc, are affected and deteriorate depending on its intensity and severity of the disaster. It may be necessary therefore to say that not all disaster can be avoided but proactive measures can be taken to avert large scale casualties. I strongly believe that past national disasters could have been avoided if appropriate actions were taken by relevant agencies and well-informed public on proactive approach even when the incidence took place.

The new concept of partnering with National Youth Service Corps members and the introduction of Emergency Management Vanguard to promote volunteerism are right steps towards the right direction. The new thinking would only require more training, capacity building for those to be involved.

Early Warning Signs Ignored

Early warning signs (EW} is a proactive measure to curbing flood hazard menace. Curbing menance, is made possible with joined efforts from the government, disaster related agencies and the people. In the wake of the flood events in these various states mentioned earlier, the Nigerian Emergency Management Agency (NEMA) accused State Governments of overlooking early warning messages of flooding across the country especially from the Nigerian Meteorological Agency (NIMET).

Most of the affected states recorded over 60% increase in the volume of rainfall in the period. This disaster caught most of the State governments unawares and they could therefore not do much to prevent, mitigate or prepare for the humanitarian crisis that ensued from the disaster. Lagos and Oyo who recorded the highest number of victims from flood in the months of July and August respectively exhibited the inability of both federal and state governments to be proactive in responding to issues of possible humanitarian crisis despite having early warning signs on the possibility of flood in the states. The cumulative risk index for urban planning, urbanization and industrialization disadvantages and poor drainage management is high in most of the affected states which increased the effects of the disaster on the people (NIMET). Also, the issue of poor urban planning and blockage of drainage and water ways came to the fore as the proximate condition for the fatality of the flood in especially Oyo state which accounted for about 905 of the total deaths from flood in the month. Sadly successive governments in the state have not done much to tackle the issue of proper urban planning and development. The importance of sensitizing government at all levels on the reality of climate change and on possible risk reduction strategies is crucial in preventing a similar crisis in the nearest future. The need for an effective community based early warning system for disaster prevention and mitigation for Nigeria cannot be over emphasized. The existing one like the National early warning system of WANEP – should be funded and improved for efficiency.

SUBSTANTIVE CAUSES OF HAZARD/DISASTER IN NIGERIA

From my findings putting together this paper, I discovered that the substantive causes of flooding and the increase in the severity and intensity of this hazard/disaster generally in all states in the country have been categorized under the following headings:

- Illegal Structure on / across Drainage Channels
- Land Reclamation/ Encroachment
- None-heeding to weather forecast
- Poor Physical Planning
- Inadequate Drainage Channel
- Global Warming
- Blockage of Canals/ Drains
- Negligence
- Government Policies and Programs
- Collapsed Dams/ Embankments
- Nature of Terrain
- Torrential Rain Storms
- Base water flow
- Water Flows from Car Wash Centers
- Spring Water Flow
- Watering Flowers/ Farm Irrigations
- Construction & Reconstruction
- Illegal channelization of Drainage Channels
- Social/Cultural Activities
- Ocean/Lagoons surge
- Poor Waste Management

SOLUTIONS AND RECOMMENDATIONS

If the cities and towns will stand the test of time, meet the Millennium Development Goals and perform their functions to the admiration of the populace, the challenges of floods must be faced head on. It should not be forgotten that global warming and climate change is a world-wide challenge. All hands must be on deck to face this global phenomenon. In averting the annual tragedy of floods often caused by torrential rains and living habits of the people, the three tiers of government need not be reactive but proactive with fighting disaster and they should step up rehabilitation activities that would ensure major roads and drainage channels are opened and widened across the cities and towns to ensure that they are free of obstacles and are serviceable. The Ministries of Works and Infrastructures of each state should be tasked by their respective state Governments to ensure adequate free drainage channels and erosion passages in all nooks and cranny of the states.

It is imperative that the national monthly environmental sanitation exercise is enforced thoroughly and sustained. It could be advised that the exercise may be observed at more regular intervals to rid homes, work environments and others of filth made up of waste waters, plastic junks, waste foods, abandoned/condemned home utensils and appliances, unserviceable tools and equipment and others, and properly disposed of them to allow free flow of water along erosion passages, gutters, drainage channels and canals. It is better late than never.

The management of River Basin Development Authorities should be magnanimous enough to provide the engineering and technical solutions that can effectively take care of floods in their respective areas of jurisdiction. Weather forecasts and forecasters should be more informed and proactive to weather events that could lead to torrential rainstorms that may subsequently result in floods, and cause hazard/disaster. There must be legislations that must deal with the use of plastic bottles and cans, sachets, such as pure water sachets, carrier bags and other non-degradable materials both in the urban and rural areas in the country. For the affected areas to achieve improvement in environmental management and quick development for sustainable growth, there must be more effective efforts to improve flood control and management, climate change mitigation and adaptation initiatives. Environmental Education is imperative and it is now a must. At every level of education; on radio, television, newspapers and magazines (the media is a power house), and in every public forum, these must be stressed; and research and development efforts should not relent. Nigeria as a country, should intensify efforts in getting early and fast response to the victims.

CONCLUSION

It is evident from the study that floods had forced millions of people away from their homes while thousands of people lost their lives to flooding at different times and locations in various states in the country. Since some parts of Nigeria are situated in low-lying areas, which can be submerged during heavy rains, such areas can adequately be coped with rather than being controlled or mitigated. However, studies could be geared toward flood control measures in low-lying areas. The citizens need to be aware of disasters that could be avoided by knowledge, skill acquisition and best practices. Furthermore, there is an urgent need for a collaborative effort of both government and stakeholders to support town planning, engineering and other professional agencies to combat/curb flooding in Nigeria to avoid its long-range consequences. For every individual, they must be advised to develop their plots of land with deep or wide drainage system, and must also be told not to don't dump wastes in the waterways (it's perilous to you in your living domain). "The environment remains our most valued possession and legacy which we must all strive to protect. Let us all join hands in protecting our common interest,"

REFERENCES

Adelekan, I. (2011). Vulnerability assessment of an urban flood in Nigeria: Abeokuta flood 2007. *Natural Hazards*, *56*(1), 215–231. doi:10.1007/s11069-010-9564-z

Aderogba, K. A. (2012a). Global warming and challenges of flood in Lagos Metropolis, Nigeria. *Academic Research International*, *2*(1), 448–468.

Aderogba, K. A, (2012b, October). Qualitative Study of recent floods and sustainanble growth and development of cities and towns in Nigeria. *International Journal of Basic and Applied Sciences*, 200-216.

Agbola, B. O., Ajayi, O., Taiwo, O. J., & Wahab, B. W. (2012). Anthropogenic causes and consequences. *International Journal of Disaster Risk Science*, *3*(4), 207–217. doi:10.1007/s13753-012-0021-3

Akintola, F. (1982). *Flooding problem at Ibadan*. Ibadan, Nigeria: Department of Geography, University of Ibadan.

Alexander, D. (2000). *Confronting Catastrophe: New Perspectives on Natural Disasters*. New York: Oxford University Press.

Bates, P. D., & De Roo, A. P. J. (2000). A simple rasterbased model for flood inundation simulation. *Journal of Hydrology (Amsterdam)*, *236*(1-2), 54–77. doi:10.1016/S0022-1694(00)00278-X

Christopherson, R. W. (1997). *Geosystems: An Introduction to Physical Geography* (3rd ed.). London: Prentice-Hall.

Dow, K., & Dowing, T. E. (2006). *The Atlas of Climate Change: Mapping the World's Greatest Change*. Earthscan.

Durotoye, B. (1999). Human Occupation of Hazard Areas in Nigeria. In A. Oshuntokun (Ed.), *Environmental Problems of Nigeria*. Lagos: Friedrich Ebert Foundation.

Etuonovbe, A. K. (2011). *The Devastating Effect of Flooding in Nigeria*. Academic Press.

Ezemonye, M. N., & Emeribe, C. N. (2014). Flooding andhousehold preparedness in Benin City, Nigeria. *Mediterranean Journal of Social Sciences*, *5*, 547–553. doi:10.5901/mjss.2014.v5n1p547

Folorunsho & Awosika. (2001). *Flood Mitigation in Lagos, Nigeria, Through Wise Management of Solid waste*. Academic Press.

Ibitoye, M. O. (2006). *Geospatial data. A panacea for Flood Management*. Paper presented at the Nigerian Institution of Surveyors (NIS) Conference, Akure.

IPCC. (2014). *Climate Change 2014: Impacts, Adaptation and Vulnerability*. IPCC Working Group II Contribution to AR5. Summary for Policymakers. Intergovernmental Panel for Climate Change.

Komolafe, A. A., Adegboyega, S. A. A., Anifowose, A. Y. B., Akinluyi, F. O., & Awoniran, D. R. (2014). Air pollution and climate change in Lagos, Nigeria: Needs for proactive approaches to risk management and adaptation. *American Journal of Environmental Sciences*, *10*(4), 412–423. doi:10.3844/ajessp.2014.412.423

Messner, F., & Meyer, V. (2006). Flood Damage, Vulnerability and Risk Perception-Challenges for Flood Damage Research. In J. Schanze, E. Zeman, & J. Marsalek (Eds.), *Flood Risk Management: Hazards, Vulnerability and Mitigation Measures* (pp. 149–167). Springer Netherlands. doi:10.1007/978-1-4020-4598-1_13

Preet, V. D. (2006). *Natural Hazard and Disaster Management*. Delhi: Shakarpur.

UNCHA. (2012). *Floods Situation Report* (Vol. 2). UN Office for the Coordination of Humanitarian Affairs.

UNISDR. (2009a). *Risk and Power in a Changing Climate: Invest today for a safer tomorrow. United Nations International Strategy for Disaster Reduction*. UNIDR.

UNISDR. (2013a). *Global Assessment report on Disaster Risk Reduction: From shared risk to shared value: the business case for disaster risk reduction*. United Nations Offices.

White, G. F. (1964). *Choice of Adjustment to Floods*. Research paper. Department of Geography, University of Chicago.

White, P., Pelling, M., Sen, K., Seddon, D., & Russell, S. (2005). *Disaster risk reduction. A Development Concern*. DFID.

Chapter 6
Internally Displaced Persons and Emergency Management in Nigeria:
Management of Internally Displaced Persons Nigeria

Franca Chitoh Attoh
University of Lagos, Nigeria

ABSTRACT

Internally displaced persons (IDPs) are men, women, and children who are uprooted from their ancestral homes as victims of natural disaster or manmade occurrences for reasons often beyond their control and comprehension. The Boko-Haram insurgency in the North-East of Nigeria has caused over two million Nigerians to be internally displaced. The crisis has created management problems for the National Emergency Management Agency (NEMA) due to paucity of funds. Secondary data and human needs theory were used for the analysis. The incessant use of improvised electronic devices has created a security lacuna in the IDPs camps, which have become targets for terrorists. The concomitant is confidence deficit between the IDPs and NEMA resulting in accusations of neglect and corruption. The chapter concludes that the failure to manage IDPs is tantamount to human rights abuse and security lapse.

INTRODUCTION

Since the return to civilian rule in 1999, thousands of people have been killed in recurring inter-communal clashes and politically motivated violence that have also caused large waves of displacement. Clashes over natural resources in the country's oil-rich Niger-Delta and the displacement they caused have declined significantly since the government granted an amnesty to militant groups in 2009. However, ethno-religious conflicts have been on the rise especially in Southern Kaduna and the ubiquitous herdsmen have been on the rampage killing and maiming farmers in a bid to graze their cattle in both the South-east and North-central states especially Agatu in Benue state.

DOI: 10.4018/978-1-5225-3194-4.ch006

Many of these conflicts are politically, ethnic and religiously motivated. For instance, the conflict over Sharia law in Kaduna in 2000 and the 2001 clashes in Jos, the capital of Plateau State, which led to the displacement of thousands of people. Also, the victory of former President Goodluck Jonathan in the April 2011 election fuelled violence, having religious and ethnic undertones, which led to the displacement of around 65,000 people (Campbell, 2014). Violence between nomadic herdsmen who are seen as settlers and Christian farmers, who are seen as indigenous citizens, led at least 40,000 people in the Middle Belt region, to flee from their homes in 2012.

Information on the military and humanitarian situation in all three states in the north was scarce, as the government suspended mobile phone networks in an attempt to disrupt communication between the militants. This made information about the displaced people in these states inaccessible but as at 18 June, 2013, more than 9,000 people had fled the north-east to neighbouring countries like Cameroun, Niger and Chad.

In the context of mass poverty, unemployment and inadequate law enforcement, Boko-haram has gained support by playing on people's frustrations and using religion to further its ends by recruiting many young men and women into its fold resulting in the sacking of many villages and more displacements. In addition, forced evictions have forced more than two million urban Nigerians, mainly slum dwellers and other marginalized people to be forcibly evicted from their homes since 2000, most notably in Abuja, Lagos and Port Harcourt. Evictions are sanctioned by state governments and carried out in the name of security and urban renewal, but they have taken place without adequate consultation, notice, compensation or offers of alternative accommodation resulting in displacement (IDMC, 2014).

Internally displaced persons face a range of threats to their physical safety and restrictions to their freedom of movement as shown by the May, 2013 military crackdown in three north-eastern states. A preliminary UN Children's Fund (UNICEF) assessment mission to Gombe State found that IDPs were forced to pay fees to pass through checkpoints. The fact that IDPs do not carry ID cards when on the move can complicate their flight further, even though most Nigerians do not have such documents. The chapter interrogates the phenomenon of internally displaced persons in Nigeria and the effectiveness of NEMA in managing this national crisis. It equally aims at suggesting ameliorative measures to ensure that IDPs are managed in line with global best practices.

Definitional Issues of Internally Displaced Persons and NEMA

Any attempt at defining the term 'internally displaced persons' throws up a number of complex, interrelated issues. One of the principal difficulties encountered in establishing a more systematic approach to the plight of internally displaced persons is the debatable nature of the concept itself. If there is to be a special legal regime for IDP's, then its beneficiaries would have to be clearly defined and identified. Any definition would have to avoid the twin pitfalls of being overly all-encompassing or constricted. The former case tends to employ the concept in relation to all those people who have moved within their own country for reasons that are not entirely voluntary (Mooney, 2005). This includes, for example, changes of residence induced by environmental and industrial disasters, as well as the forcible relocation and population distribution programme which governments often employ to counter security threats and to implement large-scale development projects. In this case, practically anyone would qualify as an IDP. Thus the definition of internal displacement generally excludes from its scope those situations in which people are obliged to move as a result of environmental disasters, development projects and infrastructural schemes. For although such people often suffer from material and psychological hardship, they may also continue to benefit from the protection of the state, and even receive some form of compensation from it.

Alobo and Obaji (2016) were of the view that the term should be limited to people who have left their usual place of residence in the context of involuntary movements, and in circumstances similar to those which create refugees. They however cautioned that care must be taken to ensure that any definition adopted is not extremely constricted so as not to leave too many people outside the protection net. In that case, the very purpose of having a separate legal regime would be lost.

A special regime would also need to address the question as to when an individual ceases to be a displaced person. At present, there is no internationally agreed definition of who is an internally displaced person. Achieving one is essential both for the development of accurate statistics and information and for comprehensive and coherent action (Alobo and Obaji, 2016).

The UN's working definition of IDPs is that of persons who have been forced to flee their homes suddenly or unexpectedly in large numbers, as a result of armed conflict, internal strife, systematic violations of human rights or natural or man- made disasters, and who are within the territory of their own country. The quantitative and time qualifiers in this definition make it unduly narrow. Restricting the internally displaced to those forced to leave 'suddenly or unexpectedly in large numbers' would exclude serious cases of internal displacement- such as in Colombia, where the displaced often flee in small numbers, making them less conspicuous; or in Iraq, where the government organized the uprooting of Kurds over a period of years in the late 1970's, 1980s and early 1990s. Even in Nigeria, IDPs escape in drips and drabs to places considered safe and not necessarily in droves. United Nations' Guiding Principles affords a working definition which has tried to cure this deficiency. It defines them as people or groups of people who have been forced or obliged to flee their homes or place of habitual residence, in particular as a result of or in order to avoid the effects of armed conflict, situations of generalized violence, violations of human rights or natural or human-made disasters and who have not crossed an internationally recognized state border". Taking these observations into consideration, the internally displaced persons should be understood as persons or groups of persons who have been forced to flee, or leave, their homes or places of habitual residence as a result of armed conflict, internal strife and systematic violations of human rights, as well as natural or man- made disasters involving one or more of these elements, and who have not crossed an internationally recognized State border.

When an internally displaced person ceases to be displaced also needs clarification. Ideally, the voluntary return of the displaced to their homes or their re- integration elsewhere should mark the end of internal displacement. However, if protection is largely lacking in these areas and their land and homes are occupied by others, internal displacement cannot be said to be over. In Nigeria, for instance, groups of internally displaced persons voluntarily transported back to their home areas found that they could not remain there because all infrastructure had been destroyed and they had no means of sustaining themselves. The mere act of return therefore did not end their internal displacement. Determining when an internal displacement is ended should go beyond merely registering whether return or relocation has taken place. It should include whether the returns or relocations are reasonably viable and whether basic security and survival are assured. Internally displaced persons in Nigeria have been exposed to more direct physical threats. In a number of cases, camps and settlements for displaced persons have been the target of attacks by the insurgents.

Unlike refugees, who have fled across a border and are therefore no longer under the protection of their country of origin, IDPs are still citizens of their country. Their Government is legally responsible for their protection and welfare. IDPs do not always end up in camps – the majority could be taken in by host families, some find temporary shelter on the move and others may settle in urban areas. IDPs often move several times during their displacement. These varied and dynamic patterns admittedly create challenges for tracking IDPs. As a result, figures are based on estimates.

The Geneva-based Internal Displacement Monitoring Centre estimates that in 2008 there were at least 26 million IDPs worldwide due to conflict (IDMC, 2014). Many more millions were displaced as a result of natural disasters, including in the context of climate change, or large-scale development projects such as dams. Having fled their homes without time to pack food, belongings or identity papers, and often in circumstances of serious trauma and violence, displaced persons can be particularly vulnerable and in need of humanitarian assistance and protection. National authorities are at the core of any response to internal displacement, yet they may be unable or unwilling to meet IDPs' needs. In such circumstances, the international community supports and complements national authorities' efforts, at their request.

The National Emergency Management Agency (NEMA) is a creation of the Nigerian state via Act 12 amended by Act 50 0f 1999 to tackle disaster related issues in Nigeria. Its core mission is to coordinate resources towards efficient and effective disaster prevention, preparation, mitigation and response in Nigeria (Retrieved June 24, 2017 from www.nema.gov.ng). However, the paucity of funds and infrastructure has hampered the activities of the agency. For instance, in 2016, the agency in collaboration with the National refugees Commission was allocated the Sum of N3.4 billion for its activities (National Planning Commission, 2016). In the 2017 budget, the sum of N45 billion was allocated for the feeding 2.6 million IDPs which the spokes-person of the Presidential Committee on North-east Initiatives (PCNI) considered grossly inadequate considering the magnitude of the humanitarian crisis (International Centre for Investigative Reporting, 2017). The chapter adopts the UN definition of IDPs as a working definition and is organized into five sections. Section one comprises the introduction and issues of definition. Section two discusses the threats and challenges of managing internally displaced persons in Nigeria whereas section three is the theoretical anchorage. Sections four interrogates the future approaches while section five concludes the chapter with relevant recommendations to enhance the efficiency of NEMA.

The Current Status of IDPS in Nigeria

According to the Internal Displacement Monitoring Centre (IDMC), as at 2013, 3.3 million people were displaced in Nigeria essentially as a result of insurgency. It also recorded 1,538,982 IDPs in Nigeria as of April 2015. As at December 2015, the total number of IDPs identified in Adamawa, Bauchi, Borno, Gombe, Taraba and Yobe amounted to 2,152,000 people. The vast majority of IDPs identified in the states have been displaced because of the insurgency (91.98%), a smaller number was forced to leave their place of origin because of community clashes (7.96%) or natural disaster (0.06%). In Borno, 24.2% of the population was displaced in 2015 following the increase in violence in this state since the beginning of the year (IDMC, 2015). The above figures are based on the Displacement Tracking Matrix conducted by the International Organization on Migration and NEMA as at December, 2015.

It is important to mention that many of those displaced do not live in the IDPs camps as a good number of IDPs in the north east of live with relatives and friends in town. For instance, 87% of IDPs live with host communities and 13% live in camps or camp-like settings. The situation in other states are quite different, while there are no identified camps in Yobe, Bauchi and Gombe, 12% of the IDP population in Adamawa and 18% of IDPs in Borno live in camps or camp-like setting. A total of 43 displacement sites have been identified in Adamawa, Borno and Taraba States. The number of individuals residing in these sites consists of 154,008 individuals within 19,348 households. The bulk of these victims are women and children who are also the most vulnerable (Internal Displacement Monitoring Centre, 2015). Research shows that 62% of the IDPs populations in sites are female while 38% are male. Half of the total number of individuals residing in sites is children under 17 years old. The vulnerability identified that within the IDP population, 3.33% are single headed-households, and 2.26% are breastfeeding mothers while 1.22% is pregnant women. In addition, 1.61% of the IDPs have been identified as unaccompanied and separated children (Internal Displacement Monitoring Centre, 2015).

The challenges faced by IDPs in Nigeria are not entirely different from the general problems associated with IDPs as observed above but it is necessary to identify specific difficulties IDPs encounter in Nigeria. First, there is improper understanding of the rights of IDPs as set out in the UN Guiding Principles on Internal Displacement and the Kampala Convention. These rights include the right to life, freedom of movement, association, dignity of human persons, personal liberty, right to private and family life etc. IDPs usually suffer the worst violation of their fundamental human rights. In Nigeria, their lives are threatened as a result of the violence by the Boko Haram insurgency. They are forced to flee from their homes to seek shelter in schools, churches, mosques and informal settlements and in the process the situation gravely affects their security and health. There are recorded cases where IDPs are killed in their make-shift camps by Boko- haram insurgents (Retrieved June 21, 2017 from www.thecable.ng).

Similarly, IDPs along with the majority of the affected population, have suffered numerous atrocities such as losing family members and succumbing to the insurgents' campaign of terror. Many IDPs therefore fear that they will be pursued by the insurgents and may be perceived by the authorities as supporters of the insurgents or supposed by the insurgents or other elements as informants for the authorities. Remaining silent and hidden for IDPs becomes the unpalatable option. Besides, Women and children remain the most vulnerable to sexual and gender based violence as there are reported cases of rape, sexual harassment, forced marriages, child marriages, sexual diseases and uncontrolled births occasioning high infant and maternal mortality in make-shift IDPs camps in Nigeria. The needs of children are highly disregarded in armed conflict situations and this is the case in Nigeria. Children are exposed to enhanced risk of abuse, forceful conscription by insurgents as child soldiers, suicide bombers, sex slaves and abrupt discontinuation with their education.

According to the National Emergency Management Authority (NEMA), there are over 750 unaccompanied and separated children of school age in Adamawa, Borno and Yobe whose opportunities for schooling have been severely constrained. The destruction of schools and indiscriminate killing of students and teachers by Boko-haram insurgents has left many schools deserted and deprived children of their right to education. Most of the displaced persons are camped in schools thereby interrupting learning and other school activities for even the indigenes of those communities. The education of displaced persons is virtually non-existent in some camps due to dearth of learning materials and the right ambience to learn. It is on record that 338 schools have been damaged or completely destroyed by insurgent attacks. Apart from such destruction, the most common types of accommodation used by IDPs in Nigeria are schools and government buildings thus further depleting the resources for learning. In addition, many existing shelters have been damaged or destroyed resulting in many IDPs living in churches, mosques, town halls, abandoned and uncompleted buildings and where available, other forms of make shift camps which are grossly inadequate and unsuitable for accommodating those displaced. The shelters are often overcrowded and lacking in basic amenities and often devoid of privacy. There is often no proper waste management and electricity. This explains the regularity of hygiene based epidemics in camps. Things are made worse by the fact that access to health care in Nigeria generally and the North East in particular is severely constrained for both the IDPs and host communities as a result of the destruction of health care facilities and insufficient health care personnel. Outbreak of disease has increased in areas affected by dislocation and there is also the general lack of access to vaccinations. The number of cholera cases among IDPs is seriously on the increase. More ailments are undiagnosed as there is no health checks carried out in these camps (author's assessment working for NEMA Consultant in 2014, 2015 & 2016).

Some IDPs have access to food distribution every day while others receive irregular food distribution. On the other hand, some IDPs never receive food distribution resulting in severe hunger and malnutrition. This is mainly because no actual statistics of IDPs exist in Nigeria as not all displaced persons are accounted for. For those persons who have the resources, at the occurrence of violence, although they become displaced by virtue of the attaining circumstance, they migrate to other places where there is peace and stability with their available resources. Others flee to the homes of their extended families or friends in other parts of the country. All these factors make it nearly impossible to obtain an accurate number of displaced persons in Nigeria.

THREATS AND CHALLENGES OF MANAGING INTERNALLY DISPLACED PERSONS IN NIGERIA

Managing internally displaced persons has become a major challenge to most countries. The situation is not different in Nigeria. The National Emergency Management Agency (NEMA) is empowered by law to manage and provide direct material assistance to displaced persons. NEMA's core mandate includes organizing, providing and coordinating emergency relief to victims of national disasters throughout the federation and matters incidental thereto. Inferring from its statutes NEMA has little to do with prevention, protection and return or resettlement of IDPs. NEMA responds to the occurrence of disasters in the country by sending relief materials and supplies to affected communities and facilitates the ability of the displaced to survive the immediate problems of displacement and dispossession. Unfortunately, these efforts are not sustained and the victims are soon left on their own (Adebowale, 2013). In Nigeria, there are several challenges associated with managing IDPs. The government does not respond to the plight of this category of persons because it assumes that the relief materials given are adequate to meet their needs, even though these materials may not get to the target end users as a result of corruption. These items are sometimes diverted by the people who are expected to distribute them, and sold for profit (Mohammed, 2013). Olagunju (2006) in his findings criticized government efforts towards effective management of IDPs because relief materials often to do not get to displaced persons. He argues further that the Nigerian government; both at local and federal levels do not have adequate machinery in place to address IDPs issues. There are organizations created by government but their capacity to handle IDPs related problems is minimal. After many years of various degrees of violence including insurgency, it is inexcusable that a country with the size of Nigeria does not have a blueprint for managing internal displacement. Most IDPs in Nigeria flee to neighboring communities that are relatively safe, usually taking refuge in temporary shelters such as public buildings, schools and places of worship among others. Oduwole & Fadeyi (2013) argue that government throws money at the problem of IDPs welfare without concrete measures to address their basic needs. In addition, government's aid gets diverted and never gets to the IDPs themselves. This raises the issue of corruption which the country has been battling with for a while. UNICEF's emergency manager in Nigeria, Mr. Soji Adeniyi was of the view that lack of a credible data base and information management, weak legislation and poorly funded state emergency management agency as well as inadequate health facilities and over population was among the challenges bedeviling effective emergency management in the country (Retrieved August 10, 2016 from www.thisdaylive.com). As a result of this, most IDPs prefer to stay with their families, relatives and religious members rather than staying in the camps. For (OCHA, 2015) many of the IDPs are hosted by private individuals who opened their homes to them as a mark of charity. But some of these temporary living arrangements lack sufficient access to food, water and sanitation, health and non-food items. As evidenced in the numerous displacement crises across Nigeria in recent years, assistance for return and reintegration is one of the major needs of IDPs, but government assistance rarely extends beyond emergency response. The national response is generally constrained by lack of experience in dealing with IDP issues, which has resulted in inefficiencies and support gaps to affected populations, but also by competing mandates. Confusion exists at the federal level over who has the mandate to respond to and assist IDPs – especially between NEMA and the National Commission for Refugees, migrants and internally displaced persons (NCFRMI), informally mandated in 2002 to also cover IDPs) – which results in competition for resources. NEMA appears to be more of an operational agency, and a natural partner for operational UN agencies in Nigeria such as UNICEF, while NCFRMI has appeared more occupied with policy issues albeit on paper.

Effects of Displacement

There is no doubt that the Boko Haram insurgency in Nigeria has led to the displacement of huge human population in the affected states of Borno, Yobe, Bauchi and Adamawa. The increasing activities of the insurgents in Northern Nigeria have been largely responsible for the spate of the Internally Displaced Persons (IDPs) in the country in recent times. As at August 2015, with about 3.3 million displaced persons (IDPs), Nigeria has the largest population of persons displaced by conflict in Africa. The number of internally displaced persons in Nigeria is put at approximately a third of IDPs in Africa and 10 percent of IDPs in the world (UNHCR, 2016). The total number of IDPs identified in Adamawa, Bauchi, Borno, Gombe, Taraba and Yobe states due to insurgency as shown in the Table 1 as at February 2015 is 1,188,018 IDPs (149,357 households). Borno state with 672,714 IDPs has the highest number, followed by Adamawa with 220,159 IDPs and Yobe recorded 135,810 IDPs.

The plight of those displaced in their various places of refuge can best be described as critically threatening. This is due to the fact that they have been dislocated from their families and social capital bases (Alkassim, 2013) leaving them disillusioned and despondent. According to the United Nations

Table 1. Displacement matrix in the north east geo political zone

Current Location	IDP Individuals	IDP Household	Average HHs Size
Adamawa	220,159	25,807	8.5
Bauchi	60,555	9,881	6.1
Borno	672,714	76,842	8.8
Gombe	24,655	3,335	7.4
Taraba	74,125	11,599	6.4
Yobe	135,810	21,893	6.2
Grand Total	**1,188,018**	**149,357**	8.0

Source: UNHCR (2015).

Office for the Coordination of Humanitarian Affairs (UNOCHA) 2014 report, there has been a steady rise in the number of internally displaced persons from Borno, Yobe and Adamawa states. From 2009 to 2010, IDPs rose to 100,000 and from 2010 to 2011 it increased to 130,000. From 2011 to 2012, the number of IDP's rose to 200,000. From 2012 to 2013 IDP's grew to 290,000 and from May 2013 to March 2014 it decreased slightly to 250,000. From May-June 2014, it rose again to 436,608 and from August to December IDP's drastically rose to over 600,000 persons (OCHA, 2014). In 2015, IDPs arising from ongoing insurgency rose to 1, 385,298 (Oji, 2015) as a result of the military operations in the North East. This is evidenced by the estimated 30,000 Nigerian refugees in Northern Cameroun, 1,000 Nigerian refugees in Chad (on Lake Chads Choua Island) and more than 50,000 Nigerian refugees in the Diffa region of South East Niger (United Nations High Commissioner for Refugees, March 2014 cited in Imasuen, 2015). The National Emergency Management Agency (NEMA) thereafter received about 10,000 Nigerian refugees who fled to Cameroon from Borno and Adamawa states (Duku, 2015). According to the research carried out by Displacement Tracking Matrix (2015) in the north-east, the vast majority of IDPs identified in Adamawa, Bauchi, Borno, Gombe, Taraba and Yobe have been displaced because of the insurgency (91.98%). A smaller number was forced to leave their place of origin because of community clashes (7.96%) or natural disaster (0.06%).

Apart from being displaced from their ancestral homes, IDPs have also been displaced psychologically, and emotionally, nursing the wounds and injuries sustained as a result of gunshots or other life-threatening situations they found themselves, with little or no medical attention. The poor living conditions in the IDP camps have increased the psychological trauma for many of these IDPs. Reports from both domestic and international agencies have shown that these camps lack adequate facilities in addition to poor sanitation and the IDPs are exposed to diseases such as malaria, acute diarrhea, measles and pregnancy related issues as shown in Figure 1.

The activities of insurgents also have wide range implications on the Nigerian federation. It is obvious that the intensity, scope and frequency of such conflicts have shown the fragility of the Nigerian

Figure 1. A displaced malnourished mother and her children sit on the ground waiting for food in Bama's camp for internally displaced people (IDP), Borno State, north-eastern Nigeria, 30 June 2016
AFP/STRINGER.

state. Thus, the activities of the insurgents in the North-East undermine the relative peace and stability hitherto enjoyed by Nigerians with threats to social order. The insurgency has brought to the fore the problems of human security with its attendant challenges resulting in humanitarian crises (OCHA 2014). For instance, Figure 2 shows some of the returnee displaced persons being catered for in a transit camp in Mubi, Adamawa state, Nigeria.

Food Security and IDPs

In the northeast of Nigeria, the impact of the insurgency on agriculture is estimated at USD 3.7 billion resulting from livestock losses, destruction of irrigation and farming facilities and collapse of agricultural extension services, coupled with reduced production due to access limitations. Lootings and fear of attacks have prevented many farmers from working in their fields, leading to the loss of harvests, productive assets and extremely reduced purchasing power. An estimated 80–90 percent of the populations in the Lake Chad area are dependent on agriculture, fisheries and livestock for their livelihoods and food security. Returning to farming is therefore a priority for such persons. Host communities, where possible, provide access to land to IDPs, but the lack of agricultural inputs is still a major constraint for many farmers, who continue to rely on prolonged and expensive humanitarian assistance to meet their food needs during the planting season (UN, 2015). Wider negative impact includes the lack of economic and employment opportunities with possible harmful consequences including youth radicalization and enrolment in armed groups, resulting in continued civil unrest. Displacement and dislocation contribute to limited agricultural activity and disrupted trade flows, aggravate food insecurity and engender malnutrition in Nigeria and neighboring countries. By the beginning of 2015, it was estimated that 4.6 million Nigerians were at risk of food insecurity, according to the UN. Of the 2.6 million people across the Sahel estimated to be facing this crisis 70 percent were located in Cameroon, Chad, Niger, and Nigeria. Thus, the attendant security and livelihood crises threaten the collective subsistence and survival of the affected populations. It suffices, therefore, to say that the activity of the insurgents has resulted in food insecurity with attendant malnutrition and hunger induced diseases.

Figure 2. Nigerian returnees from Cameroun receiving their meal at the NEMA transit camp in Adamawa red brick factory in Mubi

THEORETICAL UNDERPINNINGS

Maslow's human needs theory articulated by Abraham Maslow (1954) talks of the hierarchy of needs with food and safety being the most important needs which provide the base for the actualization of other needs. As shown in Figure 3, Maslow suggests that the most basic of needs must be met before the individual's other higher needs (Maslow, 1954) cited in (Villarica, 2011). However, (Villarica, 2011) in his essay on "Maslow: A New and Improved Recipe for Happiness" averred that in a global survey to replicate the human needs theory spanning 2005 to 2010 involving 60, 865 respondents from 123 countries. Respondents were asked questions on six needs modeled after Maslow's hierarchy of needs. The findings show novel nuances, for instance, in societies where basic needs are provided respondents were more concerned with love and respect. But in other societies where the base needs were not sufficiently provided people still had to grapple with physiological needs such as food and safety. Any wonder that IDPs in the North-East were more concerned with the twin issues of food and security as shown in figure 3 below. For the IDPs in the North-East of Nigeria, Abraham Maslow's human needs theory captures the inability of NEMA to sufficiently cater to their needs. Hence the current despondency which envelopes many of the IDPs camps as those in the camps perceive their situation as hopeless devoid of ameliorative measures.

The importance of administrative management is captured in the postulations of Henry Fayol's 14 principles of management as articulated by (Eneanya, 2015). According to him, good administration is a process and consists of certain conditions which include division of labour, authourity, discipline, unity of command, unity of direction … (Eneanya, 2015:57). The Chapter interrogates the above salient principles with a view to determine the effectiveness of NEMA in the management of IDPs in Nigeria. Even though NEMA is a specialized government agency but the inability of NEMA to synergize with the state emergency management agencies (SEMA) and other initiatives such as the national commission for refugees, migrants and internally displaced persons (NCFRMI) set up by the Federal government has created a dysfunctional process not beneficial to the IDPs. In terms of authourity the various

Figure 3. The most basic of needs that must be met before the individual's other higher needs
Source: (Maslow,1954 as cited in Villarica, 2011).

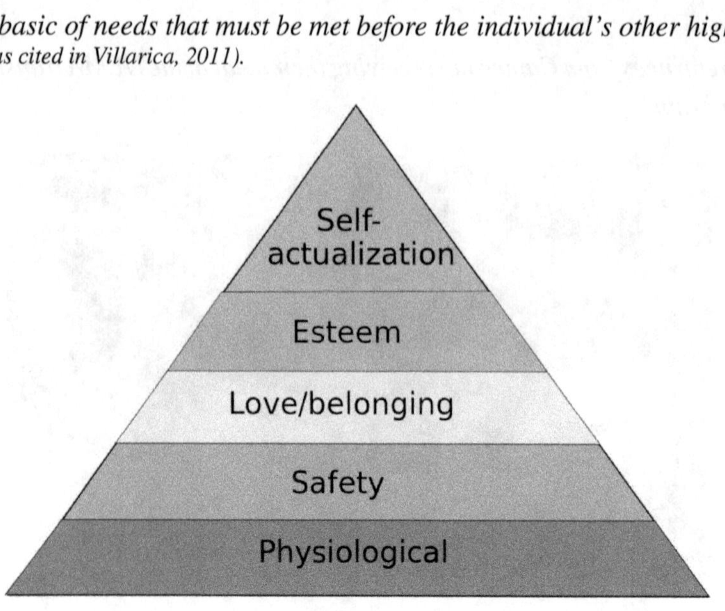

cases of diversion of items meant for those in the camps show that NEMA is not totally in control of the management of the IDPs. It is pathetic that IDPs' camps have been subjected to terror attacks, rapes, theft and other sundry crimes. This is an indication of laxity as NEMA does not even possess accurate statistics of bonafide victims in the camps thus exposing them to infiltration by insurgents. The issues of command and unity of direction are some of the challenges bedevilling NEMA considering the accusations of corruption and diversion of food stuff levelled against both the national and state officials (see News Agency of Nigeria, March 20, 2017). For instance, the May, 2017 United Nations report for Borno, Adamawa and Yobe states revealed that 8.5 million people are in need of urgent aid, 5.9 million people require emergency health care while 5.2 million people are on the edge of starvation (UN, 2017). The above information shows the absence of discipline and unity of direction within NEMA as indicative of the various lapses dire needs that exist in the IDPs' camps.

FUTURE APPROACHES

In Nigeria, the establishment of IDPs camps appears to be the most important measure taken by the various tiers of government to ameliorate the conditions of DPs considering that 8% of IDPs live in the IDPs camps (IOM, 2016). This requires little resources considering that there are often schools, Churches or Mosques to be turned into a temporary shelter. The failure to deploy long term administrative and management processes to solve such humanitarian crisis result in stop-gaps and half measures in the management of IDPs. Many managers of IDPs are often fixated with the provision of food and sundry medicine without taking a holistic view of the welfare of IDPs. To ensure a lasting solution future approaches should be anchored on the following ameliorative measures:-

Adequate Security for the Safety of IDPs

IDPs and affected communities have suffered horrendous atrocities from insurgent attacks (Protection Sector Working Group (PSWG), 2015). As a result, psychological trauma among both categories is common. According to PSWG, a major concern is the conduct of security forces, against which allegations of human rights violations have been made. Despite the Government's proclamation to investigate and bring perpetrators to justice, numerous violations continue to be reported (Protection Sector Working Group (PSWG), 2015). The escalation of the conflict in recent months has resulted in direct attacks on the IDPs camps (UNHCR, 2016) including accidental bombing by the Nigeria Air Force. In addition to physical violence that takes place in the IDPs camps, it has been widely reported that women and girls also face Gender-Based Violence (GBV) whether in camps or in host communities, particularly at night (see Attoh, 2016). The promise by the Nigeria government to reinforce security in and around IDPs Camps was reiterated by the UNHCR Assistant High Commissioner for Protection, Volker Turk during his visit to Nigeria (18-22 February 2016) focusing on the insurgency-ravaged northeast. During his meetings with federal and state authorities, Mr. Turk stressed the necessity for Nigeria to further address the protection needs of IDPs particularly with respect to the voluntary character of returns and relocations. In the same vein, he also expressed the availability of UNHCR to work with the States in the facilitation of tripartite framework for the return and reintegration processes as depicted in Figure 4 where displaced persons were fleeing a IDPs camp as a result of gunfire.

Dietrich (2015) reaffirmed that given IDPs 'increased vulnerability including the September 11, 2015 attack on an IDPs camp in Adamawa state, IDPs have called on the government to increase security measures to ensure the protection of IDPs and humanitarian actors in camps. While most of the individuals are fleeing violence, many reported that they are also fleeing a humanitarian crisis defined by food insecurity, loss of livelihoods, insufficient services, and inadequate protection. Considering the level of insecurity in IDPs camps, (Dietrich, 2015) reports that IDPs had resorted to self-help. Quoting some of the IDPs he reports: "we protect ourselves by creating well-fortified and secured camps to safeguard lives and property, conducting aggressive intelligence gathering on activities of insurgents, educating ourselves about humanitarian activities like free medical treatment, and, condemning erring persons accused of rape and sexual offences". The level of insecurity in the camps in Nigeria is alarming; it's a situation of running from insecurity to insecurity as camps are infiltrated by insurgents and women and children are vulnerable to gender based violence (Attoh, 2016). The inadequacy of security at the IDPs camps opens them to attacks from terrorists and armed robbers. In recent times, IDPs camps have been attacked by insurgents; in September 2015, the terrorist group, in a suicide mission, attacked members of IDPs Camps in Madagali and Yola killing 12 persons. In one of the attacks, bombs were reported to have been detonated inside a tent at the IDPs camp. On September 11, 2015, seven people were killed in Malkohi camp near Yola in Adamawa state, on 31st of January 2016, at least 86 people were killed in Dalori, some 12 kilometers to Maiduguri, on 10th of February 2016, another 80 displaced persons were attacked and killed in an IDP camp in Dikwa, Borno state which houses about 50,000 displaced persons. Considering these tough challenges confronting displaced persons and vulnerability of women and children, there is a need for an accelerated effort by Federal and State governments, "International Actors", Non-Governmental Organizations and the private sector to collaborate in ensuring that security is provided in the camps.

Figure 4. Villagers fleeing gunfire in a camp for internally displaced persons

Funding and Corruption

In a recent survey research carried out, respondents disclosed that although IDPs management agencies in Nigeria get funds mainly through revenue, international aids and donations, the funds they get are more often than not insufficient to meet the increasing needs of IDPs in the country (Osagioduwa & Toluwanimi, 2016). Significantly, dearth of funds results in deficiency in manpower, commodities, infrastructure, equipment and mobility. Where there is paucity of funds, needed facilities will be unavailable and the agencies will be inefficient (Obikeze and Onuoha, 2016). The under supply of funds is attributable to low budgeting for emergencies. This justifies the assertion of Osagioduwa and Toluwanimi (2016) cited in (Obikeze and Onuoha, 2016) that government in Nigeria does not have adequate machinery in place to address IDPs issues and the organizations created by the government possess minimal capacity to handle IDPs related problems as evidenced in Figure 5.

Corruption is also a major impediment to the efficient handling of IDPs in Nigeria as corrupt office holders in government, and in IDPs management agencies are often accused of diverting needed funds and relief materials meant for IDPs for their personal use and or even sold to traders for profit. A situation that reduces the efficiency of the agencies concerned in managing IDPs.

IDPs Institutions With Concurrent Juridisction

Ambiguous definitions of functions and mandates of similar institutions have been identified as a major impediment to the management of IDPs in Nigeria. For instance, the government has established several

Figure 5. FAO funding effort for IDPs in Nigeria
Source: FAO situation report, August 2016.

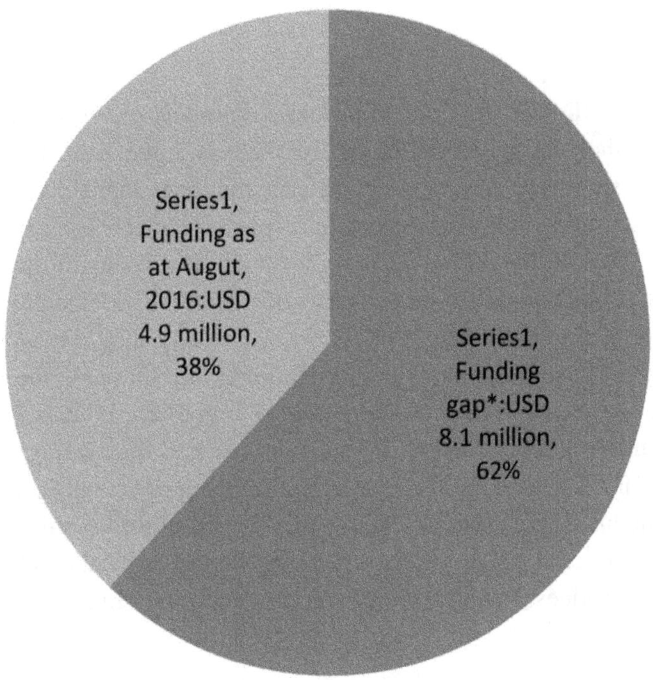

IDPs management institutions and agencies with similar mandates and structures. A good example is the establishment of the National Commission for Refugees, migrants and internally displaced persons (NCFRMI). According to (Obikeze and Onuoha, 2016) the institution was originally mandated to cater for refugees in the country and not IDPs. It is understandable that as a result of upsurge in IDPs in the country, but its mandate has since extended to include the responsibilities of NEMA/SEMA. The implication is that the similarity in functions has robbed them of operational efficiency as most of the jobs are not only duplicated but the urgent and serious jobs may be neglected or left undone. Equally of significance is that bilateral and multilateral agencies including partner NGOs may be confused as to which agency to collaborate with.

Relationship With Host Communities

Host communities may feel overburdened when swarmed with a huge population of internally displaced persons given that resources such as food, water, farmlands may be inadequate to cater for both the hosts and the IDPs. Tension may ensue between the two groups making effective local integration difficult (Kangiwa, 2012). However, due to the homogeneous nature of such communities, most host communities show love and charity to IDPs. Durosaro and Ajiboye (2011) report that in some host communities non-displaced young people usually take advantage of porous IDPs camps to access humanitarian services thus making it difficult to identify the target persons. Therefore the resultant friction between IDPs and host communities resulting from such infraction is a major hindrance to IDPs management in Nigeria. IDPs managers should establish administrative processes with which to identify and capture bonafide IDPs to avoid the diversion and distribution of aid materials to non IDPs. It is equally within their purview to initiate a strategic process of integrating the IDPs with host communities in order to reduce the trauma of displacement and dislocation.-

CONCLUSION

The ultimate desire of every IDP is to relocate to his/her community of origin, pick up the broken pieces of their lives and rebuild their means of livelihood. As Nigeria, intensifies its fight against insurgency it becomes imperative to study social trends and patterns that engender the displacement and dislocation of communities to stem the rising incidences of displacement across the nation. In the same vein government agencies such as NEMA and SEMA saddled with the responsibilities of managing and ameliorating the conditions of IDPs should endeavor to work within the parameter of global best practices. Those principles were properly articulated by (Eneanya, 2015) and should be applied to drive NEMA's processes. This is possible through the training and retraining of their personnel in order to acquire 21st Century administrative tools to enhance and sharpen their processes. In addition, staff of NEMA sent to the IDPs' camps should be a team comprising of social workers, psychologists, and sociologists not merely administrators. This is due to the fact that often IDPs require expert care by such professionals but are usually left in the lurch. One should also mention the need for organizations such as NEMA to anchor their processes on ethics. This is the only way to eliminate unethical practices such as cooperation and other unwholesome practices. In view of the foregoing the following recommendations are made:-

- The government should identify the situations that have produced IDPs in the past and take steps to prevent reoccurrence. This calls for the establishment of an early warning system to gather intelligence and nip in the bud potential conflicts.
- Build adequate temporary shelters by number and quality to accommodate all IDPs in any circumstance as was done by some states such as Lagos state.
- Involve the military and other security agencies to build tents, provide security and construct community utilities for the convenience of IDPs.
- Engage a team of medical personnel to visit the camps often to provide needed medical services such as antenatal care, immunization and the care of senior citizens.
- Enforce law and order in the temporary camps and restore orderly rhythm of life in the psyche of IDPs as a way of enhancing their quality of living.

REFERENCES

Abdulrahman, A., & Zuwaira, H. R. (2016). Effects of Insecurity on the Internally Displaced Persons in Northern Nigeria: Prognosis and Diagnosis. Global Journal of Human Social Science: F Political Science, 16(1), 3-7.

Adebowale, S. (2013). *Boko Haram: Security, Humanitarian Conditions Deteriorating UNHCR*. Retrieved from http//www.theeagleonline.com.ng/news/boko-haramsecurity-humanitarian-conditions-deterioratingunhrc/

Akowonjo, D. (2011). Ethno-religious Conflicts and National Security in Nigeria. In T. O. Ogundipe, A. Akinade, & M. Omoegun (Eds.), *Religious cum Ethnic Violence in Nigeria: Security and Socio-economic Perspectives*. Lagos: University of Lagos Institute of Security.

Alkassim, B. (2013, December 17). NEMA: 16, 470 Nigerians in displaced persons camps. *Daily Trust Newspaper*, p. 10.

Alobo, E., & Obaji, S. (2016). Internal Displacement in Nigeria and the Case for Human Rights Protection of Displaced Persons. *Journal of Law. Policy and Globalization*, *51*(2), 26–33.

Attoh, F. (2016). The Gender Dimensions of Terrorism in Nigeria: Instrumentalization of Borno Women as Gender Based Violence. Kogi Journal of Sociology, 1(3), 59-69.

Dietrich, K. (2015). *When We Can't See the Enemy, Civilians Become the Enemy: Living through Nigeria's Six-Year Insurgency*. Centre for Civilians in Conflict.

Duku, J. (2015, August 9). Boko-Haram. NEMA Receives 10,000 Cameroun Returnees. *The Nations Newspapers*.

Durosaro, I., & Ajiboye, S. (2011). Problems and Coping Strategies of Internally Displaced Adolescents in Jos Metropolis, Nigeria. *International Journal of Humanities and Social Science*, *1*(20), 256–262.

Eneanya, A. N. (2015). *Theory and Practice of Public Administration*. Lagos: University of Lagos Press and Bookshops Ltd.

Imasuen, E. (2015). Insurgency and humanitarian crises in Northern Nigeria: The case of Boko Haram. *African Journal of Political Science and International Relations*, *9*(7), 284–296. doi:10.5897/AJPSIR2015.0789

Internal Displacement Monitoring Centre. (2014). *Annual Report*. Geneva: IDMC.

Internal Displacement Monitoring Centre. (2015). *Report of the Displacement Tracking Matrix in Nigeria*. Geneva: IDMC.

International Centre for Investigative Reporting. (2017). *Report on IDPs Camps*. Abuja: ICIR.

IOM. (2015). *Displacement Tracking Matrix (DTM) Round II Report in Nigeria*. IOM.

Kangiwa, H. (2012). *National Policy in Internally Displaced Persons in Nigeria. Report of the Federal Commissioner*. National Commission for Refugees.

Ladan, M. T. (2012). Impact of insecurity in the north on internally displaced people and migration flows between Nigeria and neighbouring countries. Paper presentation made at the Forum of European Union Working Group in Nigeria and Development, Abuja.

Maslow, A. (1954). *Motivation and Personality*. New York: Harper.

Mohammed, T. M. (2013). National framework for the protection of internally displaced persons (IDPs) in Nigeria. Presentation made at a workshop for Judges and Kadis on refugee law, organized by The National Judicial Institute, Abuja.

Mooney, E. (2005). The Concept of Internal Displacement and the Case for Internally Displaced Persons as a Category of Concern. *Refugee Survey Quarterly*, *24*(3), 9–26. doi:10.1093/rsq/hdi049

National Planning Commission. (2016). *Nigeria's National Budget*. Abuja: NPC.

NCFRMI. (2009). *No End to Internal Displacement. Geneva: International Displacement Monitoring Centre (IDMC)*. Nigeria: Global Project.

NEMA. (2017). *Internally displaced persons*. Retrieved June 24, 2017 from www.nema.gov.ng/internallydisplacedpersons

Obikaeze, V. C., & Onuoha, C. B. (2016). The Nigerian State and the Management of Internally Displaced Persons 2012-2016. *African Journal of Politics and Society*, 4-21. Retrieved from www.academix.ng

Oduwole, T. A., & Fadeyi, A. D. (2013). Issues of refugees and displaced persons in Nigeria. *Journal of Sociological Research.*, *4*(1), 1–18.

Olagunju, O. (2006). *Management of Internal Displacement in Nigeria* (Unpublished Thesis). Brandeis University.

Osagioduwa, E., & Toluwanimi, O. O. (2016). Management of Internally Displaced Persons in Africa: Comparing Nigeria and Cameroon. Africa Research Review, 10(1), 193-210.

Premium Times. (2015). *Nigeria sets new record, now has Africa's highest number of displaced persons*. Retrieved from http//premiumtimesng.com/news

Protection Sector Working Group. (2015). *IDPs Protection Strategy*. Abuja: PSWG.

Thisday. (2017). *Feeding Fat*. Retrieved June 18, 2017 from www.thisday.com/index.php/feeding-fat

UNHCR. (2016). *Nigeria Humanitarian Update on the North East*. UNHCR.

United Nations. (2017). *Nigeria-North-East: Humanitarian Emergency Report No.12, 31st May, 2017*. Author.

United Nations Humanitarian Commissioner for Refugees (UNHCR). (2015). *Internally Displaced Persons Matrix in the North East*. Nigeria: UNHCR.

Villarica, H. (2011). *Maslow: A New and Improved Recipe for Happiness*. Retrieved June 21, 2017 from www.thecable.ng

Section 3

Networking Technologies and Emergency Management

Chapter 7
Prospect of Low Power Sensor Network Technology in Disaster Management for Sustainable Future

Alok R. Prusty
VIT Bhopal University, India

Ankita Mohanty
IGIT Sarang, India

ABSTRACT

Countries with better disaster management plans can better signify their existence. The earthquake is one such hazardous catastrophe that leads to heavy damage. Disaster management does not avert or eliminate the threats fully; instead, it focuses on use of technology, creating plans to lessening the effect of disasters up to significant level. Depending on the magnitude and type of waves, earthquakes crush everything, from buildings to small houses, from towers to trees, and from roads to bridges. Tsunamis and landslides follow its way as well. After the earthquake comes, the settling of things back to normal involves a lot of time, money, and manpower as well. Hence, the only solution of this is leading-edge technology support. Sensor network technology is one of the proven techniques for not only varieties of early warning, prediction, and detection systems but also post-phenomenon rescue and support. This chapter will help researchers and policy makers to think fresh on disaster management with sensor network technology.

INTRODUCTION

Disaster can be considered as a cycle and a great concern in today's developing world. Although it can be bore with the situation caused by the disaster in every disguised way it comes, the major challenge is bringing things back to normal after the diverse effects of it. Hence some technology is required to manage the scenario for which Disaster Management is the key and the plethora of technologies involved in

DOI: 10.4018/978-1-5225-3194-4.ch007

it play major role. Many Disaster Management Techniques have been developed, and are being followed by scientists and researchers for the betterment of the environment (Carter, 1991). The technology that has drawn into most researchers, in the near past is Wireless Sensor Network (WSN). This is responsible for many quick and feasible actions to be taken prior to or after any disaster. WSNs are easier, better and cheaper. Earthquake is a kind of disaster which comes without any warning, precisely speaking; the speed of onset is very sudden. This is the same traditional way of describing this disaster and yet is very difficult to predict. But, technology at this current age has widened its bed to good extent to attend both pre and post disaster circumstances. Prior to the discussion on technology support in this direction, a background study related to the earthquake is a necessity.

BACKGROUND

Earthquakes (Howell, 1986) occur naturally in almost all parts of the world but majority occurs along narrow belts which are extended up to 10-100kms wide. These are the geologically active regions which mark the boundaries. Earth's tectonic plates are in constant motion and due to this movement there is pressure on the lithosphere for which the surface of earth tends to break. This breakage occurs when the pressure or stress is large. The stress comes in the form of waves which is felt as earthquake. Vibrations due to the energy of the waves are called as seismic waves which travel with a speed of approximately 14kmph. With this speed it may take 20mins to reach other side of the earth if it goes straight through its center i.e. a distance of 13000 kilometers. The term Epicenter is the point directly above the source of the earthquake. The source is approximately 700 kilometers deep, also called as focus. This is the minimum because at high temperature and pressure rocks are not rigid and the reduced plasticity doesn't allow storage of stress. After an earthquake, aftershocks are observed that is the adjustment period for the plates. This adjustment period may last for months. Earthquake can occur in any place of the world, there is no certainty of a place being safe from an earthquake. It is only that in some places like plate boundaries, where different tectonic plates meet, the waves are frequent and nearly 80% of earthquakes occur at Circum-Pacific seismic belt, The Ring of Fire. Other than this there also occurs an intra-plate earthquake which originates at more shallow level, away from the plate boundaries.

With the developing and rushing world, researchers are going in pace with the technology developments as well. The world cannot compromise and stay back after any disaster; to keep up with the pace the scenario needs to be managed properly and efficiently. Use of various technologies in the disaster management is hence very effective, rather than any primitive practice or methodology.

Wireless Sensor Networks (WSNs) offer an interesting alternative to wired networks when infrastructures collapse, like, after an earthquake or a tsunami. In addition, WSN have extravagant features, they are known to be auto-configurable, auto-organizing, have small dimensions, etc. Diverse environment factors can be detected by sensors, such as, the atmosphere humidity and temperature for sensing forest and building fires, toxic gases instigated by an explosion, water level to notice and prevent floods, vibration level to detect chocks and earthquakes and numerous additional data. The character of a sensor node is to sense the environment, communicate and exchange sensory data with further nodes in the space, locally process its own data and make smart judgments about what it observes. Sensors can also deal with additional valuable services. For instance, a guesstimate of the victim number by counting and identifying the number of persons holding Bluetooth devices, a quick localization of incidents and victims, a quick wireless communication between sensors, major responders and command centers with such kinds of multimedia sensors. While these are the prevailing technologies, there are to be certain impeccable developments if one gives a broad way look into the future.

This chapter will focus and discuss about earth tremor, convulsion, aftershocks and their causes, also enlighten the prospect of existing and new sensor network technology's breakthrough against the nature's tremendous and destructive analog phenomenon in this digital era for sustainable future.

DISASTER AND LOW POWER SENSOR NETWORK TECHNOLOGY FOR DISASTER MANAGEMENT

Earthquake as Disaster

Disasters are calamities to the earth's surface and can change the whole scope of it. Facing the disaster, and working for the betterment and moderation of the landscape after any disaster is the main challenge which can be achieved only by good Disaster Management options. As mentioned earlier, disaster management is a cyclic process, there are various stages involved in the same. All the stages are interrelated and dependent on each other. The stages include planning, mitigation, preparedness, response and the main concern being recovery after the disaster. As of here, it has been focused on nature's most destructive calamity, Earthquake; disaster management in this also involves the same above processes. Earthquake disaster management first can be considered as the early detection the calamity to a greater extent as quickly as possible. Although the early detection processes have been developed, it's still impossible to stop the same. Yet the effects can be minimized by evacuation or similar methods, which is majorly dependent on the time factor. Now after all this, the practical thing is earthquake doesn't spare single gravel even. It affects everything in its area and can destroy anything and everything if intense. Depending on the magnitude and type of waves, it crushes everything, from buildings to small houses, from towers to trees and from roads to bridges. Tsunami and landslides follow its way as well. After the earthquake the establishment of the former scenario involves a lot of time, money and man power as well. Scientists and researchers have tried in lot of different ways of studies like varieties of animal behavior and other science for predicting earthquakes. Few of them get us idea of where an earthquake is most likely to hit, to some extent but still can't be predicted exactly when it will occur. Hence, only solution of this nature's deadly revenge on mankind is the leading-edge technology support on short term alarming.

Sensor network technology is one of the proven techniques for not only varieties of early warning, prediction and detection systems but also post phenomenon rescue and support (Rameesh, Rajan, & Pullarkat, 2014). Researchers have developed various technologies for evacuation after the quake, detection of human-beings in the area, and other rescue operations. These techniques involve sensor technology and its implementation in other devices where the whole concept lies in use of low power technology in management of the scenario before and after the event, so as to maintain a sustainable environment.

Types of Earthquakes

There are four types of earthquakes and most common type of earthquake is tectonic earthquake where the earth's crust breaks due to geological forces which occur due to movement of tectonic plates. Volcanic earthquake occurs due to the volcanic activity. Collapse earthquake, occurs in underground caverns and mines and Explosion earthquake occurs due to explosion of nuclear and chemical devices. The Seismic waves as shown in the Figure 1 are categories in to three broad categories, i.e. P-waves (primary waves), S-waves (secondary waves) and Surface waves (Frohlich & Brunt, 2013). P-waves are the fastest of all waves and the first to be detected by seismographs. These waves have the ability to move through liquid and solid. These are compressional waves i.e. they compress and expand as they move through the solid or liquid. S-waves cut the rock as they pass through them but cannot pass through liquid. Secondary waves are larger than primary waves hence are dangerous. P-waves and s-waves are called as body waves as they travel through the interior of earth's surface. The speed of these waves varies depending on the density and elastic properties of the medium they pass through but they are amplified as they reach the surface.

Surface waves are the slowest of all waves. They move either very close to the surface or outside it. This is again are of two types, Love waves: these are horizontally polarized waves and Rayleigh waves these waves move in vertical and horizontal paths in a vertical plane pointed in the direction of travel. The depth of an earthquake also classifies itself. Earthquakes occur in the depth range of earth's surface to 700km of below.

The depth is further categorized as:

1. Shallow (0-70) km
2. Intermediate (70-300) km
3. Deep (300-700) km

There is also deep-focus earthquake that occurs deeper than 700km, within segments of shallow lithosphere which are close to earth's mantle.

The size and severity of earthquake is measured from maximum heights of seismic waves within a range of 100kms from epicenter. Range in possible heights is used to construct a Richter scale. Its magnitude is measured by the estimate of energy released by it. Richter scale measures the energy released

Figure 1. Seismic waves

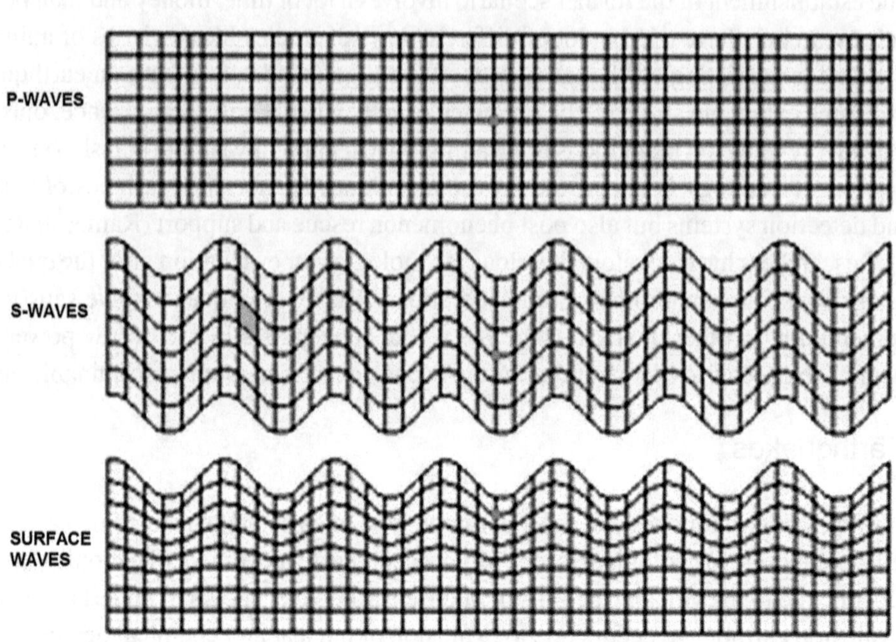

during an earthquake as a whole and is not affected but the distance from the epicenter. Larger the Richter scale reading greater is the damage. On the other hand Mercalli scale measures the intensity at a particular place. It tells us the type and amount of damage caused due to earthquake. It doesn't measure the strength of the earthquake as a whole.

Measurement of size,

1. < 3.4 measured only by seismographs.
2. 3.5-4.2 felt indoors.
3. 4.3-4.8 felt by many people and windows rattle.
4. 4.9-5.4 felt by everyone, dishes may break and door swing.
5. 5.5-6.0 there is slight damage to the buildings.
6. 6.1-6.9 there is much damage to buildings and houses move.
7. 7.0-7.4 there is serious damage to buildings, bridges and walls fracture.
8. 7.5-7.9 great damage and most buildings and structures collapse.
9. >8.0 there's complete damage, waves are seen on the ground while objects are thrown into the air as its effect.

Factors That Causes Damage

1. **Strength of the Earthquake:** According to the measurement on Richter scale the strength is known in accordance to the damage caused.
2. **Distance of the Particular Place From the Epicenter:** Places nearer to epicenter are affected more than that of the places farther from it.
3. **Type of Underlying Rocks or Soil:** If the underlying is strong and firm then the damage is minimized but in case of loose soil and presence of water, it may cause liquefaction which may then lead to loss of cohesion and slide as in case of landslides.
4. **The Building Construction or the Plan of the Area:** If the building is constructed as per good rules to be safe from earthquake, i.e. earthquake resistant, then the damage is minimized.

Types of Damage

- **Buildings Collapse:** People get trapped, but if they try to rush out the danger of casualties may increase. Buildings get knocked off their foundation. This causes serious damage.
- **Landslides:** This occurs due to liquefaction and the movement of the underlying changes the whole look of the area.
- **Fire:** Because of the earthquake the situation becomes out of control and may lead to the increased casualties.
- **Tsunami:** If earthquake occurs underneath or volcanic eruptions take place under water then it leads to tidal wave also called as tsunami. The wave may rise up to 10-20m and flood to wash away the nearby area. The given location and distance and direction of travel from the region of earthquake determines the severity of the damage.

Factors Determining the Impact of Earthquake

- **Population:** More number of people then there will be more loss.
- **Rural/urban Areas:** In rural areas there are lesser people so lesser harm and in urban areas there are more people than the impact is more.
- **Preparedness of Countries:** Less economically developed countries (LEDC's) have less time for preparations, less money and expertise to prepare for the situations. More economically developed countries (MEDC's) are prepared for situations and can limit damage. The building regulations are often ignored in LEDC's like use of cheap materials, or choice of bad ground which happens because of lack of money. They invest in less expensive materials which collapse during earthquake. Moreover the number of people in one house is more so evacuation becomes difficult. On the other hand, in MEDC's the houses are well built of good materials and they are earthquake proof and evacuation is made easy because of lesser people in individual houses. In LEDC's things are not preplanned and there is no expert help hence situation becomes difficult and people panic. Whereas in MEDC's things are preplanned and people are made aware from before so the danger is minimized. In LEDC's there are no proper roads the communication reduces. They are unable to receive proper help by charities or the government is even unable to supply. There is therefore shortage of food and medical facilities. In MEDC's the communication is good as everything is properly planned. In LEDC's the power supply is not well and if it breaks its fixation takes time due to lack of money and the minimum help of govt. and even the water supply goes down. The time taken by things to resume is more. In MEDC's the source of water is almost always available and the supply of food is even made possible. Planning makes it easy for each individual to take care of themselves and family. Due to lack of money in LEDC's they are unable to research and predict the natural disasters and even they cannot build earthquake proof houses. Even if earthquake strike the govt. is unable to fix the problems easily. But in case of MEDC's government is able to re-establish most things and as predictions up to a limit can be done, the precautions required are taken wisely. In LEDC's there is medical facilities so people may die in severe cases but that doesn't happen in MEDC's as medical facilities and hospitals are available for the victims(pp.,A1).
- **Distance from Epicenter:** The areas nearer to the epicenter are more affected than the areas farther from the epicenter.
- **Weather/Season:** In winter if earthquake occurs, the surviving people may get frost bite, hyperthermia or other illness. In case the earthquake is in summer, people may die being deficit of water.
- **Time of Day:** If the earthquake occurs at night, people will be unaware and may get trapped and panic. In day time if the same occurs, people will be in vehicles and be trapped and even breakdown may occur.

To minimize the effect, Regulations on building should be made strict to avoid casualties. There should be effective plans to evacuate the people in case of earthquake. Government help should be assured, i.e. the government should be prepared always (Goltz, & Bourque, 2016 ; Lambie, et al., 2016).

Animal Behavior to Earthquake

This is the most primitive way of detecting an earthquake. It's been prevailing in people's beliefs and practices. Like human beings express all kind of emotions like fear, or nervousness, animals even do but in a different manner which differs from species to species. Moreover these emotions are considered as a necessity for living. For animals there are some ways in which they adapt to any danger they feel, i.e., they fight, flee or freeze (Eilam, Izhar, & Mort, 2011). In some cultures earthquakes are predicted by distinguished animal behaviors, like, Buffalo in Bulgaria Boraco, Malaya and the Lesser Sunda Islands. Turtles in the North America, Serpents in the Moluccas and Sumatra, Crab in Persia (Bluestone, 2010). In their specific local areas these animals are considered to be the cause of earthquakes (Schnytzer, & Schnytzer, 2011). But what the world believes and accepts is the scientific aspect. It is known to all of us that animals have different ways or techniques to escape from predators and natural calamities. This is developed by instinct. Unlike humans, animals don't panic. Humans panic due to the secretion of adrenaline in such situations and there's nothing to do with the emotional behavior. Animals react in other way also like, Cats jump out of windows, Dogs bark furiously, Birds scream in unnatural ways, Rats are seen running out of their holes, and Bees are even noticed swarming (Tributsch, 1982).

- P-waves and s-waves, as already discussed earlier; p-waves transfer earlier than s-wave. Organisms that are sensitive enough to the p-waves become aware of the p-waves or pre-shock and in turn they respond to avoid death or escape circumstances. The time gap between p-waves and s-waves is very less. So it's natural that places nearer to epicenter its felt better but there is always less time to escape as the s-wave follows the p-waves in a short span. For p-waves the seismic response is mechanical and is thus felt by humans too. Snakes and rodents have ability to sense the effects and mark safe boundaries for themselves. This ability to sense low frequency vibrations may be the primary task of vertebrates i.e. fish to mammals. Various physical and chemical events at earth's surface occur that predict earthquake and now animals detect them. P-waves can result in other resulting signals that can be perceived by animals prior to the shaking of earth.
- Ground Tilting, the vestibular system of animals especially humans helps detection of tilting prior to the quake. Subterranean rodents sense better than those present on ground. Shallow soils under compaction add noise to signal and animal's sense better.
- Humidity Changes, Hygro-reception is the ability of humidity reception by animals. This ability is possessed by spiders and insects, because of, hygroscopic hair-like structure and can detect the humidity or temperature change. This is possible in arid environments and difficult in rainy areas. Vertebrates detect humidity by olfactory system.
- Electrical currents, Terrestrial animals are sensitive to electrical sensitivity but that is lower in case of the marine animals or freshwater animals due to resistivity of air. Sharks and rays have specialized receptor system in the ampule of Lorrenzini. They have the ability to perceive Nano volt (Eilam, Izhar, & Mort, 2011)

The unforeseen future can come up with challenges to the mankind and the civilization still would have to face it in every possible way. The circumstances as portrayed above can be a mammoth deal to science and technology. It's the humans who have developed in this field and hence bear the sole responsibility of saving the aesthetic value of the disaster struck place and the lives and lifestyle of themselves. Various researches and developments bring in ways to get a sorted scenario before and after the disaster with pre and post disaster management plans. It has been discussed some of the technologies instrumented in this field, the very importance circulating around the Wireless Sensors.

WIRELESS SENSOR NETWORK TECHNOLOGY

Wireless Sensor Networks can be referred to as a collection of sensors i.e. different types of sensors connected by a wireless communication system in a clearly good network. WSN came alive for meeting different application specific, complex, real life and practical challenges of mankind. These networks are spontaneous and unstructured with autonomous sensors communicating through radio without any prefixed infrastructure. Here the data is collected from the sensors between the sensor nodes and all together the exact information is drawn from the collected data. There's a framework i.e. the design of the whole system. It involves reliability, accuracy, flexibility, cost effectiveness, and ease of deployment (Youssef, Yousif, El-Sheimy, & Noureldin, 2007) and energy complexity (Prusty, Sethi, & Nayak, 2015) becomes a concern due to factors like geographical terrains, mobility, cost, size and weight of each deployed sensor node.

WSNs are low cost, tiny and use low power technology. For low power consumption, there are many workable technologies like solar energy and wind energy which are great supplies (Kumari, Patel, Anand, & Bhattacharya, 2013). Several kinds of technologies being used currently are high powered and battery consumption becomes comparatively more to the sensor technology thus involvement of WSN technology makes the applications low power. A sensor can be functioned in power saving mode which involves low power or the shutdown mode. But a more suitable way of attaining the same is use of digital logic in the sensors like it has been used in accelerometers, and magnetometers, etc. (freescale, 2013). Regardless of the above information provided about the wireless sensor networks, there's one thing to be know is what kind of sensor networks or what kind of wireless sensor networks that are being implemented. There are three kinds of wireless sensor networks, namely, homogenous wireless sensor networks and heterogeneous wireless sensor networks and few are hybrid one (Prusty, Sethi,, & Nayak, 2016a).

Homogenous wireless sensor network is that kind of the wireless sensor network in which the characteristics of the sensor are identical and they share similar parameters, or onecan say they work on the same parameters which include hardware system, the battery consumption as per requirement, energy level and other similar parameters. Power Efficient Gathering in Sensor Information Systems (PEGASIS) (Chen, Wang, & Lin, 2012) and Hybrid Energy- Efficient Distributed Clustering (HEED) (Kour, & Sharma, 2010) on the other hand heterogeneous wireless sensor networks are the general type of wireless sensor network that is being deployed in many places or projects. Here, the characteristics of each sensor differ from other and so are the parameters. Hetero itself means variety. The characteristics and parameters depend on which criteria they are going to function (da Silva, Almeida, Poersch, & Nogueira, 2010) There are several protocols used in low power packet routing in WSN, that makes existing technology smart and efficient (Prusty, Sethi, & Nayak, 2016b). WSN works in the following process the communication architecture and structure of an individual sensor node in WSN is as shown in the Figure 2 (Prusty, 2012). The very basic networked sensor devices in WSN are a radio, a power unit, sensor, embedded processor, memory etc. The key aim of each sensor in WSN is to route collected data to distant sink/base station for user access through internet. Occasionally, several WSN applications require an aggregate value to be reported to the observer. In this case, sensors in different regions of the It consists of various network components like, Sensor motes are the nodes which are responsible for routing various data packets for delivering the information. They are specially designed for the process control or various related components. Gateway or Access point which acts as an interface between the Host and the field devices. Network Manager designs the network and also is involved in the scheduling and monitoring of devices where Security Manager responsible for generation and storage of security keys to check both authenticity and authorization issues (Kalita, & Kar,2009).

Wireless sensor networks have applications over many fields and are well implemented by researchers. It is used in disaster aid processes, biodiversity mapping, construction and monitoring of intelligent architecture i.e. the architecture to be used for various sustainable works, machine surveillance, agriculture medicine and health care, etc. (Zennaro, Pehrson, & Bagula, 2008). The sensor nodes involve one or more sensors, embedded processors. They may be the low-power radios that can be put into use. Disaster management (da Silva, Almeida, Poersch, & Nogueira, 2010) for earthquake is an important matter of concern while the implementation of wireless sensor network ensures better certainty. Earthquake is detected by each sensor in a particular sequence depending upon the seismic frequency spectrum. As the intensity is different and every other property is different at different places, the sensors have to be managed in a good manner. Further they have to be distributed in accordance, again keeping the seismic energy and frequency spectrum in mind. WSN (Fischer, Kühnlenz, Ahrens, & Eveslage, 2009) is the

Figure 2. Wireless sensor network communication process

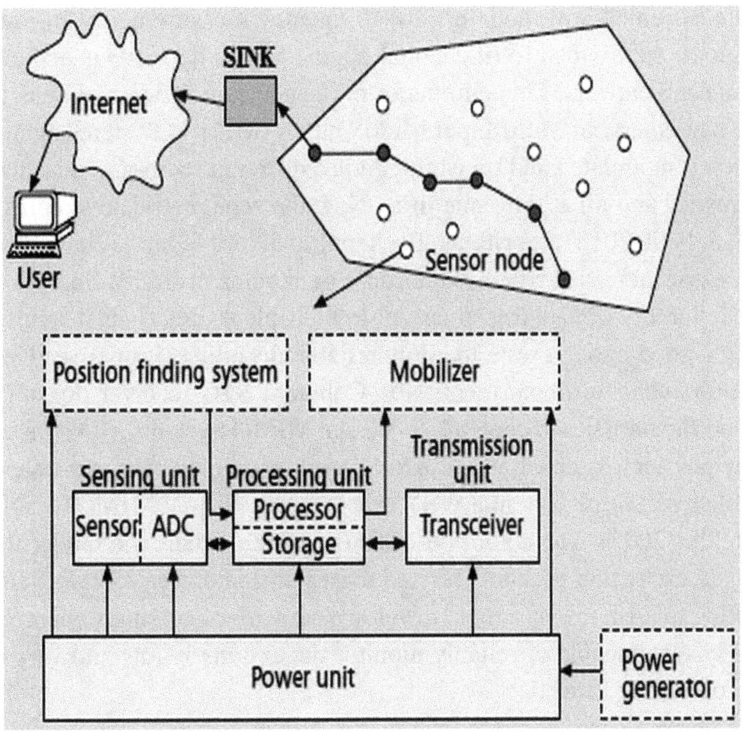

utmost favorable and emerging technologies. It offers the real time monitoring of geographical areas that are prone to the disasters. Sensor network technology has algorithms for power management at distinguished stages of the network including routing to perform at low power based on sensor type and properties along with the WSN architecture and the software modules followed by implementation (Ramesh, 2014). The Self-organizing Seismic Early Warning Information Network (SOSEWIN) (Fleming et al., 2009) represents innovative aspects with regard to existing based on the use of new, low-cost wireless sensing units. Sensors allow the performance of onsite, independent analysis of the ground motion and the real-time communication of estimated parameters. Second, following the philosophy of a decentralized network, SOSEWIN will perform early warning activities in a fully automatic way. Studies say that frequency based sensors are more effective. The working is implemented by selecting few sensors which are used to generate the information based on the seismic energy and frequency spectrum, which help generate quality information, at the base station. The whole process is crisp and can be implemented in a proper manner and is effective enough to generate the desired results (Devasena, & Sowmya, 2015).

Wireless Sensor Networks for Earthquake Early Warning Systems of Railway Lines

In the current century there are high speed railways and like everything else, it is also needed to be safe as they have to meet various safety challenges. Here comes into play the earthquake early warning system. This is because for a high speed rail even a small quake or the same from a distant epicenter can cause immense damage. A simple System model (Youssef, Yousif, El-Sheimy, & Noureldin, 2007) in this prospect includes the concept of "distributed physical layer". In the network, according to the functions, the sensors are divided into sub groups but the communication still goes on is a collective way. This has got additive noise and hence it's difficult to measure. But together it can be used to measure the accurate extremely low, and very low frequency measurements. In multi-carrier scheme, large signal bandwidth is subdivided to fit channel width. A high data rate is achieved by transmitting multiple low data rate sub-carriers. This eliminates poor spectrum efficiency, selective fading, limited bandwidth problem and others. In Multi Input Multi Output (MIMO) Systems again the data stream is divided each of which is modulated and transmitted via a different radio-antenna chain. It involves sensors with same frequency and all at the same time. Now the separated data streams are created by the use of MIMO (Patil, & Patil, 2015) algorithms. Each multipath is a separate channel and is called virtual routine. Range is likewise increased which is the cause of antenna diversity. Space Time Coding is used for improvement of reliability of data transmission. In multiple transactions it sends multiple copies so that some make it in a good enough state to allow reliable decoding. It involves Coherent STC in this case the receiver knows channel impairment. Non-Coherent STC receiver doesn't know the channel impairment but knows the statistics of channel. In Virtual MIMO Systems MIMO is created by temporal signaling and not by physical components. A WSN based on single antenna wireless sensor nodes are cooperating to establish a multiple antenna WSN that is called Virtual MIMO-WSN"(Youssef, Yousif, El-Sheimy, & Noureldin, 2007). The author descried the system where the randomly deployed sensors start communicating to each other peer-to-peer and makes an ad hoc network. Different single antennas sensors are cooperating to form this system to develop a network of antennas operating cooperatively to form multi-antenna system capable of reliably monitor the extremely low and very low frequency and high, and very high frequency as well.

Many earthquake early warnings are dependent on other kinds of sensor systems. Wireless Sensor System has been developed in the past decade and also has been employed in environmental monitoring, structure health monitoring, volcano monitoring, etc. This has got many advantages such as cost effectiveness simple maintenance, handling robustness, etc. Earthquake Early Warning (EEW)(Wang, & Ni, 2012), Where deployed sensor network with number of sensors to detect earthquake and report to the railway emergency response system. Studies say wireless communication is faster than destructive s-wave of earthquake. The primary-wave should hit the sensor so as to predict the secondary-wave before it strikes. Intrinsically, the transmission speed of wireless communication can reach 3 × 105 km/s. The transmission speeds of S-wave (destructive) and P-wave (less destructive) are 3-5 km/s and 6-7 km/s (Claudio, Yih-Min, Aldo, Abd Hiroo, 2011). Even though the wireless system may not reach the speed limit, it is understandable that the wireless system is much faster than the earthquake waves. Such difference makes it possible for the EEW. Intuitively, the more sensors can be deployed earlier can detect. In (Wang, & Ni, 2012), the author has evaluated the trade-off between the number of sensors to be deployed and the safety reporting time as shown in Figure 3. They generate timely report of earthquake and further help the report of damage.

In Railway Line Health Monitoring System (RHM) (Wang, & Ni,2012), if sensors are densely placed then the reports will be generated in a big or ample amount and hence by the aggregation of it, the cause can be located easily and impact can also be evaluated. Following is false alarms which has a big chance of generation and that is primarily needed to be eliminated. There are intrinsic problems and cannot be managed by engineering advances. Railway Structural Health Monitoring (SHM) system (Wang, &

Figure 3. An illustration of a sensor system for early earthquake warning of railway lines

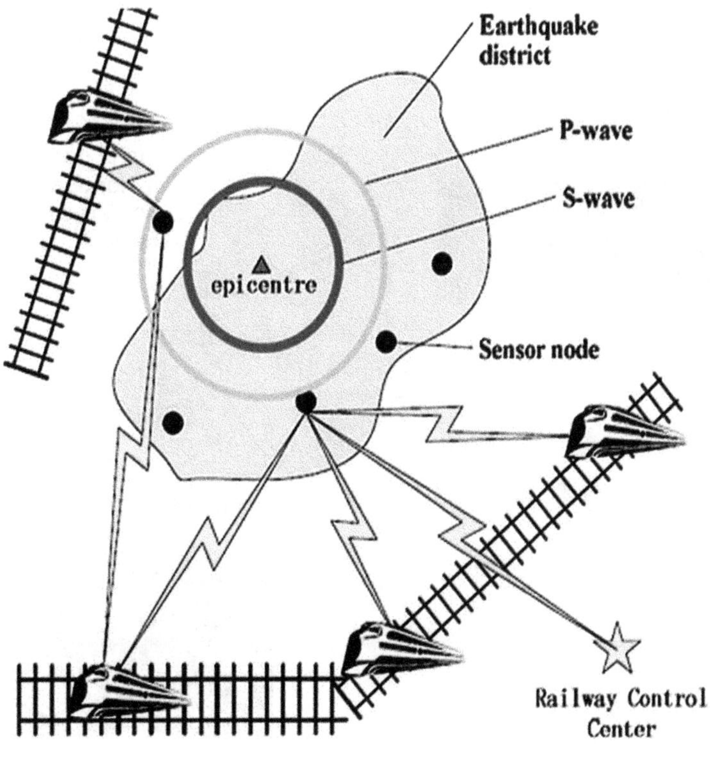

Ni, 2012), as shown in Figure 4. It can provide real-time monitoring. The EEW system will notify the sensors deployed along the railway lines. The sensors will start real-time monitoring and report OK or alarm to the control center, an OK report is necessary to show that the sensor itself is not damaged. In this system, it is required to reduce the false alarms and maintaining accurate warning. For example, reduce the speed of the train to 200 km/h; reduce speed of the train to 100 km/h; and stop the train. The outcome of these systems can rank these action levels from more critical to less critical.

Earthquake Early Warning system not only helps in the report generation of the earthquake or the studies related to it but also in accurate damage estimation and warning. Decision has to be made how to control the trains' motion as and how the magnitude of earthquake is. For this both Earthquake Early Warning and Structural Health Monitoring systems of railway lines should be strong. SHM shows the real time monitoring whereas the EEW system uses sensors. This helps in getting filtered out false alarms. To handle the false alarms there's a solution, that is, Deadline Reporting Strategy (DRS) which puts stress on behavior over sensors.

- **Vibration Tracking:** There are earthquake and Tsunami detectors which are too costly to be implemented so there has been developed a very cost effective way to do the tracking. Seismic sensors and vibration sensors use ultrasonic effect. Now this sensor then sends the tracked vibration which is then amplified but the amplifier and compared with a critical value of vibration. After some other processes, final result is displayed using Visual Basic. Vibrations of magnitude more than value four can be sensed and monitored. This process involves low cost. But better the quality of sensors used better sensitivity is achieved (Le Mehaute, Hwang, & Van Dorn, 1971; Fasihuddin, Kolhare, & Daimi, 2012).

Figure 4. Architecture of the railway structural health monitoring

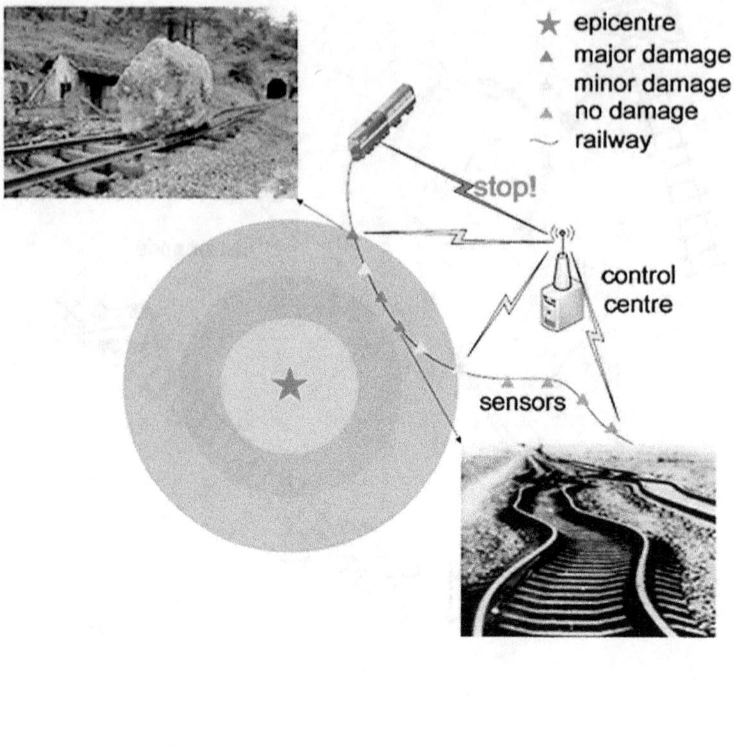

- **Geo-Sensors:** Although wireless sensor networks have made progress in this period, but something that could be more effective are Geo-sensors (Cranenbroeck, Shlyakhova, 2014). Though it's under research stage, Geo-sensor networks are best applications of WSN technology. They track, detect and monitor environmental phenomenon. Geo-sensor can be considered as an "environmental microscope" where remote sensing platform is 'telescope' and traditional sensor is the 'eye'. Sensors network helps to generate point based regional observations with the help of higher sensor node which will deliver more accurate estimation. Sensor network nodes perform local processing and filtering of sensed data and collaborate with others to detect local 'events'. WSN technology is useful for events that last just for a period of time; this is due to the battery limit of Geo-sensor elements. Geo-sensor networks are used to implement WSN. These are easy to be used but there is a disadvantage of the energy supply. Talking of sensors and micro sensors, Geoscience integrates temperature, humidity, light, and acoustic sensors. There are sensors to analyze chemical and gas as well. It has got applications over, Terrestrial Ecology Observing Systems, Geological Observation Systems, Aquatic Observing Systems and others. The scope of geoscience is more in today's world. Research and study will lead to better development of sensors and also how to perfectly use the battery life so as to get the exact information (Nittel, 2009).

Volcanic Earthquake Detection

Excluding all kinds of sensors, technology has spread its way into the radioactive elements and implementation for the execution of early warning system techniques. Radioactive decay of Radium-226 which is formerly formed from Uranium-238 is Radon-222.This is particularly found in lower levels of earth. This can be used to detect the timing of earthquake. Here also use of WSNs has been done. Numbers of sensors are deployed underground which collect data and thus is closely monitored. For early detection of the unforeseen event, it is required to determine the time span in which it is likely to arrive. That is what makes the preparedness easier. Hence Wireless Sensor Technology is used to determine the earthquake timing as well. It is a challenge to record real time seismic data due to limited WSN bandwidth and the computation of its arrival. Here, it has been introduced as a new term called the p-phase picking. Various algorithms developed to detect the same whose grounds lie in identification of changes in signal characters. STA/LTA (Short term average/ Long term average) are computed over a signal to detect slight changes. Though the process is effective but do not fit in for volcanic earthquake timing. The experiential tuning of numerous parameters is required as it's tedious to work in different environments. This timing determination uses autoregressive model, which pick the time instance. Two AR models are compared for signals before and after the picked time. Few user settings are required which are most accurate and robust. This is widely used in data-intensive WSNs (Liu et al., 2013).

- **Design Objectives:** P-phase provides information about locating hypocenter and seismic tomography. Picking p-phases of spatially distinct sensors is referenced to volcanic earthquake timing. When the network sensitivity is more and the p-wave velocities are more the difference between p-phase is small. This particular thing gives the accuracy on volcanic earthquake timing. The system model is a hierarchical network architecture which consists of sensors with limited resources and coordinators with better processing capability along with improved battery capacity. The signals are received analogously by sensors and temporarily stored in memory which follows the pattern of previous volcano monitoring WSNs. Sensor clustering, in which network is arranged in clusters, comprising a member of sensors and a coordinator called the cluster head (Liu et al., 2013).

Earthquake Detection Using UBG-04LX-F01 Laser Rangefinder

Barring the sensors and keeping them aside, it has been landed up in newer and innovative technologies that make the best use of physics for applying the detection techniques. One such device which is used in the current technology is UBG-04LX-F01 laser rangefinder which is a designed laser range finder which works on the phase shift measurement. It uses laws of optics and trigonometric equations. The total system is shown in Figure 5 and Figure 6.

The total system explains:

1. **Communication Interface:** Host computer is connected to the USB and RS-232 for range measurement.
2. **Drift Effect of the Sensor:** Also called the warm-up time, observes the distance between the target and the sensors.
3. **Effect of Intensity on the Laser Rangefinder:** The variation of various laser beams is used for the range finding. The variations are used to pass through different color mirrors to obtain different intensity of laser beam which are a help to the laser range finder.
4. **Use of Different Color Mirrors:** There are uses of various colors of mirrors as the variation they produce is different. In this case 3 colors are used. After reflection from them the laser beams, are recorded on the computer as per the vibration of the wooden bleak which is due to the strength of earthquake.

Figure 5. UBG-04LX-F01 Laser Rangefinder

Figure 6. Working principle

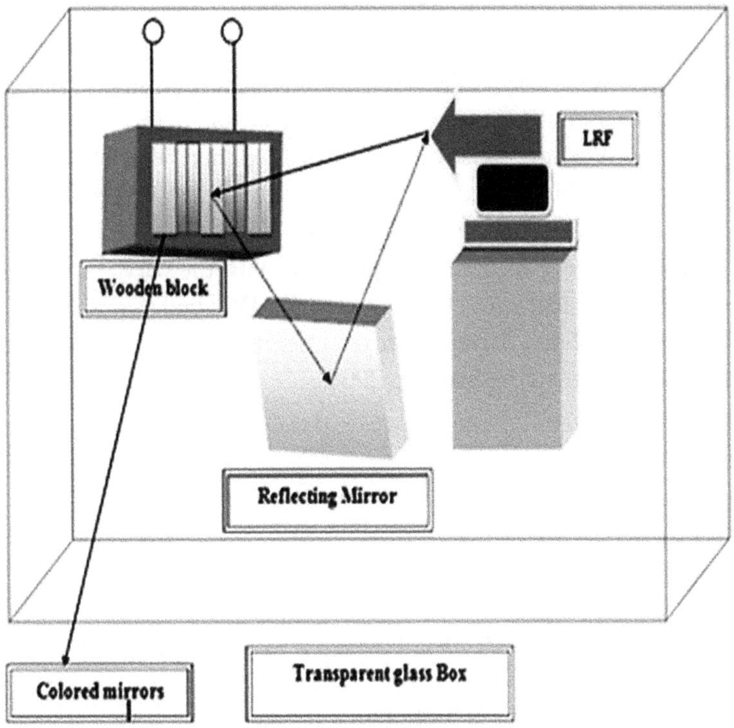

5. **Working of Proposed Model:** The beam initially strikes on the colored mirror at an angle of 45 and then also is reflected back at the same angle and after reflection finally reaches the laser range finder in the same angle. After this process if the wooden block is still, then no earthquake has occurred i.e. it shows constant reading, but in case of earthquake the reflection will change and the beam will strike different mirror to be recorded on computers (Singh, & Nagla, 2012).

Not only for early detection but these sensor techniques are used to improvise the situation after the disaster has occurred. They use various techniques for rescue operations and other tasks which follow the earthquake. PIR Sensor is one of those techniques used in rescue operation. It is used to detect humans by the naturally radiated thermal radiation. This sensor is placed within a robot that works in an earthquake prone area. By this rescue operation becomes easier and efficient. This system uses Visual Basic programming for controlling the robot. PIR Sensor is abbreviation for Passive Infra-Red Sensor. It detects live humans within a range of 10m from sensor, the average value being 5m-12m.It is advantageous as it gives accurate detection. It is also effective and safe for rescue which involves no manual work. Moreover it is a proven technology. The only disadvantage it involves is that the initial cost is more as well as the battery backup for camera used is weak (Shamroukh, & Awad, 2009; Sharad, & Thorat, 2013).

For rescue operations ICT is used as well. It is known, a small earthquake may lead to a bigger one or maybe none. Yet it is needed to be prepared for the worst. People always tend to look for social media and television networks for updates and here ICT is equipped. It informs people even on a mild quake and it saves people from falling into prey for earthquake (ur Rahman et al., 2016). Another typically used concept in rescue operations is FREQL. FREQL, that is Fast Response Equipment against Quake Load, is considered as an all-in-one seismometer. It also includes a power unit and is capable of generating the alarm in 0.2 seconds after the detection of p-wave. This all-in-one seismometer contains a network of seismometers, where one is the FREQL which manages the train operation post-earthquake, and it includes another small digital seismometer AcCo network to get the seismic details. According to Japan Meteorological Agency, it's a pseudo real-time system. The system performs repetitive calculations and is a time consuming process. Whenever a new process is called there is a chance of system failure even. Hence, installation of an on-site alarm is recommended, for gathering of precise situation after an event, and the distribution of sensors for each facility is also required. This system is practically being used in rescue operations, proper train operation by applying it with the early warning system, combining network information and onsite information for applications. The total system when works can generate the information before the arrival of s-wave earlier than before. Thus the total integrated system works best for on-site system (Saita, Sato, & Nakamura, 2008).

SENDROM

Sensor Network for Disaster Relief Operations Management (SENDROM) (Benkhelifa, Nouali-Taboudjemat, & Moussaoui, 2014), is an application designed especially for earthquake related works. It comprises of several sensor nodes which are connected together to the database called the SENDROM database as shown in Figure 7.

It has classified nodes for specific tasks. Following are the same:

- Collector nodes called as Cnodes.
- Snodes for sensing and reporting which are further classified into standalone and embedded nodes.
- Inodes used for detecting living beings, which is again further divided into standalone and embedded nodes.

Figure 7. SENDROM

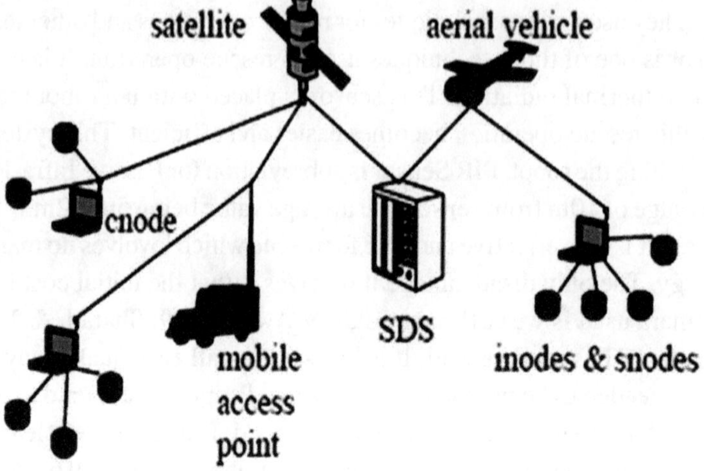

All these nodes together collect the data and in return send them to the database, namely, Central SENDROME Database. This database is regularly updated according to the data from the nodes.

Smart Phone Sensors and Applications

Smartphones are the devices that are being used by every individual on this planet. They have wide variety of applications and can work accordingly. Smartphones in today's world serve many operations. As billions of people use the same and various user convenient applications have been developed that are used in day to day life. The fact every smart phone has enormous inbuilt sensors that that are being used for several kinds of applications. Different researches have made kinds of studies to develop applications for the early warning of this massive disaster like earthquake. These applications help in keeping track of the quivers by the use of the various sensors, recording the same in a cloud based system, interpreting the same by use of various algorithms and finally act as an early warning system. There are applications called Droid-Shake and MyShake (Kong, Allen, Schreier, & Kwon, 2016) which are being used in the same (Kong et al., 2015; Finazzi, 2016). The author in (Alphonsa, & Ravi, 2016) has proposed early earthquake warning system by using the smart way for transferring the alert signal to smart phones; which has been accomplished by Internet Of Things (IOT). The hardware portion plays the role of detecting and reading the signal successfully and the software portion is to deliver the alert signal to the human on smart phones.

CHALLENGES OF LOW POWER SENSOR NETWORK TECHNOLOGY IN DISASTER MANAGEMENT

Numerous algorithms and protocols made sensor network to operate at low power with lesser communication overhead which has already been discussed in the above literature. Several researches have been performed on battery technology as well as architecture too to prolong the sensor device life time as in WSN the geographical deployments are mostly unattended. There are two kinds of work a senor in the sensor network can perform under the term "disaster management" based on its characteristics i.e. early disaster prediction and post disaster rescue operation as discussed in distinguished section above. Reliability and performance are the main challenges to the sensor technology. For the real time operations the system has to work in its particular time as the whole system is dependent on the same. While the system delivers the desired work, it has delivered it correctly. Sometimes both reliability and performance may not convey what is expected and that in turn is a lot to handle. There can be scenarios when a disaster may disturb the networked sensor devices like the sensors become out of the scope of the incidence zones by displacement and may transfer the irrelevant or no information to the destination system leads to failure of disaster management system, or the negative effects of a disaster do influence the functionality of the sensors or its network heavily by damage caused by the disaster. For instance, a serious land slide or mud landslide, forest fire may destroy many detector sensors makes the sensors not to work properly or no datasets can be transferred to the disaster management system. Another challenging scenario is the situation where the sensor network is not able to exist to perform the intended tasks can be another challenge, for instance, a vehicle carrying dangerous chemicals is drained out and exploded in a city and it thus causes the chemical leak. But, in-situ sensors for measuring the chemical conditions along with the communication module do not exist. Though, to avoid such scenarios, space-based or aerial-based sensors are being used which can sense the images of the ground in proximate real-time but in cloudy or rainy days, the taken images might not be rich enough for the ensuing interpreting task of disaster management.

The disaster scenarios are at all times emergent and have more severe impacts than expected. It is also very hard to predict those disaster locations and time since they are variable and unknown. The propinquity of the disaster management typically does not allow too much time for the Sensor network to obtain required information. Secondly, scenario contains many difficulties where it needs specific design of the Sensors at cheaper price to tackle the unpredictable anomalies that may arise in sudden arise of catastrophic situations like earthquake, landslides, forest fire, chemical leak, oil ship in deep sea, floods and others.

SOLUTIONS AND RECOMMENDATIONS

The current sensor network applications are mostly consider the normal state that reduces the usability of sensors in emergent circumstances. Hence, the networked sensor required to be robust and useful by considering both the normal state and abnormal state during design phase of the sensors which will make the sensor networks operate underground, on ground surface and everywhere. It can make stronger the defense capability which lessens the destruction probability of sensors under extreme situations. Research says that the solution to the above problem is Emergency Response(ER) system that should be much independent i.e. self-managing, self-configuring and fault tolerant. Introduction of cloud based services in sensor networks may be very useful for data analytics to predict more accurate information. The ER systems should use more advanced low power radio technology to inculcate longer range for sensor networks. In view of other extreme situations, certain sensors possibly will need precise kind of battery or power supply source to resist the extreme temperatures. Communication consistency of sensor network is often the holdup problem for the ER applications. Hence, improved design structure like presence of redundancy function and path may be used in the sensor network to resolve such communication interruption between sensors, to send real time data. The logical design of redundant application than physical one can increases the work as well as energy efficiency of the sensor network. On the contrary wireless sensor fitted drones along with high resolution onboard cameras may be very useful to monitors and transmit real time information by forming areal wireless ad hoc sensor network routing process during or prior to the disaster rescue operation.

FUTURE RESEARCH DIRECTIONS

Giving a look into the future it can be seen a vast range of forthcoming technologies. Some researchers stated that the use of Superconducting antennae and solenoid coils, Advanced Radar and Infrared Cameras, Advanced Interferometric Synthetic Aperture Radar (InSAR) and Modems and Data Processing Centers will be introduced to check for the low frequency change and would help to forecast earthquake.

The introduction of tilt meter (inclinometers) would also help in detection with even a slight movement in the earth. There are few technologies which are to take up on the near future for disaster management. SimulationDeck is one such development which would help in the public outreach via social media, Intellistreets are going to be used for alerts and evacuation process and iDAWG (Intelligent Deployable Augmented Wireless Gateway) which is to be used in the extreme scenarios. This study may lead to future realization of new and hybrid and state of the art technology.

CONCLUSION

The theme of this chapter paves its way to understand the disaster, its detection and management and rescue using low power and low cost technologies specifically sensors in wireless networks. The roles of such networks are gathering data from the catastrophe area and wirelessly communicate them to a principal database through numerous low power hybrids routing mechanism. Besides, it has been discussed that some systems are dealing with different disasters; earthquakes, landslides, and others. Some of these technologies proposed to use wearable sensors; others proposed to deploy static and dynamic deployment of sensors in structures in addition to smartphone sensors with mobile applications, etc. All of the presented techniques use wireless communication with sensor technology to be more robust in case where infrastructures breakdown occur after the disaster. Routing emergency messages with delay restrictions is one of the important tasks. In addition, tracing emergency responders and victims is also an essential aspect while the energy depletion is a fundamental concern in WSNs that should be considered when using such technology. This chapter will help distinguished reader and researchers to analyze new aspects of disaster management and emergency response with the technology of sensor networks.

REFERENCES

Alphonsa, A., & Ravi, G. (2016). Earthquake early warning system by IOT using Wireless sensor networks. *Proc. IEEE International Conference on Wireless Communications, Signal Processing and Networking (WiSPNET)*. doi:10.1109/WiSPNET.2016.7566327

Benkhelifa, I., Nouali-Taboudjemat, N., & Moussaoui, S. (2014, May). Disaster management projects using wireless sensor networks: An overview. In *Advanced Information Networking and Applications Workshops (WAINA), 2014 28th International Conference on* (pp. 605-610). IEEE. doi:10.1109/WAINA.2014.99

Bluestone, J. R. (2010). *Why the earth shakes: Pre-modern understandings and modern earthquake science* (PhD thesis). Faculty of the graduate school, University of Minnesota.

Carter, W. N. (1991). Disaster management: A disaster manager's handbook. In *Disaster management: A disaster manager's handbook.* ADB.

CGT_NetRaising_8Earthquakeconclusion. (n.d.). Retrieved from https://www.rgs.org/.../CGT_NetRaising_8Earthquakeconclusion.pdf

Chen, J., Díaz, M., Rubio, B., & Troya, J. M. (2013, June). RAISE: RAIlway infrastructure health monitoring using wireless sensor networks. In *International Conference on Sensor Systems and Software* (pp. 143-157). Springer International Publishing. doi:10.1007/978-3-319-04166-7_10

Chen, Y. L., Wang, N. C., & Lin, Y. C. (2012). Power-Efficient Gathering in Sensor Information System Architectures with a Phase-Based Coverage Algorithm in a Wireless Sensor Network. *Advanced Science Letters*, *9*(1), 620–625. doi:10.1166/asl.2012.2600

Claudio, S., Yih-Min, W., Aldo, Z., & Abd Hiroo, K. (2011). Earthquake early warning: Concepts, methods and physical grounds. *Soil Dynamics and Earthquake Engineering*, 106–118.

Cranenbroeck, J., & Shlyakhova, M. (2014). *Geosensing the world in a new generation digital earth - The new role of the surveyors*. Academic Press.

da Silva, R. I., Almeida, V. D. D., Poersch, A. M., & Nogueira, J. M. S. (2010, April). Wireless sensor network for disaster management. In *Network Operations and Management Symposium (NOMS)*, 2010 *IEEE* (pp. 870-873). IEEE.

Devasena, A., & Sowmya, B. (2015). Wireless sensor network in disaster management. *Indian Journal of Science and Technology*, *8*(15). doi:10.17485/ijst/2015/v8i15/74191

Eilam, D., Izhar, R., & Mort, J. (2011). Threat detection: Behavioral practices in animals and humans. *Neuroscience and Biobehavioral Reviews*, *35*(4), 999–1006. doi:10.1016/j.neubiorev.2010.08.002 PMID:20727909

Fasihuddin, Q., Kolhare, N. R., & Daimi, S. S. (2012). Measurement and Analysis of Vibration and Seismic Signal. *Measurement*, *7*, 8.

Finazzi, F. (2016). The Earthquake Network Project: Toward a Crowdsourced Smartphone-Based Earthquake Early Warning System. *Bulletin of the Seismological Society of America*, *106*(3), 1088–1099. doi:10.1785/0120150354

Fischer, J., Kühnlenz, F., Ahrens, K., & Eveslage, I. (2009). Model-based Development of Self-organizing Earthquake Early Warning Systems. *Proceedings MATHMOD 2009*.

Fleming, K., Picozzi, M., Milkereit, C., Kuhnlenz, F., Lichtblau, B., Fischer, J., & Ozel, O. et al. (2009). The Self-organizing Seismic Early Warning Information Network (SOSEWIN). *Seismological Research Letters*, *80*(5), 755–771. doi:10.1785/gssrl.80.5.755

Frohlich, C., & Brunt, M. (2013). Two-year survey of earthquakes and injection/production wells in the Eagle Ford Shale, Texas, prior to the Mw4. 8 20 October 2011 earthquake. *Earth and Planetary Science Letters*, *379*, 56–63. doi:10.1016/j.epsl.2013.07.025

Goltz, J. D., & Bourque, L. B. (2016). Earthquakes and Human Behavior: A Sociological Perspective. *International Journal of Disaster Risk Reduction*, *21*(2017), 251–265.

Howell, B. F. Jr. (1986). History of ideas on the cause of earthquakes. *Eos (Washington, D.C.)*, *67*(46), 1323–1326. doi:10.1029/EO067i046p01323

Kalita, H. K., & Kar, A. (2009). Wireless sensor network security analysis. *International Journal of Next-Generation Networks*, *1*(1), 1–10.

Kong, Q., Allen, R. M., Schreier, L., & Kwon, Y. W. (2016). MyShake: A smartphone seismic network for earthquake early warning and beyond. *Science Advances*, *2*(2), e1501055. doi:10.1126/sciadv.1501055 PMID:26933682

Kong, Q., Kwony, Y. W., Schreierz, L., Allen, S., Allen, R., & Strauss, J. (2015, July). Smartphone-based networks for earthquake detection. In *Innovations for Community Services (I4CS), 2015 15th International Conference on* (pp. 1-8). IEEE. doi:10.1109/I4CS.2015.7294490

Kour, H., & Sharma, A. K. (2010). Hybrid energy efficient distributed protocol for heterogeneous wireless sensor network. *International Journal of Computers and Applications*, *4*(6), 1–5. doi:10.5120/828-1173

Kumari, N., Patel, N., Anand, S., & Bhattacharya, P. P. (2013). Designing low power wireless sensor networks: a brief survey. International Journal of Advanced Research in Electrical, Electronics and Instrumentation Engineering, 2(9).

Lambie, E., Wilson, T. M., Johnston, D. M., Jensen, S., Brogt, E., Doyle, E. E., & Helton, W. S. (2016). Human behaviour during and immediately following earthquake shaking: Developing a methodological approach for analysing video footage. *Natural Hazards*, *80*(1), 249–283. doi:10.1007/s11069-015-1967-4

Le Mehaute, B., Hwang, L. S., & Van Dorn, W. (1971, September). Methods for improving tsunami warning system. In Engineering in the Ocean Environment, IEEE 1971 Conference on (pp. 329-331). IEEE.

Liu, G., Tan, R., Zhou, R., Xing, G., Song, W. Z., & Lees, J. M. (2013, April). Volcanic earthquake timing using wireless sensor networks. In *Proceedings of the 12th international conference on Information processing in sensor networks* (pp. 91-102). ACM. doi:10.1145/2461381.2461396

Nittel, S. (2009). A survey of geosensor networks: Advances in dynamic environmental monitoring. *Sensors (Basel)*, *9*(7), 5664–5678. doi:10.3390/s90705664 PMID:22346721

Patil, H. A., & Patil, V. K. (2015, July). Survey paper on multiple input multiple output (mimo) system analysis and simulations. Journal of Emerging Technologies and Innovative Research, 2(7).

Prasad, M. S. R., & Khasim, S. (2013). Structural Monitoring Of Buildings Using Wireless Sensor Networks. *International Journal of Modern Engineering Research*, *3*(3).

Prusty, A. R. (2012). The Network and Security Analysis for Wireless Sensor Network: A Survey. *International Journal of Computer Science and Information Technologies*, *3*(3), 4028–4037.

Prusty, A. R., Sethi, S., & Nayak, A. K. (2015, December). Analysis of energy complexity under mobility for cluster routing in wireless ad hoc sensor networks. In *Microwave, Optical and Communication Engineering (ICMOCE), 2015 International Conference on* (pp. 314-317). IEEE. doi:10.1109/ICMOCE.2015.7489755

Prusty, A. R., Sethi, S., & Nayak, A. K. (2016a). A hybrid multi-hop mobility assisted heterogeneous energy efficient cluster routing protocol for Wireless Ad hoc Sensor Networks. *Journal of High Speed Networks*, *22*(4), 265–280. doi:10.3233/JHS-160548

Prusty, A. R., Sethi, S., & Nayak, A. K. (2016b). Energy Aware Optimized Routing Protocols for Wireless Ad Hoc Sensor Network. Handbook of Research on Advanced Wireless Sensor Network Applications, Protocols, and Architectures, 156. DOI:10.4018/978-1-5225-0486-3.ch006

Rameesh, V. M., Rajan, P., & Pullarkat, D. (2014). Development of a resilient wireless sensor network for real-time outdoor applications. In *Proceeding on International Symposium on Technology Management and Emerging Technologies (ISTMET)*. IEEE. Doi:10.1109/ISTMET.2014.6936476

Ramesh, V. M. (2014). Design, Development, And Deployment of a Wireless Sensor Network for Detection of Landslides. *Ad Hoc Networks, 13*(2014), 2-18.

Saita, J., Sato, T., & Nakamura, Y. (2008). *What is the useful application of the earthquake early warning system?* In The 14th World Conference on Earthquake Engineering, Beijing, China.

Schnytzer, A., & Schnytzer, Y. (2011). *Animal modeling of earthquakes and prediction markets* (No. 2011-20). Working Papers, Bar-Ilan University, Department of Economics.

Shamroukh, R., & Awad, F. (2009, March). Detection of surviving humans in destructed environments using a simulated autonomous robot. In *Mechatronics and its Applications, 2009. ISMA'09. 6th International Symposium on* (pp. 1-6). IEEE. doi:10.1109/ISMA.2009.5164820

Sharad, R. P., & Thorat, P. R. (2013). Live Human Detecting Robot for Earthquake Rescue Operation. *International Journal on Recent and Innovation Trends in Computing and Communication, 1*(11).

Singh, R., & Nagla, K. S. (2012). Proposed model of an earthquake detector by using UBG-04LX-F01 laser rangefinder. *International Journal of Advanced Research in Computer Science and Software Engineering, 2*(5).

Tributsch, H. (1982). *When the Snakes Awake: Animals and Earthquake Prediction*. Cambridge, MA: MIT Press.

ur Rahman, M., Rahman, S., Mansoor, S., Deep, V., & Aashkaar, M. (2016). Implementation of ICT and Wireless Sensor Networks for Earthquake Alert and Disaster Management in Earthquake Prone Areas. *Procedia Computer Science, 85*, 92-99.

Wang, D., & Ni, Y. (2012). Wireless sensor networks for earthquake early warning systems of railway lines. In *Proceedings of the 1st International Workshop on High-Speed and Intercity Railways* (pp. 417-426). Springer Berlin Heidelberg. doi:10.1007/978-3-642-27963-8_38

Youssef, M., Yousif, A., El-Sheimy, N., & Noureldin, A. (2007, April). A novel Earthquake warning system based on virtual MIMO-wireless sensor networks. In *Electrical and Computer Engineering, 2007. CCECE 2007. Canadian Conference on* (pp. 932-935). IEEE. doi:10.1109/CCECE.2007.238

Zennaro, M., Pehrson, B., & Bagula, A. (2008, October). Wireless Sensor Networks: a great opportunity for researchers in Developing Countries. *Proceedings of WCITD2008 Conference*, 67.

KEY TERMS AND DEFINITIONS

Disaster: A calamity or disruptive event that may lead to commotion in the community or society by human, environmental, material, and economic loss. This might as well result in a situation where the community or the society becomes unable to cope with the situation and regain the previous scenario. It may be caused naturally or by humans.

Earthquake: A natural calamity that is caused by the sudden and unalarmed shaking of earth's crust. This may be violent and furious. The main reason behind earthquake is the movement of earth's tectonic plates. The earthquake when caused under the sea or in the sea bed results in a tsunami.

Energy: Energy in scientific terms is the ability to do any work. In this chapter, energy has been defined considering the catastrophic behavior of the environment and its different forms to manage the same. Further, consideration of low power is taken into account to make the wise use of the available resources and hampering the environment less.

Management: Management is how a particular thing is dealt with and how it is controlled. Management in different fields has different goals and methods. Here we discuss disaster management; it involves organization of the various available resources and taking up responsibilities to deal with the current situation created by the calamity for preparedness, response and recovery.

Technology: Practical implementation of science is called as technology. It involves science with various methodologies and processes, techniques, skills, and art.

Waves: Study of earthquake is called as seismology and hence the earthquake waves are called as seismic waves. These waves are nothing but energy which passes through the Earth's crust or in the inner layers as well, called as surface waves and body waves, respectively.

WSN: Wireless sensor networks (WSN) refers to the widely used technology of today. These are nothing but spatially spread nodes (sensors) to monitor various environmental factors and comprehensively pass the information to a remotely located data center.

Section 4

Climate Change and Disaster Vulnerability Management

Chapter 8
A Review of Psychological Resilience as a Response to Natural Hazards in Nigeria

Bolajoko Ibiyemi Malomo
University of Lagos, Nigeria

ABSTRACT

Natural hazards are potential threats to lives, properties, and the society at large. This chapter discusses psychological resilience as a strategy in overcoming the aftermath of natural hazards, which are known as natural disasters in Nigeria. Resilience consists of individual behaviours, cognitions, actions, and attitudes that enhance personal wellbeing. The ability to exhibit resilience is important in overcoming the adverse consequences of natural hazards. A review of the concept of resilience in this chapter covers an introduction to the concept of resilience and natural hazards, cases of natural hazards in Africa and Nigeria, conceptual definition and issues relating to resilience, theoretical underpinning, recommendation, and conclusion. The various factors that are critical to the development and utilisation of psychological resilience are discussed. Relevant authorities can thus be advised on the need to set up platforms on which communities can devise methods of educating individuals on methods of developing resilience in preparation for impending natural hazards.

INTRODUCTION

Hazards generally are situations or events that evoke feelings of fear, loss, pain and dejection. The occurrences associated with hazards are as old as history, but the prevalence of hazards is increasingly being experienced globally and in all spheres of life (Pryor 2012). A hazard is a threatening phenomenon that has the potential to bring about loss of lives, injuries, property damages, socio-economic disruption of life or environmental degradation (IB geography subject guide, 2011). Hazards can be described as the negative consequences of events attributable to natural or human forces. For instance, events associated with an earthquake, a meteor, and any other weather related event, could be causes of natural hazard. On the other hand, when factors associated with human activities are the major causes, then human forces are responsible.

DOI: 10.4018/978-1-5225-3194-4.ch008

Natural hazards are phenomenal events that occur naturally in the physical environment through a series of slow or forceful events. These events could be geophysical (earthquakes, tsunamis volcanic eruptions); hydrological (avalanches and floods); climatological (high temperatures, droughts and wildfires) or meteorological (cyclones, ocean surges). One of the characteristics of natural hazards is that hazards have stored energies that may get dissipated and released with a high intensity. In that state, it causes grave damage to anything along its way. These types of hazards can lead to negative consequences for the people or the environment affected.

Nigeria's typology and ecological terrain may seem to have been speared some of the world's natural hazards compared to what obtains in other parts of the world. Amongst the different types of natural hazards, the most prevalent and experienced by Nigerians is the hydrological hazards (flood), followed by meteorological (ocean surges), while geophysical (earth tremor)is experienced at a mild but worrisome level in some parts of the country. Flooding, a major hazard in Nigeria has claimed lives and properties in a magnitude that requiresurgent attention by stakeholders.

The history of flooding in Nigeria dates back to the 1900s. Flooding manifests when water flows across a land that is usually considered dry land. Floods also occur when a low land is situated adjacent to a body of water (Sada, 1988). This is because after a heavy rain fall, the likelihood of water from a stream or river overflowing its bank onto a low land is high. Notable flooding incidences in Nigeria (Ibadan, 1990; Akure, 2002, 2006; Cross River, Delta, Ebonyi, Adamawa, 2012; Niger Delta, 2012) have become a common phenomenon in the country. Flooding has recently been experienced across all regions of Nigeria and the aftermath has been catastrophic. This situation has prompted academic discuss and writings by experts (Aderogba, 2012a; Agbonkhese et al. 2014) from various fields in the physical and behavioural sciences. Usually, the area of discuss has been focused on the types of floods, typology of the areas affected, the relation of flood to sustainability and development (Adeogba, 2012), the history and effects of flooding (Tawari-Fufeyin, Megbuwe & Adams (2015); Agbonkhese et al. 2014; Okorodudu-Fubura, 2013). None has so far discussed the 30 issue of resilience and how victims can utilise it to overcome the losses associated with flooding.

Psychological resilience has been defined by Pecillo (2016), as a person's accomplishment in successfully overcoming the challenges encountered in various situations in life. It consists of behaviours that a person exhibits to achieve normalcy after a stressful experience. Resilience has been studied across domains and with various population groups such as children, sportsmen, and people living in coastal areas, victims of terrorist attacks and with patients of life threatening illnesses. In all of these situations, the resilience of the people involved was shown to be responsible for their capacity to withstand stress and overcome the challenges they faced. However, there is a paucity of knowledge on the relevance of psychological resilience in managing the aftermath of losses and trauma incurred and experienced by victims of natural hazards especially in Nigeria. The objective of this paper therefore is to review the subject of psychological resilience and how it can impact victims of national hazards in successfully managing social, emotional and economic losses caused by flooding in Nigeria.

BACKGROUND

Events associated with natural hazardsleave behind deleterious and inevitable consequences. This is because natural hazards are to a large extent, aftermaths of forces of nature that are naturally induced and out of the control of man.The incidences of natural hazards in Africa may be reported to be as common as those that occur in other continents. The situation may however be described as a much pathetic one in Africa because majority of Africans rely on natural resources for their livelihoods (United Nations Report, 2015) and these resources diminish each time a disaster occurs. Additionally, responses to incidences of natural hazards are slow and devastating. Therefore, in a situation where floods and other types of disaster affect the soil, plants and mineral resources, psychological intervention and other support will be greatly required to ameliorate the consequences experienced by victims.

There are different forms of natural hazards. Geological hazards can be described as hazards that are motivated by stored up energies that get released at a certain level after prolonged inertia. For instance, earthquakes, volcanoes, deposits of clay and sand from one location to another (Kusky, 2003). Due to the constant changes occurring underneath the surface of the earth, phenomenal events such as volcanic eruptions, earthquakes, floods and storms occur voraciously at times leading to losses especially of lives and properties. These activities have resulted in changes in sea levels, villages being wiped out and icy heaps melting at an alarming rate.

Several years of research studies have been able to create instruments that can predict the occurrences of most of these hazards. This has made it possible to leverage on the negative side effects of these hazards. For instance; the evacuation of people from environments that are likely to be affected, the cancelation of flights around lava spewing areas are possible due to early predictions and warnings that potential victims receive before a disaster occurs. Additionally, locations that are prone to natural hazards have been further fortified with buildings and roads that can stand the force presented by these hazards. These are hazards that consist of chemical substances. The chemical substances are harmful to human beings, the environment and properties. These precautionary measures are not adequate and at best not in advanced form in Nigeria. These may explain why victims of natural hazards incur and experience enormous losses and emotional trauma on perennial bases.

Biological hazards, also known as biohazards, are a type of hazard that consists of biological substances. When these hazards are handled carelessly, they could pose a threat to the existence and healthy living of all kinds of living organisms. They may manifest in the form of diseases, epidemics, viruses or toxins, pests and microorganisms. Additionally, biological hazards include medical wastes that cause harm when disposed inappropriately. The source of these hazards lies in the transformation of living organisms and may act as threat to the health and survival of these organisms and sustenance of environmental peace. The consequences of these hazards may include skin itching, allergies, infections and eventual deaths.

Meteorological hazards area form of hazard caused by severe weather conditions. For instance, floods, drought, thunderstorms and tornados are good examples of meteorological hazards that people all over the world deal with. These hazards are becoming prevalent all over the world presumably due to the climatic weather conditions occasioned by climate change. In Nigeria, and as reiterated above, flooding is common in comparison to other forms of meteorological hazards. However, despite the prevalence of flooding, efforts to prevent its occurrence is still inadequate and the response of people exposed to flooding is very poor.

Another form of hazards is chemical hazards. Chemical hazards harm and endanger human beings, properties and the environment. This is possibly because of the nature of chemicals. Chemicals are toxic, hence when they are ingested in certain amounts not recommended for human consumption over a period of time (Canadian Centre for Occupational Health and Safety, 2017), they may cause harm to the body. For example, in the Niger Delta, a region of great oil exploration in Nigeria, the toxicity of chemicals from crude oil waste sipping into streams and rivers may become a source of danger to inhabitants. The source of livelihood of the inhabitants of this region is fishing. This may become a source of danger.

When the fishes live on toxic water, this is transferred to human body when consumed. The health consequences may manifest as stunted human growth both physically and intellectually, body deformation and early death. The prevention of these hazards may not be totally possible and successful due to the nature of such hazards. Hence, victims would still be required to manage the aftermaths of the negative experiences they encounter from such events. Resilience may become useful because the individuals who possess resilience are likely to have an edge over those who do not.

A series of natural disasters in Africa and Nigeria have resulted in hazards that have caused untold hardships in the lives of the people. Some of these incidences, as posted by Eziro (online) are hereby discussed.

- In the Eastern part of Africa, a drought of great dimension was experienced between the second half of the year 2011 till the first half of 2012. Countries such as Kenya, Somalia, Ethiopia, Erithea and Djibouti were greatly affected. The consequences of the drought were food shortages, increases in the prices of food items, hunger and vulnerability to health caused by malnutrition.
- A volcanic activity was experienced in Erithea in the year 2011. This was the Nebro Strato volcano which spewed volcanic ashes that travelled over hundreds of kilometres. The consequences are: massive loss of live stocks, malnutrition, unprecedented water shortage and health challenges.
- In Southern Africa in the year 2011, Botswana witnessed severe flooding caused by incessant rains that poured for weeks. The consequences include loss of lives, closure of schools and displacement of people. This is similar to the flooding incidences that occurred in some parts of Nigeria in 2012 and in Lagos Island in July 2017. Houses, roads and business premises were flooded with properties destroyed. The severity of the situation informed the need for some victims to relocate to other places in anticipation of more rains and further flooding.
- As far back as AD 365, an earthquake was reported in the Eastern Mediterranean close to Crete. When estimated in modern day Richter scale, geologists would estimate it to be around 8 on the Richter scale. The countries affected included Greece, Northern Libya, Egypt, Cyprus and Sicily. The consequences of the earthquake in Crete were enormous with nearly all the towns affected. This was followed by another hazard known as the tsunami which spread it's devastation as far as Libya, Alexander and the Nile Delta of the Mediterranean.
- In 1960, an earthquake occurred in Agadir, a city in Western Morocco. The impact of the earthquake was intense despite the fact that it lasted for 15 seconds but it was very severe. It was reported that around 12000 people were killed while much more people were rendered homeless.
- In 1983-1985, Ethiopia experienced famine of a great magnitude. The duration of the famine was estimated at 24 months with over 350,000 human lives lost. The impact of the famine was huge because the economy of Ethiopia was hinged on agriculture which provides around 80% of the country's employment.

MAIN FOCUS OF THE CHAPTER

Numerous investigations have been conducted by experts in relevant fields on the issue of natural hazards and flooding in particular in Nigeria. Agbonkhese et al. 2014 investigated the challenges posed by incidences of flooding in Nigeria and the negative consequences of the aftermath of such occurrences. According to the authors, flooding can be traced mainly to human activities on the path of the government and the people. The role of government manifests in loop sided planning, poor and old infrastructural facilities, lack of adequate infrastructural design and poor vision. On the part of the people, the authors pointed out behavioural factors as causes of flooding. Specifically, challenges with dumping of waste were reiterated as negative behaviour associated with flooding. The authors highlighted the vulnerability of low lying areas and activities related to climate change as environmental factors of flooding. These challenges, though not created by man, ought to have received attention by way of rapid government intervention and provision of channels for excess water in the event of a flood. Additionally, the study cited early warning signals as ways of ameliorating the devastating effects of flooding. In relation to the consequences of flooding on the people affected, intervention strategies were proffered especially on the part of government. These are positive views on the topic in question. However, coping strategies on the part of the victims were ignored despite the incessant occurrences of flooding on perennial bases.

The study of Tawari-Fufeyin, Megbuwe, & Adams (2015) focused on the environmental pollutants associated with flooding in the Niger Delta region of Nigeria. The study reiterated environmental factors in the physical environment such as climate change and upsurge in the level of sea water as causes of flooding. According to these authors, flooding may be defined as "extremely high flows of river, whereby water inundates flood plains or terrains outside the water-confined major river channels". The historic flooding of the year 2012 in Nigeria was particularly mentioned as a reference point in this study. The study acknowledged flooding as a global problem. However in Nigeria, the activities of neighbouring countries have compounded the challenges of flooding in Nigeria. Tawari-Fufeyin et al. (2015) cited the decision of Cameroun to let out excess water from the Ladja dam in the year 2012 to reduce tension and avoid the dam busting and causing more havoc. This action by the authorities in Cameroun caused unfavourable consequences on the lives of the people of Nigeria in different parts of the country. The inability of the relevant authorities in Nigeria to prepare for the increase in water level that may arise thereof caused the historic flooding that was experienced nationally. The chemical properties of the water in the rivers and their effects on farmlands, aquatic animals and birds, erosion and agricultural products were discussed. The study utilized survey and experimentation of the water properties to draw conclusion on the havoc that flooding has caused the inhabitants of Niger delta region of Nigeria.

The study of Tawari-Fufeyi (2015) is similar with the review credited to Amadi (2012) on the threats posed by coastal activities to communities living along coastal lines in Lagos State. Amadi (2012), highlighted threats of erosion in Makoko and Maroko communities in Lagos State. As usual, the blame was directed at the local and State government for lapses in some areas. The truth of the matter is that the people experience these situations on yearly bases and would be better off when they are exposed to coping strategies pending the time government would come to their succour.

Studies on the applicability of resilience after a natural disaster have become very expedient. This is due to the increasing anticipation of the activities of climate change occasioned by global warming. Secondly, the increasing activities around coastal areas (Frankenberg et al. (2013) has made it important to focus on strategies that victims can adopt to ameliorate their emotional trauma after a disaster. It is pertinent to note at this point that the discussion of resilience in relation to the aftermath of a natural disaster is different from resilience relating to experiences of chronic poverty, violence or family dysfunction (Davey, Eaker & Walters, 2013). The understanding of the differences is imperative especially, due to the fact that the study of resilience is becoming popular. Also, according to Agaibi and Wilson (2005), considerable attention has shifted from the study of resilience in the aftermath of recurrent misfortune or affliction to events that are traumatic. This present paper seeks to review psychological coping mechanisms available to victims of such hazards.

Flooding in Nigeria

The history of flooding in Nigeria dates back to the 1900s with notable flooding occurring in the Niger Delta, Lagos, Ibadan and some parts of the country. Recently, in 2012, specifically, heavy flooding was intensely experienced in Kogi State claiming many lives, properties and farmlands. Other cities affected are Ibadan, Lagos, Plateau State, Delta State, Bayelsa State and others. In all of these flood events, people were displaced while lives were also lost. A major causative factor of flooding in Nigeria is behavioural (Reuters, 2012). People are in the habit of disposing trash inappropriately into drainage channels. This act becomes an impediment to free flow of rain water and in the event of a heavy down pour, rain water overflow the channels and cause flooding.

Aside from flooding, indications of earth tremors experienced in some states in Nigeria have been reported. These states include Oyo, Osun, Ogun, Lagos, Cross River, Bayelsa and Kaduna. According to Adepelumi (quoted in the Nigerian Monitor), the probability of an earthquake occurring in these states between 2009 and 2028 is 91.1 per cent. Kaduna State, a State in the Northern part of Nigeria, recently experienced a series of earth tremors.

In all of the events stated above, the consequences have been devastating so much that lives and properties were lost, people were displaced and farmlands washed away. For developing nations, this may result to economic losses because the little economic activities sustaining the people have been disrupted. This will further compound the lingering drought of funds that these nations have had to manage as aids which hitherto have been given by developed nations do not seem to come as in the past.

The severity of flooding incidences in Nigeria will be best appreciated through an analysis of the causes of floods in various parts of the country. In the Niger Delta region of Southern Nigeria, floods have become a permanent fixture in the lives of the people that reside in the region for as long as they can remember. Several incidences of flooding are experienced every year in the Niger Delta with devastating consequences. The primary cause of flooding in the Niger delta is natural and environmental while the secondary cause can be traced to human activities.

Firstly, the Niger Delta is a low-lying area in the swampy part of southern Nigeria. As described by Agbonkhese et al. (2014), the Niger Delta region serves as a reservoir to water flowing from the Niger-Benue region. This body of water becomes a serious challenge for the region to handle during heavy rains. This is because the region lacks the resources and outlets in discharging and managing the water. During the heavy rainy seasons, a period that occurs from June to September every year, the accumulation of water during these months results in the overflowing of rivers and streams onto available dry lands that are already low. During the rainy season, especially after a heavy rain, water from the Fonta Djallon Highlands from Guinea cause rivers on its way to overflow. This may be further compounded by physical objects such as trees and waste of different forms that creates cloggingandhinders the water from flowing freely. This may then cause flooding.

Secondly, heavy flooding was experienced in the Niger Delta due to activities credited to human actions. Between July and September 2012, authorities in Cameroun discharged water from Ladja dam for safety reasons. Unfortunately, Nigerians were caught unawares despite warning reports to that effect. Hence, the consequences caused by the excess water that flowed to several states in Nigeria were grave. These negative consequences of flooding in the Niger Delta region have been long standing with little hope of reprieve for the inhabitants. Other notable areas that were affected by the floods in 2012 are Kogi, Benue, Katsina, Taraba, Bayelsa, and kebbi states among others. A report on a flood incidence in Numan town (Mazawaye, Tsenbeya & Ismaila, 2014) revealed that reoccurrences of flooding annually have resulted in devastations from destruction of houses and displacement of victims. Unfortunately, inappropriate response and poor actions from the government and other concerned stakeholders have been the order of the day.

A major causative factor in flooding in Nigeria is behavioural (Reuters, 2012). People form and get used to the habit of disposing trash inappropriately into drainage channels. This act becomes an impediment to free flow of rain water. In the event of a heavy down pour, rain water overflows the channels and cause flooding. Natural hazards cause harm to human beings through deaths, injuries, diseases and stress. A typical example is the consequences of a heavy rainfall that occurred in Eleyele in Ibadan, Nigeria, in 2012. The rainfall became too heavy for the Eleyele Dam and the resultant effect caused the dam to burst. Consequently, this event resulted in the loss of over 100 lives and more than 600 people were rendered homeless (Okorodudu-Fubura, 2013).

Ibadan, a city located in the South Western part of Nigeria experiences frequent flooding due to humanactivities. The locationof Ibadan is a great distance from the coast; hence, Ibadan is not a coastal city. The rivers and streams that flow within the city can only be predisposed to flooding when the channels created for water to flow are blocked. Human activities and behaviours play a major role in the flooding that occurs in Ibadan. For instance, one or a combination of the following may be responsible: poor drainage structures and channels, inappropriate dumping of human waste, poor layout of new housing developments, attitudinal factors, general habits of inhabitants to refuse disposal and apathy or late action on the part of the government. These factors impede free flow of water in the streams and rivers in the city of Ibadan. Population explosion has further compounded this problem. Hence, any heavy rainfall renders the city susceptible to flooding because channels hitherto available through which water can easily flow have been blocked. When this happens, the water gets diverted to cause flooding.

Lagos state is a coastal city made up of islands and the Mainland. Several coastal communities are exposed to threats of flooding due to their proximity to shorelines. Hence, threats from the sea and lagoon expose inhabitants to incessant flooding with negative consequences that are sometimes beyond human endurance. Recent incidences of sea surges and flash floods have assumed increasing dimension with little hope of getting better. Globally, the consequences of climate change have been reiterated and countries have been enjoined to seek for solutions to the changing climatic conditions.

Years back in Lagos, communities that settle along the coastline and riverbeds were made up of low income earners and poor individuals who fish for a living or engage in artisan trade. Instances of such communities are Maroko, Ilaje, Makoko, Isheri areas among others. These people have had to deal with consequences of extreme weather conditions characteristic of their location. The capacity to relocate from their environment is impeded by financial constraints. Ironically, the trade they engage in is grossly hampered by perennial weather changes. This unpredictable situation is tantamount to the wellbeing of their social and economic life.

Recent reactions to climatic conditions have assumed a different dimension. This is due to the changing demographic characteristics of victims of perennial flooding usually during the raining season. In the 1990s, corporate buildings, housing banks, liaison offices of states and government buildings were the most affected by ocean surges from the bar beach in Victoria Island. Apart from the private and public residents along the bar beach in Victoria Island who have been victims of ocean surges, residents of nearby neighbourhoods in the Lekki peninsular have become victims. This may not be unconnected with inappropriate land use, inadequate government planning, dredging activities, overpopulation and lack of expert risk assessment reports prior to the influx of people to the area.

Flooding in the Lekki peninsular is common and is becoming a serious hazard to the residents. This year, 2017, enormous havoc has been experienced by residents of Victoria Island and the Lekki peninsular. Properties worth millions have been destroyed by heavy rains that poured incessantly for days. Pictures (below) depict valuables of different sorts being destroyed at great expense to the owners. The likelihood of this situation extending to human lives is high. In view of the financial capability of property owners in this neighbourhood, the ease with which they can relocate may be high, but adjusting to their new environment after such a great loss may be psychologically traumatic. The pictures below give a view of threats of flood to lives and properties.

As affirmed by DMSG (2001) reports on floods, the sources of livelihoods of flood victims are substantially affected thereby creating untold hardships economically and socially. The incidences of flooding in the Niger Delta States in the recent past resulted in loss of lives and properties. The flood, which was caused by heavy rain fall, washed away several acres of farmlands. A chain of reactions set in when natural disasters such as floods occur. The victims of flooding in the Niger Delta experienced losses in various forms. They experienced losses on their trading goods, raw materials, buildings, roads and infrastructures. For educational facilities such as schools, the need to rebuild damaged buildings and replace educational facilities may be delayed and could run into millions in economic terms. This may further cause the children in such schools to suffer delay in achievement educational advancement like their mates in secured environments.

Furthermore, the flooding in the Niger Delta affected the environment through pollution of water and air, infestation of live stocks and the spread of diseases. For instance, since floods are a common form of natural disasters in Nigeria, it can result to displacement of populations from their habitation,

Figure 1. Vehicles and buildings affected by the recent flood in Lekki near Victoria Island, Lagos State

Figure 2. Streets and residential buildings flooded by rain water in Lekki, Lagos State

contamination of water sources meant for human consumption, rise of vector borne diseases such as malaria and yellow fever. All these can result to disease epidemic among the affected population, thereby resulting in negative wellbeingof the affected population.

When natural hazards and disasters occur, survivors also experience a disengaged social life and psychological wellbeing (Ronan & Johnston, 2005). Due to various forms of losses related to livelihoods, properties and disruption to social lives, victims require some time to recover from such losses. The ability to overcome the resultant stressful situations varies across people and may be dependent on some psycho-demographic variables. For instance, in the study of Bonanno, Galea, Bucciarelli, and Vlahov, (2007), in addition to psycho-demographic variables, resilience can also be predicted and influenced by chronic diseases, social support and past experiences of traumatic situations.

Psychological Resilience

Disasters emanating from natural hazards are unavoidable phenomenon (Bonanno, Galea, Bucciarelli & Vlahov, 2007). These events sometimes occur without prior warnings and thus deal a devastating blow on people who are vulnerable and caught in the web. For people living in cities that are characteristic of coastal terrains, hazards that are attributed to weather conditions are a constant phenomenon (UN/ISDR 2002). In developed countries, and especially in places where such phenomenal events have been experienced in various degrees, history has left indelible mark on the memories of victims in those countries particularly in the location worst hit. Expectedly, such experiences should probably prepare victims on what to do should such occurrences stare them in the face in future. It will be agreed though that the general saying that once beaten twice shy may be applicable to such victims, the same may not be true in nations where the government is not proactive and ready to take action. As reiterated by Bryant (2011), prior experiences of disasters have taught or informed people on how disasters should be managed especially when it involves emotional disturbances or mental health challenges.

The behaviour of victims of natural hazards is grossly affected psychologically, mentally and emotionally (Watson, et al., 2003). These consequences extend over a long period, hence people require reprieve through psychological intervention. Therefore, this article sets out to review such interventions by discussing the definitions, models and factors enhancing psychological resilience. This is because studies have indicated that victims of natural disasters who displayed varying forms of resilience found it easier to adapt to normal life faster than those with low levels of resilience.

Resilience was originally adapted from the Latin word '*resilio'*. This means to leap, increase or jump back (Klein, Nicholls & Thomalla, 2003). In the engineering field where it was first utilised, it denotes a situation where a material for example extends to certain point, then displays the capacity to retain its original size. In the description given by (Gordon, 1978), a material is characterised by the properties of resilience when the material is able to retain strenuous pressure and yet maintain its balance or capacity without reaching a state of collapse. From the field of engineering, resilience gained grounds in psychology and was investigated to understand the reasons for individual differences in the behaviour of victims who have gone through devastating experiences.

In the field of psychology, the term resilience was first studied as a subject of investigation and discussion by developmental psychologist (Garmezy, 1971, Rutter, 1979, Bonanno, 2004) first with children's ability to overcome developmental challenges, and later with regards to ecological issues. Studies conducted on the outcome of children's interface with developmental problems revealed that while some children were unable to surmount these problems, others overcame the same problems with ease (Masters, 2013; Masten and Obradovic (2006),). This motivated developmental psychologist to investigate and understand the differences in the experiences of these children. The understanding of the factors responsible for these differences could be utilised to proffer insight on how victims of a major negative event could utilise resilience to adapt to normal life after exposure to life threatening events such as natural hazards. This assertion is related to resilience as a positive adaptation to stressful events (Hopf, 2010).

When psychological resilience is discussed, it is assumed that a devastating event has occurred and the consequences of that event have adversely affected the psychological state of a person. In that case, the emotions, mental and behaviour of that individual has been destabilised. This infers that psychological resilience is the capacity of an individual to readily overcome a life threatening situation, and to adapt despite the challenges posed by the situation. In life, people are swamped by stress inducing and adverse events such as personal, financial and work challenges (APA, 2014). However, despite the assumed positivity attached to the concept and possession of resilience as a trait, researchers in the area of resilience have affirmed the lack of a consensus operational definition of resilience (Klein, 2003) and have also expressed the ambiguity of the word. However, UN/ISDR (2002); Chaplin et al. (2000); Schulze & Mooney (1993) affirm resilience as the way to successfully manage the ecosystem.

As adults, the utility of resilience in relation to adversity occasioned by weather conditions depends on the definition. Resilience is that psychological state that enables an individual to withstand tough times and deleterious events like natural hazards. The experiences encountered due to a life threatening event require a reaction that would enable that individual to continue with normal life even after the negative experience. Adult resilience is described by Bonannoa 2004, p. 20) as "an ability of adults in otherwise normal circumstances who are exposed to an isolated or potentially highly disruptive event……, to maintain relatively stable healthy levels of psychological and physical functioning". To possess ability signifies the presence of a skill or a trait that enables the performance of a task. In essence then, this definition may seem to agree with researchers who discuss resilience as a trait instead of a process.

Luther, Cicchetti and Becker (2000, p543) defined psychological resilience as a "dynamic process encompassing positive adaptation within the context of significant adversity". In this definition, resilience is described as an on-going state of activity where the individual keeps reacting to adverse events but concurrently adapting to the situation.

A definition based on ecology defined resilience as "a measure of the ability of these systems to absorb changes of state variable, driving variables and parameters and still persist (Holling, 1973, p.17). Systems in this definition refer to natural and social systems (Bonanno, Galea & Vlahov, 2007). Environmental events that are due to climatic weather conditions have resulted in adjustments by people over the time. Hence, changes occurring thereby have required unstable reactions to suit each situation as they present themselves. The concept of resilience has therefore been developed and utilised over time. In another definition, UNISDR (2009) defined resilience as "the ability of a system, community or society exposed to hazards to resist, absorb, accommodate and to recover from the effects of a hazard in a timely and efficient manner, including through the perseverance and restoration of its essential basic structures and functions. Thus, from the aforementioned definitions, resilience is a crucial ingredient for victims of natural disasters if a return to their normal lives is desired.

Theoretical Support for Resilience

Resilience has garnered theoretical support especially from studies on health behaviours and its resultant consequences with populations at risk (Barrow et al., 2007; Isomaa et al., 2013). However, resilience was not conceived from any established theory (Richardson, 2002). Instead, it was conceived from the popularity of observations made from studies involving children and adolescents who were experiencing various risky situations occasioned by their environments. The findings emanating from the investigations on the personal characteristics of children who survived childhood traumatic experiences can be extended to the studies of victims of natural disasters.

Resiliency theory encapsulates a strength-based approach in elucidating the processes undertaken by children and adolescents during developmental stages (Zimmerman & Brenner, 2010). As reiterated by Garmezy 1991; Masten et al. 2007), resiliency theory provides concise understanding of how children and youths acquire personal skills that aid their healthy growth as adults in spite of the various risks they are exposed to. Resilience in this case relates to factors associated with positive contextual, social and individual variables. These factors influence developmental processes during human growth. Fergus and Zimmerman (2005) describe these factors as promotive factors. They are so called because of the role they play in assisting youths to overcome negative and risky health related behaviours. Additionally, promotive factors connote positivity and something of value to the individual (Luthar, Sawyer & Brown, 2006).

According to Fergus and Zimmerman (2005), promotive factors are classified into two and these are labelled as assets and resources. Assets are positive factors operating within the individual to foster positive outcomes relating to healthy development occurring on a daily basis. Examples of these personal assets are personality constructs of self-esteem and self-efficacy. The second type of promotive factors resides outside of the individual and they are known as resources. Resources can be obtained from social support, especially from parents, adults and formal programmes that are organised for youths for the purpose of creating avenues to acquire skills and learning. Assets and resources are promotive factors that aid acquisition of personal and situational characteristics for positive growth.

Another model worth considering is the challenge model of resiliency proposed by Rutter (1987). This model proposed the development of resilience through deliberate exposure of youths to challenging and risky situations for the purpose of training the youths to familiarise themselves with the dangers posed by such situations. The essence of the initial exposure is to assist the youths to overcome the initial risk so that future experiences of the same or a higher level of risk will be accommodated the same way the initial risk was dealt with. It is pertinent to mention however that the degree of the initial risk exposure must not be too intense for the youths to handle.

The person-environment view (as a process). The theoretical support for resiliency from the person-environment view proposed that resiliency is a process that commences from the beginning of experience with an adversity right through the point where the individual utilises the environment to overcome the adversity. Specifically, when an individual encounters an adversity, the person appraises the situation in line with the threats, dangers or demands the situation poses, and then builds a hedge using the environment to overcome the threatening situation. In this instance, Flach, (1980) describes resilience as the outcome of the process that the individual passes through to resist a threatening situation. The process approach was also supported by Rutter (1985). In Ruther's view, personal characteristics inherent in the individual along with experiences previously encountered make a connection with the threatening situation.

Factors Enhancing Resilience

Innumerable research studies on resilience has been conducted in relation to developmental processes (Condly, 2006), floods, natural disasters (López-Marrero & Tschakert, 2011; Goodman & West-Olatunji, 2008), bereavement (Bonanno et al., 2005), terrorist attacks (Bonanno, Rennicke & Dekel, 2005; Bonanno et al., 2007) and serious illness (Denz-Penhey & Mucdoch, 2008). These situations confirm that everyone is bound to encounter a challenging experience at different points in their lives (Ozer, Best, Lipsey & Weiss, 2003). All of these studies also signify that resilience is important as a trait or an attribute required in overcoming adversities. As reiterated by some researchers, some individuals possess degrees of resilience while some may not display as much resilience as the others. What then are these factors that enhance resilience in some people?

No one factor has a dominant predictive influence on resilience (Bonanno, Brewin, Kaniasty & La Greca, 2010).

- **Personality Characteristics:** Personality consists of the traits, attributes, values, motivations and behaviours unique to an individual and which is usually displayed on a consistent basis. When people experience the same challenging situations, their reactions are usually different behaviourally, psychologically and physiologically (Ferguson, 2013). When resilience is conceptualised as a trait (Master & Reed, 2002), it could be enhanced by personality characteristics such as self-esteem and self-efficacy which has been described as the asset component of promotive factors (Fergus and Zimmerman, 2005). Hence, the personality make up of a person would have a great influence on how resilient the person would be. In the area of sports where endurance and strenuous activities are experienced, resilience and the type of personality individuals possess plays a crucial role in winning competitions. For instance, Gould, Dieffenbach & Moffett (2002) in Fletcher and Sarkar (2012), listed confidence, the ability to focus, a hard work ethic and optimism as personal traits of champions. In relation to personality traits, the benefits of positive emotions are paramount. Tugade & Fredrickson (2004) associated positive emotions with resilience and also affirms that people with positive emotions recover faster from cardiovascular diseases compared to those who do not possess positive emotions. Examples of positive emotions are hope, interests, cheerfulness and confidence. These positive emotions enhance resilience.

- **Motivation:** Motivation is referred to as a force that pushes individuals to do what they would ordinarily not do. People set motives for staying alive, achieving a successful life and also for bringing up their children as responsible adults. Having motives in life would therefore propel people to overcome the challenges posed by their immediate environment or health. In essence then, when traumatic experiences such as floods, hurricane and drought for example occur, survivors of such events whose motives for living override the disaster devise strategies to overcome their situations. In the study of Treasure, Lemyre, Kuezka & Standage (2007), it was affirmed that for elite sports performance, a recommended psychological attribute is a high degree of motivation.
- **Social Support:** According to Cohen (2004), social support refers to the ability of a social networkto provide psychological and material resources that would assist an individual to cope with stress. Social support infers warmth, care, love, trust and emotional support. The list is endless. These can be provided at different levels; individual, family, community and the society. Social relationships are very important as social support especially when faced with loss, displacement and trauma due to natural disaster. At the community level, resilience is defined by Pfefferbaum et al. (2013) as the ability of community members to take deliberate, purposeful and collective action to alleviate the detrimental effects of adverse events. As with individual resilience, community resilience involves attitudes, thoughts, beliefs, behaviours and resources. In the event of a disaster occurring, the total community is aided to bounce back to the former state of wellbeing.

Social support from religious and educational organisations has been shown to enhance resilience among children (Marriott et al., 2004). Support for the role of religion is further enhanced by the study of Peres, Moreira-Almeida, Nasello and Koinig (2007) where findings obtained in their study revealed that half of their participants reported a relationship between religion and resilience. For social support to be effective, the spouse is not only the most important source of support. Other members of the family are equally important. Additionally, colleagues from the work place can be a source of support

CONCLUSION

Disasters occasioned by natural hazards include but are not limited to flooding, hurricane, typhoons, droughts, tsunamis, cyclones and others. These disasters will not cease to occur if the global climatic condition remains the same. In fact, the intensity and occurrence is likely to be on the increase. The activities of global warming experienced around the world are increasing and these are attributed to the behaviour of humans who ironically, are the victims of natural hazards. These aforementioned situations are inevitable and thus, people should brace up to the occasion by adopting behaviours that would ensure the existence and sustenance of life. The need to prepare people especially those who are vulnerable to these disasters is ripe. In Nigeria for example, the reaction of Nigerians to events that are due to natural hazards is still premature and in doubt. In addition to the efforts of government agencies in Nigeria through the National Emergency Management Agency (NEMA), the government should focus on the victims and other vulnerable groups of people. Due to cases of floods, droughts and earthquakes that have been reported in different locations in Nigeria, the people living in such locations should be prepared and ready for such potential disasters associated with those locations. Psychological resilience can be encouraged and adopted as a trait by such vulnerable people through educational intervention organized by government agencies involved in disaster management.

The various behaviours associated with resilience can be introduced to vulnerable people who are more exposed to natural hazards and disasters. Moreover, factors that enhance resilience such as social support can be encouraged through the different social institutions in the communities such as the family, the religious and educational institutions. Additionally, relevant personality characteristics such as extraversion, conscientiousness and openness to experience may be encouraged through educational interventions to youths in the communities.

Empirical data from the population of people who have experienced natural hazards should be sought. Furthermore, the study of psychological resilience among victims of natural hazards should be encouraged. The results obtained can serve as resource from which government can develop intervention strategies for future use.

REFERENCES

Aderogba, K. A. (2012a). Global warming and challenges of flood in Lagos metropolis *Nigerian. Academic Research International, 3*(1), 448–468.

Agabi, C. E., & Wilson, J. P. (2005). Trauma, PTSD, and resilience: A review of literature. *Trauma, Violence & Abuse, 6*(3), 195–216. doi:10.1177/1524838005277438 PMID:16237155

Agbonkhese, O., Agbonkhese, E. G., Aka, E. O., Joe-Abaya, J., Ocholi, M., & Adekunle, A. (2014). Flood Menace in Nigeria: Impacts, Remedial and Management Strategies. *Civil and Environmental Research, 6*(4), 32–40.

Amadi, A. (2012). *Brief description of the communities threatened by coastal erosion and displacement.* Retrieved from https://ng.boell.org/sites/default/files/uploads/2012/12/eko_atlantic_project_-_opinion_papers.pdf

American Psychological Association (2014). *The road to resilience.* Retrieved from apa.org/helpcenter/road-resilience.aspx

Atkinson, P. A., Martin, C. R., & Rankin, J. (2009). Resilience revisited. *Journal of Psychiatric and Mental Health Nursing, 16*(2), 137–145. doi:10.1111/j.1365-2850.2008.01341.x PMID:19281544

Barrow, F. H., Armstrong, M. I., Vargo, A., & Boothroyd, R. A. (2007). Understanding the findings of resilience-related research for fostering the development of African American Adolescents. *Child Adolescent Clinics of North America, 16*(2), 393–413. doi:10.1016/j.chc.2006.12.004 PMID:17349515

Bonanno, G., Galea, S., Bucciarelli, A., & Vlahov, D. (2007). What predicts psychological resilience after disaster: The roll of demographics, resources and life stress. *Journal of Consulting and Clinical Psychology, 75*(5), 671–682. doi:10.1037/0022-006X.75.5.671 PMID:17907849

Bonanno, G. A. (2004). Loss, trauma and human resilience: Have we underestimated the human capacity to thrive after extremely aversive events? *The American Psychologist, 59*(1), 20–28. doi:10.1037/0003-066X.59.1.20 PMID:14736317

Bonanno, G. A., Moscowitz, J. T., Papa, A., & Folkman, S. (2005). Resilience to loss in bereaved spouses, bereaved parents, and bereaved gay men. *Journal of Personality and Social Psychology, 88*(5), 827–843. doi:10.1037/0022-3514.88.5.827 PMID:15898878

Bonanno, G. A., Rennicke, C., & Dekel, S. (2005). Self enhancement among high-exposure survivors of the September 11th terrorist attack: Resilience or social maladjustment? *Journal of Personality and Social Psychology*, *88*(6), 984–998. doi:10.1037/0022-3514.88.6.984 PMID:15982117

Bryant, R. (2011). How to manage the psychological effect of natural disasters. *The Conversation*. Retrieved from http://theconversation.com/profiles/richard-bryant-161

Burton, I., Kates, R. W., & White, G. F. (1978). *The environment as hazard*. New York: Oxford University Press.

Canadian Centre for Occupational Health and Safety. (2017). *What makes chemicals poisonous?* Retrieved from http://www.ccohs.ca/oshanswers/chemicals/poisonou.htmal

Chaplin, F. S., Zavaleta, E. S., Eviner, V. T., Naylor, R. L., Vitousek, P. M., Reynolds, H. L., & Diaz, S. et al. (2000). Consequences of changing biodiversity. *Nature*, *405*(6783), 234–242. doi:10.1038/35012241 PMID:10821284

Cohen, S. (2004). Social relationships and health. *The American Psychologist*, *59*(8), 676–684. doi:10.1037/0003-066X.59.8.676 PMID:15554821

Davey, M., Eaker, D. G., & Walters, L. H. (2013). Resilience processes in adolescents: Personality profiles, self-worth and coping. *Journal of Adolescent Research*, *18*(4), 347362.

Davydov, D. M., Stewart, R., Ritchie, K., & Chaudieu, I. (2010). Resilience and mental health. *Clinical Psychology Review*, *30*(5), 479–495. doi:10.1016/j.cpr.2010.03.003 PMID:20395025

Denz-Penhey, H., & Mucdoch, J. C. (2008). Personal resilience: Serious diagnosis and Prognosis with unexpected quality outcomes. *Qualitative Health Research*, *18*(3), 391–404. doi:10.1177/1049732307313431 PMID:18235162

DMSG. (2001). *Disaster Management Support Group: "The Use of Earth Observing Satellites for Hazard Support Group." Final Report, NOAA*. Department of Commerce.

Fergus, S., & Zimmerman, M. A. (2005). Adolescent resilience: A framework for understanding healthy development in the face of risk. *Annual Review of Public Health*, *26*(1), 399–419. doi:10.1146/annurev.publhealth.26.021304.144357 PMID:15760295

Ferguson, E. (2013). Personality is of central concern to understand health: Towards a theoretical model for health psychology. *Health Psychology Review*, *7*(1), S32–S70. doi:10.1080/17437199.2010.547985 PMID:23772230

Flach, F. F. (1980). Psycho-biological resiience, psychotherapy and the creative process. *Comprehensive Psychiatry*, *21*(6), 510–519. doi:10.1016/0010-440X(80)90054-1 PMID:7428355

Fletcher, D., & Sarker, S. (2012). A grounded theory of psychological resilience in Olympic champions. *Psychology of Sport and Exercise*, *13*(5), 669–678. doi:10.1016/j.psychsport.2012.04.007

Frankenberg, E., Sikoki, B., Sumantri, C., Suriastini, W., & Thomas, D. (2013). Education, vulnerability and resilience after a natural disaster. *Ecological Sociology*, *18*(2), 16. Retrieved from http://www.ncbi.nlm.nib.gov/pmc/articles/PMC4144011/

Garmezy, N. (1971). Vulnerability research and the issue of primary prevention. *The American Journal of Orthopsychiatry*, *41*(1), 101–116. doi:10.1111/j.1939-0025.1971.tb01111.x PMID:5539483

Garmezy, N. (1991). Resilience and vulnerability to adverse developmental outcomes associated with poverty. *The American Behavioral Scientist*, *34*(4), 416–430. doi:10.1177/0002764291034004003

Geography Subject Guide, I. B. (2011). *The characteristics of killers: Earthquakes, typhoons, droughts and nuclear accidents.* Retrieved from http://geographylaunchpad.weebly.com/1 characteristics-of-hazards.html

Glantz, M. D., & Johnson, J. L. (1999). *Resilience and development: Positive life adaptations*. New York: Kluwer Academic/Plenum Publishers.

Goodman, R., & West-Olatunji, C. A. (2008). Transgenerational trauma and resilience: Improving mental health counselling for survivors of Hurricane Katrina. *Journal of Mental Health Counseling*, *30*(2), 121–136. doi:10.17744/mehc.30.2.q52260n242204r84

Gordon, E., & Johnson, J. L. (1994). Epilogue: Educational resilience: Challenges and prospects. In E. Gordon (Ed.), *Educational resilience in inner-city America: Challenges and prospects* (pp. 199–194). Hillsdale, NJ: Erlbaum.

Hopf, S. M. (2010). *Risk and resilience in children coping with parental divorce*. Dartmouth Undergraduate Journal of Science.

Isomaa, R., Väänänen, J.-M., Fröjd, S., Kaltiala-Heino, R., & Marttunen, M. (2013, August). is low? Low self-esteem as an indicator of internalizing psychopathology in adolescence. *Health Education & Behavior*, *40*(4), 392–399. doi:10.1177/1090198112445481 PMID:22872582

Joseph, J. (1994). *The resilient child.* New York, NY: Insight Books.

Klein, R. J. T., Nicholls, R. J., & Thomalla, F. (2003). resilience to natural hazards: How useful is this concept. *Environmental Hazards*, *5*(1), 35–45. doi:10.1016/j.hazards.2004.02.001

Kusky, T. M. (2003). *Geological hazards: A sourcebook.* Westport, CT: Greenwood Press.

López-Marrero, T., & Tschakert, P. (2011). From theory to practice: Building more resilient communities in flood-prone area. Environment and Urbanization. *International Institute for Environment and Development*, *23*(1), 229–249.

Luthar, S. S., Cicchetti, D., & Becker, B. (2000). The construct of resilience: A critical evaluation and guidelines for future work. *Child Development*, *71*(3), 543–562. doi:10.1111/1467-8624.00164 PMID:10953923

Luthar, S. S., Sawyer, J. A., & Brown, P. J. (2006). Conceptual Issues in Studies of resilience: Past, present and future research. *Annals of the New York Academy of Sciences*, *1094*(1), 105–115. doi:10.1196/annals.1376.009 PMID:17347344

Machida, M., Irwin, B., & Feltz, D. (2013). Reliance in competitive athletes with spinal cord injury: The role of sports participation. *Qualitative Health Research*, *23*(8), 1054–1065. doi:10.1177/1049732313493673 PMID:23771633

Marriott, C., Hamilton-Giachritis, C., & Harrop, C. (2014). Factors promoting resilience following childhood sexual abuse: A structural, narrative review of the literature. *Child Abuse Review*, *23*(1), 17–34. doi:10.1002/car.2258

Masten, A. S. (2013). Risk and resilience in development. In P. D. Zelado (Ed.), Oxford Handbook of Developmental Psychology: Vol. 2. Self and others (pp. 579-607). New York, NY: Oxford University Press.

Masten, A. S., Cutuli, J. J., Herbers, J. E., & Reed, J. M.-G. (2007). Resilience in development. In The Oxford Hand Book of Positive Psychology. Oxford, UK: Oxford University Press.

Masten, A. S., & Obradovic, J. (2006). Competence and resilience in development. *Annals of the New York Academy of Sciences*, *1094*(1), 13–27. doi:10.1196/annals.1376.003 PMID:17347338

Masten, A. S., & Reed, M. J. (2002). Resilience in development. In C. R. Snyder & S. S. Lopez (Eds.), *Handbook of positive psychology* (pp. 74–78). New York: Oxford University Press.

National Ocean Service. (2016). *What is coastal resilience?* Retrieved from http;//oceanservice.noaa.gov/ecosystems/resilience/welcome.htm

Nigerian Monitor. (n.d.). *List of Nigerian States at risk of experiencing massive earthquakes before 2020.* Retrieved from www.nigerianmonitor.com/nigerian-states-risk-experiencing-massivarthquake-2020-see-full-list/

Okorodudu-Fubura, M. T. (2013). *Environmental disasters in Nigeria: Policy and management.* Paper presented at the National Conference on the Environment (House of Representative Committee on the Environment), Transcorp Hilton Hotel, Abuja.

Ozer, E. J., Best, S. R., Lipsey, T. L., & Weiss, D. S. (2003). Predictors of posttraumatic stress disorder and symptoms in adults: A meta-analysis. *Psychological Bulletin*, *129*(1), 52–71. doi:10.1037/0033-2909.129.1.52 PMID:12555794

Pęcillo, M. (2016). The concept of resilience in OSH management: A review of approaches. *International Journal of Occupational Safety and Ergonomics*, *22*(2), 291–300. doi:10.1080/10803548.2015.1126142 PMID:26652938

Peres, J., Moreira-Almeida, A., Nasello, A., & Koinig, H. (2007). Spirituality and resilience in trauma victim. *Journal of Religion and Health*, *46*(3), 343–350. doi:10.1007/s10943-006-9103-0

Pfefferbaum, R. L., Pfefferbaum, B., van Horn, R. L., Klomp, R. W., Norris, F. H., & Reissman, D. B. (2013). The communities advancing resilience toolkit (CART): An intervention to build community resilience to disasters. *Journal of Public Health Management and Practice*, *19*(30), 250–258. doi:10.1097/PHH.0b013e318268aed8 PMID:23524306

Pryor, P. (2012). *HaSPA (Health and Safety Professional Alliance). The core body of knowledge for generalist OHS Professionals. Tullamarine.* VIC Safety Institute of Australia.

Reivich, K. J., Seligman, M. E. P., & McBride, S. (2011). Master resilience training in the U.S. Army. *The American Psychologist*, *66*(1), 25–34. doi:10.1037/a0021897 PMID:21219045

Reuters. (2012). *Nigeria floods kill 137, displace thousands*. Retrieved from https://en.wikipedia.org/wiki/2012_Nigeria_floods

Richardson, G. E. (2002). The metatheory of resiience and resiliency. *Journal of Clinical Psychology*, *58*(3), 307–321. doi:10.1002/jclp.10020 PMID:11836712

Ruther, M. (1985). Resilience in the face of adversity: Protective factors and resilience to psychiatric disorders. *The British Journal of Psychiatry*, *147*(6), 598–611. doi:10.1192/bjp.147.6.598 PMID:3830321

Ruther, M. (1987). Psychosocial resilience and protective mechanism. *The American Journal of Orthopsychiatry*, *57*(3), 316–331. doi:10.1111/j.1939-0025.1987.tb03541.x PMID:3303954

Rutther, M. (1979). Protective factors in children's response to stress and disadvantage. In M. W. Kent & J. E. Rolf (Eds.), Primary prevention of psychopathology: vol. 3. Social competence in children (pp. 49-74). Hanover, NH: University Press of New England.

Schulze, E. D., & Mooney, H. A. (1993). *Biodiversity and ecosystem function*. Berlin, Germany: Springer. doi:10.1007/978-3-642-58001-7

Tawari-Fufeyin, P., Megbuwe, P., & Adams, O. G. (2015). Some aspects of a historic flooding in Nigeria and its effects on some Niger-Delta Communities. *American Journal of Water Resources*, *3*(1), 7–16.

The Conversation. (n.d.). Retrieved from theconversation.com/black-americans-may-be-more-resilient-to-stress-than-the-white-americans

Treasure, D. C., Lemyre, P. N., Kuezka, K. K., & Standage, M. (2007). Motivation in elite-level sport. In Intrinsic motivation and self-determination in exercise and sports (pp. 153-165). Champaign, IL: Human Kinetics.

Tugade, M. M., & Fredrickson, B. L. (2004). Resilient individuals use positive emotions to bounce back from negative emotional experiences. *Journal of Personality and Social Psychology*, *86*(2), 320–333. doi:10.1037/0022-3514.86.2.320 PMID:14769087

UN/ISDR. (2002). *Living with risk: A global review of disaster reduction initiatives. Preliminary version prepared as an inter-agency effort co-ordinated by the*. Geneva, Switzerland: ISDR Secretariat.

UNISDR. (2009). *Terminology on disaster risk reduction. United Nations international strategy for disaster risk reduction*. Retrieved from http://www.nat-hazards-earth-syst-sci.net/13/2707/2013/nhess-13-2707-2013.pdf

United Nations. (2015). *Strengthening resilience to natural disasters crucial, speakers stress as second committee takes up sustainable development*. Seventieth Session, 13th& 14th Meetings. Retrieved from http://www.un.org/press/en/2015/gaef3427.doc.htm

Watson, P. J., Friedman, M. J., Gibson, L. E., Ruzek, J. I., Norris, F. H., & Richie, E. C. (2003). Early intervention for trauma-related problems. In R. J. Uesano & A. E. Norwood (Eds.), *Trauma and disaster: response and management*. Washington, DC: American Psychiatric Publishing.

Zimmerman, M. A., & Brenner, A. B. (2010). Resilience in adolescence: Overcoming neighbourhood disadvantages. In J. Reich, A. J. Zautra, & J. S. Hall (Eds.), *Handbook of Adult Resilience* (pp. 283–308). New York, NY: Guildford Press.

KEY TERMS AND DEFINITIONS

Emotional Trauma: A strong, negative, and painful feeling caused by stress and losses.

Flood: A situation where excess water from the sea, river, or stream flows to areas that are usually dry land.

Hazards: Negative consequences of weather conditions.

Natural Forces: Events that occur in the environment due to nature and one that is not under the control of human beings.

Natural Hazards: Events that are potentially dangerous and are caused by natural forces in the physical environment.

Nigeria: The name of a country situated in West Africa.

Psychological Resilience: Behaviours that a person exhibits to achieve normalcy after a stressful experience.

Wellbeing: A state of happiness and positive health.

Chapter 9
Preparing for Refugee Exodus in Crisis:
Poland Case Study

Magdalena Denham
Sam Houston State University, USA

Scott Vautrain
Memorial Hermann Hospital System, USA

ABSTRACT

Using macro-, exo-, meso-, and micro-level framework for assessing interdependencies in crisis management system in Poland, this chapter addresses expansion of planning and preparedness to novel hazards and threats such as precipitated mass movements of populations due to environmental changes. The Ukrainian refugee crisis served as backdrop for Poland's crisis management system testing. Benefits of planning and preparedness for mass displacements highlighted (a) an emerging all-hazard approach to crisis management, (b) proactive threat identification and assessment, (c) issuance of acts, laws, and regulations based on novel threats, (d) comprehensive multi-jurisdictional exercises, and (e) increasing importance of mutual aid agreements.

INTRODUCTION

I once had a record of whale sounds, I swear I understood. It didn't matter what world they were under, what language, what depth of water divided, the song went on and on. What I mean is: none of this is chaos. Immigration, cross the river, the blood of us. It goes like this: water, land, water, like a waltz. – From the Whale and the Waltz Inside of It by Ada Limón

The objective of this study was to determine whether and in what ways situating mass migration hazards and threats within a country's domestic crisis management system could enhance resilience to address future migratory spikes. To that end, the chapter illustrates the integration of mass migration policy into the national framework of crisis management in the Republic of Poland. Using the conflict in Ukraine

DOI: 10.4018/978-1-5225-3194-4.ch009

between 2013 and 2014, the authors analyzed planning and preparedness processes undertaken in Poland to face challenges of potential refugee exodus out of Ukraine across Shengen lines into European Union territory. Planning and preparedness processes are explored within the framework of domestic crisis management and at the intersection of macro-level, supranational structures.

Data were drawn from structured interviews conducted with officials from the Polish National Police (PNP) Office of Commander-on-Chief, Bureau of International Cooperation, and operational command responsible for operational planning for mass refugee exodus from Ukraine in 2014. To triangulate data (for example, Anfara, Brown, & Mangione, 2002), documents including news articles, governmental reports, legal acts, regulations, and after-action reports served as additional collection sources.

The Bronfenbrenner's model (Bronfenbrenner, 1995) framed our study. The model proposes four subsystems (a) micro, (b) meso, (c) exo, and (d) macro, all of which comprise the overall ecological environment for human development. The micro-system can be described as set of relations between the entity and the environment in an immediate setting. Interactions and linkages immediately outside of proximate setting or community constitute the meso-system. Meso-system resides within wider cultural contexts represent an exo-system. Finally, larger global environment constitutes the layer called macro-system. Because understanding ecological interdependence means understanding relationships (Capra & Luisi, 2014), dividing research space into micro-, meso-, exo-, and macro-systems is useful when examining organizational contexts; this is especially true within the domain of crisis management across domestic/internal, supranational, and human security arenas.

BACKGROUND

Mass migrations have been exacerbated by ecological factors, subsequent conflicts, and destabilization. Mass population displacements are no longer slow-developing phenomena, but are precipitated by greater interconnectedness, mobility, communication, and transportation pathways. Management of ongoing and future displacements necessitates an adoption of internal, supranational, and human security perspectives. When discussing differences in crises and approaches in managing them, Quarantelli (1993) contended that *consensus crises* (brought about by natural and technological agents) and *conflict crises* (brought about by international human agency such as riots and civil disturbances) differed from *ecological crises* (such as land subsidence, climatological pollution, famines, or draughts). The dissimilitude resulted from their slow onset/slow-progressing nature, as such "Chronic ecological problems very seldom create an immediate emergency or crisis" (Quarantelli, 1993, p. 26). Subsequently, slow-progressing phenomena foster social familiarity and tend to be relegated to local level social problems like poverty or homelessness rather than viewed as potential unexpected crises. Thus, ecological issues have historically taken back seat in traditional approaches to emergency management that have relied on the distinction between natural, or technological hazards and man-made threats. However, cumulative sum of slow progressing phenomena peak eventually. Chronic ecological issues can, and do create immediate emergencies and crises. For example, the Big Smoke wrapped Londoners in December of 1952. The five-day catastrophic event caused thousands of immediate respiratory-related deaths and triggered mortality and morbidity effects extending over time (Bell & Davis, 2001). The crisis reminds us about consequences of unanticipated convergence between weather and chronic pollution. Today, seemingly slow evolving hazards and threats may gain culminating urgency much faster due to increasing mobility and interconnectedness. This in turn has dramatically changed the way ecological issues spread beyond a specific and local social domain issue (e.g., Capra & Luisi, 2014).

Moreover, slow progressing crises may act as *hazard and threat multipliers.* Lanicci, Ramsay, & Murray (2017) underscore that chronic environmental challenges such as reduced access to fresh water, impaired food production, health catastrophes, or population and land loss could trigger escalation of resource-based conflicts and mass migration. Inarguably, migration and population displacement have been sempiternal to human history (see Gatrell, 2013; Haddad, 2008). However, global exploitation affects the scale of human migration both internally and beyond borders. Displaced populations have become more mobile, more interconnected, and better informed. Hence, they are quicker to mobilize and make better decisions such as risk assessments regarding danger during travel, sustained communication with families left behind, or access to critical networks such as transportation. Projections based on the worldwide threat assessment to the U.S. intelligence community (Clapper, 2016) underscored:

... historic increase in mobility and displacement. The growing scope and scale of human displacement will probably continue to strain the response capacity of the international community and drive a record level of humanitarian requests. At the same time, host and transit countries will struggle to develop effective responses. (p.15).

Viewed through this lens, it may be beneficial to view mass migration spikes as a new type of permanent crisis to manage. Hence, the "new normal" may call for strategies incorporating mass migration flows into general domestic emergency management frameworks. This may also be the first step to systematic approaches eventually situating population displacement along the crisis management cycle from planning and preparedness, through response, recovery, and mitigation.

Human migration raises deep human security and humanitarian dilemmas because "People fleeing serious economic and social distress or environmental degradation, stranded migrants, trafficked migrants, and people who have developed needs as a result of suffering trauma or violence during transit all have protection rights" (Betts, 2009 p. 162). Meanwhile, the European Council of Foreign Relations (ECFR) admitted that "European efforts at crisis management in 2014 were overshadowed by the EU's inability to develop a coherent, effective, or humane policy on the thousands of migrants trying to cross the Mediterranean" (2015, para. 1). Similarly, Sherwood, Smith, Davies, and Grant (2014) reported that the United Nations High Commissioner on Refugees (UNHCR) deplored the abandonment by Brussels to find solutions to massive deaths of refugees from Syria, Libya, and Sahel at sea. The 2014 report for the International Organization for Migration (IOM), cited 3,027 deaths in the Mediterranean region alone as compared to 700 in 2013; global estimates of refugee deaths approximate the number to 4,077 even though due to challenges in tracking actual numbers might have been higher (Brian & Laczko, 2014). When discussing worldwide refugees and their fate in the international landscape, Haddad (2008) summarizes that "they are the human reminder of the failings of modern international society" (p. 3).

Markedly, mass migrations transcend concerns about human security and trigger supranational as well as internal security concerns. Population flows may threaten community cohesion in destination countries and cause social, cultural, political, and economic tensions (Sasnal & Uzelac, 2014, Szpak, 2016). In that perspective, they call for policy formation that converges on the nexus of supranational, internal/homeland, and human security elements. The displaced population of approximately 10 million Syrians and Iraqis across the Middle East constituted "not only the moral dilemma for Europeans, it was a future security threat in the making" (Sasnal & Uzelac, 2014, p. 1). For example, in case of Lebanon, the country with a 4.5 million population admitted from 1.1 to 1.2 million refugees (Haase, 2014; Sasnal & Uzelac, 2014). Reportedly, Lebanon lacked crisis management structures to plan refugee camps, create temporary places to stay, provide basic medical care, employment, education, and the overall infrastructure for the mass influx of refugees (Haase, 2014; Sasnal & Uzelac, 2014). Concurrently, *FRONTEX* estimated 280,000 foreigners crossed over to European Union in 2014 (Majchrzak-Czarnecka, Rafalik, & Zasadzińska-Baraniewska, 2015).

Cianciara (2015) admitted that low effectiveness of peacemaking coupled with debatable effects of military interventions exacerbated situations encountered by the EU in 2015 and 2016 constituting merely a prelude of impending exodus. From the sending countries, ongoing hazards stem from successively and radically worsening climatological, economical, and social conditions magnified by increasing insecurity threats in the Middle East and North Africa. On the receiving end, "the refugee brings to the fore the clash between pluralism and solidarism, communitarianism and cosmopolitanism, sovereign rights and human rights" (Haddad, 2008, p. 3). The clash breeds prejudice and hostile social stance towards refugees and foreigners in general. Beliefs that migrant waves represent temporary and local crisis centered mainly on southern Europe are naïve; they manifest tendency to apply quick fixes such as patrols of ships on the Mediterranean represented by Frontex Operation Triton, barbed wire fences in Hungary, revocation of Shengen, or blockage of Eurotunnel. All the while the UNHCR most modest predictions anticipated more than 1,000,000 individuals relocating to EU in 2016 (a 22% increase from 2015) – scale unprecedented in postwar Europe (Cianciara, 2015).

MAIN FOCUS OF THE CHAPTER

The focus of this chapter is to assess sudden migration flows within the function and structure of the crisis management system of Poland. This investigation is worthwhile for several reasons. First, in supranational terms "the more that policy makers and practitioners in different countries can learn from others' strategies and approaches, the greater will be the shared pool of knowledge and this knowledge will ultimately make people safer" (Morag, 2011, loc. 241). Second, Poland occupies a strategic defense-line border position to the east of the EU – a factor related to supranational security. Third, the development of country's crisis management system has been a relatively new phenomenon – allowing for developmental internal/homeland security lens. Finally, Poland has a long-standing tradition of democracy and protection of human rights; indeed, its constitution of 1791 was the first in democratic constitution in Europe and only the second in the world.

Poland as a Shengen Signatory Issues

Shengen agreements of 1990 (European Commission [EC], 2014; Grabbe, 2000) by selected European Union (EU) member states introduced *acquis communautaire*, a mechanism for eliminating internal border controls while regulating external boundaries through border control systems, asylums, visas, immigration, and cooperation on civil justice practices. Most EU countries and selected non-EU states (that is, Iceland, Norway, Switzerland, and Lichtenstein) have joined the Schengen zone to date (EC,

2014). Poland which entered the Schengen Area on December 21, 2007 (for example., Zdanowicz, 2009) became a holder of the longest strip of the eastern frontier of the zone protected by a singular member state; as illustrated by Figure 1, its 746 miles-long border extends from the northern Baltic coast to the southern Carpathian Mountains (Iglicka & Ziołek-Skrzypczak, 2010).

Thus, Poland emerged as principal frontier country between the EU and the defense line for migrations from the legacy countries of the former Soviet Union as well as from more distant Asian regions. The Polish Minister of Interior Affairs described this position of Poland as "location on the tectonic plate" (Czuchnowski & Wroński, 2014). Specifically, the Polish border with Ukraine (not an EU state itself) occupies 329 miles (Figure 2); of the 16 administrative regions of Poland (voivodships), 2 are located along the Ukrainian border. Those regions are central-eastern Lubelskie (voivodship capital Lublin – location of regional administrative authorities) and Podkarpackie (voivodship capital Rzeszów). On the Ukrainian border, there are eight points of entry (POEs) for pedestrians as well as 6 POEs for trains.

Crisis Management System in Poland Issues

With respect to Polish policy on national safety, crisis management represented the greatest challenge throughout the 1990s and better part of 2000s (Bynander, Chmielewski, & Simons, 2008; Dworzecki, 2012; Matczak & Abgarowicz, 2013; Sharpak, 2011). There was a general lack of agreement on systematic solutions to public safety in the context of non-military crises (Gierszewski, 2011). The Millennium Flood of 1997 that affected 2,592 towns, of which 1,362 were totally flooded, destroyed 480 bridges, led to displacement of 162,500 people and left 54 dead in its wake revealed the existing system of crisis management in Poland was ineffective and dysfunctional (Dawidczyk, Gryz, & Koziej, 2006; Dworzecki, 2012). From the emergency management perspective, catastrophic flooding in 1997 could be described as what Butler termed *focusing events* (Butler, 2012). It revealed lack of administrative coordination structures and reactive improvisation during response (Cielecki, 2009). In the direct aftermath

Figure 1. Map of Poland within the Shengen Zone
History of the European Union, 2017.

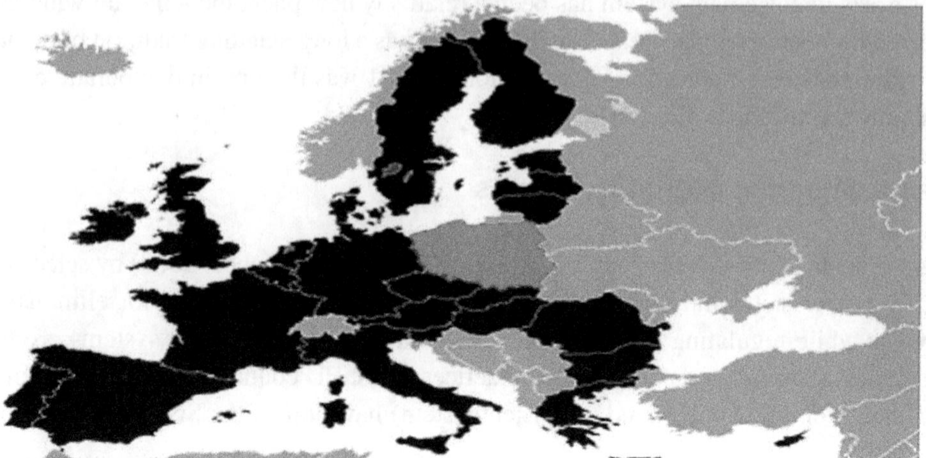

Figure 2. Map of Poland and Ukraine illustrating shared border demarcation
Poland Ukraine Locator, 2018.

of the Millenium Flood, the Ministry of Interior and Administration created the Office of Emergency Management and Civil Protection in 1998 (Dworzecki, 2012). The Office was short lived and resolved by early 2000, but its placement under the Ministry of Interior delineated an important transition in addressing civil security. Specifically, the focus shifted beyond national civil defense to a non-military model (Śnierzyński, 2009). Bynander, Chmielewski, and Simons (2008) explained that:

Polish legal system defines the aim of civil defense as protecting people, factories, public facilities, and cultural property, as well as rescuing and giving help to victims in times of war and fighting against and cleaning up the effects of elemental disasters and threat to the environment. (p. 35).

Throughout 2000s continuous constitutional challenges prevented adoption of a bill on citizen safety and crisis management. Then, in the winter of 2006, the country was rocked by the largest structural disaster in its modern history whereby 700 visitors were affected, of whom 170 sustained injuries and 65 including 13 foreigners perished. The collapse occurred during the International Trade Fairs of Carrier Pigeons hosted in Katowice at a large exhibition hall which collapsed on participants (Chmielewski, 2007; Chmielewski & Kuźmiński, 2006). Poor coordination of response efforts was heavily scrutinized (Krajewski-Siuda & Romaniuk, 2006). The catastrophe underscored mounting criticism that "during seventeen years of Polish transformation, until present moment, there is no - even on the level of legal act proposal - institutional structure of national crisis management system" (Chmielewski & Kuźmiński, 2006, p. 17).

Inarguably, the tragedy put the lack of integrated crisis management system back on the public agenda (e.g., Krajewski-Siuda & Romaniuk, 2006) and was a catalyst for the endorsement of crisis management bill (prepared during the 2005 session of the Polish Sejm in 2005, [Dworzecki, 2012]) by the Council of Ministers in June of 2006. Hence, the Act on Crisis Management was passed and signed by the President of the Republic of Poland on April 30, 2007 and ratified August 22, 2007. It is noteworthy that under new Polish system *crisis management* is a preferred term (versus emergency or disaster management) to frame the approach and the management of domestic incidents and the protection of civil security.

Policy Formation to Address Migration Flows From Ukrainian Crisis

Within the canvass of increasing population dislocation lurking south of the EU defense lines in 2014, the Republic of Poland started mobilizing for a strategy adequate to meet the challenges of potential humanitarian crisis lurking in the east. The destabilization of Ukraine in the latter part of the 2013 and its escalation since February of 2014 after the *EuroMajdan* upheaval (Adamczyk, 2014) caused concerns of potential influx of refugees; it ushered intensified preparations to address issues of mass immigration resulting from an armed conflict into eastern Poland.

Risk Assessment and Monitoring

Poland remained optimistic about relative increases in security since the fallout of the Soviet Union, globalization, and liberalization. Concurrently, it did not underestimate the risks brought about by aggressive resolution of conflict among countries through sudden military action so close to the border of the EU and NATO (Bocharova & Giza, 2015; Skusiewicz, p. 275). EuroMajdan events in Kiev between February 18 to 21, 2014 left 82 killed and 622 wounded. Those numbers reported by the Health Ministry of Ukraine were questioned compared to 780 dead (of which many unidentified or "disappeared") by accounts of volunteers and medics from the scene (Benedyczak, 2014). Subsequently, asylum seeking numbers among Ukrainian citizens increased dramatically from 46 petitions in 2013 to 1,265 petitions in 2014 (Polska Agencja Prasowa [Polish Press Agency], PAP, 2014). Numbers reported by the UNHCR (2014) and illustrated by Figure 3, pointed to 1,565 asylum petitions in Poland from January 2014 to September 2014 and the overall displacement of 275,489 people from the onset of the Ukrainian crisis. Mass Ukrainian migration was noted mainly into Russia with UNHCR (2014) reporting 814,000 entrants of whom 121,000 seeking refugee or asylum status and Russian media (RIANovosti, 2014) estimating 875,000 entrants of whom 300, 000 seeking refugee status or asylum. Nonetheless, the westward migration was anticipated based on risk assessments of potential geographical shift of Russian military operations within Ukraine. Resource-based exploitation of Crimea offered Russia strategic access to oil resources and transportation routes and westward aggression remained probable. Moreover, Poland relied on European think tank experts (for example., European Policy Centre) predicting large refugee flows would self-direct across the border with Poland, Slovakia, or other EU countries rather than seeking refuge in Russia (Benedyczak, 2014, p. 16).

Strategic Operation Plan (SOP-10) Issues

By the end of January 2014, the Border Management Team, an advisory body to the Chief of the Polish Council of Ministers, comprising among others the Commander-in-Chief of the Straż Graniczna (Border Guard [BG]) prepared a concept of operations in the event of a sudden flow of large numbers of foreigners onto the territory of Poland. Based on ongoing monitoring, risk assessments, and analyses of the Rządowe Centrum Bezpieczeństwa (Government Security Center [GSC], 2014), the Minister of Interior Affairs - in charge of the Crisis Management Task Force at the Ministry of Interior Affairs (Zespół Zarządzania Kryzysowego w Ministerstwie Spraw Wenętrznych) - approved the Strategic Operation Plan (SOP) addressing organizing, admitting, transporting, and managing of Ukrainian refugees in March of 2014. The Plan was based on the concept of operations (CONOPS) submitted by the Border Management Team and expressed though the SOP - 10 (*SPO-10): Operations in the Event of Mass*

Figure 3. Map of Ukraine illustrating interior displacement of people in numbers as well as cross-border migration in numbers as of September 18, 2014
Source: United Nations High Council on Refugees, 2014.

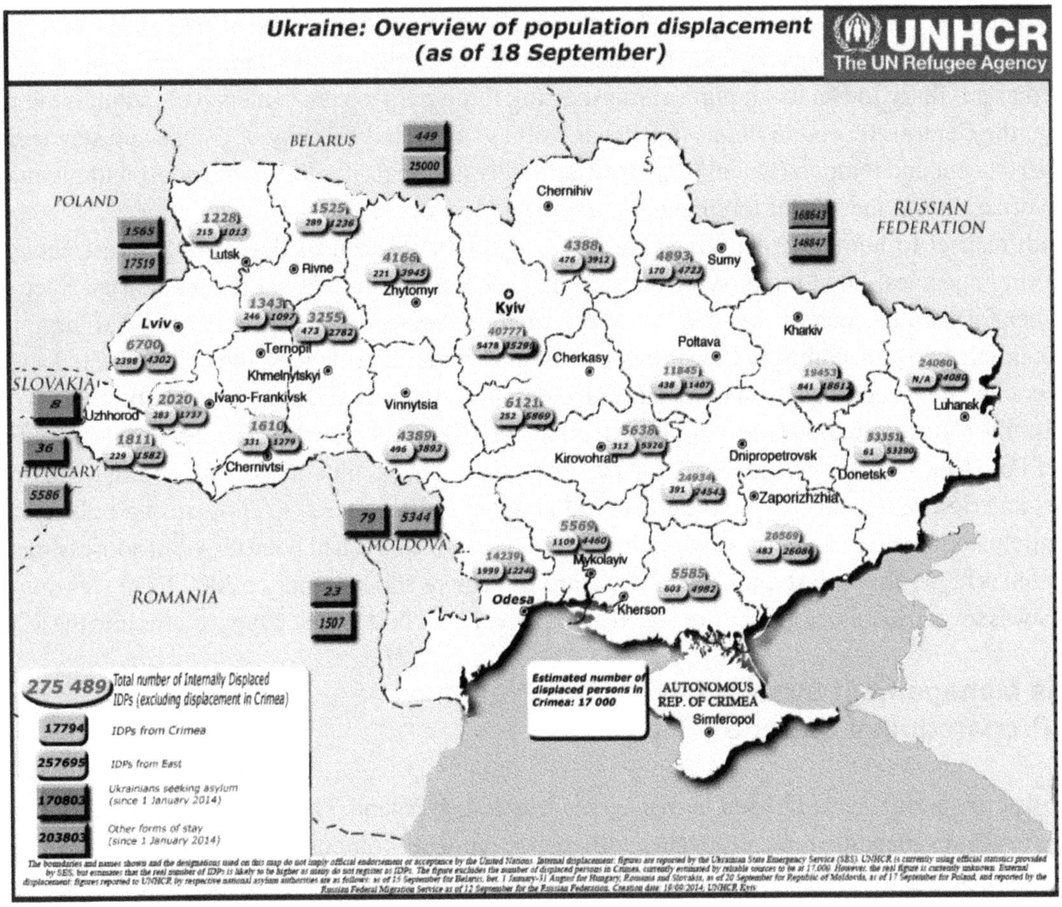

Influx of Foreigners onto the Territory of the Polish Republic. The SPO-10 became part of the National Crisis Management Plan legislated by the Polish Council of Ministers on July 23rd, 2013 (GSC, 2014). The plan delineated responsibilities and authorities of main crisis management entities in the event of escalating migration flows. It considered establishment of new refugee centers, to include camps in Lubelskie and Podkarpackie regions (territories bordering Ukraine).

Operational Authorities for Mass Influx of Refugees

Under the SPO-10, standard operations related to the protection of national borders, border traffic control, and prevention of illegal immigration in Poland fall under the jurisdiction of the BG. Thus, the BG jointly with the Urząd do Spraw Cudzoziemców (UdSC, [Bureau for Foreigners]) function as lead national agencies (comparable under the U.S. National Incident Management System [NIMS] to Emergency Support Functions [ESF]). In terms of safety and security, the scope of the role and engagement of the Polish National Police (PNP) in the planning and management processes related to mass influx of Ukrainian refugees, was defined by the Minister of Interior Affairs as auxiliary to BG. Under general preparedness and prevention missions areas of the PNP in the event of the mass influx of refugees from Ukraine the following capabilities had been addressed (a) the increased activity of organized trans-border crime; (b) potential corruption among the Polish uniformed services; (c) forced attempts to cross the border; (d) participation of Ukrainian services in facilitating illegal activities across the national border; and (e) potential for border crossings beyond designated POEs. Consequently, PNP forces and resources would be used as auxiliary to BG to support (a) transporting foreigners to stay centers; (b) maintaining public order at the Central Reception Points (CRPs) as well as safety and security of camps and stay locations both permanent and temporary; and (c) protecting safety and security during transport and maintaining public order during the stay of foreigners in designated locations.

With respect to humanitarian relief, the role of UdSC consisted of developing a vast network of supporting agencies, large majority of them NGOs such as Caritas, Polish Humanitarian Section, or volunteer formations such as Association Malta Medical Services. Because of potential large public appeal anticipated to accompany humanitarian crisis of Ukrainian refugees, the UdSC created a special email connection for private businesses and corporations that handled donations and other forms of aid. In 2014, the Centrum Badan Opinii Społecznej (CBOS, [Center for Public Opinion Polling) confirmed 63% of Polish citizens supported Ukrainian protest in Kiev, 75% sympathized with Ukraine seeking EU access, and 64% underscored Ukraine as vital to national Polish interests. Thus, strong public support was anticipated in the event of mass migration. Moreover, the UdSC had been tested in addressing needs of sudden refugee flows (last quarter of 2007 from Chechnya and summer of 2009 from Georgia). The agency was severely stressed and incorporated its experiences as best practice suggestions into the SOP-10.

Crisis Management System Levels and Planning and Preparedness Under SOP10

Figure 4 portrays four levels of the crisis management system in Poland. Those levels represent components of emergency response agencies integrated with respective-level components of public administration in Poland for national crisis management (with attention to border crisis management during mass migration). The assessment of planning and preparedness processes with respect to expected Ukrainian exodus in 2014 was conducted based on efforts devised at macro-, exo-, meso-, and micro-levels respectively.

Macro Level Planning and Preparedness

Regionally, the situation in Ukraine was addressed during the special meeting of the Visegrad Group (V4) comprising countries of the Central European region (Czech Republic, Hungary, Poland and Slovakia). The Ministers of the Interior of the V4 met in Budapest on February 25, 2014 to sign a declaration of cooperation under the circumstances of massive inflows of Ukrainian refugees. They assesesed potential for increased activity on the border of three of those countries with Ukraine (Poland, Slovakia, and Hungary) and planned for lodging and medical assistance. In addition, Czech Republic (not sharing the border with Ukraine) offered additional humanitarian help if needed (Klepner, 2014). The collaboration on the Polish-Ukrainian Border specifically included the System Wczesnego Ostrzegania (SWO [Early Warning System]) about crisis situations. It was implemented in December of 2013 to allow for continuity of operations planning (COOP) and for sustained throughput during potential escalation. The BG Commander-in-Chief and the Office of the Border Ombudsman facilitated 375 joint meetings between Polish and Ukrainian BG and Customs forces in late 2013. Moreover, 809 joint patrols of Polish and

Figure 4. Chart depicting the flow of administrative levels, coordinating elements, and operational elements applicable in the sudden influx of Ukranians into the territory of the Republic of Poland
Source: Authors.

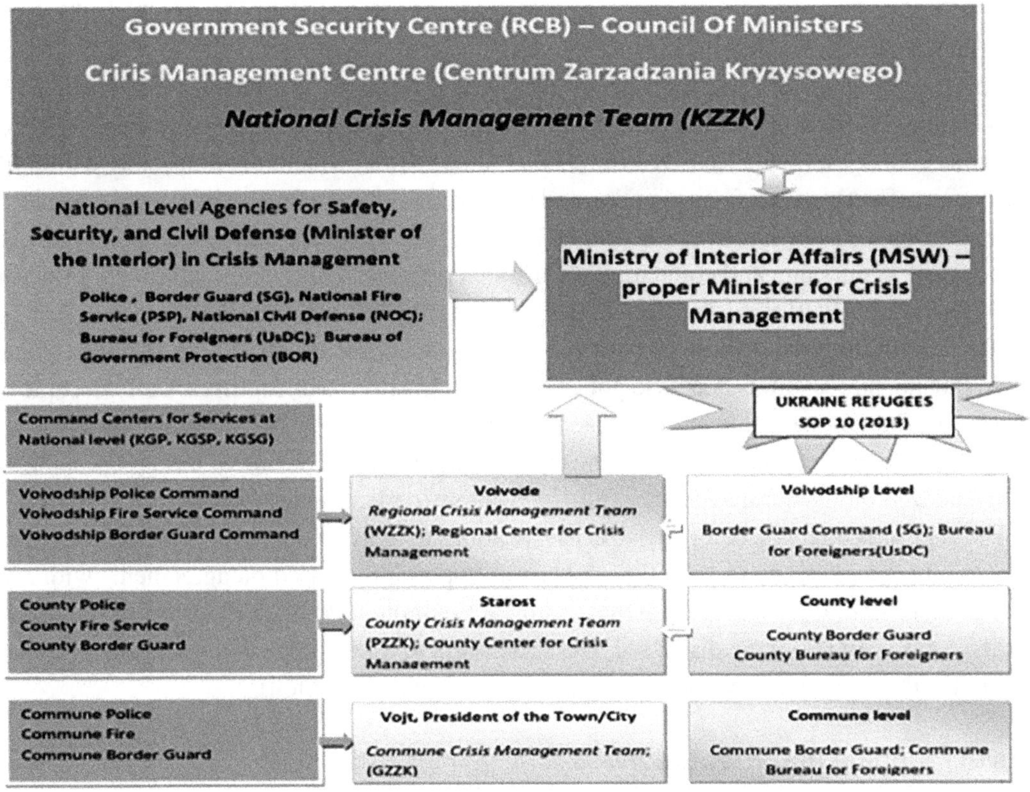

Ukrainian BG were conducted in 2013 of which 467 on the Polish side and 342 on the Ukrainian side. Finally, the information sharing points on two border crossings Dorohusk-Jagodzin (3,090 received and 2,877 transmitted reports) and Korczowa-Krakowiec (2,805 received and 3,799 transmitted reports) were heavily involved in intelligence gathering, processing, and dissemination.

Beyond that, under the North Atlantic Treaty Organization (NATO), the cooperation and collaboration of Poland in crisis management as a member state are delineated in the NATO Crisis Reaction System (NCRS). NCRS defines actions and procedures for crisis management system, to include external threats to security along five dimensions: (a) alarm levels; (b) prevention options; (c) procedures mitigating surprise actions; (d) procedures against aggression; and (e) crisis response measures. Ministers in charge of specific government sectors, Directors in charge of appropriate bureaus under the ministerial levels as well as Voivods (equivalent to Governors) had been tasked to specify and adapt actions and procedures to the scope of their own authorities. The Director of GSC had been coordinating those planning and engagement processes (classified information).

Information and intelligence sharing of the PNP, BG, and UdSC with macro-level entities focused on threat assessments and analysis of the situation in the Ukraine to account for potential escalation of conflict. The Komenda Główna Policji (KGP [Central Command of the PNP]) conducts international cooperation based on national legislations, agency-specific legal acts, and agreements facilitating local cooperation in sub-border regions. At the PNP, the Bureau of International Cooperation at the KGP - described as "one stop shop" - is a point of contact (POC) and a fusion center for all channels of international policing information exchange. The structure coordinates and monitors all international exchange activities (i.e., operational, non-operational, and training). In addition, it fulfills the role of the National Bureau of Interpol in Poland.

The cooperation entailed the exchange of information through: (a) Shengen Information System (SIS) and National Bureau of Supplementary Information Request at the National Entries (SIRENE) for Shengen zone countries; (b) EUROPOL (European Police Office) which function solely within EU and aims to enhance effectiveness of coordination and collaboration among member states in preventing and combating transnational organized crime); (c) INTERPOL (international policing organization assisting law enforcement in pursuing and combating all criminal activity); (d) coordination within the network of Liaison Officers of PNP and BG active in the EU (France, Holland, Germany, or Great Britain) and beyond EU (i.e., Belorussia, Russia, Ukraine) as well as cooperation with foreign Liaison Officers accredited in Poland; and (e) direct access to police databases (lost persons, fugitives, dactyloscopic charts/AFIS, DNA profiles, stolen vehicles, and stolen documents).

During the engagement phase of addressing crisis related to sudden influx of foreigners at the macro-level, PNP and BG could request assistance from Interpol's expert groups called Interpol Response Teams (IRTs). Those groups might be delegated to crisis locations to assist and to coordinate access to Interpol resources to all engaged responders in the crisis area. Importantly, based on agreements with *Frontex,* resources such as *RABIT,* resources from the Technical Equipment Pool (TEP) or use of European Border Guard Teams (EBGTs) were discussed. BG had access to intelligence and information analysis by Frontex via ICONet (i.e., information and monitoring of illegal migration).

Exo-Level Planning and Preparedness

Preparedness and planning actions of the PNP, BG, and UdSC at the macro-level were realized through (a) the participation of the Commanders-in-Chief in the meetings of the National Crisis Management Team under the Ministry of the Interior; (b) analysis and assessment of existing forces and resources of the PNP, BG, and UdSC discussed during the meetings of the PNP Commanders-in-Chief Crisis Management Teams; (c) national level coordination and collaboration of designated Central Commands with appropriate entities within the scope of the crisis management system (for example, GSC, ministries, institutions and services with authority defined by the legislature on crisis management and which implement the elements of the National Crisis Management Plan at the national level). The planning and preparedness included involvement of agencies' forces and resources in national exercises (discussed further) organized at different levels and whose best practices and lessons learned may be used towards addressing the sudden influx of foreigners onto the territory of Poland.

The Commanders-in-Chief are members of the Crisis Management Team at the Ministry of Interior (ZZK MSW); the Team is a coordinating structure tasked with planning and management related to the mass influx of Ukrainian citizens to Poland. Thus, the Commanders-in-Chief were directly involved in the planning process of the Ministry (e.g., advised on legal acts and participated in the preparation of two policies *The Concept of Operations in the Event of the Sudden and Large Influx of Foreigners from Ukraine into the Territory of the Polish Republic* approved February 28, 2014 and *The Operational Plan of the Ministry of the Interior Concerning the Organization, Reception, Transport, and Residence of Foreigners/Refugees from Ukraine into the Territory of Poland.*

Planning and preparedness at the PNP and BG Central Commands was conducted through the Crisis Management Teams (CMT) of the PNP and BG Commanders-in-Chief. The CMTs were instituted to function as counsel and advisory board at central level for PNP and BG initiation and coordination of actions with respect to crisis management. The management of crisis events at Central Commands was supported by the 24/7 Crisis Management Centers of the PNP and BG. The national events were monitored by on-duty PNP and BG Commands and crisis management elements under the General Staff for PNP and BG Commands (during escalation of events the Operation Centers of PNP and BG Commanders-in-Chief would be activated). With respect to threat levels, threat analysis and capacity assessments were discussed regularly at CMT meetings of the PNP and BG. With respect to information sharing at the national level, the CMTs relied on the Kriminalny System Informacyjny (Criminal Information System – comparable to the NCIC in the USA).

Meso-Level Planning and Preparedness

Markedly, the most activities for preparedness and engagement were conducted at the voivodship (province/region) level. The Voivode being the principal official responsible by law for the safety and security serves as the Chief of the regional civil defense. The Voivode achieves strategic crisis management goals through the coordination of voivodship-level crisis management services. Thus, in case of Ukrainian refugees, the Voivode coordinated with BG, PNP, or UdSC at voivodship levels. In practice, Voivodship Crisis Management Centers (VCMCs) are established at the voivodship level and house representatives of services, institutions, and agencies involved in crisis management.

Meanwhile, independently from VCMCs and depending on the level of engagement, PNP and BG established Incident Command Posts (ICPs) for Incident Commanders in charge of the police or border patrol action or operation. It is at the voivodship level that the PNP and BG were the most engaged in regional planning and operational strategy implementation. Voivodship Commanders-in-Chief/Metropolitan Commanders-in-Chief (a) possess complete and current information about forces and resources at their disposal and forces and resources engaged in the implementation of planned activities and activities in progress; (b) possess current threat assessments in their region; (c) can manage both PNP and BG forces and resources directly under their voivodship jurisdiction and those under the jurisdiction of county/city/and commune (micro-level); (d) coordinate activities directly with Voivods and other engaged agencies and services; (e) issue direct requests to the Commander-in-Chief for additional and necessary PNP and BG forces and resources (beyond voivodships); and (f) can use the PNP and BG forces and resources most efficiently. It is noteworthy that Voivodship/Metropolitan Commanders-in-Chiefs of the PNP and BG can be tasked both by the PNP and BG Commanders-in-Chief as well as by the Voivodes.

Voivodship-level Incident Command Posts (ICPs) of PNP and BG were to be activated based on police and border patrol operation needs. The activation of the ICP consists in expanding the organization of the existing full-time general staff by additional units and specialist positions in criminal investigations, prevention, and support staff relevant to specific operations. The number of ICP positions is predicated on maintaining and securing the COOP in the event of prolonged engagement. Hence, it was planned that PNP and BG forces and resources under the jurisdiction of Voivodship Commanders-in-Chief of the PNP and BG would be utilized in the event of the sudden influx of Ukrainian refugees into Poland's territory.

For example, if local PNP forces and resources in locations of temporary camps and housing would be insufficient, special forces and resources of PNP Voivodship Prevention Brigades as well as Collateral Brigades and Sub-brigades of Prevention Police would be dispatched from Voivodship PNP Commands.

Prevention Police – a specialized function within PNP can be best described as Riot Police (BG operates similar collateral structures). For PNP, those structures are formed at Voivodship PNP Commands/ Metropolitan Command to augment operations of regular Prevention Police Brigades and Sub-brigades. Voivodship Commanders-in-Chief/Metropolitan Commander-in-Chief form the Collateral Prevention Brigades and Sub-brigades based on the number of regular Police Prevention Brigades existing on the respective territory of the voivodship. Police officers and border patrol selected to collateral structures in their parent agencies are recruited among those working in prevention units under the voivodship jurisdiction. At the micro-level (county PNP), police officers are assigned to collateral duties individually by the county PNP Police Chief based on their fitness level and necessary qualifications to assume required job functions.

Collateral, auxiliary structures may also be directed to operations in a separate voivodship or to conduct operations of larger scope. In such event, it is the Commander-in-Chief of the PNP or the Director of the General Staff at the KGP make decisions about their allocation. The Collateral Brigades and Sub-brigades are equipped and armed just like regular Prevention Police Brigades. However, they lack additional technical support such as water cannons, armored vehicles, and spot lights. Overall, at the national level, the total number of police officers within those structures is 9,351 spread across 98 companies under control of Voivodship Commanders-in-Chief. With respect to two voivodships bordering Ukraine, the Commander-in-Chief in Lublin may mobilize 5 companies (475 officers) and the one in Rzeszów may activate 6 companies (4,568 officers).

Micro-Level Planning and Preparedness

Collaboration and coordination in crisis management at the poviat level (i.e., local comanders in the county with mayors or city presidents) was predicted based on local need and was based on the national Center Command situation monitoring.

Towards Resiliency: Developing Capabilities

To address immediate human security issues (e.g., sheltering), the Ministry of National Defense used the National Infrastructure Inventory and feasibility study to suggest refugee camp setup at the old military installation in Nowa Dęba (Podkarpackie). It was estimated that approximately 7,500 Ukrainian refugees could be housed there temporarily. There were 160 additional facilities at the installation that could be transformed into refugee centers. Such housing management would offer capacity for 27,000 individuals. Specific catering companies and food vendors were considered for contracts in addition to agreements with the Polish Red Cross for sustenance needs of refugees. Coordinating activities for sheltering and feeding operations were conceived under the auspices of the UdSC.

Moreover, at the Voivodship level, the UdSC was coordinating with security departments and CMCs to gain additional information about potential facilities that could be used as immigration centers. It was also at the Voivodship level (meso-level) that administrative authorities in charge of crisis management had to identify Central Reception Points (CRPs). CRPs were conceived to process necessary paperwork, to conduct initial medical assessment and to serve as departure points from which refugees would be directed to temporary residence areas/facilities.

Ukrainian citizens seeking international protection in the Republic of Poland, after the border crossing, at the POE or beyond (Figure 5) were to be transported by the BG to the closest BG stations to conduct standard procedures connected to identification, securing basic medical aid, and issuance of proper documents legalizing their stay. Each person applying for the refugee status would obtain social assistance based on the existing regulations. From those locations, refugees would be transported to CRPs (managed by the UdSC) located in the first instance in border areas where temporary stay had been planned. After necessary procedures, the refugees would be transported to stay locations designated by the Voivods on the first come, first serve basis. In the event of overcrowding at CRPs and stay locations, temporary camps were planned. Voivodship Commanders-in-Chief (BG, PNP) of affected territories as well as UsDC officials advised Voivods in the selection of the camps.

Based on the National Plan, the Voivodship Commander-in-Chief of the PNP of the affected territory would be responsible for assistance in transport of foreigners to stay centers, convoy security, maintenance of public safety and public order at the CRPs, security and public order of camps and places of stay (temporary and permanent), help in foreigner transport to temporary refugee centers and subsequent convoy security. Preparations included three preparedness levels based on flow. Level I planned for 0 to 20 persons a day, Level II for 50 to 100 persons a day, and Level III for more than 100 persons a day. Under Level III, significant PNP involvement was anticipated.

In case of necessary engagement of PNP or BG in Lubelskie and Podkarpackie regions, forces and resources were to be utilized from county commands (micro-level county and commune) located along the routes of refugee transport or at designated camps and stay locations and under the jurisdiction of the Voivodship PNP and BG Commands in Lublin or Reszów, respectively. Thus, the large temporary camp in Nowa Dęba would first utilize local (county level) forces and resources. If the PNP and BG forces and resources at CRPs, camps, or stay locations proved insufficient, PNP Prevention Police forces and reserve BG resources would be dispatched from neighboring voivodships. In case of escalation of the mass influx of refugees, police and border patrol operations would be conducted at exo-level and controlled by the Central Commands in Warsaw.

Figure 5. Polish border crossings with surrounding countries
Border Guard, 2017.

In the event of the sudden influx of Ukrainian refugees, the coordination and collaboration between PNP and BG at the voivodship and county levels was based on the memorandum of collaboration signed June of 2004 between the PNP and BG Commanders-in-Chief. The agreement included (a) combating and preventing criminal activity through assessment of potential terrorist threats; (b) protection of the national border; (c) preventive protection of the security and maintenance of public order; (d) enhancement and development of methods and service activities; and (e) logistical support. The cooperation was to be realized through (a) information sharing; (b) coordination of efforts in mission areas such as joint exercises; (c) identification of individuals and infrastructure of interest to joint operations by the PNP and BG; and (d) necessary temporary integration of activities to improve operational effectiveness; (e) shared specialized technical support; (f) exchange of experiences gained during the conduct of operational tasks and service-related activities; and (g) mutual logistical cooperation.

Because of the highly hierarchical structure of policing in Poland, police officers and PNP employees are obligated to follow the chain of command. Thus, information at the voivodship level is submitted to the proper Incident Commander at the voivodship ICP, then to the on-duty staff at the KGP/Metropolitan Command. Then, the information is channeled based on protocols and on the need to know basis to proper sectors such as BG. Consequently, to further enhance information sharing (and based on lessons learned during exercises discussed below) the Central Investigations Bureau (Investigative Section of the PNP) and BG signed a memorandum of understanding (MOU) in July, 2014.

In preparation for the sudden influx of Ukrainian refugees, a full-scale exercise (the biggest in 40 years) was conducted in April, 2014. It included 3,000 participants (BG, PNP, UdSC, National Fire Service, and military). Overall, 700 police officers, 500 Fire Service, 400 BG and canine units as well as 1,000 military personnel, several helicopters and two MIGs -29 were utilized. A fake border to be protected and patrolled was delineated in two states (albeit not border regions) in central Poland (Mazowieckie and Świętokrzyskie). Responders were required to react to incidents and serious accidents. The 3-day exercises comprised episodes with elements of surprise testing dozens of scenarios (illegal border crossings, conflict between the community and the residents of the refugee camps, emergency landing of helicopters, and mass influx of immigrants attempting to ask for asylum). The common core to all scenarios was the element of foreign immigrants.

The exercise tested coordination, communications, and application of new technologies. The Ministry of Interior Affairs monitored the information flow among the services at different levels. In both regions involved, VCMCs and Voivodship PNP and BG commands in Kielce and Warsaw were engaged. Activities at the Mazowieckie (central) VCMC revealed several lessons learned to include (a) issues of information flow within the Center; (b) poor integration of county and commune (micro-level) CMCs and their respective local services poorly integrated with respect to communication; (c) poor utilization/integration of voluntary organizations (e.g., Water Rescue Services) in support of operations at CMCs; and (d) deficiencies in effective utilization of NGOs.

Moreover, in preparation for the sudden influx of refugees the PNP, BG, and the UdSC participated in a joint voivodship level exercise organized under larger full-scale initiative conducted by the Polish Military Forces (and related to the war in Ukraine) from September 27 to October 2, 2014. Those exercises were run in realistic settings (e.g., container area of the planned Ukrainian refugee overflow camp). The functional exercise called *Anakonda* 14 was managed by the Deputy Voivodship Commander-in-Chief of the BG (Podkarpackie voivodship). It tested coordination capabilities for speed and effectiveness in handing mass immigration flows and for safety and security of the refugee camps. For example, one exercise scenario addressed dealing with 50 refugees from the territory of conflict all of whom applied for refugee status. Tasks were to process applications, vet incomers through various databases, provide information on rights and responsibilities in refugees' native language, or conduct criminal processing (photos, fingerprinting).

The UdSC engaged in exercises such as identification of suspicious contagious disease, local resident demonstrations targeting new-comers, refugee riot related to poor camp conditions, or fires. Upon processing, BG and PNP transported refugees to the Temporary Refugee Center (i.e., Nowa Dęba military polygon). Additional scenarios were conducted there: (a) detention of individuals suspected of terrorism; (b) public demonstration and restoration of security and public order; (c) property theft from the area located at the temporary refugee camp; (d) potential anthrax release; and (e) drug detection at the temporary refugee camp.

Towards Resiliency: Identified Weaknesses

In theory, administrative support and coordination and operational command and control mirrored the tiered-level approach extant in the United States (i.e., tribal, local, regional, state, and federal/EU). Hierarchical and centralized national control at the ministerial levels in Poland evidenced a shift towards decentralization in crisis management. However, in practice, administrative coordination as well as operational command and control seemed to occur predominantly at the voivodship- level (meso-level). This was underscored by voivodship-level concentration of specialized teams (i.e., Prevention Brigades, Sub-brigades, or Anti-terrorist Groups). Concentrated voivodship-level responsibility for disaster management could raise questions and concerns about communal resilience at the micro-level. For example, the concept of activation of Collateral Prevention Police Brigades remains highly controversial at local levels. At the local commune and county level police officers otherwise engaged in community policing functions or criminal investigations may be assigned to perform social control while lacking proper training, proper personal protective equipment (PPE) and with unclear rules of engagement for the use of force. In the case of a massive refugee flow these issues may be challenging for the police at the community level to combat.

Further, the analysis of lessons learned from the full-scale exercise conducted in April 2014 confirmed poor integration of county and commune (micro-level) CMCs and their respective local services with respect to communication. Those findings echo previous research reporting lower micro-level resilience with respect to information flows in Poland (Gudzbeler, Nepelski, & Urban, 2011). The current administrative system in Poland is relatively new; local level governance by the elected official at the lowest communal level was established in 1990, followed by county and voivodship levels in 1999. Thus, even though the devolution of authority has been evolving, its history is relatively short (Matczak & Abgarowicz, 2013). Hence, the new political system, coupled with barely decade old crisis management system present ongoing challenges. The transition from military model of civil defense to the non-military civil protection necessitates increased involvement of local communities in crisis preparation and planning.

The assessment of command and control structures points to the practice of coexisting operational commands in crisis management. Specifically, ICPs are established separately by agencies with specific primary legal jurisdiction for the event. Thus, for operational command and control in the event of the sudden flow of refugees into the territory of Poland, the PNP will rely on its Command and General Staff, the BG will conduct operations through its central Command and General Staff, and the UdSC will rely on its central command and control structures. Similarly, at the voivodship-level ICPs will be established separately by the PNP, BG, and the UdSC. Clearly, the concept of Unified Command as it functions in the United States post 9/11 has not been evidenced in Polish structures. Moreover, lines of authority at both the meso-level and the micro-level underscore reporting ambiguities exist which is problematic given it is unity of command that fosters efficient crisis management (e.g., Copeland, 2008).

SOLUTIONS AND RECOMMENDATIONS

Vetting the crisis management system of Poland against the Ukrainian migration case stimulated capability, redundancy, and resourcefulness building – all of which strong predictors of resiliency. Subsequently, planned capabilities served as springboard in PNP, BG, and UdSC preparation for intensified migration from southern Europe in 2015. Low micro-level preparedness and reliance on state/national and supranational (for example, Frontex) resources reinvigorated discussions about effects of force reduction in BG defensive positioning after Shengen admittance (i.e., from 20,000 to 14,300, Zieliński, 2015) and the overall diffusion of realistic defensive presence not only at POEs but what is called a *green border* (e.g., Zapałowski, 2016). In effect, appeals have been made for reassessment of positioning, assignments of BG units as well as their arms, equipment directly available, and training standards to reduce over-reliance on meso-, exo-, and macro-levels of response (Zieliński, 2015).

Adaptations were considered that addressed streamlining activities based on new and improved model estimates under three influx levels*: moderate* - defined as influx beyond regular reaction capacity at approximately 3,600 per 6 months or 140 a week, *significant* – defined as significantly exceeding regular reaction capabilities from 3,600 to 18,000 per 6 months or averaging 690 per week, and *massive* – defined as exceeding 18,000 persons per 6 months (Majchrzak-Czarnecka, Rafalik, & Zasadzińska-Baraniewska, 2015). The number of patrols on roads leading close to Polish borders increased significantly.

FUTURE RESEARCH DIRECTIONS

Future research adopting the emergency management cycle to assess the integration and recovery of the displaced (for example, Balicki & Chamarchuk, 2013; Bocharova & Giza, 2015) coupled with mitigation practices aiming at lessening environmental degradation (for example, Lanicci et al., 2017) may contribute to more comprehensive approaches to population migrations in the future.

CONCLUSION

Indubitably, planning and preparedness for Ukrainian migration flows tested proactively the national crisis management system in Poland. Planning and preparedness processes revealed (a) emerging all-hazard approach to crisis management in Poland; (b) proactive threat identification and assessment; (c) issuance of acts, laws, and regulations based on evolving threats; (d) comprehensive multi-jurisdictional exercises for capability testing; and (e) increasing importance of mutual aid agreements. Planning and preparedness processes underscored the shift from the reliance on focusing events to drive political decision on safety and security to proactive measures. Importantly, internal domestic crisis management structures were applied and tested to address ecological supranational issues such as population displacements.

Lessons learned from exercises highlighted the importance of capability testing across a wide range of all-hazard scenarios (for example, technological plane crash, social unrest, and terrorism-driven biological event). Moreover, challenges such as communication breakdowns, unclear lines of authority, or poor integration of NGOs into the crisis management system revealed during exercises can illuminate on corrective actions in the future. Moreover, under hierarchical and sector-specific nature of crisis response entities collaboration as well as implementation of mutual aid agreements (for example, PNP's Central Investigative Bureau and Boarder Guard on information sharing) are invaluable. Clearly, planning for "all-hazards" to include environmental issues (both of slow and more rapid onsets), allows for the ongoing clarification of the role and function of crisis management structures in supranational, internal, and human security realms. The chapter's objective was to gauge whether and in what ways situating mass migration hazards and threats within a country's domestic crisis management system could enhance resilience to address future migratory spikes. Indubitably, given reactive stance of the EU with respect to Syrian refugee catastrophe, value of strategic planning and preparedness to all threats and hazards especially those rooted in ecological and environmental stimuli cannot be overemphasized. Ongoing research on what measures have been implemented by the EU within its crisis management strategy and based on lessons learned from the Syrian crisis to date is sorely needed. Moreover, evaluation of response capability and resilience building at local community levels present opportunities for future comparative research. This may be particularly important in countries with (a) strong focus on national versus civil defense; (b) large centralization of public and administrative services; and (c) strong reliance on supranational structures such as that may have weakened local response capabilities.

REFERENCES

Adamczyk, A. (2014). *Ukrainian immigration to Poland during the political crisis in Ukraine*. Poznań, Poland: Adam Mickiewicz University. doi:10.14746/ssp.2014.3.2

Anfara, V. A., Brown, K. M., & Mangione, T. L. (2002). Qualitative analysis on stage: Making the research process more public. *Educational Researcher*, *31*(7), 28–38. doi:10.3102/0013189X031007028

Balicki, J., & Chamarczuk, M. (2013). Wokol problematyki migracyjnej: Kultura przyjecia [Around migration policy: The culture of reception]. Uniwersytet Kardynała Stefana Wyszyńskiego, Międzyinstytutowy Zakład Badań nad Migracją.

Bell, M., & Davis, D. L. (2001). Reassessment of the lethal London Fog of 1952: Novel indicators of acute and chronic consequences of acute exposure to air pollution. *Environmental Health Perspectives*, *109*(s3), 389–394. doi:10.1289/ehp.01109s3389 PMID:11427388

Benedyczak, J. (2014). *Polska granica w czasie rewolucji ukrainskiej* [Polish border during the Ukrainian revolution]. Warsaw, Poland: Fundacja im. Stefana Batorego. Retrieved from http://www.batory.org.pl/upload/files/Programy%20operacyjne/Otwarta%20Europa/Granica_W_czasie_kryzysu.pdf

Betts, A. (2009). *Forced migration and global politics*. Malden, MA: Wiley-Blackwell. doi:10.1002/9781444315868

Bocharova, O., & Giza, T. (2015). Ludzie w sytuacji granicznej – cazus Donbasu [People in borderland situation- the case of Donbas]. In S. Jakonowska (Ed.), *Czlowiek w sytuacji granicznej. Filozoficzne, kulturowe I historyczne wymiary refleksji I jej implikacje pedagogiczne* (pp. 234–248). Radom, Poland: Wydawnictwo Naukowe Instytutu Technologii I Eksploatacji.

Border Guard. (2017). *Headquarters of the Border Guard*. Retrieved from https://www.strazgraniczna.pl/

Brian, T., & Laczko, B. (2014). *Fatal journeys: Tracking lives lost during migration* [Report]. Geneva, Switzerland: International Organization for Migration.

Bronfenbrenner, U. (1995). Developmental ecology through space and time: A future perspective. In P. Moen, G. H. Elder Jr, & K. Luscher (Eds.), *Examining lives in context: Perspectives on the ecology of human development* (pp. 619–647). Washington, DC: American Psychological Association. doi:10.1037/10176-018

Butler, D. (2012). Focusing events in the early twentieth century: A hurricane, two earthquakes, and a pandemic. In C. Rubin (Ed.), *Emergency Management: The American Experience 1900-2010* (pp. 13–50). Boca Raton, FL: CRC Press. doi:10.1201/b11887-3

Bynander, F., Chmielewski, P., & Simons, G. (2008). *The politics of crisis management in transitional Poland from 1990-1999*. Vällingby, Sweden: Swedish National Defense College.

Capra, F., & Luisi, P. L. (2014). *The systems view of life: A unifying vision* [Kindle]. New York, NY: Cambridge University Press. doi:10.1017/CBO9780511895555

Centrum Badan Opinii Spolecznej (2014). Stosunek do protestow na Ukrainie (21).

Chmielewski, P. (2007). Errare humanum est, but sometimes costs are extremely high. The biggest constructional catastrophe in Polish history: The collapsed building in Chorzow-Katowice. *Polish Psychological Review*, *3*(159), 321–339.

Chmielewski, P., & Kuźmiński, L. (2006). *The biggest constructional catastrophe in Polish history: The collapsed building in Chorzów-Katowice*. Retrieved from https://dlc.dlib.indiana.edu/dlc/bitstream/handle/10535/4850/The%20biggest%20constructional%20catastrophe%20in%20polish%20history%20the%20collapsed%20building%20in%20chorzow%20katowice.pdf?sequence=1

Cianciara, D. (2015). Kryzys zwiazany z uchodzcami nie jest chwilowy [Refugee crisis is not temporary]. *Hygeia Public Health, 50*, 427-434. Retrieved from http://www.h-ph.pl/pdf/hyg-2015/hyg-2015-3-427.pdf

Cielecki, T. (2009). Koordynacja systemu bezpieczeństwa wewnętrznego w Polsce [Coordination of the system of internal security in Poland]. In, J. Supernat (Ed.), Między tradycją a przyszłością w nauce prawa administracyjnego [Between tradition and future in law administration education] (pp. 75-91). Wrocław, Poland: Wydawnictwo Uniwersytetu Wrocławskiego.

Clapper, J. R. (2016). *Statement for the record Worldwide Threat Assessment* [Report]. U.S. Intelligence Community Senate Armed Services Committee. Retrieved from http://www.armed-services.senate.gov/imo/media/doc/Clapper_02-09-16.pdf

Copeland, J. (2008). *Emergency response: Unity of effort through a common operating picture* (Master's Thesis). Carlisle Barracks, PA: U.S. Army War College.

Czuchnowski, W., & Wroński, P. (2014, March 21). Szef MSW: Jesteśmy przygotowani na eskalację konfliktu wokół Ukrainy [Chief of MSW: We are prepared for the escalation of conflict around Ukraine]. *Gazeta Wyborcza*. Retrieved from http://wyborcza.pl/1,75478,15659536,Szef_MSW__Jestesmy_przygotowani_na_eskalacje_konfliktu.html

Dawidczyk, A., Gryz, J., & Koziej, S. (2006). Zarządzanie strategiczne bezpieczeństwem. Teoria-Praktyka-Dydaktyka [Strategic security management. Theory-Practice-Pedagogy]. £ódź, Poland: Wydawnictwo Wyższej Szkoły Humanistyczno-Ekonomicznej w £odzi.

Dworzecki, J. (2012). Crisis management system in Poland. *The Science for Population Protection*, *2*, 1–15.

European Commission. (2014). *Shengen area*. Retrieved from http://ec.europa.eu/dgs/home-affairs/what-we-do/policies/borders-and-visas/schengen/index_en.htm

European Council of Foreign Relations. (2015). *Multilateral issues and crisis management*. Retrieved from http://www.ecfr.eu/scorecard/2015/issues

European Union. (2014). *A Common European Asylum System*. Publications Office of European Union. Belgium, Brussels.

Gatrell, P. (2013). *The making of the modern refugee*. Oxford, UK: Oxford University Press. doi:10.1093/acprof:oso/9780199674169.001.0001

Gierszewski, J. (2011). Pozamilitarne ogniwa ochronne panstwa i ich miejsce w systemie obronnym [Non-military protection systems and their place in the defense system]. In M. Chrabkowski, C. Tatarczuk, & J. Tomaszewski (Eds.), *Bezpieczeństwo w administracji i biznesie we współczesnym świecie [Security in administration and business in modern world]* (Vol. 1, pp. 49–67). Gdynia, Poland: Wyższa Szkoła Administracji i Biznesu.

Grabbe, H. (2000). The sharp edges of Europe: Expanding Schengen eastwards. *International Affairs*, *76*(3), 519–536. doi:10.1111/1468-2346.00149

Gudzbeler, G., Nepelski, M., & Urban, A. (2011). Japanese support system of crisis management services. *International Security*, *3*(1), 107–127.

Haase, T. W. (2014, October). *Administrative resilience in Lebanon*. Presentation at Sam Houston State University, Huntsville, TX.

Haddad, E. (2008). *The refugee in international society: between sovereigns*. Cambridge, UK: Cambridge University Press.

History of the European Union. (2017) Retrieved from https://pl.wikipedia.org/wiki/Historia_Unii_Europejskiej

Iglicka, K., & Ziołek-Skrzypczak, M. (2010, Sept.). *EU membership highlights Poland's migration challenges*. Migration Policy Institute. Retrieved from http://www.migrationpolicy.org/article/eu-membership-highlights-polands-migration-challenges

Klepner, T. (2014). Wspolpraca transgraniczna panstw grupy Wyszehradzkiej na rzecz zapewnienia bezpieczenstwa granic. *Poliarchia, 2,* 99-122.

Krajewski-Siuda, K., & Romaniuk, P. (2006). *Can pigeons change the emergency system in Poland? European Journal of Public Health.* doi:10.1093/eurpub/ckl065446

Lanicci, J. M., Ramsay, J. D., & Murray, E. H. (2017). Re-conceptualizing environmental security as resilience: Strategic planning for human and national security. *Journal of Human Security and Resilience, 1*, 1–37. Retrieved from http://www.thinkhumansecurity.org/sft965/re-conceptualizingenvironmentalsecurityasresilience-strategicplanningforhumanandnationalsecurity.pdf

Majchrzak-Czarnecka, M., Rafalik, N., & Zasadzińska-Baraniewska. (2015, October 8). Masowy napływ cudzoziemców – wyzwanie humanitarne I polityczne. 5-9. Rzadowe Centrum Bezpieczeństwa (RCB). Centrum Operacyjno-analityczne, Warsaw, Poland.

Matczak, P., & Abgarowicz, G. (2013). *Analysis of civil security systems in Europe. Country study: Poland* [Report]. Poznań, Poland: University of Adam Mickiewicz. Retrieved from http://www.anvil-project.net/wp-content/.../01/Poland_v1.0.pdf

Morag, N. (2011). *Comparative Homeland Security: Global lessons* [Kindle]. Hoboken, NJ: John Wiley & Sons, Inc. doi:10.1002/9781118048252

Poland Ukraine Locator. (2008). Retrieved from https://commons.wikimedia.org/wiki/File:Poland_Ukraine_Locator.png

Polska Agencja Prasowa (Polish Press Agency). (2014, August 21). UDSC: w 2014 r. już 1200 wniosków obywateli Ukrainy o status uchodźcy [Bureau for Foreigners: In 2014 already 1200 asylum petitions by Ukrainian citizens]. *Gazeta Wyborcza.* Retrieved from http://wyborcza.pl/1,91446,16508207,UDSC__w_2014_r__juz_1200_wnioskow_obywateli_Ukrainy.h

Quarantelli, E. L. (1993). Community crises: An exploratory comparison of the characteristics and consequences of disasters and riots. *Journal of Contingencies and Crisis Management, 1*(2), 67–78. doi:10.1111/j.1468-5973.1993.tb00009.x

RIANovosti. (2014, September 24). Over 300,000 Ukraine Refuges seek Asylum in Russia: Federal Migration Service. *RIANovosti.* Retrieved from http://en.ria.ru/russia/20140924/193243399/Over-300000-Ukrainians-Seek-Asylum-in-Russia-Federal-Migration.html

Rzadowe Centrum Bezpieczenstwa (Government Security Center). (2014). *Działania resortu spraw wewnętrznych w związku z sytuacją na Ukrainie* [Activities of the interior affairs sector related to the situation in Ukraine]. Author. Retrieved from rcb.gov.pl/wp-content/uploads/biuletyn/6.pd

Sasnal, P., & Uzelac, A. (2014). Unaided refugees make ISIS stronger: Europe and the worst humanitarian crisis of the 21st century. *Polish Institute of International Affairs, 102*(19), 1-8. Retrieved from https://www.pism.pl/files/?id_plik=18847

Sherwood, H., Smith, H., Davies, L., & Grant, H. (2014, June 1). Europe faces 'colossal humanitarian catastrophe' of refugees dying at sea. *The Guardian.* Retrieved from http://www.theguardian.com/world/2014/jun/02/europe-refugee-crisis-un-africa-processing-centres

Skusiewicz, K. (2016). The "Ukrainian" crisis as a chance for the certain sectors of Polish economy. In Przedsiębiorczosc i zarzadzanie: Bezpieczenstwo i zarzadzanie kryzysowe – nowe wyzwania bezpieczenstwa narodowego. Lodz, Poland: Wydawnictwo Spolecznej Akademii Nauk.

Śnierzyński, M. (2009). Struktura bezpieczenstwa narodowego w Polsce [The structure of national security in Poland]. In, M. Włodarczyk & A. Marjański (Eds.), *Bezpieczeństwo i zarządzanie kryzysowe – aktualne wyzwania* [Security and crisis management: Current challenges] (pp. 57-68). Stołeczna Wyższa Szkoła Przedsiębiorczości i Zarządzania w £odzi, £ódz, Poland. Retrieved from http://www.swspiz.pl/wydawnictwo/pliki/2009/X-6.pdf

Szarpak, L. (2009). Aspekt prawny zarzadzania kryzysowego w Polsce (Legal aspect of crisis management in Poland). In M. Chrabkowsi, C. Tatarczuk, & J. Tomaszewski (Eds.), *Bezpieczeństwo w administracji i biznesie we współczesnym świecie [Security in administration and business in modern world]* (Vol. 1, pp. 145–155). Gdynia, Poland: Wyższa Szkola Administracji i Biznesu.

Szpak, A. (2016). Kryzys migracyjny w Europie a bezpieczenstwo – podloze spoleczno-ekonomiczne, zagrozenia, pomoc rozwojowa [Migration crisis in Europe and security – socio-economic background, threats, and development aid]. *Sprawy Miedzynarodowe, 1,* 111-132. Retrieved from https://www.ceeol.com/search/article-detail?id=425293

UN High Commissioner for Refugees (UNHCR). (2014). *Ukraine: Overview of population displacement.* Retrieved from http://www.refworld.org/docid/542a9cc04.html

United Nations High Commission on Refugees. (2014, September 2). *Number of displaced inside Ukraine more than doubles since early August to 260,000.* United Nations High Commission on Refugees. Retrieved from http://www.unhcr.org/540590ae9.html

Zapałowski, A. (2016). *Gra wokół ochrony południowo-wschodniej granicy Polski* [Play around the defense of the south-eastern border of Poland]. European Center of Geopolitical Analysis. Retrieved from http://www.geopolityka.org/analizy/andrzej-zapalowski-gra-wokol-ochrony-poludniowo-wschodniej-granicy-polski

Zdanowicz, M. (2009). *Polska w Shengen* [Poland in Shengen]. Białystok, Poland: Temida 2.

Zielinski, J. (2015). *Interpelacja poselska w sprawie zabezpieczenia polskich granic przez Straż Graniczną oraz braku wyposażenia funkcjonariuszy Straży Granicznej we właściwe elementy uzbrojenia* [Appeal about securing Polish borders by Border Guard and lack of proper arming of BG]. Directed to Polish Senate, Warsaw, Poland. Retrieved from http://ciechanowiec.info/downloads_2/IT%20227.76%20Straz%20Graniczna.pdf

KEY TERMS AND DEFINITIONS

Acquis Communautaire: (From French *shared acumen*) Involves all laws, treaties, rulings, and fundamental rights provisions of the European Union (EU) by which all member states must abide.

Euro-Majdan: Ukrainian revolution started in Kiev at the Independence Square in November 2013 and ended by bloody riots in February 18-20, 2014. Ukraine's destabilization in Kiev was a contributing factor to opportunist separatist movements in West Ukraine and subsequent annexation of Crimea by the Russian Federation.

Focusing Event: In emergency management focusing events refer to disasters that were so significant in their scope and breadth of disruption that they created high and widespread levels of public discontent. In turn, the discontent influenced political will as well as research and ultimately led to changes in policy and laws.

FRONTEX: Fronières Extérieures (from French *external borders*). European Agency created in October 2005 for the management of operational cooperation at the external borders of the member states of the European Union.

Hazard and Threat Multipliers: Hazard refers to risk of insecurity from natural and technological sources that are unintentional; threat refers to risk of insecurity from intentional and criminal actions of humans. Hazard and threat multiplier is a condition that creates additional vulnerability for hazard or threat to become a larger risk.

RABIT: Rapid Border Intervention Teams. Involves tactical resources requested from FRONTEX by member states and deployed in case of local/internal resources being overwhelmed. Approximately 500-600 border guard agents constitute "RABIT pool."

Shengen Zone: An agreement signed by 26 European countries that eliminated border controls and allowed free movement across all state members. The Schengen Zone was established in 1995.

UNHCR: United Nations High Commissioner on Refugees. The UNHCR was established by the United Nations in 1950 with the goal of providing aid to refugees who had fled their homes during World War II. Its functions were preceded by the United Nations Relief and Rehabilitation Administration (UNRRA) in charge of displaced persons immediately after WWII.

Unity of Command: Management principle suggesting that each individual who participates in a crisis operation reports to only one supervisor. This eliminates opportunity for conflicting orders, confusion, and redundancy of effort.

Visegrad Group (V4): Originally a Triangle of Poland, Czech, and Slovak Federal Republic (respectively, later Czech Republic and Slovak Republic), and Hungary member states who entered in collaborative agreement in 1991 in Bucharest to support collaboration towards the integration of security systems in the aftermath of the break of the Soviet Union; its first mission was removal of Soviet occupying forces in territories of member states.

Chapter 10
Environmental Security Threats and Policy Response in the Niger Delta, Nigeria 1990–2016

Luke A. Amadi
University of Port Harcourt, Nigeria

Henry Alapiki
University of Port Harcourt, Nigeria

ABSTRACT

In recent years, environmental security in the coastal Niger Delta has had a growing scholarly interest from divergent perspectives seeking for broader elucidation and understanding of State policy response. These security threats notably oil spill, water and land pollution, gas flaring, acid rain, mangrove deforestation, etc are linked to the Multinational Oil Corporations (MNOCs)and oil resource exploitation. These have been perverse, resulting in ecological breakdown, vulnerability, emergency and environmental insecurity challenges since at least the 1970s when oil in the region became the main stay of Nigeria's economy. This conceptual paper builds on the political ecology framework which discusses the impact of global power asymmetry on natural resource extraction and extensive body of work in the broad field of environmental security to explore salient indicators which demonstrate the evidence of environmental insecurity threats and poor State policy response and made some recommendations.

INTRODUCTION

The environment perhaps has been a neglected component of the security debate as security was initially conceived largely as a military affair with emphasis on protecting the territorial integrity of the State against nuclear warfare (Deudney, 1990). This pointed to the debate on redefining security beyond its narrow militaristic scope. Brown (1977) reopened the debate on "redefining national security" from energy and ecological perspectives. Ullman (1983) further advanced this debate within the environmental context. Perhaps the debate did not gain much scholarly attention till approximately the end of the Cold War in 1989.

DOI: 10.4018/978-1-5225-3194-4.ch010

Remarkable fluidity and openness had emerged in security studies since the end of the Cold War. The non-military security threats persisted, resulting in scholarly re-engagement with debates on "redefining security" as new wars and local conflicts persisted in the global South (Kaldor, 1999). This took divergent dimensions including environmental concerns (Mathews, 1989; Myers, 1989; Klare, 1996;Paris, 2001). The Copenhagen school of security studies provided a new security framework within the "security securitization" thesis aimed at sectorial security analysis including societal, economic, environmental, etc. (Buzan et al., 1997).

The persistent ecological threats including environmental degradation saw the emergence of the field of environmental security (ES) as a distinct field of enquiry. Thus, ES studies aim to examine some of the foundational questions of ecological concern, including anthropogenic such as climate change and global warming and non-anthropogenic issues including deleterious environmental resource extraction and its effects on both the ecology and human being. In particular, human- centric perspectives on environmental security provides plausible linkages between the environment and human survival (Dalby, 2013).

To define ES requires an exploration of the interaction of humans and the natural environment, including the causes and effects of environmental degradation and vulnerabilities. Hence, ES integrates a number of issues such as climate change, emergency management, the impact of policies and human interaction with nature including ecological factors etc. and how policies or remediation strategies are deployed to cope or withstand the threats. Thus, environmental security could be defined as absence of threats such as risks or hazards.

The question of environmental policy response in this chapter aims to draw attention to the coastal Niger Delta region noted for historic struggles against environmental degradation, resource marginalization and persistent environmental pollution since at least the 1970s when crude oil became the mainstay of Nigeria's economy. Although there have been several environmental policies in Nigeria since the discovery of oil in 1956. A number of evidence point out the prevalence of environmental security threats in the region such as oil spill, gas flaring, acid rain, land and water pollution, mangrove deforestation, black soot etc. These have become intense to the extent of threatening the livelihoods and the very existence of the people (Amadi, 2013).

Against the background of these problems, the central objective of this chapter is to provide evidence of these environmental security threats in the Niger Delta and how they have negatively impacted the region and the need for policy response. Although it is argued that environmental degradation arising from crude oil extraction by MNOCs and social injustice is expressed through youth restiveness including vandalization of oil equipment, militancy and taking of oil workers into hostage, this chapter suggests that this might leave these problems unresolved as policy dialogue is always a preferred strategy to resolving such problems.

BACKGROUND

Environmental security(ES) could be seen from a number of perspectives which include the absence of ecological threats which taint the environment and has adverse effects on humans(Dalby,2002;Chelecki,2002) or the ability to withstand resource scarcity, threats or shock. Deudney (1990) re-inscribed the need for delineation of what constitutes national security threats. He contends that "national security from violence and environmental habitability have little in common. The rising fashion of linking the risks creating a conceptual muddle rather than a paradigm or world view shift –a de-definition rather than a redefinition of security".

Generally, in the neo liberal era with perverse commodification of natural resources and quest for capitalist accumulation, there have been development 'crises' (Harvey, 2005), and in particular, environmental insecurity challenges reappear in most recent scholarships as a central problem in emancipating the developing societies (Deudney, 1990). Essentially, the environment has become part of the 'new' activism, particularly in most developing societies. A number of key articles have recently been written in the field of environmental security, reflecting on its emergence, reviewing key trends in its consolidation and suggesting future research agenda that might advance knowledge in environmental security studies (Dalby, 2013; Flyod & Mathew,2013).

This orientation has given rise to a number of interesting scholarship debunking the dominant neo Malthusian perspectives in the mainstream debates since the 1990s popularized by the Toronto School led by Thomas Frazer Homer Dixon which argues that environmental scarcity is necessitated by population growth in the poor societies of the global South resulting in competition over scare resources which gives rise to violent conflicts (Homer Dixon, 1991) .On the contrary, in the 2000s, the OSLO school led by Indra de Soyza emerged and argued that environmental security emerged from violence over and the struggle by the global power to control abundant resources in the South (de Soyza, 2000, p. 5).

The terminological shifts and changing nature of environmental security which include "environmental emancipation", "environmental risks", "vulnerability", "ecological insecurity", "eco literacy" etc. entail the need for collaborative engagement of core stakeholders at various levels to recreate the dominant orientations and to take an emancipatory approach. Booth (1991, p. 319) argued that security should be equated with emancipation: "Emancipation is the freeing of people (as individuals and groups) from the physical and human constraints' which stop them from carrying out what they would freely choose to do. Security and emancipation are two sides of the same coin. Emancipation, not power or order, produces true security. Emancipation theoretically is security."

Since crude oil became the fiscal basis of the Nigerian State, and as a critical element in economic and political power, the Niger Delta became ever more prone to environmental security threats and increasingly neglected(UNDP,2006). Thus, environmental insecurity which largely results in degradation has unquestionably tainted the Niger Delta environment (Obi, 2000). It was further compounded by the character of the Nigerian State which Ake (1985) described as a modality for class domination.

The foregoing combined to fuel the resurgence of militancy, poverty and livelihoods vulnerability. For instance, the findings of the UNEP (2011, p. 6) report on environmental degradation in Ogoniland a community in the Niger Delta, reveal that "there are, in a significant number of locations, serious threats to human health from contaminated drinking water to concerns over the viability and productivity of ecosystems". Thus, environmental degradation in the region has had an unprecedented scale pointing out the basis of Nigeria's democracy.

Democracy in Nigeria thus appears not to have transformed the environmental problems of the region, creating feelings of marginalization and exclusion from both direct access and control of the natural resources and a healthy living environment. In a society where critical policy issues as environmental degradation and remediation strategies are matched with policy quiescence of the government and political elites in ways which point to ecological injustice calls for scholarly investigation. Here, what is at stake is not simply ' the formulation of various environmental policies ', rather, ' implementation'. For instance, the UNEP (2011, p. 12) report shows that; "The Nigerian Government agencies concerned, lack qualified technical experts and resources". In the particular instance of National Oil Spill Detection and Response Agency (NOSDRA), the report points out that: "In the five years since NOSDRA was established, so few resources have been allocated that the agency has no proactive capacity for oil-spill detection".

ES studies provide the basis to challenge the persistent problems of environmental degradation and argue instead for the adoption of effective policy response to redress ecological breakdown, environmental risks and disasters which pose security threats particularly in the societies of the global South. For instance, in the Niger Delta, the debilitating effects of environmental insecurity notably oil spill, gas flaring, acid rains, land and water pollution, deforestation etc have often been part of this new research agenda as they become central problems to the survival of the people, affecting their everyday lives. To capture the extent of State policy inertia in this regard, it is necessary to examine the evidence of prevailing environmental insecurity problems in the Niger Delta to provide a broader elucidation of background to the problem of the chapter.

Since the 1950s when crude oil was discovered in commercial quantity in Oloibiri a community in Bayelsa State, a number of seminal data suggest that oil spill has caused havoc to this community and several others. Most affected in the State include, Batan, Odidi, Nembe and Aleibiri whose farmlands and rivers are contaminated with oil (Frynas, 2000). For instance, in Botem village an Ijaw community, the incidence of oil spill has affected soil quality for the cultivation of food crops which are important source of subsistence for the communities.

Pollution in Oloibiri, Imiringi, Otuasega, and Anyama in Ogbia community in Bayelsa has been a consistent ecological hazard to many local people in the area, including terminal diseases and birth defects (Ashton-Jones et al., 1998; Eweje, 2006). Osuide (1990) and Iacciano (2015) have reported the impact of pollution on the rural households resulting in disempowerment and poverty.

The incidence of oil spill persists in parts of the region including the Forcados terminal oil spill of 1979, (Bayelsa), the Funiwa No 5, well blowout of 1980, the Oyakama oil spill in 1980, Oskika oil spills in 1983 all in Rivers State. The System 2c Warri/ Kaduna pipeline rupture at Abudu in 1982 in Delta State, the Idoho oil spills in 1998 in Akwa Ibom State,the Jones creek oil spills of 1998 and the Jesse creek oil spills of 1998 both in Delta State.

The Ogoni community in Rivers State remains an environmentally degraded community. This includes Bodo, Bomu, K-Dere, B-Dere in Gokana Local Government Area (LGA) known for cultivation of staple crops such as vegetables, cassava, plantain, maize and legumes (Nriagu et al., 2016). Some community members at Sugi Bodo, in Gokana, Ogbe Ijaw and Burutu in Delta State and Southern Ijaw areas in Bayelsa State drink water from river and wells that are contaminated with benzene, a known carcinogen, at levels over 900 times above the World Health Organization (WHO) guideline (Adeola, 2000). The incidence of oil spill is equally prevalent in most neighbouring rural communities such as Ejama-Ebubu and Nisisioken Ogale in Eleme LGA in Rivers State.

In April 1997, samples taken from water used for drinking and washing by local villagers were analysed in the U.S. A sample from Luawii community in Ogoni, which has not experienced oil production for four years, had 18 ppm of hydrocarbons in the water, 360 times the level allowed in drinking water in the European Union (E.U.). A sample from Ukpeleide, Ikwerre, contained 34 ppm, 680 times the E.U. standard (Naanen & Tolani, 2014).

In the late 2000s, the highest reading of dissolved hydrocarbon in the water column, of 7,420 μg/l, was detected at Ataba-Otokroma, bordering the Gokana and Andoni LGAs in Rivers State (Naanen & Tolani,2014). This has grave consequences for both human health, aquatic creatures and rural subsistence as water contamination results in the death and migration of various aquatic species including fish.

Fish samples from Bodo creek (Rivers State) have the highest levels of TPH and PAH contamination of 0.526 mg/kg and 0.044 mg/kg. It is followed by Forcados river (Bayelsa State section) which has 0.495 mg/kg of TPH and 0.038 of PAH; and the New Calabar river at Choba (Rivers State), which has 0.338 mg/kg and 0.022 mg.kg of TPH and PAH, respectively (Naanen & Tolani,2014:12).

Other rural communities faced with environmental insecurity of oil spill include Ogale and Bile in Rivers State. In early 2016, the Ogale community was reported to have been denied access to clean drinking water for more than 25 years due to ongoing spills (Gaffey, 2016). Similar oil spill impacted communities in Okrika LGA in Rivers State include Ogan Ama,Kalio Ama, Oba Ama, Dumo Ama,Okuru Ama, Abuloma Bolo,Wakama Ogu,Ikpokiri Iyokiri and Okerekana .

Related incidence persisted in some rural households in coastal communities in Akwa Ibom particularly in Eket, Onna, Ibeno, Mbo and Oron LGAs (Nriagu, et al;2016). For instance, it is reported that in Akwa Ibom, the Idoho oil spill dispersed from the coastal State to Lagos State. This culminated in the presence of sheen of oil on the coastal areas of Cross River, Akwa Ibom, Rivers, Bayelsa, Delta, Ondo and Lagos States(Mogborukor,2014).

Similar experience associated with flaring of gas in communities in the Niger Delta have been documented. The experience with communities such as Mgbede, Rumuekpe (Rivers State), Imiringi, Gabrain, Eremah (Bayelsa State), Akala-olu, Idama, Iwherekan (Delta State), and Eket in Akwa Ibom State have been reported (Ekpo, 2004; Ubani & Onyejekwe, 2013; Diukwu et al., 2013).

Anyanwu and Tanee (2008) observed dramatic reductions in cassava yield parameters (growth, fresh weight of shoot and tubers, total fresh weight, etc.) in the Niger Delta due to spill and gas flaring. Gas flaring remains a key problem in the region as it heats up soil temperature against food crop cultivation including vegetables, legumes, cassava, maize and yam and makes the environment uninhabitable for human and non- human species (Korvenoja, 1993; Adeola, 2000).

Most rural communities such as Mgbede, Rumekpe, Umuechem in Rivers State, have been subjected to incidents of gas flaring which adversely affect them (Ajugwo,2013).

Mangrove deforestation is another environmental insecurity threat which affects the region. This takes several dimensions. The mangrove forests and creeks constitute the main areas of oil exploration and exploitation activities in the Niger Delta. Beyond failing of trees for domestic use, there is the challenge of mangrove deforestation arising from the prospecting of oil resource by the MNOCs. The Niger Delta people rely on forest resources for their survival which include tress, various animals that inhabit the mangrove which are both sources of food and income, woods for building of bridges, canoes and houses and fish stock which inhabit the mangrove (Akachuku, 2006; Mmom & Arokoyu, 2010). For instance, in Oyorokoto a fishing pot in Rivers State, fisher men and women rely on mangrove trees to build their fishing settlements, deforestation has accounted for challenges of distant exploration for woods.

Studies have shown that 60% of fishes in the Gulf of Guinea breed in the mangroves of the Niger Delta(Irikana,2015). There are several rural mangrove forest dependent communities in the Niger Delta such as Buguma, Burutu, Kuruama, Opobo,Bonny, Nembe, Abua communities etc. These rural communities are located on beach ridge forest Islands within the mangrove ecosystem.

. In communities such as Abonnema, Bonny, Buguma, Brass, Nembe etc. there have been deforestation resulting in loss of plants and marine species as a result of deforestation and oil pollution for over 50 years (Ashton-Jones, 1998). Other impacted communities include Buguma, in Rivers State, Burutu in Delta State, Opobo, Bonny in Rivers State, Nembe in Bayelsa State, Abua in Rivers State, Baranyonwa Dere, Kegbara Dere, Kpor, Bodo, Mogho, Bomu, Lewe all in Gokana LGA in Rivers State, others are Kuruama in Okirika LGA, Ejama-Ebubu, Nsisioken Agbi, Nsisioken Ogale all in Eleme LGA in Rivers State.

Illegal or artisanal refining equally causes both mangrove deforestation and pollution. Illegal refining largely takes place in some creeks around the region namely river Nun creek in Bayelsa State, the creeks of Ojobo in Burutu LGA, Delta State, Nembe and Brass creeks in Bayelsa State, Bodo West in Gokana Local Government Area, close to Bonny, Okrika and Ogu/Gbolo Local Government Areas, Mbiama in Ahoada West, Rivers State. Other sites include Oteghele, Otegbene-Agbara, Bennett Island and Jones Creek, all in Warri South-West Local Government Area of Delta.There are others namely; Koko, Tagola Creek, Lolomu Creek, Ugbuwangue, etc.

Against the background of these environmental insecurity problems, there is need for an investigation of policy response by the government. The chapter considers how sustained policy response might be helpful in the context of environmental security transformation and sustainable development. In particular, it calls for environmental policy review and environmental security reform. The rest of the chapter is structured as follows; theoretical framework, the concept of environmental security, overview of environmental policies in Nigeria, environmental security threats and policy response nexus: the Niger Delta experience, recommendations, future research directions and conclusion.

THEORETICAL FRAMEWORK: THE MARXIAN POLITICAL ECOLOGY

The Marxian political ecology arose in opposition to the neo-Malthusian views of complexities of the interaction between human beings with nature. Neo-Malthusian views are based on an eco-scarcity argument originally put forward by Thomas Malthus which argues that, an ecological 'crisis' erupts when the demands of a growing human population overtakes the capacity of an environmental system to support it. The Malthusian perspective emphasizes the need for population control to tackle ecological degradation, without putting into consideration the issues of unequal global distribution of power and economic (Brown & Jacobson, 1986; Robbins, 2004: 7-8).

Against this misconception, Neo-Marxists draw attention to political and economic factors in clarifying how material power (e.g. capital, wealth, military power) mediate human society and nature relations (Bryant & Bailey, 1997, p. 28; Bryant, 1998, p. 80; de Soyza, 2000; Biersack, 2006, p. 3; Robbins, 2004, pp. 7-8).

Thus, ecological Marxism emerged as a critique of the neo- Malthusian theory which examines the primacy of inequality and deleterious implications in the interactions between humans and their natural environment (Shiva, 1988; Harvey, 1996; Speake & Gismondi, 2005, p. 55). Ecological Marxists argue on the structure of the global power including inequality in resource distribution and asymmetrical structure of the international capitalist system and its effects on natural resource access.

The field of political ecology has grown dramatically, and in so many directions including the study of power and inequality in natural resource use, conflict and struggle for natural resource exploitation, unsustainable resource consumption etc. (Peet & Watts, 1996).

The fundamental question this chapter seeks to answer stems from environmental insecurity arising from oil spill, gas flaring, acid rains, deforestation etc. giving rise to deleterious oil resource extraction in the Niger Delta. Roggers (1997) had argued on the understanding of *ecological security* as state in which the physical environment of a community satisfies its needs without depleting natural capital. This may adequately account for the basis of suitability of the Marxian political ecology framework.

Peet and Watts (1996, p. 6) demonstrate its broader academic reach encompassing geography, development studies, environmental studies, ecological, social sciences, political economy etc. Robbins (2004,

p. 14) contends that political ecology has vast themes such as environmental consumption, pollution, degradation, exploitation, marginalization, environmental resource extraction and conflict, resource equality, ecological justice, conservation and control etc. For instance, Speake and Gismondi (2005, p. 55) argue that political ecology is a critical neo Marxist theory which studies inequalities in the use of nature, such as water access.

This "inequality" in resource consumption has been a central thesis of the neo Marxist political ecology debate informed by the inevitable contradictions of global power and struggle on environmental resource control. This is reflective of the assumption that the natural environment within capitalist mode is commodified and exploited within the logic of value augmentation.

Though political ecology thinking has proliferated in development studies, it has not been sufficiently deployed in the study of environmental security particularly in most developing societies. The particular case of the Niger Delta is important as the region is known as a minority area (numerically less) and importantly, a coastal region. In particular, the region is rated one of the most environmentally insecure areas in the world (Bruce, 2003; Jike, 2004; Eweje, 2006; Emordi, 2008; UNEP, 2011).

Orientations towards "political ecology" in development studies have increased in recent decades and point to its saliency (Peet & Watts, 1996; Bryant, 1998; Foster, 1999; Robbins, 2004; Speake & Gismondi, 2005; Biersack, 2006). Although many studies have rarely examined the structures and patterns of environmental security issues in the Niger Delta from the political ecology prism, this does not only make this work a novel and original contribution, rather it provides seminal insights in understanding salient issues raised in the dynamics of deleterious environmental resource consumption and need for alternative policy response.

CONCEPTUAL CLARIFICATION: THE CONCEPT OF ENVIRONMENTAL SECURITY

There are a number of conceptual issues raised on environmental security suggesting its saliency on both environmental management and development studies. Environmental security forms part of a wider debate that illuminates the ecological and social relations which shape both the environment and human existential realities. Beyond this, there is a widespread assumption that the term 'environment' has divergent understanding. Thus, ES has been a contentious term among scholars and policy makers partly on the basis of definitional goals and at arriving at various policy issues. Aspects of this contention are largely linked to the boundaries of environmental security, i.e., the sets of events and chain of causes that constitute threats to the environment such as environmental degradation, uncertainties of climate change, global warming etc.

While there is a broad agreement on the prevalence of environmental degradation in most societies both in the global North and South, conceptual debates on the term "environmental security" have been scant and less clear. This suggests the basis for conceptual clarity, by providing a clear scholarship on ES and importantly, linking ES to the divergent threats arising from ecological breakdown. More than this, to offer a way of looking at the problems and issues of the environment as an integral part of human survival. This is against the dominant capitalist conception which centers on natural resource extraction, profit accumulation and reliance on technological solutions to environmental problems propagated by the ecological modernization theorists (Spaargaren, 2000).

In the 1960s the environment gained consciousness in the United States following the publication of the book *Silent Spring* by Rachael Carsons (1962).This was bolstered with the writings of Garret Hardin particularly the *Tragedy of the Commons* (Hardin, 1968) resulting in the emergence of environmental movements such as the Green Movements in the United States in 1970s.

In 1968 a group of scientists later to be known as the Club of Rome in their report *Limits to Growth* (Meadows et al., 1972) provided evidence of non-military threats posed by the economic oil crises of the 1970s which gave greater impetus to environmental security consciousness resulting in the United Nations Conference on Human Environment held in Stockholm in 1972. In this vein, environment and security discourse had emerged as central policy concerns for countries across the world.

Before the end of the Cold War, Brown (1977) had identified the increasing need for 'redefinition of national security' within environmental contexts. Ullman (1983) re-echoed the 'redefining national security' thesis to include environmental issues such as natural resource shortages. Similarly, Mathews (1989) further elaborated the debate on "redefining security" to encompass environmental issues and particularly argued on the 'bleak future' of the environment. Other scholars in this line of debate called for a more holistic definition of security that moves away from competitive models of State behaviour (Mische, 1989; Myers, 1989).

Myers (1989) advanced debates on "environment and security" identifying the fundamental relevance of the environment within security contexts. Equally, Westing (1989) examined environmental security from a human perspective espousing the negative effects of the role of military on the natural environment, he contends that environmental security is critical both in war times and post war periods.

The consequences of the neglect of the complex twists of environmental degradation and pollution grew worse at the end of the Cold War following the collapse of the Soviet Union and increasing environmental disaster which resulted in a reconsideration of the dominant traditional notion of security as new "ecological awareness" was on the increase (Brown, 1989).

Post -Cold War discussions on security took a novel turn following the reports of the 1992 Earth Summit which provided a rethinking of the concept of security and mobilized scientific evidence of global environmental change into a global policy agenda (Floyd & Mathew, 2013). In particular, Agenda 21 of the summit – *The Blue Print for the implementation of sustainable development*, gave renewed impetus to environmental security in the context of sustainable development. Agenda 21 came into force to bolster the actualization of the principles and overarching objectives of sustainable development.

Global trends and changes on environmental security such as ongoing debates on climate change vulnerability saw the increasing global attention given to environmental security in the Earth Summit of 1992 as environmental insecurity was perceived to be a threat to sustainable development. Concerns for climate change not only led to the creation of the Intergovernmental Panel on Climate Change (IPCC) to assess scientific, technical and socio-economic aspects of human impact on climate change but also to the signing of the UN Framework Convention on Climate Change in Rio in 1992 (UN, 1994). According to UNFCCC (1994) climate change is often referred to specifically as change caused by human activity, as opposed to changes in climate that may have resulted as part of Earth's natural processes (Amadi and Ogonor, 2015).

ES has been argued to be more of a debate than a concept encompassing divergent threats to the environment (Buzan et al., 1997; Floyd, 2008:). There are two broad classifications of environmental security threats namely anthropogenic and non-anthropogenic (Amadi et al., 2015). Anthropogenic environmental security threats include threats to the environment arising from the activities of man such as wars, conflicts, oil spill, acid rains, gas flaring, deforestation, etc. while non-anthropogenic environmental security threats include natural environmental disasters such as earth quakes, hurricanes, cyclones etc.

In the 1990s, several neo Malthusian debates on environmental security emerged (Homer Dixon, 1991; Kennedy, 1992; Kaplan, 1994; Klare, 1996) notably the Toronto School led by Homer Dixon (1991) popularised the environmental scarcity debate or the eco- scarcity thesis which argues that when scarcity of *renewable* resources (such as cropland and river water) interacts with harsh social effects (for example, population, displacement or economic decline) particularly in the poor societies, it can lead to intrastate conflict. The environmental resource scarcity and conflict debate provided a new insight into environmental security studies. Homer-Dixon (1991) identified "gaps" among the poor countries to meet such environmental demands which could result on conflicts unlike rich countries, stating that the rich countries have the potential for a widening gap between demands on the state and its financial ability to meet these demands -a gap according to him that could lead to internal conflict between competing ethnic groups, or significant out-migration to countries better able to cope with environmental stresses. In a related account on environmental security threats, Kaplan (1994) underscores this growing concern in the context of a "coming anarchy" that threatens 21st century Africa including a vortex of factors such as deforestation, population growth, crime, migration, diseases, poverty and social security.

Changing discourse in recent literature provides a critique of the scarcity debate and incorporates human security or security of the individual in all ramifications. The most plausible articulation of "human security" comes from the United Nations Development Program (UNDP) in its 1994 Human Development Report. It includes environmental factors as one of its human security themes. Human security shifted the focus of security from the State to the individual and 'the legitimate concerns of ordinary people who sought security in their daily life' (UNDP, 1994). Krokfors (1995) provides an insightful analysis of perceptible linkages between human and environmental security notably poverty, stress and migration. Krokfors (1995) contends that these insecurity triggers are germane in the dilemma of environmental security threats at individual level analysis.

The Copenhagen School of Security Studies provided the widening and deepening discussion of security (Buzan et al., 1997). This is considered in recent environmental security scholarship the leading study that argued that security concept should be broadened from military and political to societal, economic and environmental security and threats to identity.

Since the 2000s some researchers had come up with the debate that the environmental security debate had evolved within four phases. Dalby (2002) contends that the first was an initial conceptual work aimed at a broader understanding of security other than the Cold War discourses. The second phase was an attempt by theorists to sketch out how to specify links between environment and insecurity in order to establish a practical research agenda for scholarly analysis. The third encompassed a search for empirical verification or refutation of the initial postulates (Dalby, 2002). While in the 2000s, the fourth school of thinking largely linked to the International Peace Research Institute, Oslo (PRIO), provides a critique of the scarcity debate of the Toronto school suggesting eco efficiency (de Soyza, 2000).

This later perspective which has much relevance with the Niger Delta scenario became easily associated with this chapter as it unveils aspects of ecological breakdown arising from competition among the MNOCs in the Delta for access and control of the oil resources.

Against the background of the foregoing issues and controversies, the field of ES attempts to remedy environmental insecurity problems in various ways. Morel and Igor (2006) have reinforced this tendency at both local and international environmental and resource conflict levels including the sustainability interface as they argue that the "concept of "environmental security" has emerged as one basis for understanding international conflicts" and that the question is "what should be done today to preserve the quality of the environment in the future".

Dalby (2013) and most proponents of human centric perspectives focus on issues of ecological justice, human rights, the effects of globalization and the increasing challenges of Northern consumption patterns on the global South. With the surge in environment and security challenges, the General Assembly officially introduced the concept of security and environment at its 42nd Session (Amadi &Ogonor, 2015).

Amadi and Ogonor (2015) argue that Nigeria is committed to several global Multilateral Environmental Agreements (MEAS), summits and protocols. They contend that the Stockholm conference of 1972 became a spring board on which subsequent global environmental summits were organized. In a general way, these concepts are integral to the overall scope of ES which partly include the dynamics of prevention and remediation of environmental degradation and related threats and justification of environmental protection. Perhaps such justification compels implicit quest to investigate Nigeria's environmental policy response in relation to the oil rich Niger Delta. Barnett (1997) argues that "environmental security is the proactive minimization of anthropogenic threats to the functional integrity of the biosphere and thus to its interdependent human component".

Beyond these contestations, the concept of environmental security captures a wide range of issues which underlie and inform the patterns and dynamics of environmental threats notably the concerns about the adverse impact of human activities on the environment (resource extraction, migration, conflicts, displacement, degradation, etc.). The concept remains rather sensitive to human and natural resources.

Overview of the National Policy on the Environment

Since the return to democracy in 1999, environmental policy response in the Niger Delta appears ineffectual. The complex insecurity problems explicated suggest the saliency of policy response. Considering the complexity of the prevailing environmental security threats, it is useful to have an overview of existing national policy on environment to identify policy gaps and how the gaps could be filled.

Nigeria has a number of environmental policies and statutes most of which were mentioned in the introductory section. However, the chapter focuses on the National Policy on the Environment.

Before 1988, Nigeria never had any national policy on the environment. This does not only point to environmental policy inertia since independence in 1960 but suggests the level of awareness and commitment of the government to environmental issues. A genealogical mapping of environmental policy in Nigeria suggests that the National Policy on the Environment was set up 1989. This was a direct consequence of the 1988 toxic waste incident at the small port town of Koko Delta State, which perhaps drew the attention of the federal government to the critical issues of environmental protection. The National Policy on the Environment launched by the Government on the 27th day of November 1989, two years after the Brundtland Commission report following the World Summit on Sustainable Development lays down the guidelines and strategies for achieving the policy goal of sustainable development, as well as defining the strategies for its implementation.

Government's policy response was to establish an agency for environmental protection- The Federal Environmental Protection Agency (FEPA). That Agency was created by Decree No. 58 of 1988- Federal Environmental Protection Agency Act, Cap 131, 1990 LFN as it was then. The Agency is now the Ministry of Environment. The duty of the Agency was to "prepare a comprehensive national policy for the protection of the environment and conservation of natural resources, including procedure for environmental impact assessment for all development projects" (Section 4(a) of FEPA Act as amended by Decree No. 59 of 1992).

The National Policy affirms Nigeria's commitment to a "policy that ensures sustainable development based on proper management of the environment". It also recognizes the interdependent linkages between "development processes, environmental factors as well as human and natural resources." The policy, which is in essence a policy of sustainable development aims at securing for all Nigerians a quality of environment adequate for their health and well- being amongst other things; seeks to adopt an integrated, holistic and systemic view of environmental issues. It covers various sectors like the Human Population, Culture, Housing and Human Settlements, Biological Diversity, Natural Resources Conservation, Land Use and Soil Conservation, Agriculture, Water Resources, Forestry, Wild life and Protected areas, Marine and Coastal Area Resources, Mining and Mineral Resources, Industry, Energy, Oil and Gas, Construction, Health, Education, Trade, Transport and Communication, Tourism and Science and Technology. The Policy recognized the importance of public participation, institutional and inter-governmental arrangements, legal framework, monitoring and evaluation, for a successful achievement of the laudable goals of the Policy.

The document also identifies the various means for implementing the National Agenda 21, including among others: integrating environment into development planning and decision making; strengthening the legal basis for sustainable development creating and improving capacity for sustainable development; and environmental education, information and public awareness

ENVIRONMENTAL SECURITY THREATS AND POLICY RESPONSE NEXUS: THE NIGER DELTA EXPERIENCE

Despite a plethora of policies outlined in the previous section, environmental policy response in the Niger Delta remains an interesting engagement which perhaps has not been given adequate scholarly attention. The significance of this dynamic lies both in the economic and resource relevance of the Niger Delta in Nigeria as oil from the region accounts for about 75% of the total revenue to the country. Allen (2010) had observed that there has been ineffectual implementation of oil related environmental policies in Nigeria. Thus, the nexus between environmental security threats and policy response creates useful insights in exploring the linkages between access, resource allocation and ecological justice and how policy framings in this direction guide these fundamental principles.

Furthermore, it sets the scene for the persistent clamor for environmental transformation of the region resulting in youth restiveness and militancy within the context of resource struggles and vandalization of oil resource tools of the MNOCs pointing out the palpable policy schism and the clamor for environmental sustainability in the region.

The counter-argument as put forward by the MNOCs is that they observe green corporate social responsibility(GCSR) and ensure that the environment is protected in the cause of resource extraction (Shell Bulletin, 2012). Beyond these, the debate on environmental degradation partly forms the collective basis for agitation on impoverishment of the region. Ibeanu (2000) argued that poverty in the area contrasts the resource rich region. Obi (2001) had argued that the: "analogy of the producers (land owners) being alienated from the products of their land (oil and gas) by the oil multinationals and the state, fits into the identity that the oil minorities have constructed in the quest for self-determination and liberation from their expropriators". This and similar concerns point out the critical need for an analysis of the nexus between environmental security threats and policy response in the Niger Delta

Environmental policy contains rules and regulations which have as their object or effect the protection of the environment. According to FEPA (1998) environmental policy in Nigeria "demands positive and realistic planning that balances human needs against the carrying capacity of the environment". This requires that a number of complementary policies, strategies and management approaches are put in place which should ensure, among others, that environmental concerns including environmental remediation costs are integrated into major economic decision- making process among others

Against the background of these provisions and as useful as this recent spate of environmental policies have been in setting new policy agenda and reforms, there is hardly any evidence of effective environmental remediation or mitigation strategy in the Niger Delta. In a comprehensive environmental assessment of Ogoniland which covers contaminated land, groundwater, surface water, sediment, vegetation, air pollution, public health, industry practices and institutional issues, UNEP (2011) report in their findings underline that; "there are, in a significant number of locations, serious threats to human health from contaminated drinking water to concerns over the viability and productivity of ecosystems. In addition that pollution has perhaps gone further and penetrated deeper than many may have previously supposed". The helpless situations arising from the challenges of environmental insecurity was reinforced with water pollution as UNEP (2011) recounts that hydrocarbon contamination was found in water taken from 28 wells at 10 communities adjacent to contaminated sites. The study concludes that the environmental restoration of Ogoniland is possible but may take 25 to 30 years (UNEP 2011: 12). The soil contamination exceeds Nigerian national standards, as set out in the Environmental Guidelines and Standards for the Petroleum Industries in Nigeria (EGASPIN). There are issues with institutional capacities to confront growing environmental problems. For instance, the UNEP 2011 report found that, "the Department of Petroleum Resources (DPR) and NOSDRA have differing interpretations of EGASPIN" (UNEP, 2011, p. 12). These contradictory interpretations suggest agential policy inertia and enervated grasp of the intensity of environmental degradation in the delta. This has resulted in the persistence of environmental degradation. Beyond this, suggests the urgency of policy engagement with environmental policy issues in the Niger Delta.

The forgoing points out that the question of the relation of State policy to environmental security has never been satisfactorily resolved. This has serious consequences for environmental security transformation and —indeed, sustainable development of the Niger Delta.

RECOMMENDATIONS

What the chapter has primarily demonstrated is that environmental policies for both emergency management and public safety have been ineffectual in addressing environmental insecurity threats in the Niger Delta. The analysis of Nigeria's environmental policy reveal that it was only in 1989 that the federal government took a bold initiative at putting into force a national policy on the environment. Beyond this effort, there have been subsequent environmental statutes. However, these have been both eclectic and peace meal, hence its inability to have a strong bite on the core environmental security challenges identified in this study. It is useful to have a novel environmental policy model, one which would specifically address salient environmental security issues peculiar to the Niger Delta region being a coastal and resource rich area. There is need for urgent review of the provisions of the existing environmental policies to redirect and redefine their focus on environmental challenges in the coastal oil producing areas. Issues of responsibility for environmental contamination beyond oil spills and issues of clean up should be given a distinct attention and assigned to a separate government Agency.

Beyond environmental impact assessment (EIA), periodic environmental due diligence assessment should be undertaken to critically examine patterns of resource extraction in-line with extant laws and global best practices. Harmonized approach to environmental insecurity mitigation with an all-inclusive, participatory bottom-top approach is suggested.

Firstly, Nigeria should evolve official definition of environmental security and policy response that unifies policy and action. Such novel policy framework should be a simplification of reality that focuses and clearly defines specific aspects of the environmental security problems distinctively and possibly as emergency issue to attract the much needed urgent policy attention to the persistent problem of deleterious environmental resource extraction which from this study has become a systemic extermination of the oil bearing communities of the Niger Delta. For instance, since the National Emergency Management Agency (NEMA) was established at the national level with Act No. 12, 1999 and State Emergency Management Agency at the State level, the response to emergency situations have not gained a boost and the local levels.

Local Emergency Agency (LEA) is suggested the aim is to step down emergency response to the local level through capacity trainings on emergency management and recruitment of local emergency management volunteers. This could be useful at the local levels. Although emergency management are often non sustainable remedial strategies that could bring a permanent end to disaster and environmental security threats rather could help reduce the immediate effects of the threats to both humanity and the ecosystem. For instance, the recommended clean- up of oil pollution in Ogoniland does not imply that there could not be possible spill in the future. The policy gap here is how to device strong measures that could serve as both deterrent and sustainable strategies of remediation in case of future occurrences.

There is need to engage civil society organizations (CSOs), public resource managers, relevant Ministries, Departments and Agencies(MDAs), practitioners, Non-Governmental Organizations (NGOs), individual volunteers, researchers, professionals and students in the dynamics of policy response and strategies to emergency management. Novel re-awakening on environmental security challenges and mainstreaming of these threats into development planning and similar policy response is critical.

In the clear case of the Niger Delta, environmental degradation such as water and land pollution, acid rains, gas flaring, deforestation, etc. should form part of emergency policies through which the vulnerability of the oil bearing communities could be mitigated with protection against environmental hazards such as the persistent problems of oil spill. Mitigation strategies through which the affected areas could cope with such disasters are poorly operationalized. The State plays various roles which are relevant to the process of degradation. It can intervene for better resource management. It can regulate resource use or make more resource efficient policies to resource users. The actions of the Nigerian Government can impact on the range of choices of the resource users and managers both in the Niger Delta and elsewhere in Nigeria.

Secondly, one of its most important role is defining the distribution of resources and access to the resources. The State is also responsible for formulating and implementing macroeconomic policies which, depending on its substance, focus and scale, has the potential to transform the environment and its resources. Ultimately, effective environmental security policy aimed at natural resource management such as crude oil, land management, forest reserve, etc. rests directly with the government as it makes the decisions concerning resource use practices (Blaikie and Brookfield, 1987). This study puts the perceptions, actions and patterns of behaviour of the resource users and the managers at the centre of ES and resource efficiency. In this approach, oil resource becomes, inextricably related to the people and society that use them. This also means avoiding a single hypothesis explanation of both environmental degradation and remediation (Blaikie and Brookfield, 1987). Thus, efficient and equitable resource allocation model will meet the needs of the resource bearing communities, the derivative formula in Nigeria remains a ruse.

Lastly, in enunciating a national policy on the environment, there is need for "eco literacy" which aims to provide an all- inclusive and participatory understanding of the complexity of environmental security challenges. This includes multi-stakeholder engagement including institutional capacities and varying historical, social, cultural and legal contexts which shape policy response and outcomes particularly affecting environmental security issues and management which have been at issue in the identification and implementation of measures designed to solve local environmental problems.

FUTURE RESEARCH DIRECTIONS

This chapter attempted to examine environmental security and policy response in the Niger Delta. While a number of policy interventions were identified, the chapter demonstrated that there has been evidence of ineffectual policy response. Future research direction should examine a wide range of issues linked to environmental insecurity in the Niger delta which include the impact on both the environment and humanity particularly the rural households whose subsistence derive directly from the natural environment and options for policy remediation.

Again, future research should investigate environmental renewal, efficient natural resource management, youth restiveness as a reaction to poor government policy response to environmental insecurity and how new policies could be formulated with direct response to ecological breakdown, the plights of the rural communities, women and youths including linkages with unemployment and inclusive bottom top policy initiatives.

Eco literacy research agenda is critical as a new option for the understanding of inclusive resource efficiency for ecological justice, equality and sustainable development.

CONCLUSION

The foregoing discussion points out that since the return to democracy in 1999 and despite series of environmental policies in Nigeria, environmental security threats in the Niger Delta have not been given adequate policy attention. ES thus remains a key issue in the Niger Delta, which prompts the chapter to critically investigate the complex problems as concrete categories in both environmental risk management and development Studies. The insecurity threats briefly highlighted included oil spill, gas flaring, acid rain, deforestation, black soot among others. These are largely anthropogenic environmental security threats linked largely to the MNOCs in the region and oil resource exploitation and exploration.

Broader theoretical elucidation of these environmental security threats were examined from the Marxian political ecology framework which provided a foundational theoretical analysis of the contexts including the nexus between deleterious natural resource extraction by the international capitalist MNOCs and inequality in the Niger Delta.

Conceptually, ES was examined from divergent scholarly perspectives. This also gives the impression that the concept has both policy and scholarly relevance. However, despite its robust offshoot, ES and policy response in the Niger Delta have been ineffectual as the literature reviewed suggested. Scant studies have adequately interrogated the patterns of government policy response to environmental insecurity problems in the Niger Delta. This does not only make this study a useful contribution that draws policy attention to the complex problems of environmental degradation in the Niger Delta, rather it adds value to policy makers, researchers and new comers in the field of ES. Essentially the exploration of ES policy in Nigeria provides a tendency to engage in a peculiar manner with analysis of policy inertia in which the chapter justifies with evidence of divergent environmental insecurity problems in the Niger Delta

Rather than an assumption on the existence of environmental policies in Nigeria, the impact of such policies on environmental insecurity remediation remains at issue. This is primarily an aspect of the objective of this chapter which points out that future research agenda should be directly focused on distinct factors linked to environmental insecurity including the vested capitalist interests of the MNOCs as causal factor of environmental insecurity. This suggests neglect of the oil bearing Niger Delta communities whose environments are degraded. The chapter suggests that it is possible—indeed, desirable—to define novel policy strategies by the government to specifically meet the environmental problems of the region largely conceived as volatile. This include enforceable policy response that should be bidding on key stakeholders. This can be strengthened through legislation and inclusive multi-stakeholder engagement involving the host communities, the policy makers, environmental activists, civil society organizations (CSOs), MNOCs and community leaders with insistence on the distinction between policy formulation and policy implementation as regards sustainable mitigation of threats posed by environmental insecurity in the region.

REFERENCES

Adeola, F. (2000). Endangered community, enduring people: Toxic contamination, health, and adaptive responses in a local context. *Environment and Behavior, 32*(2), 209–249. doi:10.1177/00139160021972504

Ajugwo, A. (2013). Negative effects of gas flaring: The Nigerian experience. *J. Environ. Pollut. Hum. Health, 1*, 6–8.

Akachuku, A. (2006). *Disappearing Forests, the consequences and the challenges of Sustainable development of Nigeria*. Proc. of Forestry Ass. of Nigeria.

Ake, C. (Ed.). (1985). *Political Economy of Nigeria*. London: Longman.

Allen, F. (2010). *Implementation of Oil-Related Environmental Policy in Nigeria: Government Inertia and Conflict in The Niger Delta* (Unpublished PhD Thesis). University of Kwazulu Netal.

Amadi, L. (2013). Climate Change, Peasantry and Rural Food Production Decline in the Niger Delta Region: A case of the 2012 flood disaster. *J. Agric. Crop Res., 1*(6), 94-103.

Amadi, L., & Ogonor, M. (2015). Climate Change, Environmental Security and Displacement in Nigeria: Experience From the Niger Delta. *African Journal of Environmental Science and Technology, 1*(9), 53–64.

Annan, K. (2005). *Towards Development, Security and Human Rights for All*. Available at www.ohchr.org/Documents/Publications/A.59.2005.Add.3.pdf

Anyanwu, D., & Tanee, F. (2008). Tolerance of Cassava (Var. TMS 30572) to different concentrations of Post-planting cruse Oil Pollution. *Nigeria Journal of Botany*, *21*(1), 203–207.

Ashton-Jones, N. (19998). *The Human Ecology of the Niger Delta. An ERA Handbook*. Environmental Right Action (ERA).

Barnett, J. (1997). Environmental Security: Now What? In *Seminar proceedings*. Department of International Relations, Keele University.

Basedau, M. (2005). *Context Matters – Rethinking the Resource Curse in Sub-Saharan Africa*. German Overseas Institute (DÜI) Responsible Unit: Institute of African Affairs. Working Papers Global and Area Studies.

Biersack, A. (2006). Reimagining political ecology: culture/power/history/nature. In A. Biersack & J. Greenberg (Eds.), *Reimagining political ecology* (pp. 3–40). Durham, NC: Duke University Press. doi:10.1215/9780822388142-001

Blaikie, P., & Brookfield, H. (1987). *Land Degradation and Society*. London: Methuen.

Bogardi, J., & Brauch, H. (2005). Global Environmental Change: A Challenge for Human Security Defining and Conceptualising the Environment Dimension of Human Security. In A. Rechkemmer (Ed.), *UNEO-Towards an International Enviornmental Organization-Approaches to a Sustainable Reform of Global Environmental Governance* (pp. 85–109). Baden-Baden, Germany: Nomos.

Booth, K. (1991). Security and Emancipation. *Review of International Studies*, *17*(: 4), 319.

Brauch, H. (2005). *Environment and Human Security. Towards Freedom from Hazard Impacts, UNU-EHS, Intersection 2*. Bonn, Germany: UNU-EHS.

Brauch, H., Oswald, U., Grin, J., Mesjasz, C., & Kameri-Mbote, P et al. (Eds.). (2008). *Facing Global Environmental Change: Environmental, Human, Energy, Food, Health and Water security Concepts*. Berlin: Springer-Verlag.

Brown, L. (1977). *Redefining National Security*. Worldwatth Inst.

Brown, L., & Jacobson, J. (1986). *Our Demographically Divided World*. Worldwatch Paper 74. Washington, DC: Worldwatch Institute.

Brown, N. (1989). Climate, ecology and international security. *Survival*, *31*(6), 519–532. doi:10.1080/00396338908442497

Bruce, E. (2003). Nigeria: The Ogoni: Oil, Blood, and the Death of a Homeland. An Encyclopedia of Indigenous People and Environmental Issues, 1-11.

Bryant, R. (1998). Power, knowledge and political ecology in the third world: A review. *Progress in Physical Geography*, *22*(1), 79–94. doi:10.1177/030913339802200104

Bryant, R., & Bailey, S. (1997). *Third world political ecology: an introduction*. London: Routledge.

Buzan, B. (1991). New Patterns of Global Security in the Twenty-First Century. *International Affairs, 67*(3), 431-451.

Buzan, B., Waever, O., & Wilde, J. (1997). *Security a New Framework for Analysis*. Lynne Rienner Publishers.

Chalecki, E. (2002). *Environmental Security: A Case Study of Climate Change*. Pacific Institute for Studies in Development, Environment, and Security. Retrieved from www.pacinst.org

Dalby, S. (2002). *Environental Security*. Minneapolis, MN: University of Minosota Press.

Dalby, S. (2002). Security and Ecology in the Age of Globalization. *ECSP Report, 8*(95), 95-108.

Dalby, S. (2013). Environmental dimensions of human security. In R. Floyd & R. Mathew (Eds.), *Environmental Security Approaches and Issues*. London: Routledge Taylor and Francis.

de Soyza, I. (2000). The resource curse: Are civil wars driven by rapacity or paucity? In M. Berdal & D. M. Malone (Eds.), *Greed and grievance: Economic agendas in civil wars* (pp. 113–135). Boulder, CO: Lynne Rienner.

Deudney, D. (1990). The Case Against Linking Environmental Degradation and National Security. *Millennium, 19*(3), 465. doi:10.1177/03058298900190031001

Diukwu, I., Ijaiya, M., Mohammed, M., & Egila, A. (2013). The Effect of Gas Production, Utilization, and Flaring on the Economic Growth of Nigeria. *Nat. Resour, 4*, 341–348.

Ekpo, U. (2004). *The Niger Delta Oil and Politics*. Lagos: International Energy Communications Limited.

Emordi, E., Osaghae, E., & Nwaokocha, M. (2008). *Niger Delta Crisis and the State's Response Up to 2007*. Paper presented at the 1st Annual Conference of the Society of Peace Studies and Practice, Institute for Peace and Conflict Resolution.

Eweje, G. (2006). Environmental Cost and Responsibilities Resulting from Oil Exploration in Developing Countries: The Case of the Niger Delta of Nigeria. *Journal of Business Ethics, 69*(1), 27–56. doi:10.1007/s10551-006-9067-8

Federal Environmental Protection Agency (FEPA). (1998). *Revised National Policy on the Environment*. Federal Republic of Nigeria.

Floyd, R. (2008). The environmental security debate and its significance for climate change. *The International Spectator, 43*(3), 51–65. doi:10.1080/03932720802280602

Flyod, R., & Mathew, R. (2013). *Environmental Security Approaches and issues*. London: Routledge.

Foster, J. B. (1999). Marx's Theory of the Metabolic Rift: Classical Foundations for Environmental Sociology. *American Journal of Sociology, 105*(2), 366–405. doi:10.1086/210315

Frynas, G. (2000). *Oil in Nigeria: Conflicts and Litigations Between Oil Companies and Village Communities*. Hamburg, Germany: LIT Verlag Munster.

Gaffey, C. (2016). Niger Delta Communities Sue Shell Over Oil Spills. *Newsweek*. Available at http://www.newsweek.com/niger-delta-communities-sue-shell-over-oil-spills 432300Accessed 10/7/2017

Haque, M. (1999). The Fate of Sustainable Development under Neo-liberal Regimes in Developing Countries. *International Political Science Review*, *20*(2), 197–198. doi:10.1177/0192512199202005

Harvey, D. (1996). Justice, Nature and the Geography of Difference. Oxford, UK: Blackwell.

Harvey, D. (2005). *Brief History of Neoliberalism*. New York: Oxford University Press.

Homer-Dixon, T. (1991). On the Threshold: Environmental Changes as Causes of Acute Conflict. *International Security*, *16*(2), 76–116. doi:10.2307/2539061

Homer-Dixon, T. (1999). *Environment, Scarcity and Violence*. Princeton, NJ: Princeton University Press.

Iaccino, L. (2015). Nigeria's oil war: Who are the Niger Delta militants? *International Business Times*. Available at http://www.ibtimes.co.uk/nigerias-oil-war-who-are niger-delta militants-1520580.accessed12/21/2015

Irikana, G. (2015). The Niger Delta Ecosystem and Oil production. An Analysis of Capitalist Incorporation of A Natural Economy. In Conflict and Development in the Niger Delta Sociological Perspectives. Enugu: ABIC Books.

Jike, V. (2004). Environmental Degradation, Social Disequilibrium, and the Dilemma of Sustainable Development in the Niger Delta of Nigeria. *Journal of Black Studies*, *34*(5), 686–701. doi:10.1177/0021934703261934

Kaldor, M. (1999). *New and Old Wars: Organized Violence in a Global Era*. Cambridge, UK: Polity Press.

Kaplan, R. (1994, February). The Coming Anarchy:How security crime, overpopulation, tribalism and disease are rapidly destroying the social fabric of our planet. The Atlantic Monthly, 49.

Kennedy, P. (1992). *Preparing for the 21st century*. New York: Random House.

Klare, M. (1996, November). Redefining Security: The New Global Schisms. *Current History*.

Korvenoja, T. (1993). The Environmental Problems and Politics of Power Review on the African Elite. *Nordic Journal of African Studies*, *2*(1), 140–153.

Krokfors, C. (1995). Povery, Environmental Stress and Culture as Factors in African Migrations. In J. Baker & T. Akin Aina (Eds.), *The Migration Experience in Africa* (pp. 54–64). Nordiska Afrikainstitutet.

Mathews, J. (1989). Redefining Security. *Foreign Affairs*, *68*(2), 2. doi:10.2307/20043906 PMID:12343986

Meadows, D. (1972). *The Limits to Growth: A report for the Club of Rome's project on the predicament of mankind, Part 1 Club of Rome*. Potomac Associates.

Mische, P. (1989, February 18). Arctic Expedition Finds Chemical Threat to Ozone. *New York Times*.

Mmon, P., & Arokoyu, S. (2010). Mangrove forest depletion, biodiversity loss and traditional resources management practices in the Niger Delta, Nigeria. *Research Journal of Applied Sciences, Engineering and Technology*, *2*, 28–34.

Mogborukor, J. (2014). The Impact of Oil Exploration and Exploitation on Water Quality and VegetalResources in a Rain Forest Ecosystem of Nigeria. *Mediterranean Journal of Social Sciences, 5*(27).

Morel, B., & Igor, L. (Eds.). (2006). *Environmental Security and Environmental Management: The Role of Risk Assessment*. Springer. doi:10.1007/1-4020-3893-3

Myers, N. (1989). Environment and Security. *Foreign Policy*, *74*(74), 23–41. doi:10.2307/1148850

Naanen, B., & Tolani, P. (2014). *Private Gain Public Disaster Social Context of Illegal oil Bunkering and Artisanal Refining in the Niger Delta*. NIDEREF.

Nigerian Environmental Study/Action Team (NEST). (2004). Available at www.climatenetwork.org/profile/.../nigerian-environmental-studyaction-team-nes

Nriagu, J., Udofa, E., Ekong, I., & Ebuk, G. (2016). Health Risks Associated with Oil Pollution in the Niger Delta. *NigeriaInternational Journal of Environmental Research and Public Health*, *13*(3), 346. doi:10.3390/ijerph13030346 PMID:27007391

Obi, C. (2000). Globalised Images of Environmental Security in Africa. *Review of African Political Economy*, *27*(83), 47–62. doi:10.1080/03056240008704432

Obi, C. (2001). *The Changing Forms of Identity Politics in Nigeria under Economic Adjustment: The Case of the Oil Minorities Movement of the Niger Delta*. Research Report Nordiska Afrikainstitutet No.11.9.

Ogata, S., & Sen, A. (2003). *Human Security Now: Commission on Human Security*. Final Report. New York: United Nations.

Okoko, K. (2002). The Politics of Resource Extraction, Distribution and Sustainable Development. Nigerian Journal of Oil and Politics, 2(1), 118-128.

Osuide, S. (1990). Environmental pollution in Nigeria. *Habitat International*, *14*(1), 5–15. doi:10.1016/0197-3975(90)90014-R

Paris, R. (2001). Human Security: Paradigm Shift or Hot Air? *International Security*, *2*(26), 87–102.

Peet, E., & Watts, W. (1996). Liberation Ecology: Development, Sustainability and Environment in an Age of Market Triumphalism. In R. Peet & M. Watts (Eds.), *Liberation Ecologies: Environment, Development, Social movements*. London: Routledge. doi:10.4324/9780203286784

Pirages, D. (1991). Social Evolution and Ecological Security. *Bulletin of Peace Proposals*, *22*(3), 329–334. doi:10.1177/096701069102200310

Ramsay, J., & O'Sullivan, T. (2013). There's a Pattern Here: The Case to Integrate Environmental Security into Homeland Security Strategy. *Homeland Security Affairs, 9*, Article 6. Retrieved from https://www.hsaj.org/articles/246

Robbins, P. (2004). What is political ecology? In P. Robbins (Ed.), *Political ecology: A critical introduction* (pp. 5–16). Malden, MA: Blackwell Publishing.

Roggers, K. (1997). Article. *Ecological Security and Multinational Corporations*, *2*, 7–17.

Shiva, V. (1988). *Staying alive: Women, ecology, and development*. London: Zed Books.

Spaargaren, G. (2000). Ecological modernization theory and domestic consumption. *Journal of Environmental Policy & Planning, 2*(4), 323-335.

Speake, S., & Gismondi, M. (2005). Water: A Human Right. In Consuming Sustainability Critical Social Analysis of Ecological Change. Fernwood Publishing.

Ubani, E., & Onyejekwe, I. (2013). Environmental impact analyses of gas flaring in the Niger delta region of Nigeria. *American Journal of Scientific and Industrial Research*, *4*(2), 246–252. doi:10.5251/ajsir.2013.4.2.246.252

Ullman, R. (1983). Redefining Security. International Security, 31, 129.

UNDP. (2006). *Niger Delta Human Development Report.* Oxford Press.

UNEP (United Nations Environment Programme). (2011). Environmental Assessment of Ogoniland. UNEP Report.

United Nations Development Programme. (1994). *Human Development Report: Redefining Security.* Oxford, UK: Human Dimensions.

United Nations Framework Convention on Climate Change. (1994). *UNFCCC Report.* Author.

Westing, A. (Ed.). (1986a). *Global Resources and International Conflict: Environmental Factors in Strategic Policy and Action.* Oxford, UK: Oxford University Press.

World Commission on Environment and Development (WCED). (1987). *Our Common Future.* Oxford Press.

KEY TERMS AND DEFINITIONS

EGASPIN: Environmental guidelines and standards for the petroleum industries in Nigeria.

EIA: Environmental impact assessment.

ES: Environmental security.

FEPA: Federal environmental protection act.

LEA: Local emergency agency.

LGA: Local government area is the smallest level of government in Nigeria's federal system, which includes the federal, state, and local government.

MNOCs: Multi-national oil corporations.

NOSREA: National oil spill detection and response agency.

WSSD: World summit on sustainable development.

Chapter 11
Climate Change and Sustainable Development in Nigeria:
Limitations and Opportunities

Ighodalo Bassey Akhakpe
Lagos State University, Nigeria

ABSTRACT

The chapter assesses the nature and effects of climate change on sustainable development in Nigeria. It observes that climate change has a multifarious effect not only on the environment but also on the socio-economic life of the people. Therefore, if sustainable development is to be realized in the country, climate change should be properly managed through extant public policies. However, if government track records on policies and program implementations is anything to go by, the future of sustainable development is gloomy. This makes an interrogation of the interface between climate change and sustainable development germane. The chapter observes that while government has shown willingness to manage climate change for the sustainability of the environment and its people, certain limitations stand on its path. These include poor policy or program implementation, inadequate funding of climate change management, poor sensitization program on environment management, among others. However, there are opportunities that can be harvested at the state and individual levels.

INTRODUCTION

Nigeria is in urgent need of transformation in its socio-economic and political life. The most populous nation and biggest market in Africa ironically houses some of the poorest of the poor in Africa. This development is not unrelated to several challenges that characterize its match to human and societal development. While many of its challenges are not unique to the country in particular and Africa in general. however, its capacity to manage them and possibly overcome them has been low and in some cases, non-existence. This has made the quest for socio-economic transformation of the continent an herculean task inspite of efforts by successive governments in the various African countries to achieve its goal.

DOI: 10.4018/978-1-5225-3194-4.ch011

One of these challenges is the phenomenon of climate change. Why climate change is a global phenomenon that is affecting different countries of the world, the capacity of the individual countries to attenuate its effects through well-articulated policy instrument and well managed policy implementation mechanisms differs from one country to another. While policies on climate change have not been in short supply, across the continent the human and technical capacity gap have constrained efforts at tackling this challenge. Climate change negatively affects both human and material resources of every society. It has led to early death of most living things, destroys the ecosystem and depletes finite and non-finite resources.

With this development, a chance of achieving sustainable development in Nigeria becomes increasingly remote. The quest for sustainable development is hinged on the desire of using and preserving the resources of the present generation without endangering their usage by future ones. It is against this backdrop the United Nations declared in 2015, a decade and half for the pursuit of sustainable development in the globe. However, the sub-sets of sustainable development are presently threatened by climate change which makes it imperative to investigate its effects on the overall well being and sustainable socio-economic development of the people and society.

To address the central issues in the study, the paper is divided into the following parts. The first part engages in conceptual and theoretical indulgence, the second explores the nature and character of climate change and sustainable development. The third discusses the effects of climate change on sustainable development in Nigeria with particular reference to the socio-economic concern of the people and society. The forth analyzes challenges facing efforts at managing the effects of climate change on sustainable development. The fifth provides a road map for managing climate change for sustainable development through some remedial policy proposals.

Conceptual Clarification and Theoretical Indulgence

The concept of climate change has become a global phenomenon concept and process. Ojo and Oni (in Adejobi and Akhakpe (2012) define climate as "the total experience of weather many place over some specific period of time". This perspective to the concept of climate rests on its dynamic nature and coverage which defer from one period of time to the other. It denotes the average deviations that occurs and allows for intensive and extensive evaluation of climate data (Adejobi, and Akhakpe, 2012). Similarly, climate change is a process of alteration of climate conditions covered by factors such as increased frequency of heat and ground level ozone. However, climate change should be seen in the context of many other factors in the global setting including the process of urban development. Other climate change inducing factors are desertification, deforestation, exploration finite and non-finite resources, reduction of biodiversity and soil fertility.

There are many perspectives to defining the concept of development. The United Nations define it as a process that increases people's opportunity of choice. For Rudeback (1997), development is the ability of a people to use their cultural values to bring about progressive change in which the present condition is better than the previous one. Therefore, a people must make a choice of the developmental paradigm to apply or follow. In the same vein, sustainability could mean maintaining and retaining a chosen and viable developmental path for a long time in order to preserve both finite and non finite resources. Therefore, sustainable development demands the use of world resources to meet the needs of the present and future generation. The World Commission on Environment and Development (WCED) (in Oyeshola, 2005) defines sustainable development as "development that meets the needs of the present without compromising the ability of the future generations to meet their own needs".

This perspective to sustainable development is very germane to the quest for equity, fairness and social justice in the exploitation of natural resources and their management. It is instructive to note that the present politico-economic elite have tended to exploit and use the country's natural resources in a wasteful and reckless manner to disempowered the people socio-economically and endanger global security through climate change and such negative effects.

The Basic Need Theory hinges its arguments on the desirability and inevitability of state institutions to pursue and realize the provision of social services that promote and enhances the well-being of the people. Public policy should be directed at providing food, housing, water supply, education, clothing for the people, among others. However, the instruments of public policies as tools for dealing with social service delivery and climate change management have been largely ineffective, as they frequently suffer causalities (Ayee, 2008; Oyeshola, 2008 and Olawaye, 2010; Iyoha, 2008). While the Basic Need Theory (BNT) may not be the most elastic theory to underpin the subject matter of the chapter, it nonetheless, prioritizes the imperative of concentrating on the most pressing and demanding concerns of the people across the continent of Africa – food, shelter, education, moral leadership, among others. These concerns are affected by climate change. Unless they are approached and dealt with cooperatively, their effects on people and society would not be achieved.

The Nature and Character of Climate Change and Sustainable Development

Climate change is a portent force that is affecting every facets of societal development. In the quest for sustainable development, climate change cannot be underestimated or undervalued. For example, "the Federal government has observed that the present high rate of population growth is already contributing substantially to the degradation of the ecology of the country (Adejobi and Akhakpe, 2012). Also, Olawole (2010) has shown that climate change is not just about environmental, scientific or technological issues but embraces social issues especially, those pertaining to gender realities.

Just as climate change affects different areas and aspects of human lives so also sustainable development should be observed, planned and analyzed in its social, economic, environmental and even moral dimensions (Ayeni, 2010; Amue and Igwe, 2016). While conceptualizing the various dimensions of sustainable development can segregate or disentangled, in terms of process they should be configurative pursued so that no one part should be left stranded. Socio-economic and developmental sustainability entails:

An economically sustainable system (that is) able to produce goods and services on a continuous basis and maintain manageable economy without jeopardizing sectoral balances of economic activities of the country. Environmentally, sustainable system must maintain a stable resource base avoiding over-exploitation of renewable resource system or environmental sink functions depleting non-renewable resources only to the extent that investment is made in adequate substitutes ... The last factor must achieve distributional equity, adequate provision of social services including health care, education, gender equity and political accountability, transparency and participation. (Harris, 2000, p. 7)

To these primary factors could be added a moral one that places/demands on leaders in both the public and private space to take responsibility and responsiveness to the demands and needs of the people. As a factor that combines other dimensions of sustainable development and climate change, it should lead to a process that places emphasis on due diligence and commitment to the course of societal development and the citizens' welfare and well-being.

Meanwhile, in the Niger-Delta region of the country, oil exploration and exploitation activities are being carried out relentlessly with little or no control from the various governments in the country. While government at different levels have set up agencies to manage issues relating to climate change, however, their operations have tended to fall short of the requirements for dealing with climate change and sustainable socio-economic life (Olojede, Fajonyomi and Akhakpe, 2013; Amue and Igwe 2016).

Climate change has had different variational effects ranging from depletion of the ozone layer, excessive heat, flooding and erosion, air and water pollution, etc. However, it could take a long time before one can conclude that there is climate change. Sustainability of some features of climate change should be present such as: the alteration of the concentration of greenhouse gases aerosol both of which influences and are influenced by climate change. Similarly, on a global scale climate change has been caused by greenhouse gas emission including cases of global warming and rise in sea level (Adejobi and Akhakpe, 2012).

Meanwhile, it has been argued that "Nigeria like other developing countries, contributes insignificantly to greenhouse emission. However, no matter the level of its contribution, climate change brings among other outcomes: flooding, rise in sea level, erosion, rise in temperature and nomadism (Olojede, Fajoinyomi and Akhakpe, 2013; Akhakpe, 2008 and Olawoye, 2010).

These developments affect the basic needs of the people in particular and society in general. Yet, it is not clear how these development have been manifested in concrete terms across the country. This demands some identifications and analysis. To this, we now direct our attention in the next part of the chapter.

Climate change has continued to exert immense impact on not only the global environment but also human livelihood in most urban and rural settlements. The deleterious activities of natural forces, hazardous human habits and unwholesome practices of national and international corporations in some parts of the world, affect the delicate balance of the ecosystem. In the global south including countries like Nigeria logging, deforestation and mining activities are rampant (Dauda, 2017). In the Northern part of Nigeria, desert encroachment is progressing almost inhibited. While soil and river erosion have been ravaging towns, cities, villages and communities in Southern part of the country. All these activities have made the environment vulnerable to climate change and unsustainability of the development drive of the country.

Arguably, the most fundamental driver of climate change is Global Environmental Change (GEC). According to Simon (2007):

The burning of fossil fuel as an energy source and various industrial processes that generate heat are contributing to an increase in turn affect rainfall and weather pattern in the short term and climatic condition in the long run. Direct heat released into the atmosphere is one contributor but the principal mechanism is through the build up of Greenhouse Gases (GHG) in the atmosphere. As the same implies, these gases of which carbon dioxide is the most important – hence its use as the key benchmark indicator – act like a thermal blanked by retaining heat. This inhibits albedo (reflected solar radiation) and heat generated by human activities from escaping through the earth's atmosphere, so that combient atmospheric rise.

The emission of GHG is likely to increase unless urgent steps are taken to stem the tide. For example, it is expected that "by the end of the century, unabated emission discharges would raise the level of CO_2 equivalent to around 750ppm watts 50 percent change of 5^0C temperature increase (Simon, 2007). However, all hope is not lost as the worst impact of climate change could substantially reduce if GHG levels were stabilized within the range of 450-550ppm" (Simon, 2007).

Also, some environmentally harmful activities and actions have further made the environment vulnerable to pollution and consequently climate change. Indiscriminate felling of trees, sand filling of water ways, poor refuse disposal and lack of proper lay out for construction of houses particularly in large cities are contributors or drivers of climate change in this part of the world. People need to be reminded that nature has a way of supplying us with oxygen therefore; nothing should be done to upset the balance of nature which these activities tend to engender in our environment. Unfortunately, such activities are likely to increase with the present socio-economic challenges facing the country as people struggle to make ends meet.

A major affront on climate and environmental stability comes from the activities of Multinational Corporations and Transnational Corporations (MNCs and TNCs). This is exemplified in the Niger-Delta region of Nigeria where the oil-ecology nexus is well played out! In this region, environmental abuses arising from gas flaring is a common feature in their oil production activities. Gas flaring began when Shell Petroleum Company first struck oil in commercial quantity at Oloibiri in the present day Bayelsa State. Thereafter, oil has been discovered in other areas in the Niger-Delta region making it the most prolific oil producing area of the country. However, gas flaring has continued for over 55 years with deleterious effects on the environment and people of the area in particular and Nigeria in general. In the face of gas flaring, the state through its government appear, handicap to decisively deal with the environmental hazard this produces due probably to the huge revenue they derive from the oil. Given the weak legal and policies frameworks for managing oil exploration and exploitation in the country, environmentalist and international agencies like the World Bank, United Nations Development Programme (UNDP) Climate Justice Programme (CJP) and Environmental Right Action (ERA) among others, have launched series of campaign to draw world attention to the environmental injustice and threat they pose to environment and Sustainable Human Development (SHD).

Also, environmental stress is further created by the over exploitation of forest resources by those engaged in logging activities and other forest related activities. The tropical rain forest habors forest resources which for years have been exploited for profit by National and Multinational Corporations with little respect for global environmental standard. These activities destroy the high species diversity thereby upsetting the tropical forest constantly that is warm and humid environmental conditions of the area. These create enabling context for climate change to occur with serious effect on environmental sustainability.

Climate Change and Environmental Vulnerability

Today, the environment has become more vulnerable to natural and human events and activities which predispose it to global environmental change. Natural disasters can and have indeed occurred in our times. However, such events occur not often and when they do, they are of short duration, spontaneous and with different degrees of impact on the environment and human life. They take the form of among others: earthquakes, volcanic eruption, tsunamis and hurricanes (Simon, 2007). Another category of this so called "natural disaster" is caused by human activities that have significantly changed the environment most times negatively. Simon (2007) identifies these activities as: "severe flooding due to the destruction of coastal mangrove swamps, river flooding due to upstream deforestation and the destruction of roadbeds along riverbanks lands and mudslides, down steep mountain slopes following deforestation, the blockage of natural drainage channels and the construction of homes on vulnerable slopes". All these human activities and their outcomes expose the fragile environment to direct impact of global climate change such as Greenhouse emission and depletion of the ozone layer and its accompanying global warming.

Perhaps, no better place has the effect of human activities demonstrated than in the Niger-Delta region where gas flaring for example has resulted not only in lost of profit by the Nigeria state but also poses health hazard to the people in areas where these activities take place. According to ERA and CIP (in Akali, 2007):

More gas is flared in Nigeria than anywhere else in the world. This is equal to 40% of all Africa's natural gas consumption in 2001 while the annual financial loss to Nigeria is about USD 2.5 billion. The flare have contributed more to Greenhouse gas than all Sub-Sahara Africa combined. And the flare contain a cocktail of toxin that affect the health an livelihood of local communities exposing Niger-Delta residents to increase risk of premature death, child respiratory illness, asthma and cancer.

Beyond the enormous loss of revenue from gas flares, the healths of the people are in constant and serious danger. It would seem lack of political will on the part of successive governments is at the heart of these environmental and human injustices. This explains why the Human Development Index (HDI) and Sustainable Human Development (SHD) of the country's human and material resources have been on the downward slide over the years.

Besides, the policy response capacity of government has been everything but efficient and effective. While institutions have been created and policies rolled out to regulate and cushion the effects of exploration activities on the environment such as: the Niger-Delta Development Commission (NDDC), Environmental Protection Agency (EPA), Climate Justice Program (CIP) and other environmental right bodies, the policies of environmental governance has not been felt by the people and environment concerned.

In 2011, the Climate Change Vulnerability Index (CCVI) published by the UK based Risk Company – Maplecroft classifies Nigeria as being of high risk. Yet, Nigeria rank low in the area of environmental control performance. For example, "The 2012 Environmental Performance Index (EPI) ranks the country as 112th of 119 countries surveyed in the world with the score of 40.1. Although this shows some improvement over the 2008 performance in which Nigeria scored 56.2, the low failure still puts Nigeria behind many other African countries" (FGN, 2012). This shows that there is lack of sustainability in the management of climate change and environmental vulnerability. Given the low resistance and fragile nature of the environment in areas like Lagos and the Niger-Delta region, more concerted effort are needed to sustain the management of the backlashes of climate change (FGN, 2012).

However, the global level, efforts have been stepped up to address issues emerging from climate change. The 2015 Paris agreement has brought the major industrialized countries together in fashioning out action plans to manage issues arising from climate change. Though, the United States of America under President Donald Trump appears to be reluctant in signing the Paris agreement, some States in the US have started its implementation. It should be emphasized that this is an issue that requires global commitment and concerted efforts by all concerned.

Also, in 2015, the United Nations came up with the Sustainable Development Goals (SDGs) which have several of its goals directly and indirectly related to issues of climate change. These goals include:

Goal 1: "Call for an end to poverty in all its ramifications";
Goal 2: "End to hunger, achieve food security and improve nutrition and promote sustainable agriculture";
Goal 3: "Ensure healthy lives and promote well-being for all at all ages";
Goal 4: "Ensure inclusive and equitable quality education and promote life long learning opportunities for all";
Goal 5: "Achieve gender equality and empower all women and girls";
Goal 6: "Ensure availability and sustainable management of water and sanitation for all";
Goal 7: "Ensure access to affordable, reliable, sustainable and modern energy for all";
Goal 8: "Promote sustained, inclusive and sustainable economic growth, full and productive employment and decent work for all";
Goal 9: "Build resilient infrastructure, promote inclusive and sustainable industrialization and foster innovation";
Goal 10: "Reduce inequality within and among countries";
Goal 11: "Make cities and human settlements inclusive, safe, resilient and sustainable";
Goal 12: "Ensure sustainable consumption and production patterns";
Goal 13: "Take urgent actions to combat climate change and its impacts";
Goal 14: "Conserve and sustainably use the oceans, seas and marine resources for sustainable development";
Goal 15: "Protect, restore and promote sustainable use of terrestrial ecosystems, sustainably manage forest, combat desertification and halt and reverse land degradation and halt biodiversity loss";
Goal 16: "Promote peaceful and inclusive societies for sustainable development, provide access to justice for all and build effective, accountable and inclusive institutions at all levels"; and
Goal 17: "Strengthen the means of implementation and revitalize the global partnership for sustainable development" (United Nations, 2016).

Given the importance of issues involved in the implementation of these goals, it brings to the fore challenges of policy implementation that has for decades affected national development in the country. If the experience of the country with national plans is anything to go by, it would appear that the implementation of the sustainable development goals as it relates to climate change may run into some difficulties given the poor policy implementation records of the federal, State and Local governments in Nigeria. This point is well made by Gambari (in Maduabum 2014) that:

Nigeria's development efforts have over the years been characterized by lack of continuity, consistency and commitment (3Cs) to agreed policies programmes and prospects as well as an objective of a long term perspective. The culminating effort has been growth and development of the Nigerian economy without concomitant improvement in the overall welfare of Nigerian citizens.

It is not enough to initiate a policy, implementing such policy is key to achievement of its goals. Therefore, all hands should be on deck to implement the climate change policies at all levels. Sustainable Development Goals present another challenge on how well public policy can be implemented in the country. Will the goals relating to climate change in general and environmental sustainability in particular go in the way of previous ones?

The Effect of Climate Change on Sustainable Development: An Overview

In the preceding part of this discourse we examined the nature and characters of sustainable development in Nigeria. In this part, we will analyze the different areas where the occurrence of climate change has affected the sustainability of both human and material resources.

It is important to state at this juncture that climate change is now a global phenomenon. But the activities of man have intensified its occurrences and outcomes. In the Niger-Delta region of Nigeria, the feverish search for finite and non-finite resources has left the area and community devastated with negative implication for farming, drinking water, gas flaring, greenhouse emission, among others. The oil producing companies and other mineral exploration activities have left their areas of operation seriously devastated with adverse effect on the eco-system.

Most farmers in these areas have left their sources of livelihood for other things like migrating to the cities in search for white-collar jobs. This has reduced drastically the number of those involve in farming. The immediate effect of these atrocities on the ecosystem is the migration of able bodied men and women who have abandoned their farms and consequently create room for food crisis and or food insecurity in the country. Agriculture products are almost priced out of the reach of the average Nigerians. Farmers who refuse to join the bandwagon and remain in agriculture (farming) are made to contend with adverse conditions caused by heavy torrential rain or drought caused by climate change. However, these conditions are sometime caused by unfriendly environmental habits. The resultant effect of this is what has become known as food crisis or insecurity in the country. As Idachaba (in Olaniran, 2000) put it "Nigeria, blessed as it is with abundant agro-ecological resources and diversity, has become one of the largest food importer in Sub-Saharan Africa".

The effect of climate change on the social well-bring and security of the people is devastating is equally. It has generated excessive heat that depletes the ozone layer thereby subjecting human beings to unprotected heat. Presently in Nigeria, over sixteen states have been affected by the outbreak of cerebral meningitis leading to the death of hundreds of people. Death toll is expected to rise in the coming days as vaccines are in short supply while distribution of existing ones is facing logistic difficulties in some areas (The Punch Newspaper, 2017).

Rapid urbanization in most towns and cities and overcrowding it brings on its heals, generate heat that is partly the cause of the outbreak of some diseases such as: cholera, polo, malaria fever, lasa fever, malnutrition, river blindness and so on. Studies have shown that that urbanization structure in the area of population density have caused the urban environment to deteriorate (Adejobi and Akhakpe, 2012). Also, the size of housing plots has been reduced with the growth in population thereby increasing density and the potentials for traffic congestion. It would seem deficiencies in development control have serious consequences for the urban climatic conditions (Adejobi and Akhakpe, 2012).

From the perspective of vulnerability to the danger of climate change, studies have equally shown that women are more likely to be affected by such occurrence. This is particularly true of "societies that are m ore inequitable, men are likely to receive preferential treatment in rescue efforts and women are likely to suffer more from shortages of food and other resources in the aftermath of disasters. These gender inequalities usually render climate change events to be so devastating to women" (Olawole, 2010). The same can be said of other vulnerable groups in the society. The ability of these groups including children and students to contend with climate change has reduced drastically in recent times due to economic recession, unemployment and growing poverty.

As the poorest of the poor continue to increase in most urban and rural areas of the country, social unrest and upheaval have intensified. Violence and impunities have become the order of the day in most towns and cities across the country. Some of their extreme manifestations are reflected in insurgency in the northern and southern parts of the country, child trafficking, killings by Herdsmen, street trading, school drop outs, kidnapping, among others. On the part of government, human right abuses have continued unchecked as it seeks to suppress the opposition to its policies and programs. Many Nigerians are detained without trial and state apparatuses are covertly and overtly used to intimidate and harassed both perceived and imaginary "enemies" of the government in power.

In mineral bearing and producing regions of the country, mineral exploration and exploitation, particularly in the oil rich region of the Niger-Delta, has brought about enormous ecological damage that has exacerbated the phenomenon of climate changes in these areas (Olojede, Fajonyomi and Akhakpe, 2013). Land and water pollution has intensified thereby negatively affecting the economic and social life of the habitants of these areas. Oil spill, gas flaring, oil pipeline vandalization have all combined to make these areas unsuitable for human habitation. Unfortunately, not much is being done on the part of government and the oil companies to ameliorate or manage the effect of these socio-economic injustices (Olojede, Fajonyomi and Akhakpe, 2013).

Efforts of Government at Combating Climate Change for Sustainable Development

Successive governments in the country have not been oblivious of the untoward effects of climate change particularly, as it affects sustainable development. The policy regimes of government at the Federal and State levels are directed towards checkmating the effect of climate change. In this regard, institutions have been set up to provide policy guidelines and frameworks for addressing issues arising from climate change. In order to promote climate resilient environmental sustainability efforts have been put in place at all levels of governance. These can be gleaned from the following areas:

1. "Institutional and capacity building;
2. Promoting an enabling climate change policy environment;
3. Political and legal framework;
4. Participation in global initiatives;
5. Partnership building for climate change;
6. Climate change financing;
7. Climate change information;
8. Other climate change initiatives" (FGN, 2012).

The federal government has set up some national bodies with specific mandate to address issues relating to and arising from climate change. In December, 2011, the Federal government created the Department of Climate Change (FGN, 2012). It is the Designated National Authority (DNA) with the mandate to ensure development mechanism, and also to work with a number of ministries through the inter-ministerial council on climate change (FGN, 2012). Another major player in the climate change scenario are the Nigerian Meteorological Agency (NIMET) and the Center for Climate Change and Freshwater Resources, Federal University of Technology, Minna has joined the group of climate change actors at the national level. At the state and local government levels, similar agencies and departments are operational to manage the issue of climate change in their respective areas.

In spite, of these efforts at the various levels of governance to address and manage the negative effects of climate change on socio-economic and environmental life of the people and society, Nigeria still score low in the management of issues arising from climate change. Beyond institutional structure and public policy formulation, not much seem to have been achieved in the area of curtailing the adverse effect of climate change on human development and environmental sustainability. It has been observed that:

Overall, Nigeria ranks low in terms of its environmental performance. The 2012 Environmental Performance Index (EPI) ranked the country as 119th out of 137 countries surveyed in the world, with a score of 40.1. Although this shows some improvement over the 2008 performance in which Nigeria scored 56.2, the low EPI figure still puts Nigeria behind many other African countries. Nigeria was rated 19th out of 21 Sub-Saharan African countries scoring better than only Eritrea and South Africa. The country ranked 130th on Environmental burden of disease, 26th on agriculture and 81st on diversity and 41st on climate change. (Federal Government of Nigeria, 2012, pg. 57)

From the comparative analysis carried out above, it would seem Nigeria is yet to come to terms with challenges posed by climate change on the environment, economy and social well-being of the people. The fact that the country is presently in recession makes this challenge more urgent and worrisome. The point has been made FGN (2012) that:

In general, Nigeria is strongly predisposed to severe negative impact of climate change due to its fragile economy, weak resilience and low adaptive capacity. Much of the economy is dependent on climate sensitive resources. For example, the agriculture, forestry and fishing sectors employ up to 70% of the workforce. Its fossil fuel dependent economy will be particularly vulnerable to climate change induced frequency and severe extracted events, such as flood and droughts. The heavy concentration of GDP generating industries in locations that are highly vulnerable to climate change induced sea level rise e.g. Lagos and the Niger-Delta makes the country extremely vulnerable.

Given the fragile nature of the economy, environmental vulnerability and social dislocation in the country, the basic need for life sustenance is declining rapidly. Yet, it is not clear what factors promote these trends in the country. While climate change is a global phenomenon, the ability of different countries to address its emergence differs. This explains why it is more impactful in some country's people and environment than others. In the next part, we shall discuss challenges facing management of climate change for sustainable development in Nigeria.

Challenges Facing Management of Climate Change and Sustainable Development

Nigeria as a developing country is faced with the problem of resource gap that has made it difficult for it to tackle headlong challenges of climate change. While it could be diagnosed causes of climate change and put in place measures or policies to manage climate change issues, the resources needed to implement and monitor the process appear inadequate. This condition is likely to worsen with the weight of economic recession breathing down the neck of the country.

Closely related to the foregoing point is the problem of sustainable policy regime. While well articulated policies may not be in short supply, their casually rates are very high (Iyoha, 2008; Olaopa, 2016). Besides, received paradigms are common features in public policy management in this part of the world (Akhakpe, 2016). The effect is that policy made and implemented in the country lack adequate capacity to address issues of climate change peculiar to the Nigerian circumstances.

Also, human capacity needed to manage these issues remains pathetically low. Issues relating to climate change cut across a wide range of areas with each having its own peculiarities. In the same vein, the manpower needed to deal with them also varies from one sector to another. However, the country, either do not have the resources to pay available experts in these areas or they are simply non-existence. In either way, these are challenges that impact negatively on the country and the living standard of the people.

The behavioural contents of climate change actors leave much to be desired. This is not unrelated to the general perception of government business and how it affect individual employee. Government work is generally seen as no man's work and should not be taken serious. Unless material gains are in sight, public servants are unlikely to give much attention to public issues like climate change and its aftermath.

Further, inability of government to regulate activities of multi-national companies and other industries particularly, in the Niger-Delta region and other industrial areas across the country is a major challenge to the management of climate change. The primary reason for the existence of any government is welfare and security of its people. However, either due to failure on the part of regulatory agencies or convergence of interest among stakeholders in the oil industries and allied bodies, government appears incapacitated to regulate their activities in the interest of its citizens (Olojede, Fajonyomi and Akhakpe, 2013).

Bad governance and the leadership question impede the process of climate change management for sustainable development in the country. Nigeria is a blessed country with resources to address any challenge thrown at it by nature or human activities. However, failure to manage its resources – human and material efficiently and effectively is the bane of poor public management. Its developmental challenges cannot be adequately explicated without an account of its leadership deficit. No nation became great without visionary and missionary leaders. Yet, the problem of Nigeria is leadership (Achebe, 1983).

By and large, these challenges are not insurmountable! In the final part of this study, we focus on these identified challenges which can be converted into opportunities for sustainable development.

Opportunities for Climate Change Management for Sustainable Development

Addressing the developmental needs of Nigerians is a task that must be done! The 1999 Constitution notes that the fundamental objectives and directive principles of state policies are that "the security and welfare of the people shall be the primary purpose of government…". It therefore, behooves on government to manage the rampaging role of climate change for the sustainable development of the citizenry and society. Opportunities abound for these to take place some of them are: Promotion of good governance and positive leadership, with the expansion of the democratic space and a vigilant public, issues of climate change would be better addressed. With an attentive public and vigilant press, issues of climate change will always be in the front burner for redress in the interest of the people (Ingwu, Okoro and Urni-Agbor, 2012).

Prioritizing the basic needs of the people has become the clarion call of the opposition and policy advocacy group across the country. Elected governments are often reminded that they must deliver democratic dividends for them to have legitimacy to rule. Gone are the days governments ignore the voice of the people for too long. The fear of the people is the beginning of good governance and sustainable development (Amue and Igwe, 2016; Chikwonye, 2011; Gyang, 2013).

Efforts should be made to reduce gender inequalities that have became very glaring as humanity tries to tackle challenges of globalization, bad governance, pervasive poverty and social exclusion of some groups, particularly, from decision making. These are cross-cutting issues that have to be addressed holistically by development planners, actors and policy analysts (Olawoye, 2010). Selected treatment of these issues may not deliver the expected results except a cross-cutting approach is applied.

Proactive measures should be put in place to enhance the system adaptive capacity to climate change. This means optimizing the "abilities or potentials to adapt to the effect of climate. It usually depends on factors such as wealth, technology, education, information, skills, infrastructure, access to resources and management capabilities (Olawoye, 2010). These steps help to checkmate vulnerability caused by climate change and promotes sustainable development. Luckily, some of these structures are in place waiting to be harvested and domesticated for sustainable development.

Government has put a lot in place to manage climate change and promote sustainable development. However, leadership at various levels including those actors on the field must develop the political will and patriotism to leverage on the resources available for socio-economic and environmental transformation of the country. Promoting inclusive growth and development through grassroots empowerment would go a long way in promoting sustainable development irrespective of changes in climatic conditions. The people therefore, must be sensitized and mobilized towards these goals.

CONCLUSION

We have analyzed the vexed issues of climate change and sustainable development. We observed that climate change affect the socio-economic and environmental well-being of the people. The study came up with opportunities that are available to the government and the people to manage these phenomena. Some of them include: enhancing the adaptive capacities of the government and people of Nigeria to bring about the needed change in management perspective to climate change and sustainable development. Above all, positive leader is most crucial in this regard. Political leadership and other actors should embrace due diligence and global best practices in tackling this hydra-headed challenge of climate change in the country. And the time to do that is now.

REFERENCES

Achebe, C. (1983). *The trouble with Nigeria*. Enugu: Fourth Dimension Publishers.

Adejobi, S., & Akhakpe, I. (2012). Urbanization process and climate change: Lagos State, Nigeria Experience. *International Journal of Social Science and Humanities Review, 3*(1).

Adilieje, C. (2011). Public policy and environmental management in Nigeria. *The Constitution: A Journal of Constitutional Development, 11*(1).

Akali, K. (2007). The contribution of Environmental Right Action (ERA) in the struggle for ecological rights in the Niger-Delta. *The Constitution: A Journal of Constitutional Development, 7*(1).

Akhakpe, I. (2008). Food crisis, climate change and human development in Nigeria: Challenges and the Way Forward. *The Nigerian Journal of Energy and Environmental Economies, 1*(2).

Akhakpe, I.B. (2016). Public policy and sustainable national development in Nigeria's Fourth Republic: challenges and possibilities. *International Journal of Management Perspectives, 13*(1).

Amue, G. J., & Igwe, S. R. (2016). ICT resource management and sustainable development: An inquiry into the causes of Sustainable development. In A. Nkemnabe (Ed.), *Resource management for sustainable development. Awka: SCOA Heritage Nigeria Ltd.*

Aye, J. R. (2008). *Reforming the Africa public sector retrospect and prospects*. Dakar: Codesria Green Book.

Ayeni, A. (2010). Sustainable development through international cooperation in research and development. In L. Popoola & O. Olorunmisola (Eds.), *Contemporary issues in sustainable development: lessons for, and challenges to Nigeria*. Ibadan: University of Ibadan.

Echikwonye, R., & Beetseh, K. (2011). The role of public policy making and development in Nigeria. *Journal of Social Science and Public Policy, 3*.

Fajonyomi, S. O. (2012). *When the state goes to sleep: of citizen and state relationship*. Lagos: LASU Inaugural Lecture Series.

Federal Government of Nigeria (FGN). (1999). The 1999 Constitution of the Federal Republic of Nigeria. Abuja: Government Press.

Federal Government of Nigeria (FGN). (2012). *Nigeria's path to sustainable development through green economy: country report to the Rio + 20 Summit Abuja*. Federal Ministry of Environment.

Gyong, J. E. (2013). A social analysis of the transformation agenda of president Goodluck Ebele Jonathan. *European Scientific Journal, 8*(16).

Harris, J. M. (2000). *Basic Principles of Sustainable Development*. Global Development and Environment Institute, Tufts University. Retrieved from http://ase.tufts.edu/gdae/publications/working_papers/Sustainable%20Development.pdf

Ingwe, R., Okoro, J., & Uim-Agbor, U. (2012). Neo-liberal policies and regimes in Post – independent Nigeria at the dawn of post neo-Liberalism (1960 – 2007): A Research Agenda. *Universal Journal of Management and Social Sciences, 2*(12).

Iyoha, F.E. (2008). *Local government and the elusive rural development in Nigeria: Constraints on local authorities capacity and the promise of decentralized participatory development*. Ekpoma: Ambrose Ali University. Inaugural Lecture Series.

Maduabum, C. P. (2014), What has public administration got to do with it? Lagos: Inaugural lecture, National Open University of Nigeria.

Olaniran, A.O. (2002). *Review of the state of agricultural enterprise in Nigeria*. Being the 2nd LASU CDDS lecture series held at the Nigerian Institute of International Affairs, Lagos, Nigeria.

Olaopa, T. (2016). *The state of public administration and sustainable development in Africa: Towards a bold and functional public administration.* Guest lecture at the 2nd International Conference, Department of Public Administration LASU.

Olawoye, J. E. (2010). Sustainable rural development: Challenges, opportunities and lessons from Nigeria. In L. Popoola & O. Olorunmisola (Eds.), Contemporary Issues in Sustainable Development: Lessons for, and Challenges to Nigeria. Ibadan: University, of Ibadan.

Olojede, I., Fajonyomi, B., & Akhakpe, I. (2013). The Ecology – security nexus: oil, environmental pollution, community unrests and national security in Nigeria. In A. Ikelegbe (Ed.), *Oil, Enviornment and Resource Conflicts in Nigeria.* Zurich: UT VERLAG Gmb IT and Co.

Oyeshola, O. P. (2008). *Sustainable development: issues and challenges for Nigeria.* Ibadan: Daily Graphics Nigeria Ltd.

Simon, D. (2007). Global change and urban risk: The challenge for African cities. *The Constitution. A Journal of Constitution Development, 7*(1).

United Nations. (2015). *Getting started with sustainable development goals: a guide for stakeholders.* New York: United Nations.

United Nations. (2016). *The sustainable development goals report.* New York: United Nations.

KEY TERMS AND DEFINITIONS

Climate Change: The alteration in the atmospheric condition leading to negative impact on the environment.

Development: The progressive change from one state of being to another.

Environmental Vulnerability: The tendency of the environment to respond either positively or negatively to changes in human and climatic conditions.

Sustainability: The ability to maintain improvement or progress in environmental and human conditions over a long period of time.

Chapter 12
Climate Change Impact on Agriculture and Food Security

Ali Syed
Mulungushi University, Zambia

Urooj Afshan Jabeen
Government Degree College for Women, India

ABSTRACT

Research on the impact of climate change on agriculture and food security is important, especially in the agricultural economies, not only to know the severity of impact but also the policies to be adapted to halt climate change and the technology to be used to mitigate the impact of climate change. The study was conducted in Kapiri Mposhi district of Central Province in Zambia to find out the impact of climate change on agriculture and food security. The objectives of study include to know the intensity of climate change and its impact on area under cultivation, late sowing of seed and damage of seed due to lack of water, fertilizer absorption reduction, food shortage, livestock, and productivity. The chapter also focuses on the sources of credit to the farmers.

INTRODUCTION

Today, climate change is at the top of the agenda of policy makers in all countries of the world. The demand for cereals will increase by 70% by 2050 and will double in many low-income countries due to increase in population and higher per capita consumption with growing incomes (FAO, 2006). Agriculture is highly sensitive to climate change. Climate change is a threat to the humanity itself. But the actions to mitigate climate change impact on agriculture and food security are less and uneven in countries of the world.

Climate change means changes in the patterns of overall climate of the earth, i.e., changes in the earth's average temperature and precipitation patterns. There are two reasons for climate change – natural and human. The natural reasons are changes in earth's orbit, in the amount of energy coming from the sun and volcanic eruptions. The human causes for climate change are due to the combustion of fossil fuels like coal, oil and natural gas whose burning releases green house gases, such as carbon dioxide in

DOI: 10.4018/978-1-5225-3194-4.ch012

to the air. The Intergovernmental Panel on Climate Changes (IPCC) Fifth Assessment Report (2007) concluded that "the human influence on the climate system is clear and is evident from the increasing green house gas concentrations in the atmosphere, positive radiative forcing, observed warming and understanding of the climate system". 97 per cent or more of climate scientists agree that this is due to human activity- notably, emissions of carbon dioxide (Cook, 2013).

Climate change affects agriculture by destroying harvests and reducing agricultural productivity. Research on impact of climate change on agriculture and food security is important, especially in the agricultural economies not only to know the severity of impact but also the policies to be adapted to halt climate change and the technology to be used to mitigate the impact of climate change. Individuals living in rural areas in poor countries are also disproportionately affected by climate change because they are more dependent on natural resources and the environment for their livelihood (World Bank, 2011). Climate change and global warming reduces economic growth and slows economic activity. Global warming affects agriculture in two ways. Firstly, it destroys agricultural harvests and secondly, it affects on agricultural productivity permanently (Dell, et.al, 2012).

World Bank (1986) defined food security as "access by all people at all times to sufficient food for an active and healthy life". Naiken (2003) defined undernourishment as that part of the distribution lying below a minimum energy requirement level. Mauro, et.al. (2006) identified five indicators for measuring food security. The first indicator was "undernourishment", which was also adapted by FAO. The second indicator was "food intake", which measures the amount of food actually consumed at the individual or household level. The third indicator was "nutritional status", which considers dietary deficiencies. The fourth indicator was "access to food", which considers wealth status measured by total consumption or income. Income as access-to-food was the main food security indicator. Lastly, "vulnerability" indicator, which expresses ex-ante vulnerability and ex-post outcomes.

Table 1 shows the global share of greenhouse gas emissions by country.

Table 1. Global share of greenhouse gas emissions by country

Country	Share of Global	CO_2Population (tco$_2$ per Capita)	CO_2/GDP PPP (kg co$_2$ per Current International $)	GDP per Capita (Current ppp)
China	28%	6.65	0.55	12,196
United States	16%	16.18	0.31	52,980
India	6%	1.49	0.28	5,418
Russian Federation	5%	10.75	0.43	25,033
Japan	4%	9.70	0.27	36,223
Germany	2%	9.42	0.21	43,887
Korea	2%	11.39	0.34	33,089
Canada	2%	15.25	0.35	43,033
Islamic Republic of Iran	2%	6.79	0.42	16,067
Saudi Arabia	1%	16.39	0.31	52,993
Total share (10 top countries)	67%			

Source: (IMF, 2015).

Table 1 shows that the 10 largest emitters of greenhouse gases are responsible for 67 per cent of global emissions. While high income countries are big emitters in per capita terms, energy efficiency has been gaining ground in these countries. It is expected that the consumption of fossil fuels by advanced economies will come down. Whereas, emerging markets and developing countries are still heavily reliant on coal and their consumption of fossil fuels will continue to rise. China's global share is the highest. It is the largest coal consumer in the world. If the countries do not adopt relevant technology for coal-powered plants, global emissions will reach dramatic levels and in turn accelerate global warming. Poorly designed regulations in the use of coal in developing countries could also discourage technological change in the electrical power sector. As a result, the world might not benefit from the downward trend in coal usage in developed countries (IMF, 2015).

Agriculture in Zambia

Agriculture is an important t sector in the Zambian economy. The agriculture sector contributed 16.7% of the country's GDP and the annual growth rate was 7.1 percent in 2009 (MFNP, 2009). It provided livelihood for more than 60% of the population. It employed about two-thirds of the labor force. The agriculture exports as a percentage of non-traditional exports stood at 55% in 2015. Zambia has an estimated nine million hectares of land (12% of its total land area) suitable for cultivation. The country is divided into three agro-ecological zones with rainfall as the dominant distinguishing climatic factor. Zone I lies in the western and southern part of the country and accounts for about 15% of the land area. It receives less than 800 mm of rain annually. It used to be considered the bread basket of the nation but for the last 20 years it has been experiencing low, unpredictable and poorly distributed rainfall. The observed meteorological data suggest that it is currently the driest zone, very prone to drought and with limited potential for crop production. Zone II covers the central part of the country, extending from the east through to the west. It is the most populous zone, with over four million inhabitants, and has the highest agricultural potential. The soils are relatively fertile. It receives about 800-1000 mm of rainfall annually. Zone III covers the northern part of the country and has a population of over 3.5 million. It receives over 1000 mm of rainfall annually.

The mean temperatures computed for the agro-ecological zones for three time periods, November to December, January to February and March to April, indicate that the summer temperature in Zambia is increasing at the rate of about 0.6 ^{0}C per decade, which is ten times higher than the global or southern African rate of increase in temperature. The major cropping season in Zambia runs from November through April. Agriculture in this season depends on the rains to a very large extent. There is a minor cropping season from June through September. The major crops grown are maize, sorghum, millet, rice, wheat, cassava, groundnuts, sunflower, cotton, soya beans, mixed beans and tobacco. Most of these are summer crops which depend almost entirely on the rains. Significant rainfall deficits at critical stages of crop growth have frequently led to a serious shortfall in crop production. Maize is a staple grain. The maize yield in the Southern, Central and Eastern provinces showed a high positive correlation with the total seasonal rainfall. Marginal net revenue per hectare for an increase of 1 ^{0}c in the mean temperature of November and December is US$ 322.62, indicating that if the temperature rises at the beginning of the cropping season, when plants are germinating, this may have a negative effect on the crop. The marginal net revenue per hectare for an increase of 1 ^{0}c in the mean temperature of January and February is US$ 315.70, indicating that if the temperature rises during the growing stage of the plant, this may have a positive effect on the crop (CEEPA, 2006).

Statement of the Problem

The Government of Zambia has been implementing different programmes to increase agricultural production and solve the problem of food shortage. But the climate change had negative impact on agricultural production and food security. Many studies were conducted at national and international levels to find out the climate change impact on agricultural production and food security. But research at village level is needed to know the impact of climate change on agriculture and food security. There prevails uncertainty in these studies regarding crops grown which leads to considerable difficulty to the quantification and evaluation of impacts and adaptation. There prevails also the gap in data, tools and models at spatial and temporal scales appropriate to decision-making. Production impacts are often aggregated over large areas such as the country or region, and this can hide considerable heterogeneity in climatic conditions and agricultural production (Jones and Thornton, 2003). This study fills the gap in the existing literature by bringing out the clear picture at the grass-root level on the impact of climate change on agriculture production and food security.

Objectives of Study

The specific objectives of study were to:

1. Know the climate change impact on rain days, water shortage, sowing of seed, damage of seed and fertilizer absorption to the plant.
2. Find out the impact of climate change on agricultural production.
3. Explain the climate change impact on food security.
4. Research the climate change impact on live stock.
5. Ascertain the sources of borrowing and the amount borrowed.

REVIEW OF LITERATURE

Wisdom, et.al. (2015) revealed that climate change impacts are likely to become more profound in this century, particularly if mitigation policies are absent or inadequate. Due to wide variations in climatic conditions across regions and countries in the developing world, detailed impact studies require assessments at local levels.

Channing, et.al. (2015) examined the implications of climate change for economic development in Ghana until 2050 by using CGE model. The study found that climate change will reduce Ghana's national welfare or total absorption in 2050 relative to a no climate change baseline. Estimates of these damages range from 0.8 to 2.5 percent of baseline absorption depending on which climate projection is considered. However, in some sectors there are benefits from climate change, particularly within agriculture. These results therefore underscore the importance of accounting for multi-sector impact channels and climate uncertainty. Climate change's adverse impacts are also unevenly distributed across households and sub national regions in Ghana. The study suggested that investments in agricultural research and extension and in sealing road surfaces are likely to be cost-effective means of reducing the damages from climate change in Ghana.

Eric (2013) revealed that soils are intricately linked to the climate system through the carbon, nitrogen and hydrologic cycles and climate change will have an effect on soil processes and properties. Hence, study of soil-climate interactions in a changing world is critical to addressing food security. The study suggested for continued research with the particular goal of understanding the complex interactions that take place in the natural environment.

IPCC (2007) warned that the global temperatures are expected to increase between 1.1 and 6.4 ^{o}c during the 21st century and precipitation patterns will be altered. It has assessed the potential impacts of climate change on agricultural production in different parts of the world.

IRI (2006) rightly pointed out that climate information was not used adequately across sectors and from local to policy levels in Africa. It attributed the substantial gap in the provision and use of climate information to "market atrophy" associated with long-term ineffective demand by development practitioners and inadequate supply of relevant climate information services.

Challinor, et.al,, (2009) reported that there has been much progress in recent years in combining climate models with crop models in order to understand and project climate impacts. In spite of the inherent uncertainties, robust responses of yield to climate change have been found using both empirical and process-based crop models.

The study by Jain (2006) on impacts of climate change on agriculture in Zambia depends on Ricardian theory which measures the effect of climate on the value of agricultural land. A multiple regression model with net farm revenue as response variable was fitted with climate, hydrological, soil and socio-economic variables as explanatory variables. The results showed that most socio-economic variables were not significant. Whereas some climate variables and the corresponding quadratic variables were significant in the model. Further findings showed that an increase in the November-December mean temperature and a decrease in the January-February mean rainfall had negative impacts on net farm revenue. Whereas an increase in the January-February mean temperature and mean annual run off had a positive impact.

Riche (2007) undertook a climate change vulnerability assessment based on community consultations in seven locations in Zambia and came to the conclusion that Zambia faced increased risks due to climate change due to a rise in the frequency and severity of extreme vents, including droughts, floods and high temperatures and a decrease in the length of the rainy season. In the words of Riche (2007):

The rise in extreme climatic events is negatively affecting the natural, physical, financial, and human resources that are crucial for people's livelihoods, and is leading to increased food insecurity and health issues. When facing climate hazards, small scale farmers (which are negatively affected by disruption of their normal farming cycles) rely heavily on access to alternative natural resources from forests and wetlands.

According to Kundzewicz, et.al.,(2007) climate change will impact the delivery and effectiveness of irrigation. The predicted increase in precipitation variability, coupled with higher evapo-transpiration under hotter mean temperatures, implies longer drought periods and would therefore lead to an increase in irrigation requirements, even if total precipitation during the growing season remained constant.

According to OneWorld.net (2010) deforestation was the key agent in the linkages between poverty reduction, food security and climate change in Zambia. Smith (2009) reported on the devastation brought on to Zambia as climate change brought on early flooding and that the Red Cross had warned that global warming will lead to more disasters along the Zambezi river basin. UNDP (2010) also reported the climate change implications and the need to increase awareness on this matter in Zambia.

Couroche (2010) concluded that Zambia was one of the most affected countries by climate change. Despite ambitious policy statements, Zambia had no hope of getting out of poverty due to negative impacts of climate change.

Ericksen (2009); Ingram (2009); Liverman and Kapadia (2010) reported that the impact of climate change on food security should consider both direct impacts on local food production and also the fuller set of interactions with the whole food system. According to Easterling, et.al. (2007) changes in the mean and variability of climate affects the hydrological cycle and crop production.

According to Godfray, et.al. (2010) climate change will bring further difficulties to millions of people for whom achieving food security is already problematic and perhaps humanity's most pressing challenge as we seek to nourish nine billion people by 2050. Thornton, et.al. (2009) revealed that uncertainty in rainfall is not always a factor that limits the predictability of yield; temperature is more important in a number of cases. Gregory, et.al. (2009) pointed out that climate change was influencing the distribution and virulence of crop pests and diseases. The study by Kundzewicz, et.al. (2007) found that climate change impacted the delivery and effectiveness of irrigation.

The study by Thornton (2010) linked climate change to livestock production . The impacts of climate change on livestock production will be both direct, i.e., productivity losses (physiological stress) owing to temperature increases and indirect, i.e., changes in the availability, quality and prices of inputs such as fodder, energy, disease management, housing and water. Couroche (2010) reported that climatic situations have a direct effect on animal populations. Cattles, for instance, have reduced in numbers due to the increased temperatures and the lack of food and water. The reduced cattle population will also directly affect the farmers that depend on them for productivity. The animals are suffering from the lack of food and getting more diseases.

Low (2005) addressed negative effects of climate change, specifically in Zambia, on agriculture and food production. The most comprehensive report on climate change in Zambia is Zambia's National Adaptation Programme of Action (NAPA) of 2007, which provides a process to identify priority activities that respond to immediate needs to adapt to climate change.

METHODOLOGY

The study was based on multi-stage random sampling technique to collect the data. In the first stage the Central Province was selected out of ten (10) provinces in Zambia. In the second stage Kapiri Mposhi district was identified and it consists of six (6) Blocks, i.e., Mulungushi, Chipepo, Lukanga, Changonda, Nkole and Lounchu. In the third stage Mulungushi Block was selected which had six (6) Camps, namely, Luanshimba, Kakulu, Imansa, Kaunga, Lukanda and Kambosha. In the fourth stage Luanshimba Camp area was selected which had six (6) Zones – Luanshimba, Nchembwe, Mukabe, Ngololoeni, Njanji and Lyuba. In the fifth stage, Luanshimba Zone was selected which had 28 villages. In the sixth stage nine (9) villages – Njalamimba, Lukanga Kateneka, Kakulu Mabwe, Mukabe, Fumbelo, Makaiva, Njanji and Satulo- were selected. In the last stage 214 households, out of 900 households, were selected for the study. The primary data were collected in October and November, 2016 for the immediate preceding agricultural season.

In this study, agricultural production meant the production of maize. The food scarcity was estimated on the basis of number of times the household members ate in a day and the number of days of food shortage in a week. The impact of climate change on agricultural production was measured by using the following econometric model.

$$Ap_i = \beta^i{}_+ \beta^1{}_{aci} + \beta^2{}_{rdi} + \beta^3{}_{lsi} + \beta^4{}_{dsi} + \beta^5{}_{fri} + \beta^6{}_{wsi} + \mu$$

where

Api = Agricultural Production during i_{th} season
aci = Area under cultivation during i_{th} season
rdi = Rain days reduced during i_{th} season
lsi = Late sowing of seed in days during i_{th} season
dsi = Damage of seed during the i_{th} season
fri = Fertilizer absorption to the plant reduced during i_{th} season
wsi = Water shortage during the $_{ith}$ season
u = Error term

Table 2. Econometric model used to measure impact of climate change on agricultural production

Dependent Variable: API
Method: Least Squares
Date: 12/19/16 Time: 16:39
Sample (adjusted): 1 213
Included observations: 213 after adjustments

Variable	Coefficient	Std. Error	t-Statistic	Prob.
C	16.69985	4.671562	3.574789	0.0004
ACI	0.870435	0.496824	1.751999	0.0813
RDI	0.020281	0.075059	0.270197	0.7873
LSI	0.005847	0.067179	0.087033	0.9307
DSI	0.179410	6.425578	0.027921	0.9778
FRI	5.955331	5.423800	1.098000	0.2735
WSI	-0.523779	4.521644	-0.115838	0.9079

R-squared	0.046287	Mean dependent var	25.23944
Adjusted R-squared	0.018509	S.D. dependent var	9.343116
S.E. of regression	9.256244	Akaike info criterion	7.320785
Sum squared resid	17649.68	Schwarz criterion	7.431250
Log likelihood	-772.6637	Hannan-Quinn criter.	7.365428
F-statistic	1.666334	Durbin-Watson stat	1.801985
Prob(F-statistic)	0.130802		

The model revealed that the independent variables, sci, rdi, lsi, dsi, fri and wsi were not significant. They did not have significant impact on Api, since the p values of all the independent variables were greater than 0.05 (5% level of significance).

The R^2 was 0.046, which was 4.6 percent, This means that the model explained only 4.6 percent of the variations in Api.

The overall significance of the model shows that the probability of F statistics was 0.130 which was greater than 0.05. This means that the model was not statistically significant. These results confirmed the results of study by Jain (2006), which revealed that most of the socio-economic variables were not significant in explaining the impact of climate change on agriculture in Zambia.

MAIN FOCUS OF THE CHAPTER

Impact of Climate Change on Rain Days, Water Shortage, Late Sowing of Seed, Damage of Seed and Reduced Fertilizer Absorption to the Plant

Table 3 shows the impact of climate change on rain days, water shortage, late sowing of seed, damage of seed and reduced fertilizer absorption to the plant.

Table 3 shows that due to climate change 31.52 rain days were reduced during the last agricultural season. Out of 214 farmers, 199 farmers (93%) reported that there was water shortage during the season. Sowing of seed was delayed by 23.82 days due to delay in rainy season. 193 farmers (90.18%) reported that seed was damaged due to shortage of water and 189 farmers (88.31%) explained that fertilizer absorption to the plant was reduced.

Impact of Climate Change on Agricultural Production

Table 4 shows the impact of climate change on agricultural production.

Table 3. Impact of climate change on rain days, water shortage, late sowing of seed, damage of seed due to lack of water and reduced fertilizer absorption to the plant

Sl. No.	Details of Impact	Number of Days/Percentage
1.	Rain days reduced	31.52
2.	Water shortage:	
	Yes No	93.00 7.00
3.	Late sowing of seed (days)	23.82
4.	Damage of seed due to lack of water (percent)	
	Yes No	90.18 9.82
5.	Reduced fertilizer absorption to plant (percent)	
	Yes No	88.31 11.69

Source: Primary data.

Table 4. Impact of climate change on agricultural production

Sl. No.	Details	Before	After	Change	Percentage
1.	Area Cultivated (in hectares)	2.98	1.71	-1.27	42.61
2.	Quantity Produced in 50 Kg bags (per hectare)	109	25	-84	77.0

Source: Primary data.

Table 4 shows that the average area cultivated during the last agricultural season was affected negatively due to climate change. The average area cultivated when there was no climate change effect was 2.98 hectares but it was decreased to 1.71 hectares, i..e., decreased by 1.27 hectares (42.61 percent) due to climate change effect during the last agricultural season.

The average quantity of production per hectare during the last agricultural season was also affected due to climate change. The average quantity of production when there was no climate change effect was 109 bags per hectare but it was decreased to 25 bags per hectare, i.e., decreased by 84 bags per hectare (77 percent) due to climate change effect during the last agricultural season.

The results of this study did not confirm the results of the studies by Thornton, et.al. (2009); Cobell and Bruke (2008), which revealed that un-certainty in rainfall was not a factor that limits the yield per hectare. Whereas, this study agreed with the results of study by Dell, et.al. (2012) and Couroche (2010), which revealed that uncertainty in rainfall affected the yield per hectare.

Impact of Climate Change on Food Security

Table 5 shows the impact of climate change on food security.

Table 5 shows that there was negative impact of climate change on food security. More than half farmers (56.54%) revealed that there was food shortage during the last agricultural season due to climate change. The average number of days of food shortage was 1.08 days in a week. The results of this study confirmed the results of the studies by Mauro, et.al (2006); Erickson (2009); Ingram (2009); Liverman and Kapadia (2010) and Godfray, et.al. (2010), which revealed that there was negative impact of climate change on food security.

Impact of Climate Change on Livestock

Table 6 shows the impact of climate change on livestock.

Table 5. Impact of climate change on food security

Sl. No.	Food Shortage			No. of Days of Food Shortage in a Week
	Yes	No	Total	
1.	121	93	214	1.08
	(56.54)	(43.46)	(100)	

Source: Primary data; Figures in the parenthesis are percentages.

Table 6. Impact of climate change on livestock

Sl. No.	Details	Number/Percentage
1.	Livestock sick	
	Yes No Total	144 (67.28) 70 (32.72) 214 (100)
2.	Livestock sold due to sickness	
	Yes No Total	74 (34.58) 140 (65.42) 214 (100)
3.	Amount spent on treatment of livestock	231.52 ZMW

Source: Primary data; Figures in the parentheses are the percentages.

Table 6 shows that there was negative impact of climate change on livestock. Out of 214 farmers, 144 farmers (67.28%) reported that their livestock was sick due to climate change and 74 farmers (34.58%) revealed that they were forced to sell sick livestock due to climate change. The average amount spent on the treatment of livestock was ZMW 231.52. These results confirmed the results of the study by Thornton (2010).

Sources of Borrowing and Amount Borrowed

Table 7 shows the sources of borrowing and amount borrowed by the farmers.

Table 7 shows that out of 214 farmers, 103 farmers (48.14%) only borrowed and the average amount was ZMW 554.35. The amount financed by banks was nil. Only one farmer borrowed ZMW 1800 (3.15%) from micro finance, 22 farmers (21.35%) borrowed an average of ZMW 242.72 (9.35%) from Co-operatives; 26 farmers (25.24%) borrowed an average of ZMW 577.30 (26.28%) from Relatives and 54 farmers (52.42%) borrowed an average of ZMW 647.18 (61.22%) from their friends.

SOLUTIONS AND RECOMMENDATIONS

There are many issues related to climate change. There is no single solution to the problem of climate change impact. However, the following solutions/ recommendations could be considered.

Table 7. Sources of borrowing and amount borrowed in ZMK

Sl. No.	Source	No. of Farmers	Total Amount	Average Amount
1.	Bank	-	-	-
2.	Micro Finance	1 (0.97)	1800 (3.15)	1800
3.	Co-operatives	22 (21.35)	5340 (9.35)	242.72
4.	Relatives	26 (25.24)	15010 (26.28)	577.30
5.	Friends	54 (52.42)	34948 (61.22)	647.18
6.	Total	103 (48.14)	57098 (100)	554.35

Source: Primary data; Figures in the parentheses are percentages.

- Interactions are required between decision-makers and researchers in all sectors, greater collaboration among climate, agriculture and food security departments and consideration of interdependencies across whole food systems and landscapes.
- Investment in crop improvement should be increased to address specific characteristics of a progressively changing climate, i.e., heat, drought, water logging, pest resistance.
- National policy on subsidies and incentives should be formed for crop substitution or expensive farming inputs.
- The private sector and public-private partnership should play their role in generating knowledge in the agriculture sector that is needed to deal with food security and climate change challenges.
- Research in agriculture, food security and climate science is needed to understand uncertainties, to allow more confident decision-making and allocation of limited resources towards new climatic conditions.
- Investment in data base and tools should be increased in the spheres of agricultural risk-management and adaptation to develop capacity to tackle climate change impacts on farming and food security.
- The involvement of farmers, policy makers, the private sector and civil society in the research process is vital to identify and solve the issues relating to climate change impact on agriculture production and food security.
- New varieties of crops, which mature faster and are heat resistant, should be developed. Traditional crops should be diversified to other types of crops which can withstand drought and higher temperatures.
- Land use between crops and livestock should be rotated to replenish soil nutrients.
- Rural credit facility should be provided to enable subsistence farmers to buy new varieties of seeds and fertilizer.
- Information on various adaptation options should be disseminated through extension services.
- Appropriate policies for marketing agricultural input and output products should be formulated to face the negative impacts of climate change.
- Investment should be made on research for climate resistant crop varieties, water harvesting, irrigation schemes.
- Alternative employment opportunities should be provided in non-farming activities during the times of drought conditions to enable rural farmers to make a livelihood..
- Adequate funds could be provided to meteorological department to increase its capacity in climate data collection, storage, analysis and forecasting. Climate forecasts should be disseminated to the farmers in their local languages.
- Policies should be formulated to encourage the NGOs, private sector and civil societies to complement the government's efforts to implement adaptation policies.
- The national government and all other stakeholders should respond to climate change by formulating and implementing adaptive measures to minimize the negative effects of climate change on agriculture which may pose a serious threat to food security.

FUTURE RESEARCH DIRECTIONS

Uncertainties exist in the direction and magnitude of climate change, which in turn leads to uncertainty in food production and food security. Looking to the future, Zambia's climate is expected to see an increase in temperatures in Regions I and II as well as an increase in alternating occurrences of drought and floods throughout the country (Couroche, 2010). This will increase the risks of drowning, climate-sensitive diseases, outbreaks of infectious plant, zoonotic and human diseases, the destruction of agricultural crops which will result in food shortages and an increase in malnutrition or starvation. All other sectors of the economy that depend on Zambian food resources are at risk. The maize industry, for instance, will see difficulties as most of its varieties will not mature because of too short growing seasons. Given that agriculture is such an important part of Zambia's economy, the resulting food shortages will greatly affect the whole country.

Future research areas include: incorporation of global climate models with regional climate models for adaptation processes; mitigation of greenhouse gas emissions; linking climate change with other key drivers of agricultural systems and food security; international, national and local policy and program options for improving environmental benefits, etc.,

CONCLUSION

The study revealed that climate change reduced rain days, delayed sowing of seed, reduced fertilizer absorption to the plant and the area under cultivation The study also revealed that the climate change affected negatively on yield per hectare. These results do not confirm the results of the study by Thornton, et.al. (2008), which revealed that un-certainty in rainfall was not a factor that limits the yield per hectare, but it confirmed the results of the study by Dell, et.al. (2012) and Couroche (2010), which revealed that uncertainty in rain fall affected the yield per hectare. The results of this study revealed that climate change had negative impact on food security, which confirmed the results of studies by Mauro, et.al. (2006), Erickson (2009), Ingram (2009), Liverman and Kapadia (2010), which revealed that there was negative impact of climate change on food security. Climate change also had negative impact on livestock, which confirmed the results of study by Thornton (2010) that climate change had negative impact on the health of live stock. The study also revealed that the sources of finance to the farmers were limited and the main source was relatives and friends. The solutions to this problem included: development of new varieties of crops which mature faster and are heat resistant ; provision of rural credit facility to buy them; investment on research for climate resistant crops varieties, water harvesting and irrigation schemes; provision of alternative employment opportunities in non-farming activities; allocation of adequate funds to meteorological department to increase its capacity in climate data collection, storage, analysis, forecasting and dissemination to the farmers in their local languages

REFERENCES

Akpalu, W., Arudt, C., & Matshe, I. (2015). Introduction to the Special Issue on Economics of Climate Change Impacts on Developing Countries: Selected studies on Sub-saharan Africa and South-East Asia. *Sustainability*, *7*(9), 12122–12126. doi:10.3390/su70912122

Arndt, C., Asante, F., & Thurlow, J. (2015). Implications of Climate change for Ghana's Economy. *Sustainability*, *7*(6), 7214–7231. doi:10.3390/su7067214

Centre for Environmental Economics and Policy in Africa (CEEPA). (2006). Climate Change and African Agriculture. *The economic impacts of climate change on agriculture in Zambia.* Discussion Paper No. 27, CEEPA, University of Pretoria.

Challinor, A. J., Ewert, F., Arnold, S., Simelton, E., & Fraser, E. (2009). Crops and climate change: Progress, trends, and challenges in simulating impacts and informing adaptation. *Journal of Experimental Botany*, *60*(10), 2775–2789. doi:10.1093/jxb/erp062 PMID:19289578

Cook, J., Nuccitelli, D., Green, S. A., Richardson, M., Winkler, B., Painting, R., & Skuce, A. et al. (2013, June). Quantifying the Consensus on Anthropogenic Global Warming in the Scientific Literature. *Environmental Research Letters*, *8*(2), 024024. doi:10.1088/1748-9326/8/2/024024

Dell, M., Jones, B., & Olken, B. (2012). Temparature shocks and Economic growth: Evidence from the last half century. *American Economic Journal. Macroeconomics*, *4*(3), 66–95. doi:10.1257/mac.4.3.66

Easterling, W. E., Aggarwal, P. K., Batima, P., Brander, K. M., Erda, L., Howden, S. M., . . . Tubiello, F. N. (2007). Food, Fibre and Forest products. In Climate Change 2007: Impacts, Adaptation and Vulnerability. Contribution of working Group II to the Fourth Assessment report of the intergovernmental Panel on climate Change. Cambridge University Press.

Eric, C. B. (2013). The Potential Impact of Climate Change on Soil Properties and Processes and Corresponding Influence on Food Security. *Agriculture*, *3*(3), 398–417. doi:10.3390/agriculture3030398

Erickson, P. J. (2009). Conceptualizing Food Systems for global environmental Change research. *Global Environmental Change*, *18*(1), 234–245. doi:10.1016/j.gloenvcha.2007.09.002

Food and Agriculture Organization. (2006). *World Agriculture: Towards 2030/2050.* Rome: Food and Agriculture Organization of the United Nations.

Godfray, J. C. J., Beddington, J., Cute, I., & Haddad, L. (2010). Food Security: The Challenge of Feeding 9 billion people. *Science*, *327*(5967), 812–818. doi:10.1126/science.1185383 PMID:20110467

Gregory, P. J., Johnson, S. N., Newton, A. C., & Ingram, I. S. I. (2009). Integrating Pests and Pathogens in to the Climate Change/Food Security Debate. *Journal of Experimental Botany*, *60*(10), 2827–2838. doi:10.1093/jxb/erp080 PMID:19380424

Ingram, J. S. I. (2009). Food System concepts. In R. Rabbinge & A. Linneman (Eds.), *ESF/cost Forward Look on European food Systems in a Changing World.* Strasbourg: European Science Foundation.

Inter-governmental Panel on Climate Change (IPCC). (2007). *Summary for Policymakers. In climate Change 2007: The Physical Science Basis. Contribution of Working Group I to the Fourth Assessment report of the Intergovernmental panel on Climate Change.* Cambridge, UK: Cambridge University Press.

International Monetary Fund (IMF). (2015, December). Who are the biggest emitters of greenhouse gases? *IMF Research Bulletin*, *16*(4), 4.

International research Institute (IRI). (2006). *A Gap Analysis for implementation of Global Climate Observing System Programme in Africa.* IRI Technical report No. 0601, International Research Institute for Climate and Society.

IPCC (Intergovernmental Panel on Climate Change). (2007). *Climate Change Impacts, Adaptation and Vulnerability. Contribution of Working group II to the Fourth assessment Report of the IPCC.* Geneva, Switzerland: IPCC.

Jain, S. (2006). *An Empirical Economic Assessment of Impacts of Climate Change on Agriculture in Zambia.* Pretoria, South Africa: University of Pretoria, Centre for Environmental Economics and Policy in Africa (CEEPA), CEEPA discussion paper No. 27.

Jones, P. G., & Thornton, P. K. (2003). The potential impacts of climate change in tropical agriculture: The case of maize in Africa and Latin America in 2055. *Global Environmental Change, 13*, 51–59. doi:10.1016/S0959-3780(02)00090-0

Kalantary. (2010). Climate Change in Zambia: Impacts and Adaptation. *Global Majority e-Journal, 1*(2), 85-96.

Kundzewicz, Z. W., Mata, L. J., Arnell, N. W., Doll, P., Kabat, P., Jimenez, B., . . . Oki, T. Z-Sen and I.A. Shiklognov (2007). Fresh water Resources and their Management. In Climate Change 2007; Impacts, Adaptation and Vulnerability. Contribution of Working group II to the 4th Assessment report of the Intergovernmental Panel on climate Change. Cambridge University Press.

Liverman, D. M., & Kapadia, K. (2010). Food Skystems and Global environment: An Overview. In J. I. Ingram, P. J. Erickson, & D. Liverman (Eds.), *Food Security and Global Environmental change.* London: Earthscare.

Lobell, D. B., & Bruke, M. B. (2008). Why are Agricultural Impacts of Climate Change so uncertain? The importance of Temperature Relative to Precipitation. *Environmental Research Letters, 3*(3), 034007. doi:10.1088/1748-9326/3/3/034007

Low, P. S. (Ed.). (2005). *Climate Change and Africa.* New York, NY: Oxford University Press.

Migotto, M., Davis, B., Carletto, G., & Kathleen, B. (2006). *Measuring Food Security using repondents' perception of food Consumption Adequacy.* Research paper No. 2006/88, UNV WIDEr.

Ministry of Finance and National Planning (MFNP). (2009). *Annual Progress Report of the Fifth National Development Plan.* Lusaka, Zambia: MFNP.

Naiken, L. (2003). FAO Methodology for Estimating the Prevalence of Undernourishment, Measures and Assessment of food Deprivation and Under nutrition. In *Proceedings of the International Scientific Symposium.* Rome: FAO.

OneWorld. (2010). *Climate Change in Zambia.* London, UK: One World, Briefing Guide. Available at http://UK.Oneworld.net/guides/Zambia/climate-change

Riche, B. (2007). *Climate Change Vulnerability Assessment in Zambia.* Geneva, Switzerland: The World Conservation Union.

Smith, D. (2009, July 6). Devastation in Zambia as Climate Change Brings Early flooding. *The Guardian.*

Thornton, P. K. (2010). Livestock Production: Recent Trends, Future Prospects. *Philosophical Transactions of the Royal Society of London. Series B, Biological Sciences, 365*(1554), 2853–2867. doi:10.1098/rstb.2010.0134 PMID:20713389

Thornton, P. K., Jones, P. G., Alagarswamy, G., & Andresen, J. (2009). Spatial Variation of Crop Yield Responses to Climate Change in East Africa. *Global Environmental Change, 19*(1), 54–65. doi:10.1016/j.gloenvcha.2008.08.005

United Nations Development Programme (UNDP). (2010). *Climate Change: the Zambian Story.* New York: UNDP News Report.

World Bank. (1986). *Poverty and Hunger: Issues and Options for Food Security in Developing countries.* World Bank Publications.

World Bank. (2011). *Gender and climate Change: 3 things you should know.* Washington, DC: World Bank Group.

KEY TERMS AND DEFINITIONS

Climate Change: Changes in the patterns of overall climate of the earth (i.e., changes in the earth's average temperature and precipitation patterns).

Food Security: Food security is measured in terms of eating food three times a day by the family members and number of days of food shortage in a week.

Human Causes of Climate Change: The combustion of fossil fuels like coal, oil, and natural gas whose burning releases greenhouse gases such as carbon dioxide in to the air.

Impact on Agricultural Production: The impact on agricultural production is measured on the basis of production of maize before climate change and after climate change.

Impact on Livestock: The impact of climate change is measured in terms of sickness of animals and the amount spent on their treatment.

Natural Causes of Climate Change: Changes in Earth's orbit, in the amount of energy coming from the sun and volcanic eruptions.

Section 5

Ecosystem Management, Environmental Media, and Public Health

Chapter 13
Impact of Sea Level Rise on Coastal Regions and Strategic Responses

Pedro Miguel Gomes
New University of Lisbon, Portugal

Francisco Sacramento Gutierres
Eurecat – Technology Centre of Catalonia, Spain

ABSTRACT

This chapter includes an assessment of physical vulnerability of the coast, including a coastal vulnerability index composed of 9 physical variables—elevation, distance to shore, tide amplitude, significant wave weight, erosion/accretion rates, geology, geomorphology, ground cover vegetation, and anthropogenic actions—followed by a quantification of coastal recession and the data of special report on emissions scenarios (SRES) developed by the Intergovernmental Panel on Climate Change (IPCC) on the rise in average sea level. It includes an estimate of the economic value of an area of recreation based on the travel cost method. Finally, a bibliographic review is made to assess strategies and responses to the impacts of sea level rise in order to make comparisons and to develop a road map of interventions for shoreline protection. The proposed methodology was applied to a case study on the Portuguese coast corresponding to the beaches of Costa de Caparica, Almada.

INTRODUCTION

The rise in mean sea level is considered one of the most significant results of global warming (Gilman, 2005). To date, few studies have been conducted that considered damage from sea level rises, and those who did so have been exposed to potential damage, such as reductions in the area of coastal habitats (Nicholls, 2006 in Anthoff, 2001). Throughout the geological history, changes in sea level have resulted in a rise of about three hundred meters due to changes in the shape and size of ocean basins, the amount of water in the oceans and the average density of seawater. Studies conducted since 1993 based on satellite altimetry indicate that the rate of sea level rise at the global level has been higher in the last decade,

DOI: 10.4018/978-1-5225-3194-4.ch013

with a rise of more than 3 mm per year (Church et al., 2001 in Holgate & Woodworth, 2004). In the last century, sea level rose about 20 cm, to an average of 1 to 2 mm per year (Nerem & Mitchum, 2001; Leuliette et al., 2004 in Antonov et al, 2005).

Coastal areas are an unstable system conditioned by processes, such as the balance between the accretion and erosion of sediments, which will influence the formation of different types of coasts (Gomes, 2009). The natural processes create a normal sediment migration due to wind, tides and storms, but the anthropic actions can accelerate these processes at the long term, while may reduce it at short term.

Vulnerability is a fundamental concept in the investigation of the human environment. It is the potential for loss and can be defined as the social and ecological context that adjusts the capacity to ensure well-being when confronting a population with situations of variability (Downing & Patwardhan, 2003). Its assessment to coastal systems becomes necessary in the sense that from an ecosystem point of view is not a problem but a natural response to a new equilibrium created by Earth's biogeochemical mechanisms. However, Man is embedded in the various ecosystems and derives benefits such as housing, food and raw materials for economic activities; therefore, it is justified by the possibility of identifying areas with high potential for loss.

In industrialized countries, most populations are protected from flooding by building structural coastal protection measures such as the United Kingdom, the Netherlands and Japan. In developing countries, defenses are less developed and populations are more often subject to floods, economic losses and in the worst cases, loss of life. Among the potential impacts of sea level rise are erosion, coastal recession, increased severe flooding, and contamination of groundwater by seawater intrusion (Small & Nicholls, 2003).

There are many stakeholders in coastal areas that have different preferences for actions to be adopted, including economic and social partners, organizations representing coastal residents, NGOs and the business sector. This shock of preferences are often seen among those whose interests lie in maintaining natural values and progressing towards sustained development and those whose interests are purely economic (McFadden & Green, 2007).

BACKGROUND

It is estimated that 1.2 billion people, approximately 23% of the world's population, live 100 km from a coastline (Nicholls & Small, 2002; Nicholls, 2003). The population density in coastal regions is about three times the world average (Nicholls, 2003). Coastal areas are of great economic value (Schernewski & Loser, 2004) nonetheless, they are susceptible to hazards such as sea level rise.

Rising sea levels will have significant impacts on a range of marine and terrestrial ecosystems, many of which are already threatened by human activities (Feagin, 2005). Even small saline intrusions in coastal lagoons may cause disruption of the zooplankton community regarding its structure and abundance (Schallenberg et al., 2003).

In Portugal, 28.5% of coastal areas were at risk of erosion (Costa, 2007). Coastal shoreline intensification has coincided with the development of several interventions in the catchment and coastal areas, including sand extraction, construction in sensitive areas and the construction of more than 100 large dams in the national territory (Costa, 2007). As a consequence, sediment transport from the mainland to the coast decreased by about 90%. The construction of infrastructures to protect the coastline, such as spurs, consolidation of cliffs and artificial feeding of beaches altered the coastal dynamics (Crooks, 2004).

Portugal, with large stretches of rocky coastline, is not one of the most vulnerable countries to sea level rise. It has, however, a great extension of sandy beaches, which will surely reflect this elevation. The areas where the consequences of sea level rise will be most significant are the regions of Aveiro, Costa da Caparica and Ria Formosa (Ferreira et al, 2005). They are separated from the ocean by sandy coasts, which may disappear or become very narrow, thus making an assessment of vulnerability to sea-level rise in these regions necessary.

MAIN FOCUS OF THE CHAPTER

The assessment of vulnerability to sea level rise in coastal areas is fundamental since the knowledge of the associated level of risk is an important aid for planning and decision-making for the implementation of coastal protection measures aimed at protecting natural, human and cultural heritage.

This chapter intends to link the natural component with the economic component in coastal areas. The benefits that can be obtained are grouped into the various components of the total economic value, namely the value of use, non-use and option. The main focus will be on the analysis of the value of use as a recreational area.

The objective of this work is to propose an approach for a fast assessment of the potential impacts of rising mean sea level in a coastal region, namely the beaches of Costa de Caparica in Portugal; with the integration of physical assessment of the coastal zone regarding its vulnerability, based on a tripartite methodology including a vulnerability index developed by Coelho et al (2006) based on the estimation of coastline parameters, the quantification of coastal recession in three different periods (2020, 2050 and 2080), with the application of the Bruun Rule, using the SRES scenarios developed by the IPCC as an estimate of sea level rise and the estimation of its economic value as a recreation area, based on the travel cost method. The tripartite methodology concerning estimations will in fact be a four part methodology since it will be complemented with a road map of coastal interventions in the study area, detailing the total costs of intervention so that it can be compared with the values obtained with the travel cost method.

SEA LEVEL RISE

It is estimated that the rise in sea level rise in the next century will be between 0.2 and 0.6 meters, with this range representing the uncertainty of processes affecting sea level due to factors not determined exclusively by the climate, such as the morphology of the oceans (IPCC, 2007). Two primary processes contributing to sea-level rise are thermal expansion with increasing water volume as a result of heat absorption and land transfer to the ocean (Houghton et al., 2001).

Changes in sea level at any coastal location depend on the sum of factors at global, regional and local levels, and are called relative sea level changes. Therefore, a global increase in sea level does not translate into a uniform rise around the world (Church et al., 2001):

- **Global:** Refers to a rise in sea level due to an increase in the global volume of the ocean. In the 20th century, the global rise was due to the thermal expansion of the upper ocean, due to the fusion of small glaciers caused by anthropogenic induced global warming (Church et al., 2001).

- **Regional:** Due to factors such as spatial variation in thermal expansion, long-term changes in wind, atmospheric pressure and ocean circulation such as those occurring in the Gulf Stream. This component has been largely ignored in the impact studies carried out to date (Church et al., 2001).
- **Vertical Land Movement:** Due to geological processes such as tectonic movements and glacial-isostatic adjustment (Emery & Aubrey, 1991).

The IPCC concludes that the impacts will be "mostly negative", including wetland flooding, increased coastal erosion, increased flooding, increased salinity of rivers, bays and groundwater. Rising sea levels will have implications for many coastal populations and the systems that depend on them. The consequences for a population of a given locality will depend on some factors such as (Brooks, Nicholls & Hall, 2006):

- The rise in sea level recorded at local level;
- Interactions with other effects caused by climate change, such as changes in the severity of storms, and their frequency;
- The geomorphological response of coastal zones, which in turn depend on natural and human processes;
- Physical exposure to the population and systems associated with the impacts of the dangers of rising sea levels;
- The capacity of the population to cope with these impacts.

Coastal erosion is a natural process, but it is often provoked or amplified by anthropic actions. Currently, 20% of the European coast is eroded, despite the implementation of coastal protection works (Fortunato et al, 2008). It can represent, among other consequences, loss of natural or built heritage, or an increased cost in the protection and reconstruction of the same, as well as the displacement of people and economic activities, thus representing an inconvenience to the tourist activity (Valiela, 2006).

All European coastal states are in some way affected by coastal erosion. Most of the affected areas are in retreat despite the defense works. It is estimated that 15 km2 are lost or severely affected annually. Studies by a United Nations international panel on climate change have estimated that the annual number of victims with current erosion and flooding trends could rise to 158 000 by 2020, while half of the European wetlands may disappear due to (EC, 2004).

The extension of new coastal engineering interventions in European maritime fronts totaled 934 km. Of these, 875 km were built in places that in 1996 showed no signs of erosion but in 2001, 63% of which were located within 30 km of artificially stabilized sea fronts. In the remaining 37% of new erosion areas, it tends to be more intense in areas where average sea level has risen by more than 20 cm over the last century and could potentially rise by over 80 cm over the course of this century (EC, 2004).

Response of Coastal Habitats

The coastal zone behaves like an unstable environmental system in function of a series of continental and marine geological processes that are determinant in the formation of different types of coast (Silva, 2004). The imbalance in the sedimentary balance can induce serious changes in the coastline, often induced by sediment retention by engineering works, readjustment of the equilibrium profile to a rise in sea level and modification of the wave climate, which together or alone cause the reduction of the sedimentary sources, being responsible for the erosive processes and consequent changes in the coastline (Muehe, 2004).

Under normal conditions there is a balance between deposition and removal of coastal habitats. Rising sea levels can change the look and feel of a beach, tending to migrate with rising sea levels, being one of the main factors contributing to the loss of coastal habitats. These are constantly changing due to wind, tides, storms, and anthropogenic actions. Natural migration as a response to the relative rise in sea level can be obstructed by coastal defense structures by reducing the area of wetlands and beaches and consequently reducing the available area.

The concept of equilibrium profile was evidenced from the theory known as the Bruun Rule (1954) and was initially applied to a geological scale process, with the gradual adjustment of the profile to different sea level situations. At present, this concept is applied in the smaller scale processes, adopted to follow the studies of morphodynamics (Machado, 2007).

Coastal Intervention Measures

In a sustainable development policy, priority must be given to maintain the capacity of coastal systems to respond to the pressures, to which they are subject, basing integrated management strategies on a social and economic logic (Bodungen & Turner, 2001).

Coastal intervention measures include traditional methods of coastal protection, such as adherent longitudinal defenses, spurs, breakwaters and artificial feeding. Adherent longitudinal defenses are used in emergency situations where assets at risk of erosion or flooding must be protected at all costs. The construction works of these interventions may lead to beach downswing and induce erosive effects on the underlying sandy shores and adjacent contiguous zones due to the added turbulence they cause (Martin et al 2005 in Airoldi & Beck, 2007).

In a stretch of coast, when there is an imbalance of its alluvial balance, i.e. when more material comes out than enters, the construction of spurs may allow a reordering of the coast with the reduction of the intensity of the sedimentary transport. The sedimentary saturation of the spurs with sands by means of an artificial filling can minimize the eventual temporary negative impacts of the sedimentary cut caused by the works. The use of spores has side effects depending on their use as retention or containment works. Breakwaters are most applicable to areas with moderate or reduced sea shaking energy.

These works modify the longitudinal and transverse transport processes, and the control and reversion of a given erosion situation can be achieved through their use. The beach itself is the best way to dissipate wave energy. However, in case of alluvial deficit, the best technical solution involves the artificial feeding of the beaches, which is a viable process technically and economically when four conditions are met: the region must be a zone of great importance for tourism with an high population density, a moderate regime of marine agitation, existence of adequate sand reserves in nearby places and not having great environmental and ecological constraints (Dinis & Tavares, 2005).

METHODOLOGY

It will be used a four part methodology, consisting in the application of the Vulnerability Index developed by Coelho et al. (2006) to a beach, in order to provide an estimate for the vulnerability of the natural systems. The Bruun Rule will be used to quantify the value of coastal erosion of the region in study according to the Special Report on Emission Scenarios (SRES) scenarios for 2020, 2050 and 2080. The Travel Cost method was used to provide an estimate of the benefits generated by the study region. Finally it will be detailed a road map of the shoreline interventions made on the beach and its surroundings in order to understand the economic values already invested to maintain the equilibrium in the coastal ecosystems and protect the recreational benefits.

Vulnerability Index

In this index there's the evaluation of nine physical parameters (Table 1). The classification depends on the users' own experience as well as data availability concerning the parameters. The classification ranges from 1, corresponding to "very low" to 5 which is "very high" (Gomes, 2009).

In Table 2 there are four more vulnerability parameters, related with the traits (natural and anthropic) of the coastal region of study.

Table 1. Vulnerability classification of 5 physical parameters

Vulnerability Parameters	Very Low	Low	Moderate	High	Very High
	1	2	3	4	5
Elevation related to datum (meters) (TE)	> 30	> 20 < 30	> 10 < 20	> 5 < 10	< 5
Distance to shore (meters) (DS)	> 1000	> 200 < 1000	> 50 < 200	> 20 < 50	< 20
Tide average (meters) (TA)	< 1,0	> 1,0 < 2,0	> 2,0 < 4,0	> 4,0 < 6,0	> 6,0
Significant wave height (meters) (WH)	< 3,0	> 3,0 < 5,0	> 5,0 < 6,0	> 6,0 < 6,9	> 6,9
Coastal erosion/ accretion rate (meters/year) (EA)	> 0 (is accretion)	> -1 < 0	> -3 < -1	> -5 < -3	< -5

Source: Adapted from Coelho et al., 2006.

Table 2. Vulnerability classification (II)

Vulnerability Parameters	Geology (GL)	Geomorphology (GM)	Green Cover (GC)	Anthropic Actions (AA)
1 – Very Low	Rock (magmatic)	Mountains	Forest	With Interventions
2 - Low	Rock (metamorphic)	Plains (Rocky)	Green cover	With Interventions
3 - Moderate	Rock (Sedimentary)	Plains (Eroded)	No green cover	With Interventions
4 - High	Non-consolidated sediments (rough)	Beaches	Urban or Rural	No intervention
5 – Very High	Non-consolidated sediments (thin)	Either dunes, rivers or estuaries	Either urban or Industrial	No intervention

Source: Adapted from Coelho et al., 2006.

Bruun Rule

The Bruun Rule still is the most used method to quickly provide an estimate regarding changes in shoreline as a result of the rise of sea level (Bruun, 1988; Hennecke & Cowell, 2000). The value of shoreline recession (usually in meters) can be quantified with the expression:

$$Recession = Sea\ Level\ Rise \frac{Profile\ Width\left(W\right)}{Active\ profile\ depth\left(h\right) + berm\ height\left(B\right)}$$

SRES Scenarios

Greenhouse gas emissions are the result of constraints such as demographic, socio-economic development and technological evolution. Scenarios are alternate images, allowing analysis of how driving forces can influence future emissions. They assist the analysis of climate change, including the assessment of the impacts and effectiveness of adaptation and mitigation measures (IPCC, 2007).

In 1992 the IPCC published six scenarios on future emissions of major greenhouse gases (CO2, CO, CH4, N2O, NOx and SO2) for the period 1990 to 2100. These scenarios were known in the literature as IS92 (variants A, b, c, d, e, f). From the time of its publication, IS92 scenarios were considered as the maximum state of the art in terms of scenarios (Leggett et al, 1992).

These are reference scenarios that seek to exclude the effects of change and climate policy on society and the economy. Its scientific base is based on a set of quantified narratives using different models. The field of use of SRES scenarios covers climate modeling, impact assessment, vulnerability assessment and analysis of adaptation and mitigation policies. None of the SRES scenarios represent an estimate of a central tendency for all driving forces or emissions and none should be interpreted as such. In summary, the four scenarios combine two sets of divergent trends: one set varies between strong economic values and strong environmental values and the other between increased globalization and increased regionalization (IPCC, 2001):

A1: A world of very fast economic growth, low population growth and rapid introduction of new and more efficient technologies. The main underlying themes are economic and cultural convergence and capacity building, with a substantial reduction in regional differences in per capita income. In this world people pursue personal health more than environmental quality.

A2: A differentiated world. The underlying theme is the intensification of regional cultural identities, with emphasis on family values and local traditions, high population growth, and less concern for rapid economic development.

B1: A converging world with rapid change in economic structures, and introduction of clean technologies. Emphasis is placed on global solutions to environmental and social sustainability, including efforts in rapid technological development, dematerialization of the economy, and increased equality.

B2: A world in which the emphasis is on local solutions to economic, social and environmental sustainability. It is a heterogeneous world with less rapid technological changes, and greater and diverse technological changes, but with a strong emphasis on community initiative and social innovation to find local instead of global solutions.

Travel Cost Method

The travel cost method pairs values in order to determine the demand function of the study area based on the visitor's rate and the cost of travel. With the following equation it becomes possible to estimate a demand function based on the variation of the number of visits through the consumer's surplus, which will be the area below the demand function.

Visitors rate = f (cost of travel)

RESULTS

The Rainha beach (Figure 1), located in Costa de Caparica, is one of the most visited touristic areas of the region. This section is inserted in the coastal arc of Caparica – Espichel cove, composed essentially with detrital rock with a frontal dune. The natural processes are modified and the beaches are confined within rigid intervention works for stabilization preventing the impact of coastal hazards.

Vulnerability Analysis of Coastal Area

Three parameters corresponding to different weights were considered (Table 3). The first criteria gives equal weighting to all parameters, while the second criteria diminishes the importance of green cover and

Figure 1. Rainha Beach

Table 3. Three vulnerability weights

Parameters	1- First Weighting	2-Second Weighting	3-Third Weighting
TE	1	1	7
DS	1	2	8
TA	1	1	2
WH	1	0.5	5
EA	1	1	3
GL	1	2	9
GM	1	2	4
GC	1	0.5	1
AA	1	2	6
Total Sum	9	12	45

Source: Adapted from Coelho et al., 2006.

significant wave height, since these are not considered in the majority of these classifications (Coelho et al., 2006), finally, the third criteria consists in a parameter scaling from 1 to 9. All three were used in this analysis.

Regarding the range of the tides, it is given a rating of 3 (Table 4), valid for almost all Costa da Caparica shore. This region is positioned near the maximum degree of vulnerability, at a height of less than 10 meters and a distance of the coastal zone less than 20 meters. For the tidal range, Rainha beach is at the moderate level with values between 2 to 4 meters. Concerning the parameter coastal erosion / accretion rate, it is stated in the literature that while the coastal region remained stable in recent years, in the past 50 years it has lost over 160 meters. Considering the geology parameter, a sandy beach will always have the highest degree of vulnerability as well as in geomorphology with the presence of dunes. This area has anthropic defenses, in order to protect ecological and recreational benefits of the beach.

The calculated results are in the region of high vulnerability, which is between 3.5 and 4.5 (Table 5), which is a strong indicator for the vulnerability of this region of study.

Bruun Model

The Bruun Model was calculated using the bathymetric variables from a study of Coelho & Veloso-Gomes (2004) and integrated with the SRES Scenarios (Tables 6-8).

In A2, where the underlying considered subject is the substantial population increase, and decreasing concern with the a fast development in the economy, coastal recession will trend to increase, probably due to the impact of high population growth which will inevitably increase all impact indicators. The results of the Bruun Rule are static, therefore a modelling evaluation with dynamic changes of the morphodynamic elements of coastal regions would be required to assess the precision of these results.

Table 4. Vulnerability matrix of parameters

Region	Results								
	TE	DS	TA	WH	EA	GL	GM	GC	AA
Rainha	4	5	3	4	4	5	4	2	1

Source: Adapted from Coelho et al., 2006.

Table 5. Vulnerability classification of S. Jacinto beach

Region	Classification of Vulnerability					
	1		2		3	
Rainha	3.6	High	3.8	High	3.9	High

Source: Adapted from Coelho et al., 2006.

Table 6. Bruun rule results using SRES Scenarios and bathymetric variables for Rainha beach for 2020

SRES Scenarios	Sea Level Rise (m)	Profile Width (W)	Active Profile Depth (m)	Beach Berm (m)	Bruun (m)
B1	0.07	480	3	50	0.63
B2	0.20	480	3	50	1.81
A1	0.21	480	3	50	1.90
A2	0.38	480	3	50	3.44

Table 7. Bruun rule results using SRES Scenarios and bathymetric variables for Rainha beach for 2050

SRES Scenarios	Sea Level Rise (m)	Profile Width (W)	Active Profile Depth (m)	Beach Berm (m)	Bruun (m)
B1	0.13	480	3	50	1.18
B2	0.36	480	3	50	3.26
A1	0.39	480	3	50	3.53
A2	0.68	480	3	50	6.16

Table 8. Bruun rule results using SRES Scenarios and bathymetric variables for Rainha beach for 2050

SRES Scenarios	Sea Level Rise (m)	Profile Width (W)	Active Profile Depth (m)	Beach Berm (m)	Bruun (m)
B1	0.19	480	3	50	1.72
B2	0.53	480	3	50	4.80
A1	0.58	480	3	50	5.25
A2	1.04	480	3	50	9.42

Travel Cost

The application of this method resulted in a bibliographic revision of the methodology produced by Silva (2012), considering Costa da Caparica beaches with valuation of direct and indirect use, assuming that travel costs to the beaches are an "entry fee" to provide an estimate of demand. The first step was the identification of individuals via survey that considered "surfing" as a reason to visit the beach, while "waves" were the indirect use considered in the survey. 60,4% of the 432 individuals presented at least one of these motives. This study considered that the region of Costa de Caparica has a direct use valuation of 1022789, 52€ and 46635, 12€ of indirect use valuation.

It must be referred that this study only considers National visitors, disregarding foreign ones, which will inevitably translate in an underestimation of the travel cost values, however, these are reference values since it's a pioneer study on methods of economical valuation to ecosystems applied to Costa de Caparica beaches.

Protection Measures

Artificial feeding of the beaches is the fastest measure; however they imply the prohibition of beaches during the installation of the metallic pipes for the creation of a line of metallic repulsion. The use of submerged repulsion line is one of the most effective solutions taking into account the ripple characteristics. These measures were carried out on the beaches of Costa de Caparica between August 11 and October 19, 2008. This process lasted around three months, implying a value in the order of six million euros per million cubic meters of deposited sand (INAG, 2008).

The most costly coastal defense process was the construction and repair of spurs, worth eight million and two hundred thousand euros, for the repair of the spur field of Costa de Caparica and Cova do Vapor in a period between October 2004 To May 2006, and despite the repair of the spurs, still requires an artificial feeding of sand, thus contributing to an increase in the final value (INAG, 2006).

Still in front of Caparica beaches, it should be mentioned that the relocation and construction works to support fishing, parking, pedestrian access, demolition, rehabilitation of coastal defense works and artificial feeding of beaches, carried out within two years between 2004 and 2006, were valued at a total cost of 17 million euros (INAG, 2008). The costs mentioned will be raised or reduced by the value of the benefits that are intended to be safeguarded in a given region. If the regions to be safeguarded represent an engine of the economy at local level, we can say with some certainty that these measures represent an added value in the protection of resources.

SOLUTIONS AND RECOMMENDATIONS

Vulnerability assessment is vital from a social point of view, since the natural response of the ecosystems to a new equilibrium might affect the benefits that populations collect from it, such as housing, raw materials and recreation values such as tourism.

The main use of a coastal vulnerability index is to assess the impact that natural events such as sea level rise will have in a coastal region. Görnitz and Kanciruk (1989) developed one of the first attempts to include a vulnerability index in climate changes studies. The object of study was the coastal region of the United States of America and it is, until today, one of the main studies considered in the development of such indexes. Gomes (2009) developed a tripartite methodology to assess vulnerability and quantify its potential of lost in the Portuguese shore, considering the effects of the sea level rise due to climate changes.

The coastal vulnerability index used in Rainha beach of Coasta de Caparica allowed providing a fast and cost effective estimate of local vulnerability, supported by a heavy bibliographic revision and the use of visualization tools. The combination of these methods is useful for the stakeholders in terms of decision making for coastal management.

The Bruun model is included in the logic of production of fast and cost effective data to provide to stakeholders, however, there are some limitations. The Bruun Model used in a two-dimensional logic is static. When providing scenarios, it should also be considered in a more detailed analysis the variability of the morphometric variables for each period.

Regarding the travel cost method, it considered only national tourists, while excluding foreign ones and the excursionists visiting the area and with a stay less than one day, meaning that the obtained values might be underestimated regarding the total economic value generated by the beach in study. Regardless of the model limitations, the results provided by Silva (2012) are a pioneer attempt to provide an estimate for the beaches of Costa de Caparica in Portugal.

The bibliographic collection for the interventions in shoreline for the region of Costa de Caparica were an attempt to demonstrate that these do not intend only to protect the economic value generated only by recreational activities *per se* such as tourism, there's also the importance of protecting the ecosystem itself behind.

FUTURE RESEARCH DIRECTIONS

A Framework for Using GIS and Stakeholder Input to Assess Vulnerability to Coastal Regions

The "vulnerability of site" model contends that place risks and considerer the integration of biophysical events of geographic context and socioeconomic stressors (Cutter et al. 2000). Füssel (2007) mentioned some advances to this framework by stating that vulnerability results from exposure to external biophysical and socioeconomic stressors (e.g., an extreme physical event, national and local policies) and from internal characteristics of a system (e.g., local environmental conditions, household incomes, social networks, economic situation) that influence the extent of the mentioned stressors' impacts on the system.

An assessment exclusively based on a contextual framework consider all internal and external components that contribute to a site vulnerability, and can be improved with a geographical framework at more detailed scales in guiding sea level rise (SLR)-related hazard mitigation and adaptation strategies. Thus, we suggest an improvement of the presented vulnerability assessment framework, considering the premises presented by Füssel (2007) but also considering social media data, Internet of Things (IoT) and spatio-temporal Earth Observation data.

As a future development, we proposed the implementation of the framework presented by Füssel (2007), which categorizes vulnerability into four broad dimensions—the system (e.g., economic activities), the attribute of concern (e.g., human life condition), the hazard, and the temporal reference. We also intend to apply this framework to our study area by assessing vulnerability as specific not only to particular hazards and sites, but also to the multi-scalar aspects (e.g., internal vs. external) inherent in Füssel's system dimension.

We will intend to assess multi-scalar aspects of vulnerability by coupling geospatial analysis with multi-criteria collaborative decision-support methods. Therefore, geospatial analysis will involve the use of Geographic Information System (GIS) tools to integrate modeled physical data delineating potential contemporary and SLR-enhanced storm surge inundation zones with socioeconomic data characterizing coastal development. Multi-criteria collaborative decision-support methods will involve the use of focus groups to validate and expand GIS-based results with local expert knowledge and to include value-based human dynamics of geospatially dependent vulnerability, population growth, and development.

CONCLUSION

This study accomplished some goals, namely that intervention measures are important not only to protect the economic value provided by coastal areas, but especially, the ecological value. Both components should be deemed as a relevant issue while developing adaptation and mitigation policies in terms of coastal management.

The quantitative and qualitative assessment of the impact of vulnerability and sea level rise in Rainha beaches based on the Vulnerability Index and the combination of the Bruun Rule with the SRES Scenarios provide a first and fast estimate to classify this beach with a high vulnerability. The Vulnerability Index is based on the users' experience and a heavy bibliographic support based on available data, the main limitation is the lack of benchmarking for these results, due to the diversity of indexes and studies provided by several authors. The Bruun Rule continues to be a benchmark for the scientific community in terms of sea level rise impacts, however, for more elaborated studies, the third dimension analysis must be considered, since the two dimensional approach relies on beach profiles, while three dimensional models surpass this limitation by the material budget equations, providing more precise estimates.

The Travel Cost method is used to calculate economic values of environmental goods, especially for recreational value. It still has several limitations, mainly related to each visitor and the calculation of a "real entry fee" and the development of a tourist concept that allow including in official statistics the figures for the fluctuant population visiting the site for recreation and the actual population. Surveys still are the main source of information, however, the burden answer and the difficulties of validation don't provide a reliable estimate. Other methods should be considered in the future of travel cost, such as the use of real time data via tracking or via mobile applications. The results for Caparica beaches are a first attempt to quantify the recreational benefits of this coastal area, allowing users to recognize them as an important source of income for the local economy and in terms of decision making and management programs involving local communities to protect these natural values.

REFERENCES

Airoldi, L., & Beck, M. W. (2007). Loss, Status and Trends for Coastal Marine Habitats of Europe. *Oceanography and Marine Biology - an Annual Review*, 387.

Anthoff, D., Nicholls, R. J., & Tol, R. S. J. (2001). *Global sea-level rise and equity weighting*. Working Paper FNU – 136.

Antonov, J. I., Levitus, S., & Boyer, T. P. (2005). Thermosteric sea level rise, 1955–2003. *Geophysical Research Letters*, *32*(12), L12602. doi:10.1029/2005GL023112

Bodungen, B., & Turner, R. K. (2001). Science and Integrated Coastal Management, An Introduction. In *Science and Integrated Coastal Management* (pp. 1–14). Berlin, Germany: Dahlem University Press.

Brooks, N., Nicholls, R. J., & Hall, J. (2006). *Sea Level Rise: Coastal Impacts and Responses*. Berlin: WBGU.

Bruun, P. (1988). The Bruun Rule of Erosion by Sea-Level Rise: A Discussion on Large Scale Two- and Three Dimensional Usages. *Journal of Coastal Research*, *4*(4), 627–648.

Bruun, P. M. (1962). Article. *Journal of Waterways and Harbor Division*, *88*(WW1), 117–130.

Church, J. A., Gregory, J. M., Huybrechts, P., Kuhn, M., Lambeck, C., Nhuan, M. T., & Woodworth, P. L. et al. (2001). Changes in sea level. In J. T. Houghton, Y. Ding, & D. J. Griggs et al. (Eds.), *Climate Change 2001: The Scientific Basis* (pp. 639–693). Cambridge, UK: Cambridge University Press.

Coelho, C., Silva, R., Veloso-Gomes, F., & Taveira Pinto, F. (2006). *Risk Analysis Approach for the Portuguese West Coas.* Conducted at Risk Analysis 2006 – Fifth International Conference on Computer Simulation in Risk Analysis and Hazard Mitigation, Malta.

Coelho, C., & Veloso-Gomes, F. (2004). Crosshore Beach Profile Models – Application to Aveiro Coast. *Journal of Coastal Research*, *39*.

Costa, C. (2007). Introduction to Environment Géotechnique – Fundamentals of Géotechnique, Faculty of Sciences and Technology, New University of Lisbon. *Civil Engineering (New York, N.Y.)*, 1–38.

Crooks, S. (2004). The effect of sea-level rise on coastal geomorphology. *British Ornithologists Union*, *1*, 18–20.

Cutter, S. L., Mitchell, J. T., & Scott, M. S. (2000). Revealing the vulnerability of people and places: A case study of Georgetown County, South Carolina. *Annals of the Association of American Geographers*, *90*(4), 713–737. doi:10.1111/0004-5608.00219

Dinis, J. L., & Tavares, A. (2005). Geomorphological susceptibility of the Portuguese west coast to tsunamis. *Proceedings of the III Congress on Planning and Management of Coastal Areas of Portuguese Speaking Countries*. Hydric Resources Association.

Downing, T. E., & Patwardhan, A. (2003). *Vulnerability Assessment for Climate Adaptation*. APF Technical Paper 3, United Nations Development Programme.

Ecosystem Valuation. (2009). *Travel Cost Method: Ecosystem Valuation*. Retrieved Oct 12, 2008, from: http://www.ecosystemvaluation.org/travel_costs.htm

Emery, K. O., & Aubrey, D. G. (1991). *Sea levels, land levels and tide gauges*. New York: Springer-Verlag. doi:10.1007/978-1-4613-9101-2

European Comission. (2004). Living with coastal erosion in Europe: Sediment and Space for Sustainability, PART I - Major findings and Policy Recommendations of the EUROSION project. *Directorate General Environment European Commission.*, *1*, 1–57.

Feagin, R. A., Wu, X. B., Smeins, F. E., Whisenant, S. G., & Grant, W. E. (2005). Individual versus community-level processes and pattern formation in a model of sand dune plant succession. *Ecological Modelling*, *183*(4), 435–449. doi:10.1016/j.ecolmodel.2004.09.002

Füssel, H. M. (2007). Vulnerability: A generally applicable conceptual framework for climate change research. *Global Environmental Change*, *17*(2), 155–167. doi:10.1016/j.gloenvcha.2006.05.002

Gilman, E. (2005). *Assessing and Managing Coastal Ecosystem Response to Projected Relative Sea-Level Rise and Climate Change*. Honolulu, HI: Blue Ocean Institute.

Gomes, P. (2009). *Impacte da subida do nível do mar sobre o turismo* (Master thesis). Faculty of Science and Technology of the New University of Lisbon.

Gornitz, V., & Kanciruk, P. (1989). Assessment of global coastal hazards from sea-level rise. *Proceedings of the 6th Symposium on Coastal and Ocean management.*

Hennecke, W. G., & Cowell, P. J. (2000). GIS modelling of impacts of an accelerated rate of sea-level rise on coastal inlets and deeply embayed shorelines. *Environmental Geoscience*, *7*(3), 137–148. doi:10.1046/j.1526-0984.2000.73002.x

Holgate, S. J., & Woodworth, P. L. (2004). Evidence for enhanced coastal sea level rise during the 1990s. *Geophysical Research Letters*, 31.

Houghton, J. T., Ding, Y., Griggs, D. J., Noguer, M., van der Linden, P. J., Dai, X., . . . Johnson, C. A. (2001). Climate Change 2001: The Scientific Basis: Contributions of Working Group I to the Third Assessment Report of the Intergovernmental Panel on Climate Change. Cambridge University Press.

INAG I.P. (2007). *Water quality of bathing areas*. Retrieved Oct 12, 2008, from: http://www.inag.pt/

INAG I.P. (2008). *Coastal Interventions*. Retrieved Oct 12, 2008, from: http://www.inag.pt/index.php?option=com_content&view=article&id=47&Itemid=90

IPCC (2001). *Summary for policymakers climate change 2001: Mitigation*. A Report of Working Group III of the Intergovernmental Panel on Climate Change.

IPCC. (2007). *Climate Change 2007: The Physical Science Basis*. Working Group I Contribution to the Fourth Assessment Report of the Intergovernmental Panel on Climate Change. Cambridge, UK: Cambridge University Press. Retrieved Oct 12, 2008, from http://ipcc-wg1.ucar.edu/wg1/wg1-report.html

Kennedy, V. S., Twilley, R., Klepas, J., Cowan, J., Jr., & Hare, S. (2002). Coastal and Marine Ecosystems & Global Climate Change: Potential Effects on U.S. Resources. Report for the Pew Center on Global Climate Change.

Leggett, J., Pepper, W. J., & Swart, R. J. (1992). *Emissions Scenarios for the IPCC: an Update, Climate Change 1992: The Supplementary Report to The IPCC Scientific Assessment*. Cambridge University Press.

Machado, G. (2007). *Morpho-sedimentary analysis of the beach, and internal continental shelf of the coastline of the Jurubatiba National Park* (PhD dissertation). Federal University of Rio de Janeiro, Rio de Janeiro, Brazil.

McFadden, L., & Green, C. (2007). Defining 'vulnerability': Conflicts, complexities and implications for Coastal Zone Management. *Journal of Coastal Research*, 120–124.

Mozambique, D. N. G. A. (2005). *Vulnerability assessment for climate change and adaptation strategies, Ministry for the coordination of environmental action*. National Directorate of Environmental Management.

Muehe, D. (2004). Definition of boundaries and typologies of the border under morphodynamic and evolutionary aspects. In *Ministry of environment and planning, budget and management. Subsidies for a management project* (pp. 13–32). Brasília: MMA and MPO.

Nicholls, R. J. (2003). *Case Study on Sea-Level Rise Impacts*. Working Party on Global and Structural Policies, OECD Workshop on the Benefits of Climate Policy: Improving Information for Policy Makers. Ref: ENV/EPOC/GSP(2003)9/FINAL. OECD.

Nicholls, R. J., & Small, C. (2002). Improved Estimates of Coastal Population and Exposure to Hazards Released. *EOS Transactions*, *83*(2), 301–305. doi:10.1029/2002EO000216

Schallenberg, M., Hall, C. J., & Burns, C. W. (2003). Consequences of climate-induced salinity increases on zooplankton abundance and diversity in coastal lakes. *Marine Ecology Progress Series*, *251*, 181–189. doi:10.3354/meps251181

Schernewski, G., & Löser, N. (2004). *The modern tourist's perception of the beach: Is the sandy beach a place of conflict between tourism and biodiversity?* Managing the Baltic Sea. Coastline Reports.

Silva, C. G. (2004). *Coastal Sedimentation Environments and Morphodynamic Processes Acting on the Coast Line. In Introduction to Marine Geology* (pp. 175–218). Rio de Janeiro: Intersciences.

Silva, S. A. F. (2012). *Management of Beaches of Costa de Caparica. Load capacity, wave value, security and information* (Master thesis). Faculty of Science and Technology of the New University of Lisbon.

Small, C., & Nicholls, R. J. (2003). A Global Analysis of Human Settlement in Coastal Zones. *Journal of Coastal Research*, *19*(3), 584–599.

Valiela, I. (2006). *Global Coastal Change. Science Singapore*. Blackwell.

Veloso-Gomes, F. (2006). Coastal Erosion and Spatial Planning in Costa de Caparica. *Journal of Integrated Coastal Management*, *7*, 83–95.

KEY TERMS AND DEFINITIONS

Accretion: The process through which coastal sediments return to the visible portion of the beach following storm erosion.

Adaptation: Anticipating the adverse effects of climate change and taking appropriate measures to prevent or minimize the damage they can cause.

Alluvial Balance: Difference between the added and removed sediments from coastal regions.

Artificial Feeding: Regenerations of beaches by means of the artificial injection of sand.

Bathymetric: Study of underwater depth of lake or ocean floors.

Biogeochemical: Relation of biological, geological, and chemical factors.

Breakwaters: Structures constructed on coasts as part of coastal defense or to protect an anchorage from the effects of both weather and longshore drift.

CH4: Methane.

CO: Carbon monoxide.

CO2: Carbon dioxide.

Coastal Erosion: Wearing away and breaking up of rock along the coast.

Coastal Morphodynamics: Refers to the study of the interaction and adjustment of the seafloor topography and fluid hydrodynamic processes, seafloor morphologies, and sequences of change dynamics involving the motion of sediment.

Economic Valuation: Assigning monetary value to environmental factors that are normally not considered in financial valuation.

Ecosystems: Community of living organisms in conjunction with the nonliving components of their environment interacting as a system.

Elevation Related to Datum: Height above or below a fixed reference point.

Equilibrium Profile: The slope away from shore of a sea floor or lake bottom having a gradient such that waves and currents neither erode it downward nor deposit sediment upon it.

Green Cover: Natural or planted vegetation covering a certain area of terrain, functioning as protection against soil erosion, protecting the fauna, and balancing the temperature.

IPCC: Intergovernmental panel on climate change.

Mitigation: Efforts to reduce or prevent emission of greenhouse gases.

N2O: Nitrous oxide.

NOx: Nitrogen oxides.

Significant Wave Height: The mean wave height of the highest third of the waves.

SO2: Sulfur dioxide.

SRES: Special report on emission scenarios.

Sustainable Development: Organizing principle for meeting human development goals while at the same time sustaining the ability of natural systems to provide the natural resources and ecosystem services upon which the economy and society depends.

Tide Average: Vertical difference between the high tide and the succeeding low tide.

Vulnerability: Potential for loss.

Wetlands: Land area saturated with water, either permanently or seasonally.

Chapter 14
Land Use Policy and Urban Sprawl in Nigeria:
Land Use and the Emergence of Urban Sprawl

Waziri Babatunde Adisa
University of Lagos, Nigeria

ABSTRACT

Land use policy is central to the development of urban life and the emergence of cities. In many developed capitalist societies, both the planning and expansion of the cities are usually anchored on sustainable urban land policies such that the growth of urban sprawl is effectively controlled. In most developing countries, land use policies are not only disparate, they are usually not connected to the growth of cities because policy makers are after the money they could make from private investors. This chapter argues that though the coming of the Land Use Act 1978 ended the era of disparate land law regimes, it has, over the years, sealed the control of urban lands to state governors, a development that has created massive corruption and arbitrariness in the allocation and utilization of urban lands. This approach to land administration has also hindered effective and sustainable urban and regional planning in many Nigerian cities. This study suggests the review of the 1978 Land Use Act and effective utilization of modern technologies in the monitoring of urban sprawls.

INTRODUCTION

Land use policy is very central and critical to the development of urban life and the emergence of cities. Whether in the developed or less developing societies, adequate knowledge of land use and the growth of urban sprawl, are very important to the development of urban life and administration of urban services. In the United States as in other developed countries, the concern of government is on how to synergize land use policy with urban development in order effectively monitor the growth and development of the cities (Sholanke, 1990; Butler, 2012; Otubu, 2010; Mehran and Kiani, 2014). The synergy between the

DOI: 10.4018/978-1-5225-3194-4.ch014

two, when carefully tapped, can go a long way in ensuring secured property rights, stable investment climate, effective regional planning across the cities in Nigeria.

Land use policy is as important as urban and regional planning because it helps government shape its decisions on the amount of land that will be allocated to or shared by each segment of the society; human settlements, the industries, road construction, the transportation systems, the economy, urban infrastructure and rural areas (Atilola, 2010). As good as land use policy is for urban development, not many governments around the world particularly in the developing world have succeeded in pursuing effective land use policies and urban administration that are capable of reducing urban poverty and the debilitating human conditions in most urban sprawls (Sun, 2015). Although, the growth of urban sprawls, is inevitable in the evolution of cities, countries where urban sprawls are poorly managed tend to make citizens vulnerable to urban problems such as armed robbery and unsecured property rights.

This chapter discusses the role of land use policy in the emergence and growth of urban sprawls in Nigeria. The objectives of the study are to: 1. examine the challenge of urban sprawl in Nigeria 2. Interrogate the land use policy of Nigeria 3. examine the effect of Nigeria's land use policy on the nature of urban sprawl in the country.

BACKGROUND

The concept of "Land Use Policy" is often used to refer to government's blueprint or regulatory framework pertaining to procedures, methods and approaches the state will adopt to control the acquisition, allocation, distribution, management and utilization of public lands for the benefits of the citizens (Abiodun, Olaleye, Dokai & Odunaiya, 2011). Sholanke (1990) opined that a land use policy is the government's blueprint guiding the reform as well as the administration of lands for urban and rural development. In many parts of the world, land use policy often determines a number of other government policies pertaining to investment growth and economic development. No wonder, it is so central to wealth creation and the development of the city (Oluyede, 1985; Aluko, 2004; Mabogunje, 2009; Otubu, 2010).

Urban sprawl, on the other, denotes a pattern of human settlement on the fringes of the city caused by a combined forces of urban expansion, increase in population size and density, pressure on urban infrastructure, public health and traffic conditions in the city and increasing road networks to the hinterlands (Sun, 2015). In the developed world, urban sprawl is a matter of preference for those who are willing to live on the fringes of the city, in the developing countries, it is matter of necessity. It is a matter of necessity because citizens' movement to the area, is usually as a result of the pressure on urban land use, population density and access to resources in the inner part of the city (Bekele, 2005). Urban sprawl, when, considered, within the context of land law, can also be describe an abnormal spatial expansion of the city to low-density areas due to the inability of people to have access to legitimate land rights or fulfill their legal requirements as prescribed by law. In Nigeria, urban sprawl also entails the construction or erection of buildings in areas not approved by city planners or government's land bureaus.

Bekele (2005) argued that "Sprawl is the spreading out of a city and its suburbs over more and more rural land area. This involves the conversion of open space (rural land) into built up, developed land over time". Owei, Obinna & Akalo (2008) stated that;

Urban sprawl is commonly used to describe physically expanding urban areas consisting largely of a pattern of low-density expansion of large urban areas, under market conditions, into mostly surrounding agricultural areas. Sprawl in Nigeria consists of informal housing development on the urban fringe on land that is mostly privately owned, sold in single small plots and in the case of Port Harcourt and Lagos, also on marginal land along the waterfronts. Usually no development permits have been granted (Owei et al, 2008).

While many factors may have helped to explain the causes of urban sprawl, Bekele (2005) noted that it main causes are population increase and land use. One of the significant parameters for determining or measuring the emergence of urban sprawl in the city is the presence or the predominance of people living on the fringes of the city-a location between the urban area and the rural community. It should however be noted that, not all urban sprawls are illegal settlements (Amujiri, 2001).

Causes of Urban Sprawl in Nigeria

Urban sprawl is a phenomenon that could be caused by a variety of issues such as geographical, ecological, social, economic, industrial, political, spatial, and hazards. This study examines some of the factors that have caused the development of urban sprawls in Nigeria. The study does not claim to have exhausted all these factors.

Urbanization and Urban Sprawl in Nigeria

Urbanization has had one of the greatest impacts on the growth of Nigerian cities. It is one factor that has had the most significant direct link to the movement of people from the hinterland to the urban areas, and in extension to the fringes of the city. Since late nineteenth century, Nigeria's pre-colonial towns and communities have been shaped by rapid increase in population caused by modern urbanization.

Ezeani (2001) argued that urbanization was said to be responsible for the transformation of the agrarian economic system in pre-colonial African cities to predominantly industrial capitalist system introduced by the Europeans in the 19th century. One significant change that accompanied the change in the mode of production was internal migration to major cities in search of greener pasture "the White Man's Job". People moved in mass towards colonial headquarters than to African cities that merely served as transportation routes of the colonialists (Onibokun & Faniran, 1995; Nwanna, 2004).

Population Increase and Population Explosion

The growth of the population size and density of a city is one of the determinants of the growth of urban sprawl. The assumption is that, the higher the population of a city, the higher the propensity for the growth of urban sprawl. Since urban sprawl, is the movement of people to the fringes of the city, increase in population combined with limited infrastructure can drag people out of the city. At the same time, improved road networks to the rural areas can facilitate population decongestion where the city has experienced population explosion. What this implies is that, whether city planners like it or not, people will move to the periphery of the city, as the Central Business District get congested. This is one of the factors that has influenced the growth of Lagos from Lagos Island, to Apapa, Ikoyi, Lekki, Ibeju etc.

Studies have shown that high population increase in Nigeria, in Nigeria, since the colonial days, is responsible for the growth of Nigerian cities. For instance, the Nigerian population, was said to have grown steadily from 18.72 million people in 1921 to 30.4 million in 1952. By 1963, it was estimated to have risen to 55.67 people, almost doubled of the total population of the country in 1952. Although, new population estimates put the population size of Nigeria at over 180 million, as at 1991the country's population stood at 80.5 million while it rose to 140 million in 2006.Interestingly, as at 1931, less than 7 percent of Nigerians lived in urban centres with Lagos leading other urban areas in Nigeria (Onibokun and Faniran, 1995).Today, almost half of Nigeria's population, 49.9% live in urban areas.

Nwanna (2004) held that since the end of the Second World War, there has been upward movement of people towards the cities in the South and later to the North. The author, believed, this migration pattern, was fueled partly by changes in the infrastructure in the cities and the pressure caused by the Second World War. For instance, the population of people living in Lagos is reported to have grown from 230, 256 in 1950 to 267, 400 in 1952. By 1963, it accelerated to 449, 500 indicating over eighty percent increase. As at 1991, Lagos was estimated to have 6.8 million people, by 2006, the city population was put at 9 million (National Population Census, 2006). Although, the 2006 population census figures put the population of Lagos at 9 million below that of Kano, the World Population Review stated that, Lagos is one of the mega cities in the world with an estimated population size of over 21 million as at 2016 (World Population Review, 2017).

The growth of Lagos, during the colonial era, has been attributed too much influence that the colonial government exerted on the city. First, Lagos was the first colony in Nigeria to be annexed to the British Empire in London in 1861. The annexation of Lagos in 1861 sparked and brought Western-form of urbanization. Colonial institutions began to spring up in Lagos and later Bonny, Calabar, Kano, Kaduna, Jos and Ibadan. Second, Lagos happened to be the first colony to witness the growth of colonial administrative headquarters such as the first courts that were built in 1963 and the first prison which the colonialists launched in 1872.

Similarly, Nwanna (2004) argued that the influence of urbanization reverberated across the country including the Northern region of Nigeria particularly in Kaduna such that by 1952, the population of Kaduna had reached 131, 316. By 1963, the number of people living in Kaduna had grown to 295, 432 despite the fact that, colonial presence had not been fully felt in the city unlike Lagos. It should however be noted that urban growth in many Nigerian cities followed the dictates of the colonial administrators such that people moved towards the capital cities and colonial administrative headquarters than to other places in the colonies. One of the consequences of uneven urban growth in many African countries is the problem of urban sprawl.

The United Nations Habitat 2010/2011 report estimated that a total of 227 million people in the developing countries have been moved out of the slums between year 2000 and 2010. This, when measured against the targets of the Millennium Development Goal 7, exceeded the target with 2.2 times and was 10 years ahead of the 2020 deadline (United Nations Human Settlements Programme, 2010).

While there seems to a general improvement, the rate of performance varies across regions. Asia is said to have done relatively better than other regions by moving an estimated 172 million people out of slums. China and India top the hierarchy because they have been able to improve the lives of more than seventy-percent (125 million) of the total of number of people that have moved out of slums in Asia. The reverse is however, the case in Africa, except for North-Africa where 8.7 million people have been lifted out of slums and the proportion of slum dwellers have declined from 20 to 13 percent. In the case of Sub-Saharan Africa, the proportion of people living in the slums just decreased with 5 per cent ((United Nations Human Settlements Programme, 2010: x).

It is believed that the worsening human conditions in Sub-Saharan Africa, are caused by the sporadic rise in the level of urbanization and the lackluster colonial and postcolonial urban land use policies (Mabogunje, 1968; Nwanna, 2004). The United Nations Human Settlements Programme (2010) summed up the crisis this way:

More and more people both in the North and South are moving outside the city to "satellite" or dormitory cities and suburban neighborhoods, taking advantage of accommodation that can be more affordable than in central areas, with lower densities and sometimes a better quality of life in certain ways. Spatial expansion of cities is triggered not only by residents' preference for a suburban lifestyle, but also by land regulation crises, lack of control over peri-urban areas, weak planning control over land subdivisions, improved or expanded community technologies and services, as well as greater population mobility. Whether it takes the form of "peripherization" (informal settlements) or "suburban sprawl" (residential zones for high and middle-income groups), sub-urbanization generates negative environmental economic and social externalities. In developing countries, the phenomenon comes as an escape from inadequate governance, lack of planning and poor access to amenities. Rich and poor seek refuge outside the city, which generates further partitioning of the physical and social spaces (UN-Habitat, 2010: ix).

Citing the experience of Port-Harcourt, Owei, Obbinna & Akarolo (2008) noted that;

Urban sprawl in Port Harcourt is an integral part of its process of transformation from a port city, to one of the nation's most important industrial and commercial centers. It is also a testament to the failure of successive governments to manage urban growth. Over the years, government has acquired land ostensibly for public use. Oruwari & Owei (2006) have raised questions on the use of such land and suggested that the beneficiaries have been high income people. Public land acquisition has been the arena of conflict between indigenous land owners and government over the years. Thus, access to land for the majority of urban dwellers especially low and medium income people is through the informal land market. Primarily, it is the interplay of these factors and the impact on both demand and supply of land that has fuelled the pattern of urban sprawl in Port Harcourt (Owei, et al, 2008).

Statistics in Table 1 shows that, Sub-Saharan Africa has the highest of urban population compared to North-Africa, East-Asia and the rest of the world as at 2010.

With reference to the proportion of urban population living in the slum areas, Sub-Saharan Africa (61.7%) ranks number one in the world, and it also leads the whole of the region (the developing world) when compared with North-Africa (13.3%), Latin-America (23.5%), Eastern Asia (28.2%), Southern Asia (28.2%), and Oceania (24.1%). The findings indicate a bid lesson for the whole of Sub-Saharan Africa including Nigeria. It shows that, the problem of urban sprawl is greater in the region than other parts of the developing world.

Land Use, Spatial Expansion of the City, and the Growth of Urban Sprawl

Spatial expansion of the city is also a key determinant of urban sprawl. This can be influenced by land use and urban master plan. Urban sprawl does not just grow without the need for a city to expand. The

Table 1. Urban population of major region of the world living in slums

Urban Slum Population (Thousands)						
Major Region or Area	**1990**	**1995**	**2000**	**2005**	**2007**	**2010**
Developing Regions	656, 739	718,114	766,762	795,739	806,910	827,690
North Africa	19,731	18,417	14,729	10,708	11, 142	11, 836
Sub-Saharan Africa	102, 588	123,210	144,683	169,515	181,030	199,540
Latin-America and the Caribbean	105,740	111,246	115, 192	110,105	110,554	110,763
Eastern Asia	159,754	177,063	192,265	195,463	194,020	189,621
Southern Asia	180,449	190,276	194,009	192,041	191,735	190,748
South-East Asia	69,029	76,079	81,942	84,013	83,726	88,912
Western Asia	19,068	21,402	23,481	33,388	34,179	35,713
Oceania	379	421	462	505	524	556

Source: (United Nations Human Settlements Programme, 2010).

Table 2. Proportion of urban population living in slums (%)

Major Region or Area	1990	1995	2000	2005	2007	2010
Developing Region	46.1	42.8	39.3	35.7	34.3	32.7
North Africa	34.4	28.3	20.3	13.4	13.4	13.3
Sub-Saharan Africa	70	67.6	65	63	62.4	61.7
Latin-America and Caribbean	33.7	31.5	29.2	25.5	24.7	23.5
Eastern Asia	43.7	40.6	37.4	33	31.1	28.2
Southern Asia	57.2	51.6	45.8	40	38	35
South-East Asia	49.5	44.8	39.6	34.2	31.9	31
Western Asia	22.5	21.6	20.6	25.8	25.2	24.6
Oceania	24.1	24.1	24.1	24.1	24.1	24.1

Source: (United Nations Human Settlements Programme, 2010).

expansion of a city is however dependent on whether government officials in charge of city master plan enforces and monitors the growth of sprawls or the movement of people to the urban fringes. The desire of residents to live in new environments developed by government is also contingent on the degree of conduciveness of the environment for human cohabitation. Where government is effective in the administration of a city's master plan, the development of urban sprawl, is closely under the control of government and the state (Olujimi, 2009).

Urban and regional planners, have over the years, developed Geographic Information System to capture the satellite images of the growth of cities around the world. This technology has benefitted both the cities in the developed and developing countries including Nigerian cities. The GIS has worked effectively in monitoring the development of urban sprawls and land use patterns over time. This sys-

Table 3. FCT land use projections 1998-2000

Land Use	Total (ha)	% Total
Residential	12, 486	48.67
Commercial Business District	561	2.19
Institutional	891	3.44
Industrial Research/Training	920	3.59
Transportation	1,705	6.65
National Government	500	1.96
Sports and Recreation	160	0.62
Parks, Open Space	8, 435	32.87
Total	**25, 685**	**100%**

Source: Oyesiku (2010) p. 206.

tem, has been applied and used to monitor the growth of Lagos, Abuja, Port-Harcourt, Benin etc (see Olujimi, 2009); Aluko, 2011).

The GIS can be used to capture data on the spatial expansion of the city influenced by land use patterns such as residential, commercial, industrial, institutional, transportation, architectural, sports and recreational activities. Oyesokun (2010) presented land use projections of the Federal Capital Territory to show the anticipated spatial expansion of Abuja over a period of 3 years.

MAIN FOCUS OF THE CHAPTER

Land Use Policy and Land Administration in Nigeria Since Colonial Era

The history of land use policy in Nigeria dated back to the colonial era particularly to 1861 when Lagos became the colony of the British. Although, the annexation of Lagos was the beginning of colonial administration and British land use policy in Nigeria, some customary land tenure systems had existed far before the coming of the colonial masters (Mabogunje, 2009; Butler, 2012).

Acemoglu & Robinson (2012) argued that before the coming of the colonialism to Africa, the European super powers had traversed other parts of the world and colonized many nations in the 17th and 18th centuries. In addition to colonizing their captives, they succeeded in imposing European culture and land use policies (Acemoglu & Robinson, 2012). To succeed in Africa, the European colonialists had to first introduce Trans-Atlantic Slave Trade and later imposed colonial rule on the vast territory in the continent (Mamdani, 2004).

In 1884, 14 European colonialists converged in *Berlin, Germany to cement this process of colonization of Africa and the sharing African territories among themselves.* The meeting, which is otherwise called *"the Scramble for Africa"* did not end until 1885, when the British, Germany, America, France, Portugal, Spain, Belgium all agreed on the portions of Africa that would be shared and controlled by them. With 1884-1885 land act, the Western world created for itself the formal ground for the takeover or control of Africa's large expanse of lands (Touval, 1966; Boahen, 1985; Koponen, 1993; Boateng, 2010; Graven, 2015).

At the onset of British colonial administration in Nigeria, the colonial government imposed the Foreign Jurisdictions Act of 1890 to 1913 on Nigerian territories with a view of using these laws to takeover and alienate the lands of the colonized territories in the country. The *proviso* in the law which gave the British foreign missions the rights to acquire lands in the interests of British government effectively sealed Nigerian territories to the British colonialists. Aluko (2004) made this submission about the development of British land policy in Nigeria:

Following the enactment of the Foreign Jurisdiction Acts, 1890s-1913, the British Government, which included the Parliament and the Crown, had powers to legislate on Nigeria. A major reception legislation arising from the jurisdiction of the British Government was the Interpretation Act, Cap. 89, Laws of the Federation and Lagos. There was also the (Miscellaneous Provisions) Act. By section 45 of the Act, the English common law, the doctrines of equity and the statutes of the general application that were in force in England on the 1st day of January, 1900 were in force in Lagos in so far as the limits of the local jurisdiction and local circumstances permitted and subject to federal law (Aluko, 2004: 112).

Across the colonial territories, the effect of the adoption and imposition of the English Common laws were overwhelming such that existing customary laws in the South and Islamic laws in the North were subdued to British Common laws. Butler (2012) stated that, in the South, the British Government recognized the collective ownership of lands in trust for the family while in the North all lands belonged to the Crown. Because of the differences in the culture and customs of the North and the South, colonial land laws were different and couched to suit each region. For instance, in Northern Nigeria, the British Government enacted the Crowns Lands Proclamation, 1902. Aluko (2004) wrote that:

In an effort to streamline and clarify land tenure system, a Committee was set up in 1908 which made far reaching recommendations. The immediate effect of the Committee's recommendations was the enactment of the Lands and Native Rights Proclamation, 1910. By section 2 of the Proclamation, the whole of lands in the Protectorate of Northern Nigeria, whether occupied on the date of the commencement of the Proclamation was declared native lands. Six years later, the Land and Native Rights Ordinance was enacted. This was in 1916 (Aluko, 2004:115).

In Western and Eastern Nigeria, there were also different types of land and property registration laws designed to protect the interests of the colonialists and the natives. For instance, in Eastern Nigeria, there were Kola Tenancy Law of 1935 and Acquisition of Lands by Alien Law of 1957. The development was not quite different in Western Nigeria where the British Government introduced land use policies as far back of 1862 (Oluyede, 1985; Aluko, 2004). At independence, there was concern about lack of uniformity in the land use policies in the country, and the need for government to centralize and nationalize the disparate land laws spread across the country. Due to the problems of political instability caused by the 1966 military intervention in politics, nothing significant could be achieved on this nationalization policy. The outbreak of the civil war of 1967-1970 however increased nationalists' interests in policies that were capable of increasing national unity and national integration. It was this spirit that prompted the Federal Military Government under Gen. Olusegun Obasanjo to set up the Land Use panel, in 1977. Oluyede (1985) stated that the panel was charged with the following terms of reference:

1. To undertake an in-depth study of the various land tenure, land use, and land conservation practices in the country and recommend steps to be taken to streamline them;
2. To study and analyse all the implications of a uniform land policy for the country;
3. To examine the feasibility of a uniform land policy for the entire country; make necessary recommendations and propose guidelines for implementation;
4. To examine steps necessary for controlling future land use and also opening and developing new land for the needs of Government and Nigeria's population in both urban and rural areas and to make appropriate recommendations.

Based on the recommendations of the panel, and a critical review of existing land laws in Nigeria prior to 1977, the Federal Military Government promulgated on March 29th, the 1978 the Land Use Act, Decree No. 6 of 1978. Mabogunje (2009) added that, the promulgation of the Decree offered the Federal Military Government the opportunity to extend the Northern land management system of 1910 to the whole country, which was expected to give government autonomous control over the allocation of lands in the country.

Land Use, Urban Planning, and Administration in Nigeria

The way and manner people living in urban centres consume or utilize lands that are available for urban development, is a function of the nature and effectiveness of urban and regional planning policies puts in place by the government. As a natural asset, lands in the city are expected to appreciate over time, and develop in response to increase in population that is either caused by high birth rate or migration from the rural areas to the city. As more people move into the city, densities in the city are expected to push residents to the fringes of the city. This movement is also expected to be accompanied by spatial expansion of the geography of the city and corresponding increase in the demand for lands.

While the phenomenon of urban sprawl is a matter of preference in the developed world usually to escape the noise and pollution in the city, in most developing countries, most citizens move to the sprawl as a matter of necessity either due to lack of access to landed property in the urban areas or inability to bear the high cost of living in the city. This two-hedge condition that urban sprawl presents to urban dwellers around the world, justifies the reasons why urban and regional planning is very important to the development of cities (Bekele, 2005; Olujimi and Gbadamosi, 2007; Olujimi, 2009). This also brings to the fore the inevitability of urban and regional planning units, agencies and institutions in the planning and development modern city.

Evolution of Urban and Regional Planning in Nigeria

The history of urban and regional planning in Nigeria dates back to the pre-colonial era. Prior to the advent of colonialism, urban and town planning were largely guided by the customs, culture and traditions of the people of Nigeria. Traditional cities in the North were designed around the palaces of the Emirs, so also those of the Yoruba in South-West, Nigeria. Urban and regional planning in many of these societies were largely rudimentary (Aluko, 2011).

1. Urban and Town Planning Ordinances in Nigeria:
   a. **Town Planning:** The advent of colonialism and colonial penetration of Africa including Nigeria in the nineteenth century, however changed what most people in pre-colonial societies had been used to in their societies. Following the annexation of Lagos in 1861, the colonial government introduced the 1863 Town Improvement Ordinance (Aluko, 2011). This ordinance was enacted to control the development and sanitation of Lagos colony.

The 1900 land proclamation act by Lord Lugard, in Northern Nigeria was the act that provided for indirect rule in respect to land administration and settlement development in Nigeria. Aside from the 1900 act, the colonial government also introduced the 1904 Cantonment Proclamation No. 29 of 1917 which ushered in the separation of European officials from the expatriates officials. While the 1917 ordinance created different residential areas for the Europeans and the expatriates, the Ordinance No. 4 of 1914 introduced compulsory acquisition of lands for public uses, irrespective of method of ownership. This law could be regarded as the first fierce approach to compulsory land acquisition for urban planning in Nigeria (Mabogunje, 1968; Aluko, 2011).

While the 1917 ordinance classified towns into first class, second class and third class towns, the 1924, the Town Planning Schemes created Town Planning Committees to initiate planning schemes that will lead to the overall development of the city. One of the tasks of the committees was to initiate planning programs that will lead to the decongestion of highly populated towns. It was this vision that led to the creation of Apapa in 1926 as a port town aimed at reducing the pressure on Lagos Island.

One of the remarkable impacts of the 1924 Town Planning Committees set up by the colonial government, was that it opened up the way for further centralization of urban and regional planning in major cities in Nigeria. For instance, it paved way for the establishment of the Lagos Town Planning Ordinance of 1928 and the Lagos Executive Development Board created from this legislation. Although, the delivering of public services were largely done by the Lagos City Council, the LEDB helped in slum clearance, and caused the reclamation of Victoria Island. It also led to the initiation of planning schemes in Ikoyi, Iganmu and Surulere areas of Lagos (Aluko, 2011).

In 1946, the colonial government, under their ten years development plan of 1946-1955, introduced the 1946 Town Planning Ordinance No. 4. This ordinance marked the first systematic attempts at regulating urban and regional planning across Nigeria (Aluko, 2011). The law, which helped greatly in shaping urban land use provided a centralized and systematic urban and regional plan, through which the growth of Nigerian cities could be monitored.

It is however important to note that while these laws (1924, 1928 and 1946) lasted and impacted on the development of urban life, the development of colonial cities in Nigeria, was also a function of the interactions between geography, topography, economic potentiality and the colonial significance of the cities. Land use patterns across many of these cities responded to the pressure of population growth dictated by the colonial political economy, urbanization, industrialization and migration.

   b. **Independence and Urban and Regional Planning in Nigeria:** Urban and regional planning in Nigeria entered a new phase after the independence of Nigeria. Independence, as it did in other areas of Nigerian development, provided an opportunity for Nigerian leaders redesign urban plans and formulate favorable urban programs for their people.

At independence in 1960, the country retained a good number of the colonial laws, ordinances and urban policies designed for the regulation of land use, city development, building construction,delivery of public services as well as provision of urban infrastructure (Aluko, 2011). The first law to be retained, was the 1946 Town and Country Planning Ordinance. This was followed by chapter 123 of the Town and Country Planning Law of Western Nigeria of 1959. The chapter 130 of the law of Northern Nigeria and chapter 155 of the law of Eastern Nigeria were equally retained (Aluko, 2011).

Urban and regional planning in postcolonial Nigeria, however received a great boost from the Military government following the introduction of 1976 Local Government Administration Law which empowered the local governments to have Local Planning Authority that was expected to monitor and control the planning schemes at the local government level. This policy, which was later complemented with the 1978 Land Use Act, was to determine the growth of Nigerians cities in years to come (Aluko, 2004: Aluko, 2011).

Urban and Regional Planning Decree of 1992 and Urban Administration in Nigeria

In 1992, the Federal Military Government, introduced gave urban and regional planning a boost similar to that of 1946 by enacting the Nigerian Urban and Regional Planning Decree No. 88, 1992. The decree, which created different agencies at the Federal, State and Local Government levels, wanted urban administration in Nigeria to be properly monitored by government. It also wanted to monitor land use patterns across the cities in order to avoid indiscriminate buildings that could distort the Master Plans of the cities.

Although since the 1992 Nigerian Urban and Regional Planning Decree, there has not been any new law at new by Federal Government centralizing urban and regional planning law in Nigeria, different states have adapted the components of the law to initiate schemes for the development of their cities. Such laws designed by state to effectively monitor land use and urban development in their states include the 2010 Lagos State Law and Urban and Regional Tribunal etc.

The Land Use Act, 1978, Land Acquisition and the Emergence OF Urban Sprawls in Nigeria

The Land Use Act, 1978 is the most significant land reform in Nigeria's history. This is because it was the first law to centralize, nationalize and synergize land allocation, land ownership and land administration in the country (Otubu, 2010; Butler, 2012). As good as the Land Use Act is, it is not without its flaws and consequences for urban growth and development. Mabogunje (2009) argued that, the performance of the Decree can be assessed against its essential provisions which vested the all lands situated in the territory of each state in the country in the Governor of the state and that all lands control and management, including land allocation in the urban areas must be carried out by the Governor. Mabogunje added that the law also grants the Governor the sole power to issue Certificate of Occupancy and decide, which vacant land is to be acquired for public uses.

Although the land use act was promulgated with good intentions, the wide range of powers granted the Governor, is the source of many problems relating to land ownership in Nigeria (Sholanke, 1990; Otubu, 2010).

The next section of the paper takes a critical look at the effects of the land use act, 1978 on the emergence and pervasiveness of urban sprawl in Nigeria.

Effects of the Land Use Act, 1978 on the Emergence of Urban Sprawls in Nigerian Cities

The Land Use Act, 1978 was promulgated with the intention of increasing public control of lands for the benefits of the citizens but experiences over the past 38 years after the promulgation of the law do not indicate that governments have citizens at heart. The next segment of the paper discusses the impact of the land law on the growth of urban sprawls in Nigeria. These problems include:

1. **Land Acquisition, Abuse of Power of the Governor and the Growth of Urban Sprawl:** Land acquisition is the first component of every land law or land policy. This is because it defines the process to be followed by government in acquiring or securing lands for public purposes. Over the years, government has abused this absolute right accorded the state governor by acquiring lands which to the natives to their political associates and wealthy individuals who in the country. This attitude of government obviously contravenes Section 1 of the Land Use Act (1978) which states as follows;

Subject to the provisions of the Act, all land comprised in the territory of each state in the Federation are vested in the Governor of that state and such land shall be held in trust and administered for the use and common benefit of Nigerians in accordance with the provision of this Act.

Demolition of illegal structures is one of the strategies by which governments in Nigeria have been dealing with citizens who violated the land. Some of these governors have hinged their actions on the need to build new model cities comparable anywhere in the world. However, the governors do not always follow due process of the law before eviction of people from urban sprawls or urban slums.

A good example of this, under the military was the forceful seizure of the lands of allegedly belonging to late Odumegun Ojukwu, in spite of a standing court order (Eso, 2003). Another instance was the eviction in the 1990s residents of Maroko, Lagos by then Military Governor of Lagos State, Governor Raji Rasaq. Sahara Reporter (2015) painted a gloomy picture of how the plan to demolish Maroko was hatched:

The then military governor of Lagos State, Raji Rasaki saw that Maroko was good and liked it. According to sources, he intimated then head of state, Ibrahim Badamasi Babangida (IBB) of the potentials of the place. IBB, in the company of Raji Rasaki and the then newly created Eti-Osa Local Government boss, Olatunde Beecroft, visited Maroko on November 17th 1989. IBB, during the trip, described Maroko as 'a mini Nigeria' because all Nigerian languages were duly represented in the community. He also enjoined the residents to continue to embrace government programmes; promising that he would soon send Rasaki to the residents to fast track the development of the community. The promise was followed by ovations that nearly reached into the high heavens. Surviving ex-Maroko residents told City Voice they were naïve. Raji Rasaki indeed returned to Maroko but, to the chagrin of the residents, it was to demolish the community; arrest, detain, even kill anyone who dared the military invasion. The first inkling of trouble was an announcement on Radio that residents of the community had eight days to vacate their homes. Yet, by the seventh day, the bulldozers had moved in. By 8: am on that fateful morning, the first set of 30 houses had been reduced top dust by the over 30 bulldozers brought for the task (Sahara Reporter, 2015).

The second instance of house demolition by government which also triggered public reaction and public sympathy was the Abuja demolition of 2004. The demolition was preceded with a revocation order by then FCT Minister, Malam Nasir El Rufai. The Daily Independent of July 6th, 2004 exclusively reported that the presidency issued a notice for the revocation of all C-of Os issued in Abuja between 1976 and when Abuja was first mooted. In addition to the revocation order, raised down several houses and displaced thousands of people in the capital city (Daily Independent July 6th, 2004).

On Friday March 18th March, Otodo-Ogbeme fishing community in Eti-Osa LGA, Lagos State was said to have been demolished by the state government. Al Jazeerah exclusively reported further that, over 4,700 people were displaced by the government's action.

2. **Land Allocation, Abuse of Governor's Power and Urban Sprawl:** Land allocation is another source of abuse of the land use act by the Governors and government officials. Although, corrupt public officials are often involved in unscrupulous allocation of lands to individuals, the Governors' role in the issuance of double Certificate of Occupancy impacts more on the individuals than the action of the bureaucrats.

Butler (2012) held that, the deliberate delay and bureaucratic bottlenecks in the allocation of land to land applicants, are attempts by government officials to cause confusion as to which land obtains government's approval and which one does not. Owei et al (2008) averred that, the use of executive recklessness to take over the lands of private individuals is a common practice in contemporary Nigeria. Oshio (1990) went to say that; "These abuses were not limited to privately owned lands or communal lands under customary law. The allocation of state lands was also affected in no small measure. The distribution was either to government functionaries, to persons who were closely connected with them, or to others who invariably had more wealth than most" (Oshio, 1990: 50).

3. **High Cost of Processing Land and Completing Land Title Registration:** Land title registration is a very important aspect of land ownership. This is because it is the stage at which citizens make requests to government and go through the necessary procedures to secure their land holding rights. The high cost of registration lands by low income earners have pushed many to the low-density areas, where all the formalities are not "required" before one can get lands from the land speculators. The preponderance of those who buy "acquired lands" are those who cannot afford to buy lands in approved government areas (Butler, 2012). The World Bank (2012) stated that, high cost of registering property and poor regulatory framework discourages investment and slows down economic growth in the developing countries.
4. **Corruption and Informality in Land Title Registration and the Phenomenon of Urban Sprawl:** Corruption is one of the clogs in the wheel of progress of Nigeria's quest for a genuine land reform. It is a problem because it gives the land speculators the opportunity to induce government officials who should have monitored the implemented the land use act and government land use policy. Butler (2012) believed that corruption adds to the burden of poor people by ensuring that, they pay more

for land title registration. In addition to this, it leads to other problems such as; 1. Pre-cluding the development of modern land cadastre and registration systems; 2. Under-mining state collection of recurring land charges and transaction fees; 3. preventing the implementation of good land use planning and subdivision practices; 4. creating an uneven playing field for businesses.

5. **Lack of a Workable Housing Policy for Nigerians:** Housing is one of the most essential necessities of life. It is also the best indicator of a person's standard of living as well as his place in society. Without housing, many people will be unable to contribute meaningfully to their country's national development (Aluko, 2004).

The expansion of the city or the urban area is expected to be predicated on very good and sound urban housing policy but this is not always the case in Nigeria. Since independence, governments have attempted to solve the problem of housing deficit by investing in mortgage banks, building low-cost houses and engaging the private sector in the development of the housing sector, in reality, many of these policies are elite-driven, bogus and not favorable to the low-income earners (Aluko, 2004). A good and most recent example is the building of "Eko Atlantic City" in the name of building houses for Lagosians.

Aluko (2004) stated that most of the problems of housing in Nigeria are caused by lack of synergy between the land use policy of government and housing policies. Currently, Nigeria is said to be suffering from a housing deficit of about 17 million. This is attributed largely to lack of commitment to the provision of housing for the poor. In order to effectively tackle this problem, the World Bank suggested government will require about N59.5 trillion to close the gap created by housing deficit (Emedare, 2015).

6. **Political Instability, Policy Somersault, Politicization of the Implementation of the Land Use Act, 1978 and the Problem of Urban Sprawl:** The politics of land use policy is another grey area is responsible for the growth of urban sprawl. Politics involves the use of state power to determine who gets what and how. It exists all over the world but it is used in much of colonial Africa to accumulate wealth for the benefits of the political elites.

Of the 57 years of Nigeria's independence, the military spent a total of 29 years, 21 of which were spent after the promulgation of the land use act in 1978. When it handed over power to the civilians in 1999, many Nigerians expected significant changes but political climate has seen the hijack of land administration by politicians, god fathers and wealthy individuals (Mabogunje, 2009).

SOLUTIONS AND RECOMMENDATIONS

Towards the Building of a Genuine Reform in the Land Sector and the Eradication of the Problem of Urban Sprawl in Nigeria

Across Nigeria, efforts are ongoing in all the 36 states of the Federation and the Federal Capital Territory, Abuja, to address the challenge of urban sprawl. The pace at which each state, is moving varies and is determined by many factors; availability of funds, size of the city, the rate of urbanization, population explosion, migration and pressure on urban infrastructure. At the National level, the main regulatory framework for planning, preventing, monitoring and fighting the managing urban sprawl is the Nigerian Urban and Regional Act, 1992.

The Act is the regulatory framework that has informed laudable urban and regional planning and programs that have transformed urban development in Nigeria. This law has led to the establishment of several agencies and boards at the Federal and state levels. Like, the Federal Ministry of Power, Housing and Works, the Lagos State Government has Ministry of Physical Planning and Urban Development. The establishment of the Ministry dated back to 1928 when the colonial administration enacted the Town Planning Ordinance Cap., 95 of 1928.

The Lagos State Ministry of Physical Planning and Urban Development is specifically established under this Act and the 2010 Lagos State Law to undertake, among other things: vet and approve building plans, undertake the comprehensive land use, re-planning and improvement of the conditions of urban centers, embark on housing development and general development of Lagos State. The Ministry over the years, has been involved in urban reclamation, a development, which has brought physical changes to the landscape of Lagos as well as the reduction of the volume of urban slums in the state (Lagos State Ministry of Physical Planning and Urban Development, 2017).

Urban reclamation is one of the modern ways of reducing the number of urban sprawl that is available in a country at a time. While the implementation of this policy in the developed world has always been accompanied by offers of social security measures to the poor and the migrants, the displaced people in the urban slums of developing countries rarely get compensation. This, again, is the cause of the persistence and pervasiveness of the problem of urban sprawl in Nigeria.

Although, Lagos State Government has achieved considerable success in the transformation of Lagos, just like the Federal Government in Abuja metropolitan reclamation, it is important for the state to implement this policy with human-face. In a bid to protect the poor, the Lagos State Government passage into law, *the Lagos State Grabbing Law, 2016.* ThisDAY (2016) sums up the development this way:

Lagos State government has commendably, enacted a law to check the activities of unscrupulous land acquirers and extortionists, but it must courageously implement the law to make the desired impact. Lagos State Governor Akinwunmi Ambode on Monday signed a new law outlawing land grabbing and extortion of money from land developers in the state. The properties protection law was enacted as the appropriate response to the menace of locals, popularly called Omo-Oniles, who are mainly the original or earliest inhabitants of the different areas of the state. The Omo-Oniles have long posed a huge threat to land owners and property developers in the state. On paper, this new response couched in the law "law to prohibit forceful entry and illegal occupation of landed properties, violent and fraudulent conducts in relation to landed properties in Lagos State and for connected purposes", is a deviation from the traditional norm of treating the miscreants, who are generally seen as the nemesis of land developers, with kid gloves. (Thisday, 2016).

Recommendations

The study is making the following recommendations to government in order to effectively control the growth of urban sprawl in the country and towards the concept of Smart City of the 21st century.

1. The Land Use Act, 1978 should be amended in a such way that state governors will be divested of the sole powers of allocating lands in the urban areas. Caveat can be inserted in the law to check abuse by some erring executive governors. By doing this, the state can involve other stakeholders on issues pertaining to processes and procedures of making public lands affordable for the citizens.
2. The study wants governments in Nigeria no synergize land use policy and urban and regional planning such that, the growth of urban sprawl can effectively monitored.
3. Urban reclamation program of Lagos State Government, the Federal Capital Territory and other states in the country should be informed by global best practices and respect for human rights.
4. Land administration and land management must be guided by modern land tenure systems that give citizens a voice and open up the systems for all competitors. The use of e-land administration and Geographic Information System (GIS) must be paramount to government's administrators. This will ease land allocation, title registration and issuance of Certificate of Occupancy.
5. Governments in Nigeria should synergize land reforms and housing policies in order to provide low-cost housing for the poor.
6. The operations of land bureaus in the 36 states of the federation and FCT must be closely monitored and supervised by houses of assembly and the anti-corruption agencies in the country.
7. Government should streamline the cost of processing lands and work on affordable prices that average Nigerians can afford.
8. The suggestion of Presidential Technical Committee on Land Reform that more Nigerians should be sensitized on land tenure system, land titling and registration as well as land use patterns must be given needed media supports.
9. Land reforms in the states should involve the funding and implementation of cadastral survey that will capture quite a large number of information about land ownership, land uses and urban development.
10. Urban and regional planning agencies should embark on integrated data sharing system where physical planning agencies policies will work at par with those the housing development.

FUTURE RESEARCH DIRECTIONS

The idea of Smart Cities, is increasingly becoming popular among urban planners. Future studies of urban sprawl in Nigeria, can examine the extent to which land reforms and improved land use patterns, in Nigerian cities have moved towards the concept of Smart Cities.

CONCLUSION

This study has examined critically the effects of land use policy on the emergence and growth of urban sprawls in Nigerian cities. The study discovered that, over the past three decades when government had promulgated *the Land Use Act, 1978, government and government officials have not demonstrated sincere interest in making lands available and accessible to Nigerians. The result of this, is the indiscriminate building of structures on lands not approved by government. In addition to this, is the rise of the phe-*

nomenon of "Omo-Onile" land grabbing in major cities in the country which has further compounded the challenges of urban dwellers. This study has also discovered that, there is apparent abuse of absolute power granted the state Governors to allocate lands and issue Certificate of Occupancy. This abuse has been coming in form of executive recklessness or commercialization of what supposedly belonged to the general public. The study concluded by suggesting that genuine reforms of the land sector must be informed by and rooted in transparent land management and a complete overhauling of the absolute rights hitherto enjoyed by the State Governors.

REFERENCES

Abiodun, O. E., Olaleye, J. B., Dokai, A. N., & Odunaiya, A. K. (2011). Land Use Change Analyses in Lagos State: From 1984-2005. FIG Working week 2011, Bridging the Gap between Cultures, Marrakech, Morocco.

Acemoglu, D., & Robinson, J. A. (2012). *Why Nations Fail: The Origin of Power, Prosperity and Poverty*. Profile Books.

Aluko, O. (2004). *Housing and Urban Development in Nigeria*. Kins Publications Series.

Aluko, O. (2011). Functionality of the Town Planning Authorities in Effecting Urban and Regional Planning Laws and Control in Nigeria: The Case of Lagos State. *African Research Review*, *5*(6), 156–171. doi:10.4314/afrrev.v5i6.14

Amujiri, B. A. (2001). Urban Planning in Nigeria: Problems and Prospects. In *Issues in Urbanization and Urban Administration in Nigeria*. Enugu: Jamoe Enterprises.

Aribigbe, A. (2007). *Urban Land Use Planning, Policies and Management in Sub-Saharan African Countries: Empirical Evidence From Akure, Nigeria*. Fourth Urban Research Symposium, 2007. Retrieved March 9th, 2017, sitesresources.worldbank.org/INTURBANDEVELOPMENT/

Atilola, O. (2010). *Land Administration Reform Nigeria: Issues and Prospects*. FIG Congress, 2010: Facing the Challenges-Building the Capacity, Sydney, Australia. Retrieved March 9th, 2017, fine.net/pub/fig2010/papers/invo4/invo4_atilola_4758

Bekele, H. (2005). *Urbanization and Urban Sprawl*. Department of Infrastructure, Kungliga Tekniska Hogskolan.

Boahen, A. (1985). *General History of Africa: Africa under Colonial Domination 1880-1935*. UNESCO.

Boateng, O. (2010). Carving Up Africa: 125 years of the Berlin Conference. *New African*, 492.

Butler, S. (2012). *Nigerian Land Markets and the Land Use Law of 1978*. World Resources Institute & Landesa.

Emedare, W. (2015). *Housing in Nigeria: Why Nigeria Desperately Need 17 million More Houses*. Retrieved 14th March, 2017, www.sapientvendors.com.ng

Eso, K. (2003). *Further Thoughts On Law and Jurisprudence*. Ibadan: Spectrum Books.

Ezeani, E. O. (2001). *Contending Theoretical Perspectives on Global Urbanization. In Issues in Urbanization and Urban Administration in Nigeria*. Enugu, Jamoe Enterprises.

Graven, M. (2015). Between Law and History: Berlin Conference of 1884-1885 and the Logic of Free Trade, London. *Review of International Law*, *3*(1), 31–59. doi:10.1093/lril/lrv002

Jazeerah, A. (2017). *Police Displaced Thousands in Nigeria's Otodo-Gbame*. Retrieved from www. aj jazeerah.com/news/2017/3/police-displaced-thousands-otodo-gbame-slum

Koponen, J. (1993). The Partition of Africa: A Scramble for a Mirage? *Nordic Journal of African Studies*, *2*(1), 117–135.

Mabogunje, A. (2009). *Land Reform in Nigeria: Progress, Problems & Prospects*. Retrieved March 10th, 2017, siteresources.worldbank.org/EXTARD/resources/Mabogunje

Mabogunje, A. L. (2009). *Land Reform in Nigeria: Progress, Problems & Prospects*. Retrieved March 11th, 2017, siteresources.worldbank.org/.../mabogunje

Mamdani, M. (2004). *Citizen and Subject: Contemporary Africa and the Legacy of Late Colonialism*. Fountain Publishers.

Mehran, M. G., & Kiani, A. (2014). A Study on the Relationship Between Commercial Land Use and Increase of Urban Travels (Case study: City Center of Urmia). *Trends in Life Sciences: An International Review Journal*, *3*(4), 21–27.

Nwanna, C. (2004). *Rural-Urban Migration and Population Problems in Nigerian Cities. In Industrialization, Urbanization and Development in Nigeria 1950-1999*. Lagos: Concept Publications.

Olujimi, J. (2009). Evolving a Planning Strategy for Managing Urban Sprawl in Nigeria. *Journal of Human Ecology (Delhi, India)*, *25*(3), 201–208.

Olujimi, J., & Gbadamosi, K. (2007). Urbanization: Innovation Policy and Sustainable Development. *Proceedings of International Conference Organised By Covenant University*, 246-252.

Oluyede, P. A. (1985). Development in Land Law and Law of Conveyancing. In *The Challenge of the Nigerian Nation: An Examination of its Legal Development 1960-1985*. Nigerian Legal Institute of Advanced Legal Studies.

Onibokun, A. G., & Faniran, A. (1995). "Urbanization and Urban Social Problems in Nigeria." In Urban Research in Nigeria. Ibadan: IFRA, CASSAD.

Oshio, E. (1990). The Indigenous Land Tenure System and Nationalization of Land in Nigeria. *Boston College Third World Law Journal*, *10*(1), 42–43.

Otubu, T. (2010). Land Reforms and the Future of Land Use Act in Nigeria. *Nigerian Current Law Review*, *2007-2010*, 126–143.

Owei, E., Obinna, P. N., & Akarolo, C. I. (2008). Land Market Distortions in Nigerian Cities and Urban Sprawl. *44th ISOCARP Congress.*

Oyesiku, O. K. (2010). *New Cities in Urban and Regional Planning*. Lagos: Longman.

Reporters, S. (2015). *Maroko Evictees Mark 25 Years of July ""90 Massacre Over Lekki"".* Retrieved 14th March, 2017, http:// secure.saharareporters.com/2015/11/11/maroko-evictees-25-years

World Population Review. (2017). *Lagos Population, 2017.* Retrieved March 11th, 2017, worldpopulationreview.com/world-cities/lagos/population.

Sholanke, O. (1990). Is the Grant of Governor's Consent under the Nigerian Land Use Act automatic? *Journal of African Law, 34*(1), 42–52. doi:10.1017/S0021855300008184

Sun, X. (2015). Selected Enforcement of Land Regulations: Why Large-Scale Violators Succeed. *China Journal (Canberra, A.C.T.), 74*, 66–90. doi:10.1086/681938

Thisday. (2016). *Land-Grabbing Prohibition Law: All Eyes On Lagos State Government.* Retrieved March 14th, 2017, http://www. thisdaylive.com/index/2016/08/02

Touval, S. (1966). Treaties, Borders and the Partition of Africa. *Journal of African History, 7*(2), 279–293. doi:10.1017/S0021853700006320

United Nations Human Settlements Programme. (2010). *State of the World's Cities, 2010/2011*. Nairobi, Kenya: UN-Habitat.

World Bank. (2017). *Doing Business 2017.* Retrieved March 14th, 2017,www.doingbusiness.org/media/WBG/DoingBusiness/Documents/AnnualReport

KEY TERMS AND DEFINITIONS

Land Use Patterns: The utilization of the available lands in a city as dictated by urban and regional planning and also the social, economic, political and geographical conditions in a city.

Land Use Policy: The legal and policy frameworks of a state or government guiding land acquisition, allocation, utilization, and consumption by different segments of society.

Urban and Regional Planning: The overall government plans for the planning a city's infrastructural development, which usually include the development of residential, commercial, industrial and recreational areas for the benefits of the citizens. Aside from these, government usually use urban and regional planning to extend the geographical spread of the city with a view of reducing population densities in the city and also balancing development between the urban and rural areas.

Urban Sprawl: The increasing extension of a city to sub-urban areas as a result of changes in land use patterns which may be caused by deliberate urban and regional planning or sudden movement of the citizens to the fringes of the city.

Chapter 15
Assessing the impact of Transportation on Climate Change:
A Case Study of Agartala City in India's North East Region

Sohana Debbarma
Clean Air Asia, India

Geetanjali Kaushik
MGM's JNEC, India

ABSTRACT

India's North-East Region has greater demand for road and personalized modes of transport powered by fossil fuels. And due to emissions, there has been evidence of climate change. It has been found that diesel cars cause greater emissions (per kilometer travelled) as compared to petrol cars; therefore, the use of diesel should be discouraged. The chapter suggests that the emissions in case of public transport passenger-km are lesser than other modes of transport. However, in the North-East Region, there is negligible share of public transport due to poor infrastructure and service facilities. Therefore, improvements should be made with regard to public transport system so that considerable number of passengers shifts to public transport modes. Further, it is inferred that use of alternate vehicle or fuel technologies like hybrid electric vehicles, biofuel, biodiesel, hydrogen fuel need to be initiated to mitigate the climate change.

DOI: 10.4018/978-1-5225-3194-4.ch015

INTRODUCTION

Background

North-East Region consisting of eight sister states covering a geographic area of 26.2mha and a population of 40 million, is characterized by large percentage of indigenous tribal communities (34-91%) and diverse climate regimes which are highly dependent on the southwest monsoon (June-September). The region has international border with China in the north, Myanmar in the east, Bangladesh in the south-west and Bhutan in the north-west (shown in Figure 1). The western part of the region is connected to the Eastern part of the country through the Siliguri Land Corridor. Urbanization and growing economy in the region are leading to rapid increase in the growth of motor vehicles further leading to the increase in GHG emissions causing climatic variations over the years (GoI, 2010).

Transport System

Transport sector in North-East comprises of different modes of transport like roads, railways, aviation and inland waterways, which facilitate easy and efficient conveyance of goods and people across the region. Road transport is the primary mode of transport which plays an important role in conveyance of goods and passengers and linking the centres of production, distribution and consumption. It is also a key factor for promoting socio-economic development in terms of social and regional integration. Sustained economic growth, increasing disposable income and rising urbanization has led to rising demand for road transport and personalized mode of transport, particularly cars and two wheelers. According to the 2011 census figures, 30 percent of the region's population lives in urban areas and the rate of urbanization are growing steadily due to the push/pull effects of employment opportunities and the declining viability of the agricultural sector. This has resulted in a large and growing movement of goods, services and people fuelled by trade and employment opportunities, almost all of which is powered by fossil fuels (GoI, 2011-12).

Figure 1. Map showing north-east region and its shared borders

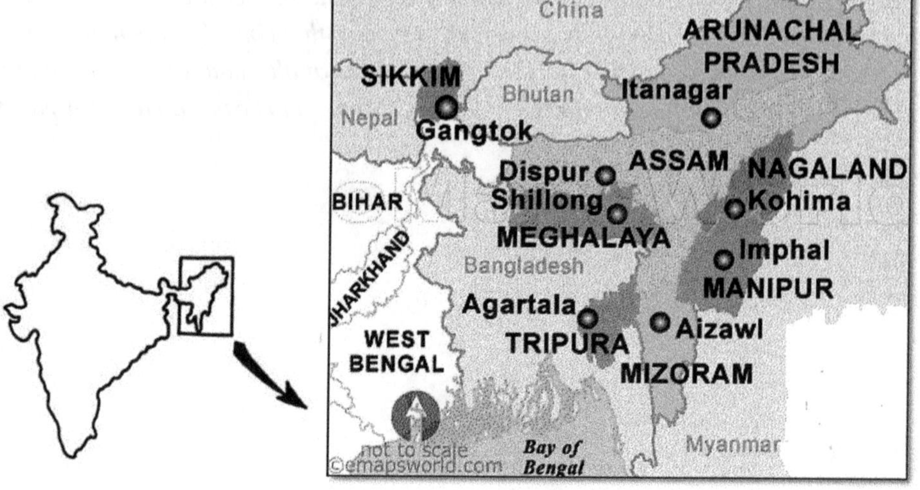

Motorization Trend

Over the years, the number of registered vehicles has recorded an exponential increase, private as well as commercial vehicles except. The state-wise average number of registered motor vehicles per 1000 population is graphically presented in the Figure 2.

Nagaland has the highest average number of registered vehicles per 1000 population. But it is growing at an annual growth rate of only 3.06%. And Tripura's motorization trend is growing rapidly at 15.67% annually, followed by Sikkim and Meghalaya. Arunachal Pradesh has the lowest motorization rate in the region (GoI, 2011-12) (shown in Table 1).

Need for Study

National air quality review carried out by the Centre for Science and Environment revealed that the states in the north east region are witnessing rapid increase in air pollution and untamed motorization.

Figure 2. Registered motor vehicles per 1000 population in north-east
Source: Transport Research Wing, Ministry of Road Transport and Highways, 2012.

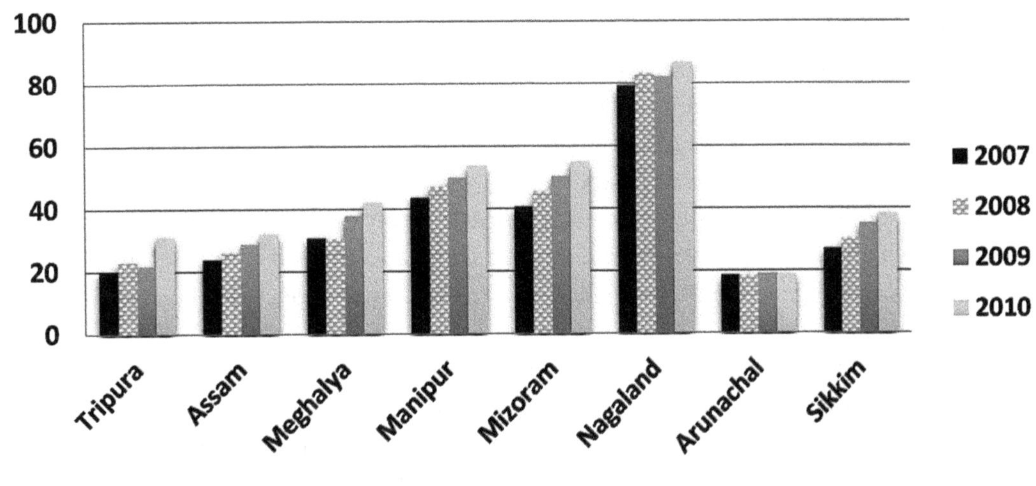

Table 1. State-wise annual motorization rate

State	Annual Growth Rate (%)
Tripura	15.67
Assam	10.07
Meghalaya	11.12
Manipur	7.18
Mizoram	10.65
Nagaland	3.06
Arunachal	0.33
Sikkim	12.22

Source: Transport Research Wing, Ministry of Road Transport and Highways, 2012.

This growth in motor vehicles is increasing greenhouse gas emissions. And due to these emissions there are evidences of climate change in the region over the years. According to the Indian Meteorological Department (IMD) study of 2011, the average temperatures are projected to increase by about 1.7 C in almost all districts of north-east. Also the World's wettest area Cherrapunji-Mawsynram recorded lowest rainfall in 34 years in 2013 (752.6cm) against the annual normal total rainfall of 11500cm. Hence, there is need for urgent action to introduce vehicle/fuel technology, scale up public transport, integrate multi-modal transport options, and encourage car restraint in the region. In this background this study is focussed on the following objectives:

1. To study the decadal climate change in the region
2. To study the existing and evolving mobility pattern of the people in Agartala city
3. To compute the emissions generated from primary study based on the vehicle kilometres travelled
4. To obtain relation between emission loads and climate change

Methodology

The steps in methodology were carried out in five stages. The stages are as explained below:

Stage 1: In this stage the need for this study, its aim and objectives were established. During this stage only the elementary scope of work was defined to give direction to this research.

Stage 2: This stage included the literature study of global and national initiatives on climate change such as IPCC, UNFCC, Kyoto Protocol and NATCOM. Also the relations between climate change and emissions from transport sector were studied.

Stage 3: In this stage surveys were conducted and data collection was done. Household survey was carried out which included information like socio-economic characteristics, vehicle kilometres of travel, fuel used, etc. Fuel consumption and climate data were collected. They have been dealt in the later chapters.

Stage 4: In this stage detailed analysis of the household, fuel consumption and climate data was carried out. From the household survey, the average distance travelled for different modes, emissions, etc were found out. The emissions were cross checked by the actual fuel consumption data. And then relation between emissions and climate change parameters were obtained.

Stage 5: The last stage consisted of developing efficient measures which will help in making policies at regional level for the mitigation of climate change due to emissions from transport sector based on all the literature search, case studies, conclusions of analysis and recommendations given by the experts.

LITERATURE REVIEW

Understanding Climate Change

According to United Nations Framework Convention on Climate Change climate change as 'change of climate which is attributed directly or indirectly to human activity that alters the composition of the global atmosphere and which is in addition to natural climate variability observed over comparable time periods'.

Also, climate can be defined as the 'average weather' and described in terms of the mean and range of variability of natural factors such as temperature, rainfall and wind speed. This system changes over time in response to internal dynamics and variations in external influences such as volcanic eruptions and solar radiations (IPCC, 2007).

Urban Climate Change

Cities are central to the climate change challenge, and their position is ever more important as the world's population is becoming increasingly urban. Climate change affects hazards, vulnerability and risk exposure in cities through a variety of direct and indirect relationships.

- **Hazards:** Are defined in the framework as the climate-induced stresses and climate-related extremes and are identified through observed trends and projections derived from global climatic models (GCMs). Extreme events affected by climate change include heat waves, droughts, inland floods, accelerated sea level rise and floods for coastal cities. Analysis of hazards specific to a city should include observed and projected data on key climatic parameters- temperature, precipitation, sea level rise, among others. Further, each hazard needs to be analyzed for variance in climate parameters over the short and long term and, where relevant, for frequency as well as intensity of extreme events. Climate change scenarios provide a reasonable understanding of potential future climate conditions. It is not expected that a single climate model will project exactly what will happen in the future, but by using a range of climate model simulations along with scenarios incorporating different atmospheric concentrations of greenhouse gases, a range of possible climate outcomes are produced and can be represented as projections that demonstrate the current expert knowledge (IPCC, 2007).
- **Vulnerability:** Is defined as the physical attributes of the city and its socio-economic composition that determine the degree of its susceptibility. The variables affecting vulnerability include flood proneness (proximity to coast or river), land area, elevation, population density, percentage of poor and quality of infrastructure. According to IPCC 2007, vulnerability is defined as the extent to which a city is predisposed to "adverse effects of climate change, including climate variability and extremes". The stakeholder concerns deal with the impacts on their inhabitants of the predicted increase in hotter days, likelihood of more intense rain events at potentially more frequent intervals, flooding that will likely result from these rain events, sea level rise and storm surge.
- **Adaptive Capacity:** Includes institutional attributes of the city and its actors that determine the degree of its capability to respond to potential climate change impacts. Thus they provide measures of the ability (institutional structure, calibre, resources, information and analysis) and willingness of actors (local governments, their constituent departments, private sector, civil society, NGOs and academics) to adapt climate change. Variables that can determine the extent of a city's ability to adapt include the structure and capacity of institutions, presence of adaptation and mitigation programs and motivation of change agents. Adaptive capacity does not assume a steady state of a city and its integrated systems; rather it measures the ability and willingness to not only copes but to respond positively to the stresses that climate change imposes.

Climate Change and Urban Transportation Systems

Transport is a critical infrastructure for development. Urban transport comprises the facilities and services to move people and materials throughout the city and its surroundings. Globally, according to the IPCC 2007 report, the transport sector accounted for 23% of the world's GHG emissions related to energy in 2004 which a reflection of local land use and mobility patterns. Automobiles and trucks are major contributors to global climate change. 17.8% of the world's energy-related greenhouse gas emissions come from road vehicles, including 10.2% from automobiles. There are over half a billion automobiles in the world today, almost all running on gasoline. Burning 1 gallon (3.8 liters) of gasoline releases 19.4 lb (8.8 kg) of carbon dioxide (CO_2) into the atmosphere, along with a number of other pollutants. An average car emits its own weight in CO_2 every year.

Emissions from automobiles are rising worldwide, as car ownership increases in the developing world and average vehicle size increases in the developed world. Increasing efficiency through decreased size, mileage increases, hybrid vehicles and diesel engines is one way of decreasing greenhouse emissions from automobiles. Also, automobile usage may be reduced through public and alternative transport, zoning cities and other settlements so as to minimize commuting and traffic jams, and other measures. Biofuels, which do not make a direct contribution to greenhouse gas emissions, are another option for reducing automobile emissions (IPCC, 2007).

Transport emissions consist mostly of CO_2, but methane (CH_4) and nitrous oxide (N_2O) are also emitted. CH_4 accounts for 0.1-0.3% of transport greenhouse gas emissions and N_2O for 2.0-2.8%. In 2004, fluorine-containing compounds (F-gases), leaked from vehicular air conditioning systems, and produced an amount of global warming equal to 5-10% of the warming produced by CO_2 emitted by vehicles.

About 95% of transport energy comes from petroleum-based fuels, with most of the other 5% coming from biodiesel and ethanol. The three major petroleum –based transport fuels are diesel, which supplies 31% of total transport energy; gasoline, which supplies 47%; and aviation fuel, which supplies 9%. The transport sector's contribution to global greenhouse gas emissions almost doubled from 1971 to 2000. Most of this increase was attributable to increased emissions from road transport. Two main factors led to this increase. One was the increase in vehicle numbers due to urbanization (increased numbers of people living in cities) and to decentralization (spreading-out) of cities. The sprawling patterns of urban transport that result from decentralization are not easily served by public transportation systems, which are best at transporting large numbers of people between a relatively small number of destinations. Two wheelers and cars have, therefore, increased rapidly in many parts of the world resulting in an increase in GHG emissions (IPCC, 2007).

Linking Emissions and Climate Change

The high natural variability and complex internal interactions create uncertainty in the way the climate will respond to increased emissions.

Global Climate Change Scenario

The global climate change scenario in terms of findings and predictions according to the IPCC, 2007 assessment report is shown in Figure 3.

Figure 3. Main findings and predictions on world's climate
Source: IPCC, 2007 Report.

Global Initiatives

There are various initiatives taken globally that address climate change. Some of the major initiatives are Intergovernmental Panel on Climate Change, United Nations Framework Convention on Climate Change and the Kyoto Protocol. These initiatives have been discussed in the following sections.

Intergovernmental Panel on Climate Change

The Intergovernmental Panel on Climate Change (IPCC) was established by the World Meteorological Organization and the United Nations Environment Programme in 1988 with the mandate to provide the world community with the most up-to-date, comprehensive scientific, technical and socio-economic information about climate change. The IPCC multi-volume assessments have since then played a major role in motivating governments to adopt and implement policies in responding to climate change, including the United Nations Framework Convention on Climate Change and the Kyoto Protocol (IPCC, 2007).

At its first session, the IPCC was organized into three working groups. The current remits of the three working groups are for Working Group I to examine the scientific aspects of the climate system and climate change; Working Group II to address vulnerabilities to, impacts of and adaptations to climate change; and Working Group III to explore the options for mitigation of climate change. The three previous assessment reports were produced in 1990, 1996 and 2001. The decision to produce a Fourth Assessment Report was taken by the 19th Session of the IPCC at Geneva in April 2002. The report was to be more focussed and shorter than before. The IPCC Fourth Assessment is intended to be a balanced assessment of current knowledge. Its emphasis is on new knowledge acquired since the IPCC Third Assessment (2001).

United Nations Framework Convention on Climate Change

The United Nations Framework Convention on Climate Change (UNFCC) is an international environmental treaty negotiated at the United Nations Conference on Environment and Development (UNCED), informally known as the Earth Summit, held in Rio de Janeiro from 3 to 14 June 1992. The objective of the treaty is to "stabilize greenhouse gas concentrations in the atmosphere at a level that would prevent dangerous anthropogenic interference with the climate system". Under the Convention, governments: gather and share information on greenhouse gas emissions, national policies and best practices; launch national strategies for addressing greenhouse gas emissions and adapting to expected impacts, including the provision of financial and technological support to developing countries; and cooperate in preparing for adaptation to the impacts of climate change.

The treaty itself set no binding limits on greenhouse gas emissions for individual countries and contains no enforcement mechanisms. In that sense, the treaty is considered legally non-binding. Instead, the treaty provides a framework for negotiating specific international treaties (called "protocols") that may set binding limits on greenhouse gases. The UNFCC entered into force on 21 March 1994. As of March 2014, UNFCC has 196 parties. The parties to the convention have met annually from 1995 in Conference of the Parties (COP) to assess progress in dealing with climate change. In 1997, the Kyoto Protocol was concluded and established legally binding obligations for developed countries to reduce their greenhouse gas emissions.

The 2010 Cancun agreements state that future global warming should be limited to below 2.0 °C (3.6 °F) relative to the pre-industrial level. The 20th COP will take place in Peru in 2014.One of the first tasks set by the UNFCC was for signatory nations to establish national greenhouse gas inventories of greenhouse gas emissions and removals, which were used to create the 1990 benchmark levels for accession of countries to the Kyoto Protocol and for the commitment of those countries to GHG reductions.

Kyoto Protocol

The Kyoto Protocol to the UNFCC is an international treaty that sets binding obligations on industrialized countries to reduce emissions of greenhouse gases. The Protocol "recognizes that developed countries are principally responsible for the current high levels of GHG emissions in the atmosphere as a result of more than 150 years of industrial activity, and places a heavier burden on developed nations under the principal of common but differentiated and responsibilities".

The Protocol was adopted by the Parties to the UNFCC in 1997, and entered into force in 2005. As part of the Kyoto Protocol, many developed countries have agreed to legally binding limitations/reductions in their emissions of greenhouse gases in two commitments periods. The first commitment period applies to emissions between 2008 and 2012, and the second commitment period applies to emissions between 2013 and 2020. The protocol was amended in 2012 to accommodate the second commitment period, but this amendment has not entered into legal force.

International emissions trading allow developed countries to trade their commitments under the Kyoto Protocol. They can trade emissions quotas among themselves, and can also receive credit for financing emissions reductions in developing countries. Developed countries may use emissions trading until late 2014 or 2015 to meet their first-round targets. Developing countries do not have binding targets under Kyoto Protocol, but are still committed under the treaty to reduce their emissions. Actions taken by developed and developing countries to reduce emissions include support for renewable energy, improving energy efficiency and reducing deforestation. Under the Protocol, emissions of developing countries are allowed to grow in accordance with their development needs.

The treaty recognizes that developed countries have contributed the most to the anthropogenic build-up of carbon dioxide in the atmosphere (around 77% of emissions between 1750 and 2004) and that carbon dioxide emissions per person in developing countries (2.9 tonnes in 2010) are, on average, lower than emissions per person in developed countries (10.4 tonnes in 2010). Countries with second-round Kyoto targets made up 13.4% of annual global anthropogenic greenhouse gas emissions in 2010. At global scale, existing policies appear to be too weak to prevent global warming exceeding 2 or 1.5 degrees Celsius, relative to the pre-industrial level (IPCC, 2007).

National Initiatives

One of the major initiatives of India addressing the aspects of climate change is the National Communication in fulfilment of commitment under the United Nations Framework Convention on Climate Change. The details of India's NATCOM to UNFCC are elaborated in the next section.

National Communication to UNFCC

India is a Party to the United Nations Framework Convention on Climate Change (UNFCCC) and the Government of India attaches great importance to climate change issues. The Communication has been prepared by a national effort and contributions of experts involving more than 200 scientists and experts belonging to various institutions countrywide, constituted into more than 120 multi-disciplinary teams to work on various aspects of climate change. The entire process has been participatory in nature involving numerous institutions and stakeholders alike. A National Steering Committee (NSC), with members from 21 Ministries/Departments of the Government of India, oversaw the implementation of the work programme (MoEF; GoI, 2012).

The Government visualizes the process of preparation of India's National Communication as an opportunity to enrich and enhance India's capabilities in identifying constraints, gaps, and related financial, technical and capacity needs to adequately fulfil the obligations under the United Nations Framework Convention on Climate Change. The Convention aims at stabilizing the greenhouse gas concentrations in the atmosphere at safer levels that would prevent dangerous anthropogenic interference with the climate system. Energy and climate change-related concerns of the Indian economy include the growing gap between the demand and supply of energy and environmental externalities associated with energy use (MoEF; GoI, 2012).

The Indian economy has been growing rapidly since the 1990s, with an even higher growth in the energy sector. This was because the economic growth was driven by energy intensive sectors, where the energy efficiency was low by international standards. High growth of these sectors has resulted in a high elasticity of energy consumption and environmental emissions with respect to GDP. Various initiatives have been taken towards establishing energy efficient technologies, energy conservation measures and regulatory frameworks, while diversifying energy sources to meet national goals as well as simultaneously address climate change concerns.

STUDY AREA

Agartala

Agartala is the capital of Tripura, one of the eight sister's states and is the second largest city in the North-East Region after Guwahati both in terms of municipal area and population.The various characteristics of the city are discussed in the later sections.

Physical Characteristics

Agartala is situated along 23° 45' - 23° 55' N latitude and 91°15' - 91°20' E longitude, in the flood plains of the Haora River. It has an altitude of 12.8 metres and an area of 62.6 sq.km. The central portion of the city is bounded by the rivers Haora in the south and Katakhal in the north. Due to its saucer shape, the city is vulnerable to recurrent flooding during monsoons. The city is located in seismic zone V. It shares border with the states of Assam and Mizoram and also shares an international border with Bangladesh (Government of Tripura, 2008).

The climate of Agartala is of tropical monsoon type. The average annual rainfall is around 220 centimetres (cm). The average no of rainy days is 100 days. The temperature varies from 4.2°c to 37.6° on the average. The winter period is from November to February, summer is from March to May and monsoon is from June to September. It has a moderate temperature and highly humid atmosphere (Government of Tripura, 2008).

Demographic Profile

Since 2004, the area of The Agartala Municipal Council (AMC) is 62.60 sq. Km. The AMC population accounts for 69% of Tripura's total urban population indicating the prominence of the city. 17% of geographical area, which is the core area, occupies over 50% of the population of AMC. The population trend and growth of Agartala city in terms of area are shown in Table 2 shows trend of decadal population growth. The growth rate indicates high migration in the city over the last decade. During 1971-1981 the growth rate was 2.4% per annum majorly due to migration and since then it has been subsequently declined slightly. But in the later years there was increase in the growth rate due to both natural increase and migration from neighbouring villages.

Socio-Economic Characteristics

The socio-economic profile of the city shows that 45% of the people have migrated to the city implying the presence of employment opportunities in the city. Among migrants, a small but significant percentage (3%) has migrated from Bangladesh. However, the maximum population (55%) of the city is originally from Agartala itself. The people residing in Agartala city are predominantly Hindus (97.7%), followed by Muslims (1.9%) and a minor share of Christians (Government of Tripura, 2008).

Table 2. Population growth in Agartala

Year	Population	Area (sq.km)
1901	6,415	3
1981	1,32,186	10.94
1991	1,57,358	15.81
2001	1,89,998	16.02
2004	3,68,493	62.6
2011	4,00,004	62.6

Source: Government of Tripura, 2008.

Figure 4. Existing land use in Agartala, 2011
Source: Government of Tripura, 2008.

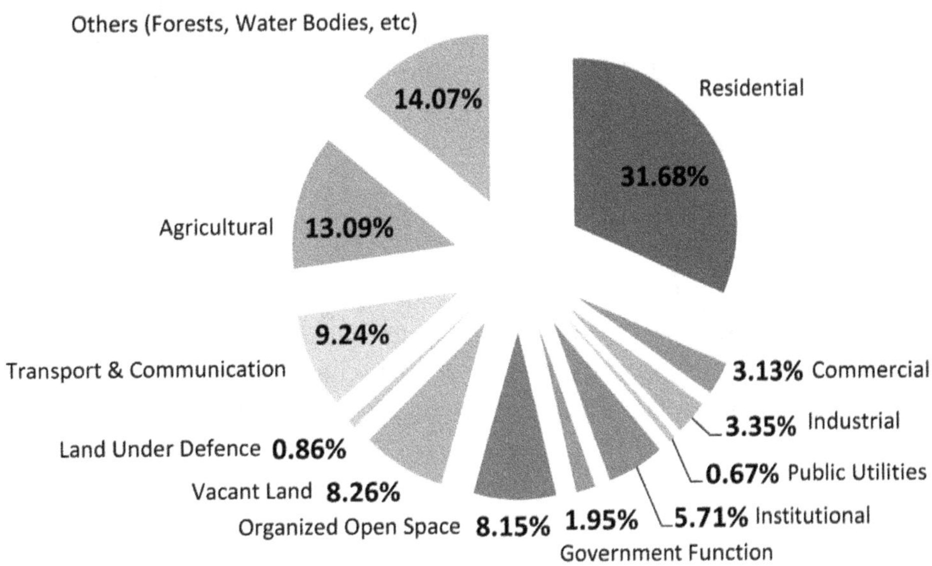

Land Use Pattern

Land under commercial and industrial use is relatively low, as these functions are still in their formative stage. The city also suffers from inadequate developed recreational area, although landform of the city and its suitability as a place of tourist attraction will justify a claim for higher proportion of land under recreation. Figure 4 represents the existing land use breakup in Agartala city.

Regional Connectivity

Agartala is well connected to other parts of the country by road, rail and air transport. The highways, NH-44 and NH-44A connect Agartala with Silchar (317km), Guwahati (599km) and Aizawl (443km). It has been connected to other cities in India by railways since June 2008. The railway station is located 5.5km from the city.There are direct flight connections to major cities of the country. Singerbill domestic airport is located 12.5km north-west from the city.It is an important international border trading town due to its trade linkages with Bangladesh.

Economy

The per capita income is rising rapidly at an annual compound growth rate of 9.6% during 2004-05 and 2012-13 at constant prices. The growth rate of primary and secondary sectors is much slower at an average annual compound growth rate of 3.38% and 5.5% respectively during the same period. The tertiary sector including transport, finance, community and social services is growing at the fastest rate of 8.12% per annum.

Motorization Trend

The total number of registered motor vehicles in Agartala is 137349 as of January, 2014. 90% of increase in the total registered motor vehicles has been observed during the 2006-2014. The highest growth rate was recorded in two-wheelers at 21.75% annually followed by cars at 11.34% and auto rickshaws at 8.24% during the same year. Growth of buses is the lowest at just 1.87% annually, indicating poor public transport (Government of Tripura, 2008)..

Modal Share

The vehicle composition analysis shows that about 75% of the total vehicles constitute of personal vehicles comprising of two-wheelers, cars, vans, jeeps, etc; which is followed by IPT modes, comprising of three wheelers, which are about 10.20% and the least is of public transport around 1.21%, which includes minibuses and omnibus (Government of Tripura, 2008). The existing modal share in Agartala is shown in Figure 5.

Road Network

The total road network of the city is 290 km and the road density is about 4.2 km per sq.km. The network follows a grid iron pattern in the central part of Agartala city. The area under circulation of roads is about 8.2% of the total area of the city. The primary road network of the city comprises of regional links such as Assam Riffle Road, Ashram Road and Bishalgarh Road as well as other roads like Secretariat Road, G.B Road and AJC Road. The NH-44 passes through the city and is the lifeline of traffic movement.

There is no distinct hierarchy of roads that can be observed. About 50% of the road length is revealed to have Right of Way (ROW) of 8-10m and 36% of the road length has more than 10m ROW. High congestion is seen in some of the traffic nodes due to improper road geometry. Majority of the existing roads in the city are about 6-10m wide with little or no scope for widening. Few stretches have 4 lane divided carriageway width with footpath on either side (NH-44 and Hari Ganga Basak Road).

Public Transport and Para Transit System

- **Bus Services:** Tripura Road Transport Corporation (TRTC), a state government owned organization, and private operators provide bus services in Agartala. TRTC and Chandrapur ISBT are

Figure 5. Existing modal share in Agartala
Source: (Government of Tripura, 2008).

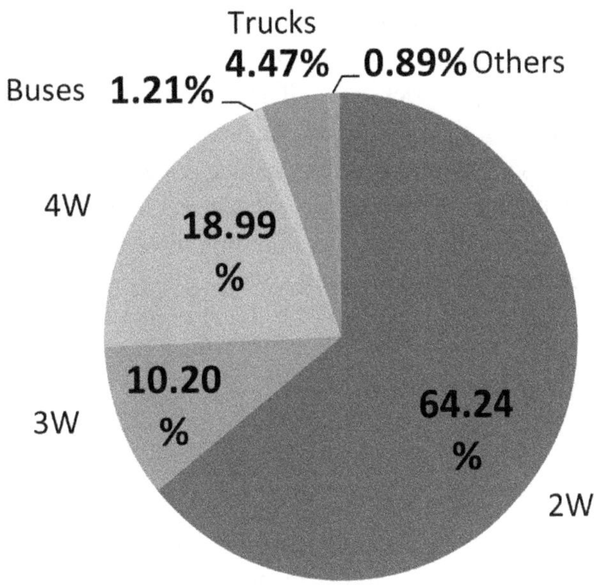

the major long distance (inter-state and inter-city) bus service provider, while private operators provide inter city services (Buses and Minibuses).

There are five bus terminals in Agartala, located along major transport corridors and based on their respective command area. While the TRTC terminal and depot is centrally located in Agartala city, the Radhanagar Terminal, Motor Stand Terminal and Battola Bus Terminal are located respectively towards north, west and south of the city. And Chandrapur ISBT is a recently introduced terminal in the city. These bus stands are operated from some of the most densely populated and economically active areas of the city, thus posing serious impediments to traffic flow and their smooth operation (Government of Tripura, 2008).

- **IPT:** IPT modes in form of 3-wheelers and maxi-cab /jeeps ply in Agartala city. Auto rickshaws cater to medium distance trips of about 5 to 10 km lead and they ply both on sharing and non-sharing passenger basis. Majority of the 3-wheelers are overloaded and carry about 6-7 passengers during peak period on sharing basis in the city. In addition maxi cab/jeeps also ply from three motor stands – Radhanagar, Nagarjala and Motor Stand and cater to passenger trips arriving/departing from the city fringe areas with an average lead of about 7-20 km. They carry about 13-15 passengers during peak period. Overloading in 3-wheelers as well as maxi-cabs is very prominent in the city due to absence of quality public transport services.

Pedestrian and NMT Profiles

- **Pedestrian Facilities:** At present footpath exist on both sides of the road Motor Stand Road, Hariganga Basak Road, Fire Brigade Road, Palace Road and Gol Bazaar Road. Construction of footpaths and zebra crossing on rest of the roads is needed to ensure safe access to pedestrians and to minimize the risk of accidents.

- **Non-Motorized Transport (NMT) Profile:** Non- motorized transport consisting of bicycles and cycle rickshaws are present in large number in the city. The Central Business District (CBD) area has a very large number of cycle rickshaws. These cycle rickshaws are registered by Agartala Municipal Council (AMC), which also issues licenses to cycle rickshaw pullers. But the AMC has stopped issuing licenses to cycle rickshaws since 2002 mainly due to the traffic congestion problems. There are about 5000 registered cycle rickshaws in the city; however it is estimated that about an equal number of rickshaws are operating unregistered, taking the total estimated figure to about 10,000 (Government of Tripura, 2008).

DATA COLLECTION AND ANALYSIS

Introduction

Household survey and secondary survey were conducted in Agartala in order to find out the average distance travelled, fuel type used and fuel consumed by different modes of transport i.e. car, two-wheeler, auto and bus. This information was required to compute the emissions from following modes. The method has been explained in the following chapters.

Agartala

The data collected from the household survey and secondary survey in Agartala are given in details in the following sections along with their analysis.

Traffic Characteristics

The study area was divided into 35 municipal wards. On an average day about 1,00,000 vehicles enter and exit Agartala, out of which 65% was motorized passenger vehicles, 6% was goods vehicles and 29% was non-motorized vehicles.

Household Characteristics

The average household size was 4.4. The occupation distribution, trip purpose, mode-wise average trip length and share of different fuel driven cars, from the household survey are given in this section.

Figure 6 shows the occupation distribution which indicates the share of students is highest in the city. Also Figure 7 shows the trip purpose and it indicates that work is the major purpose for which people travel the most.

Average travel length of each mode shows the share of different fuel driven cars which tells petrol driven cars have the highest share.

Figure 6. Occupation distribution

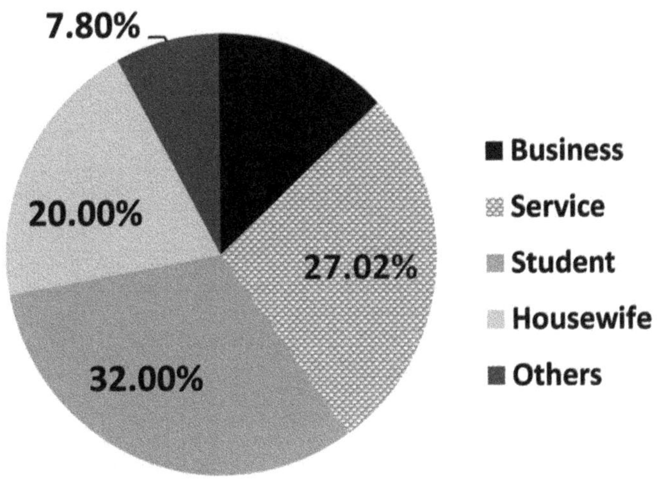

Figure 7. Trip purpose

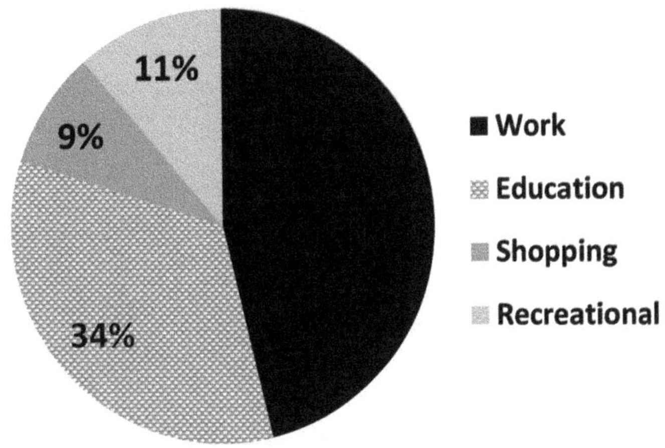

Trip Characteristics

Average trip length of different modes based on the fuel used is given in Table 3.

Using average trip length and total strength of each mode, daily vehicle-km was calculated based on the assumption that 75% of the total vehicles ply on the road daily.

Average occupancy of each mode gave the respective daily passenger-km from the daily vehicle-km. Mode-wise average occupancy and the passenger-km calculated are given in Table 5 and Table 6 respectively.

Table 3. ATL of different fuel driven vehicles

Vehicle Type	Fuel Type	ATL (km)
Car	Petrol	9
	Diesel	16
	Petrol+CNG	15
2W	Petrol	3
Auto	CNG	7
Bus	CNG	14

Source: Primary survey.

Table 4. Daily vehicle-km

Vehicle Type	Fuel Type	Vehicle Strength	Vehicle-km
Car	Petrol	19,730	1,77,570
	Diesel	2,922	46,752
	Petrol+CNG	2,194	32,910
2W	Petrol	108882	3,26,646
Auto	CNG	9526	66,682
Bus	CNG	1918	26,852

Source: Primary survey.

Table 5. Average occupancy

Vehicle Type	Average Occupancy
Car	2.7
2Wheeler	1.5
Jeep/Maxi cab	11.3
Auto	5.8
Bus	26.2

Source: Primary survey.

Table 6. Daily passenger-km

Vehicle Type	Fuel Type	Passenger-km
Car	Petrol	4,79,439
	Diesel	1,26,230
	Petrol+CNG	88,857
2W	Petrol	4,89,969
Auto	CNG	3,86,756
Bus	CNG	7,03,522

Source: Primary survey.

Emission Load

Total average distance travelled daily by all the vehicles in terms of daily vehicle-km and passenger-km were found out from the trip details in the previous section. Now emission load was computed using the formula given below:

$$E_{i} = \Sigma (Veh_j D_j)*eij \text{......... eqn} \quad (1)$$

where,

E_i = Total estimated emission of each pollutant (g/day)
Veh_j = Number of vehicles of type j
D_j = Average kilometres travelled by vehicle type j
e_{ij} = emission factor of vehicle type j (g/km)

Indian standard emission factors were used to compute the total emissions. The emission factors used are given in Table 7.

Using the above emission factors and vehicle-km from primary survey, the yearly emission load for each mode was calculated. The calculated yearly emission load emitted from the respective modes in Agartala is given in Table 8.

Data Validation

Emission load so obtained from the primary survey was validated by finding out the emission load from authentic sales data of fuel consumption in Agartala city. Yearly fuel consumption in the city during the period of 2005-06 and 2013-14 is given in the Table 9.

Trend in consumption of both diesel and petrol have been seen to decrease during the last few years which may be due to the availability of CNG in the city since 2010. Figure 8 shows the trend in consumption of diesel during the period of 2005-06 and 2013-14 and Figure 9 shows the trend in consumption of petrol during the same period.

Table 7. Emission factor used for different modes

Mode	Fuel Used	CO(g/km)	HC(g/km)	NOX(g/km)	CO_2(g/km)
Car	Petrol	3.01	0.19	0.12	150.5
	Diesel	0.3	0.26	0.49	154.56
	CNG	0.6	0.36	0.01	131.19
2W	Petrol	0.72	0.52	0.15	45.6
Auto	CNG	1	0.26	0.5	77.7
Bus	CNG	3.72	3.75	6.21	806.5

Source: (Automative Research Association of India, 2007).

Table 8. Yearly emission load (in metric tonne)

Emissions	Car	2W	Auto	Bus	Total
CO	441.65	324.85	215.35	222.65	1204.5
HC	189.8	445.3	21.9	73	730
NOX	368.65	270.1	18.25	292	949
CO_2	44318	18655	4180	12870	80023

Source: Primary survey.

Table 9. Yearly fuel consumption in Agartala (in KL)

Year	Due to Diesel	Due to Petrol	Total
2005-2006	41,185	25,854	67,039
2006-2007	47,587	26,468	74,055
2007-2008	49,571	29,531	79,102
2008-2009	56,995	31,746	88,741
2009-2010	63,570	37,408	1,00,978
2010-2011	76,485	43,302	1,19,787
2011-2012	70,801	44,180	1,14,981
2012-2013	65,874	44,831	1,10,705
2013-2014	56,290	45,005	1,01,295

Figure 8. Trend in consumption of diesel (in Kl)
Source: Annual Sales Report of Fuel 2005-2013, Indian Oil Corporation Limited.

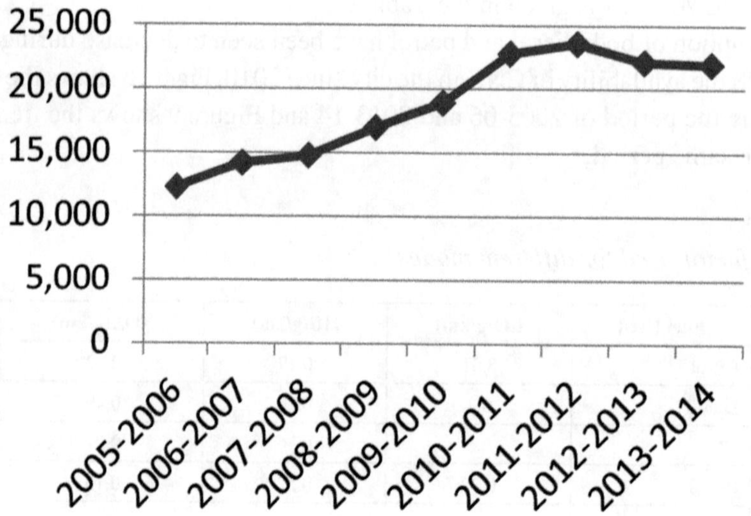

Figure 9. Trend in consumption of petrol (in KL)
Source: Annual Sales Report of Fuel 2005-2013, Indian Oil Corporation Limited.

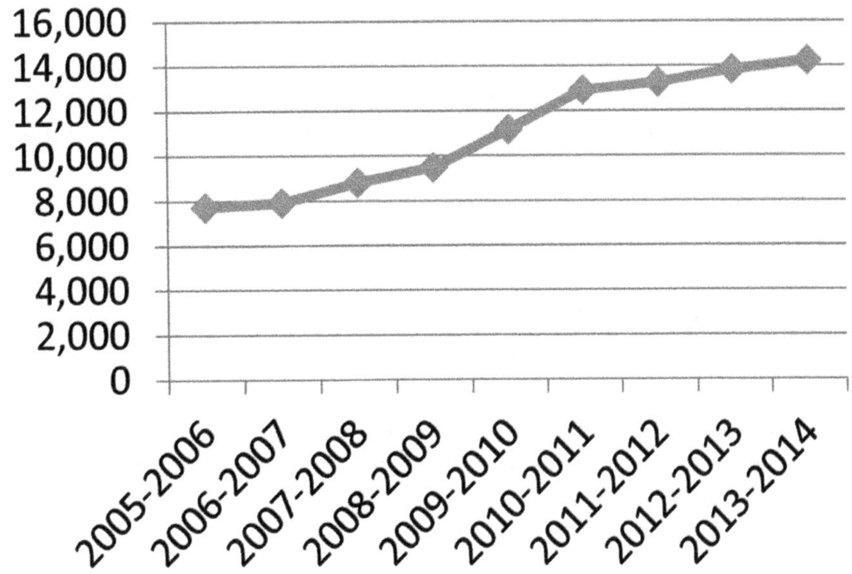

Now to calculate CO_2 equivalent emission due to Diesel, an energy density factor of 0.004375 TJ/Tonne and emission factor of 76.8 TCO_2/TJ were used. And for CO_2 equivalent emission due to Petrol, energy density factor of 0.0448 TJ/Tonne and emission factor of 74.8 TCO_2/TJ were used. The CO_2 equivalent emission so obtained during the period of 2005-06 and 2013-14 is given in Table 10.

Inference

The emission load in 2013-14 obtained from the primary data is 15% less than the emission value from the secondary data due to the inclusion of goods vehicles in the later data.

Table 10. CO_2 equivalent emission (in metric tonne)

Year	HSD (Diesel)	MS (Petrol)	Total
2005-2006	12,257.30	7,715.24	19,972.54
2006-2007	14,162.76	7,898.51	22,061.27
2007-2008	14,753.13	8,812.60	23,565.73
2008-2009	16,962.90	9,473.42	26,436.32
2009-2010	18,919.61	11,162.83	30,082.44
2010-2011	22,763.39	12,921.88	35,685.27
2011-2012	23,559.92	13,273.39	36,833.31
2012-2013	22,093.35	13,842.62	35,935.97
2013-2014	21,941.28	14,238.06	36,179.34

KEY FINDINGS

Asia Pacific Region

Key findings from the various international studies on climate change in Asia Pacific Region have been studied, the details of which is given in the following part of this section. Later these findings have been compared to the findings from this study.

Climate Model

The World Climate Research Programme (WCRP) has undertaken the largest coordinated global climate modelto provide the most comprehensive multi-modal perspective on climate change. The climate model is as shown in Figure 10.

Several steps from emissions to climate response contribute to the overall uncertainity of a climate model projection. In the Asia Pacific region, there is evidence of prominent increase in the intensity and frequency of many extreme events such as heat waves, tropical cyclones, prolonged dry spells, intense rainfall and melting of snow *(IPCC, 2007)*.

Climate models indicate temperature increase in the order of 0.5-2.0^{0}C in the region and it is expected to increase to 1-7^{0}C by 2070. Temperatures are expected to increase more rapidly in the arid areas of northern Pakistan and India and western China. Models also indicate rising rainfall concentration throughout much of the region (including greater rainfall during summer monsoon) due to the glacier melt in the Himalaya. Furthermore, winter rainfall is likely to decline in the region, suggesting increased aridity from the winter monsoon. The region will be affected by an increase in global sea level ofapproximately 3-16 cm by 2030 and 7-50 cm by 2070in conjunction with regional sea level variability. The scenario of climate change in Asia Pacific region should reflect on the climate change scenario in the North-Eastern region of India. It is discussed in details in later sections.

Figure 10. Comprehensive global climate model
Source: Fourth Assessment Report of IPCC, 2007.

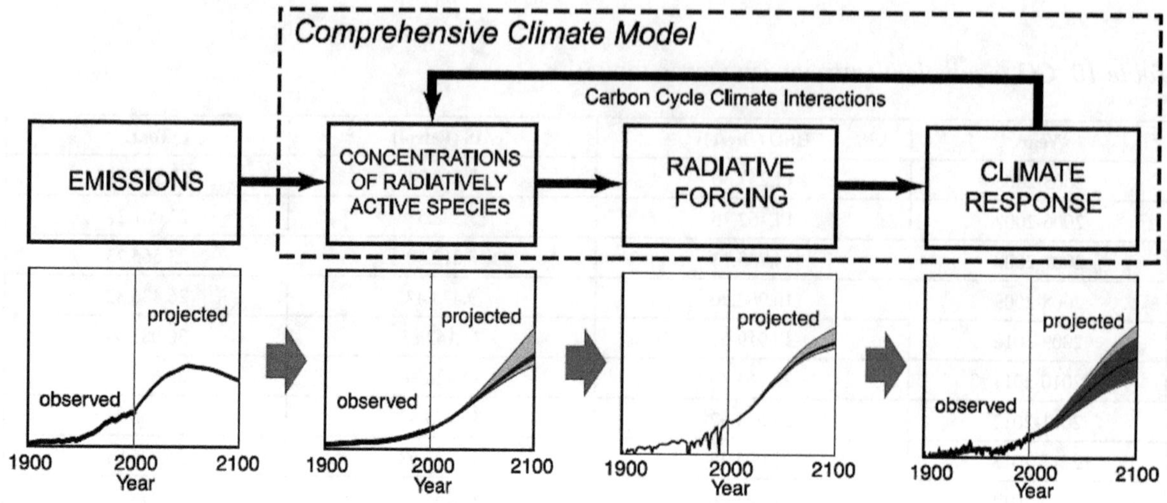

North-East India

The climate change parameters of Agartala were studied, the details of which are given in the later sections.

Climate Change Parameters of Agartala

The climate change parameters such as annual average temperature, intensity and frequency of rainfall were regressed against CO_2 equivalent pollution load in Agartala for the time period between 2005 and 2013. The observation of annual average minimum temperature is shown in Figure 11 and Figure 12 shows the observation of annual average maximum temperature over the last decade.

Figure 13 shows the observation of average frequency of rainfall in a year and Figure 14 shows the observation of average intensity of rainfall in a year.

The climate models for Agartala indicated the climate change parameters in terms of annual average maximum and minimum temperature; and intensity and frequency of rainfall are increasing with the increase in pollution load in the city. The R^2 value is greater than 0.7 in most of the case, so it is reliable. Also, the effect of pollution load on these parameters was observed to be more due to diesel than petrol.

Alternatives for Mitigating Climate Change

Biofuel

Biomass refers to recently living organisms, most often referring to plants or plant-derived materials. This biomass can be converted to energy in three different ways: thermal conversion, chemical conversion, and biochemical conversion. This new biomass can be used for biofuels. Biofuels have increased in popularity because of the rising oil prices and need for energy security. Bioethanol is an alcohol made by fermentation, mostly from carbohydrates produced in sugar or starch crops such as corn or sugarcane.

Figure 11. Annual average minimum temperature
Source: Primary survey.

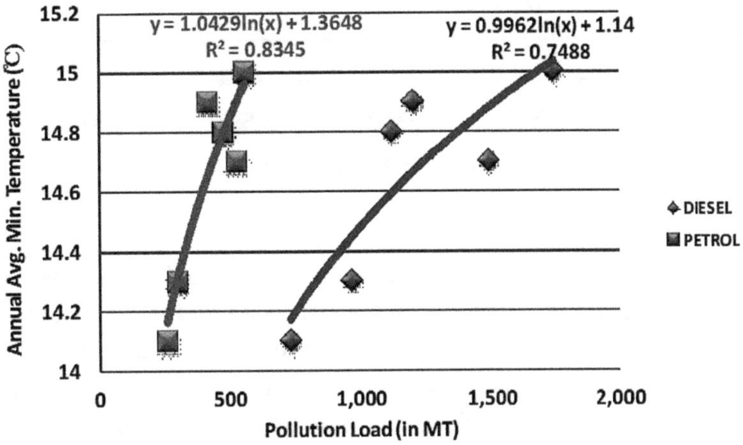

Figure 12. Annual average maximum temperature
Source: Primary survey.

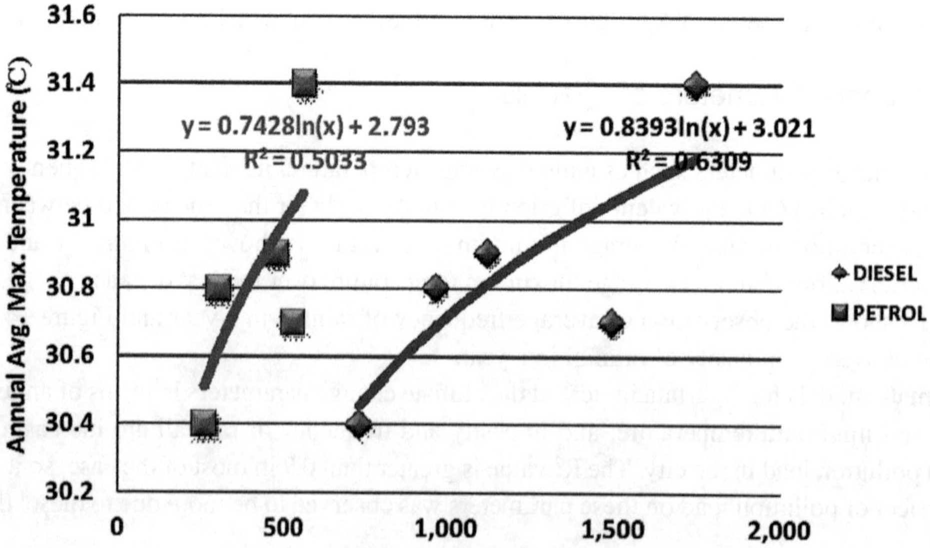

Figure 13. Number of rainy days in a year

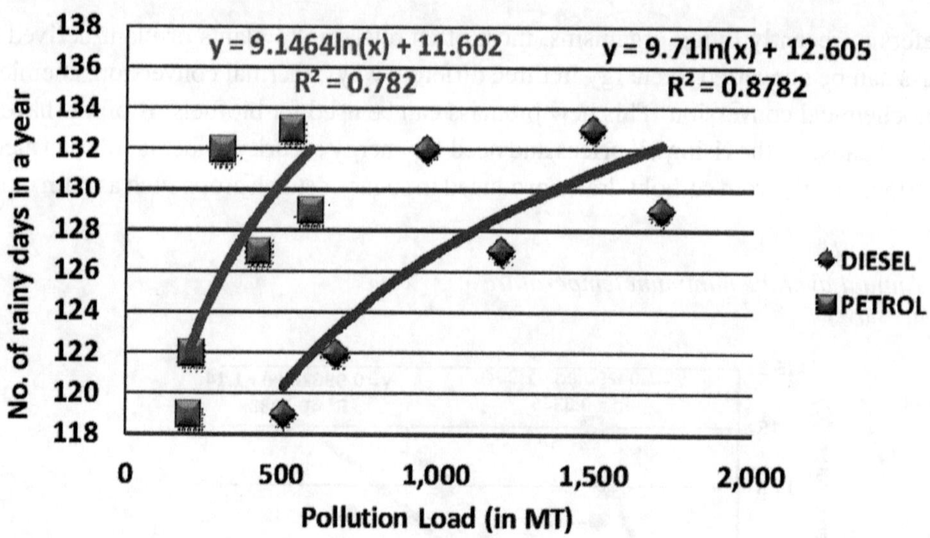

Cellulosic biomass, derived from non-food sources, such as trees and grasses, is also being developed as a feedstock for ethanol production. Ethanol can be used as a fuel for vehicles in its pure form, but it is usually used as a gasoline additive to increase octane and improve vehicle emissions. Bioethanol is widely used in the USA and in Brazil. Biodiesel is made from vegetable oils and animal fats. Biodiesel can be used as a fuel for vehicles in its pure form, but is usually used as a diesel additive to reduce levels of particulates, carbon monoxide, and hydrocarbons from diesel-powered vehicles (MoCEB, 2012).

Figure 14. Mean annual rainfall (in mm)

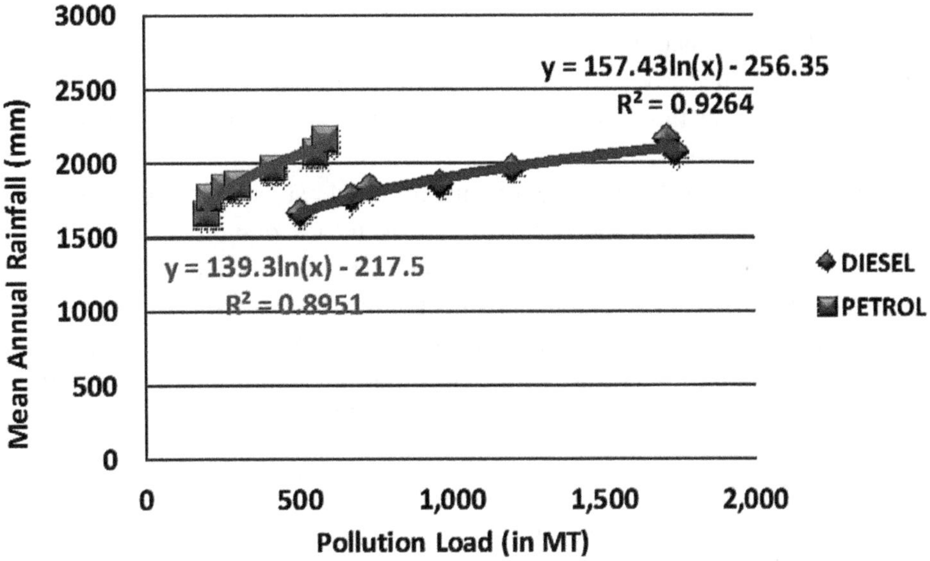

Hybrid Electric Vehicles

Hybrid electric vehicles combine the internal combustion engine of a conventional vehicle with the battery and electric motor of an electric vehicle resulting in better fuel economy and lesser carbon equivalent emissions. The power of the hybrid vehicle's internal combustion engine generally ranges from 1/10th to 1/4th of the conventional automobile. In a HEV, when the driver applies the brakes, the motor becomes a generator, using kinetic energy of the vehicle to generate electricity that can be stored in the battery for later use. A hybrid vehicle can achieve the cruising range and performance similar to advantages of conventional vehicles with low-noise, exhaust emissions and energy independence benefits of electric vehicles (DEA, 2012).

Hydrogen Fuel

Hydrogen is one of the clean fuels in the world, as it does not contain carbon compounds. Hydrogen is a clean and efficient energy carrier with the potential to replace liquid and gaseous fossil fuels. Hydrogen can be combusted directly in the IC engines, or it can be used on the fuel cell to produce electricity, which can operate the vehicle. Therefore, use of hydrogen as a transportation fuel would result in few or no emission that affects air quality (DEA, 2012).

Public Transport

Due to the lack of proper public transport service and infrastructure, the PT share is negligible in the region leading to the increase in private vehicle use (major contributor of greenhouse gas emissions). Car restraint should be encouraged in the region. To encourage public transport, the infrastructure, accessibility and frequency of public transport system have to be improved. The shift from private vehicles to public transport will reduce emissions considerably.

CONCLUSION

1. It is seen from the study that there is climate change in North-East India, and transport emissions are contributing to it. Presently, the region does not have any policy measures to reduce greenhouse gas emissions in the near future. It is therefore; felt that North-East should have some policy to mitigate the GHG emissions.
2. It is seen that the gram per kilometre emissions of diesel cars are more than that of petrol cars. Therefore, a policy should be formulated to discourage the use of diesel by deregulating the prices of diesel and by reducing the disparity between petrol and diesel prices.
3. It is found from the study that the emissions in case of public transport passenger-km are lesser than other modes of transport. But in the North-East region, there is negligible share of public transport due to poor infrastructure and service facilities. Therefore, improvements should be made with regard to public transport system so that considerable number of passengers shifts to public transport modes.
4. It is inferred that use of alternate vehicle or fuel technologies like hybrid electric vehicles, biofuel, biodiesel, hydrogen fuel, are required to be initiated to mitigate the climate change.
5. Car restraint policies should be implemented in the major cities of the region to discourage the use of private modes.Re-designing of the city structure should be considered to achieve minimum average trip length and to reduce the need to travel. Integration of land use transport planning should be an integral part of a city development programme.

REFERENCES

Attri, S.D., & Tyagi, A. (2010). *Climate Profile of India.* New Delhi: Ministry of Earth Sciences. Retrieved from http://www.indiaenvironmentportal.org.in/files/climate_profile.pdf

Automative Research Association of India. (2007). *Draft report on "Emission Factor development for Indian Vehicles" as a part of Ambient Air Quality Monitoring and Emission Source Apportionment Studies.* Pune: Automative Research Association of India. Retrieved from http://www.cpcb.nic.in/Emission_Factors_Vehicles.pdf

Danish Energy Agency. (2012). *Energy Policy in Denmark.* Danish Energy Agency. Retrieved from file:///C:/Users/mcaesar/Downloads/EnergyPolicyinDenmark.pdf

Government of Tripura. (2008). *City Development Plan-Agartala.* Retrieved from http://agartalacity.tripura.gov.in/PDF/Utilities/CityDevP.pdf

Indian Oil Corporation Limited. (2005-2013). *Annual Sales Report of Fuel.* Mumbai: Indian Oil Corporation Limited. Retrieved from https://www.iocl.com/InvestorCenter/PastAnnualReport.aspx

Ministry of Environment and Forests. (2012). *India's Second National Communication to the United Framework Convention on Climate Change*. Government of India. Retrieved from http://pib.nic.in/newsite/PrintRelease.aspx?relid=82211

New Delhi Intergovernmental Panel on Climate Change (IPCC). (2007). Fourth Assessment Report of the Intergovernmental Panel on Climate Change. *Intergovernmental Panel on Climate Change.* Retrieved from https://www.ipcc.ch/pdf/assessment-report/ar4/syr/ar4_syr_full_report.pdf

Transport Research Wing, Ministry of Road Transport and Highways. (2011-12). *Road Transport Yearbook.* Government of India.

Chapter 16
The Politics of Environmental Pollution in Nigeria:
Emerging Trends, Issues, and Challenges

Godwin Solomon Mmaduabuchi Okeke
University of Lagos, Nigeria

ABSTRACT

Pioneer efforts and early scientific and popular writings on the environment support genuine ecological concerns on the need for a clean and healthy environment. A clean environment supports good health, and good health is very important for vibrant workforce, which is necessary for a productive economy. This logical connection, in the final analysis, seems to encourage a healthy and vibrant political atmosphere. In Nigeria, the oil boom in the early 1970s led to a major boost in industrialization and the subsequent environmental pollution and the attendant undercurrent political implications. In the light of the above, the chapter investigates the environmental problems in Nigeria and the inherent political dynamics that propel and shape the calls for a cleaner environment. The authors rely on archival materials and other secondary sources of data including the library documentations and internet to query and theoretically explain the extant issues. The relevant recommendations are made on the urgent need to ameliorate the unpalatable condition and for cleaner environment.

INTRODUCTION

Environmental pollution is a matter of global concern today because of what the environment is to human survival and indeed, the survival of the earth itself. Virtually everywhere discussions focus on environment around the world. "If it is not greenhouse effect", "it is global warming". Both point to the fact that the earth is polluted. Williams (2006:88) affirms that it is well recognized that the environment constitutes one of the major challenges to health. He however points out that Nigeria suffers from both primary and secondary environmental problems resulting from under-development and growth in industrialization. As a result, human idea and activities are endangered on a daily basis. In view of the threats to the human existence, environmental pollution is one of the current Phenomena that have gained international attention, including the Nigeria. This pollution comes in different ways.

DOI: 10.4018/978-1-5225-3194-4.ch016

Environmental protection implies safeguarding all the surrounding conditions which influence growth and development (air, water, land, atmosphere and space) against pollution (Udoudo 2006). Furthermore, Udoudo (2006) observes that in some localities in Nigeria, especially among the riverine communities of the Niger Delta region, one can hardly find modern toilet facilities. In some communities land pollution is the degradation of the earth's land surface through misuse of the soil by poor agricultural practices, mineral exploitation, industrial waste dumping and indiscriminate disposal of urine waters.

Air pollution is the accumulation in the atmosphere of substances that in sufficient concentrations endanger human health or produce other measured effects on living matter and other materials. Among the major sources of air pollution are powered heat generation, the burning of solid wastes, industrial processes of air pollutants like carbon monoxide from incomplete combustion engine and others, such as hydrocarbons, nitrogen oxides, sulphur dioxide and photochemical oxidants.

Again, noise pollution is a composite of sounds generated by human activities ranging from blasting stereo system to the roar of transport vehicles. The most readily measurable physiological effect of noise pollution is damage of hearing, which may be either temporary or permanent and may cause distraction of normal activities or general annoyance. Environmental pollution has threatened human existence in diverse ways. This is because life is dependent on the environment. Environmental pollution can occur from petroleum production activities, as noted by Ugochukwu (2003) that: Since 1905, petroleum operations have caused great devastation to the Niger Delta. These have impacted negatively on fishing and farming which are no longer productive enough to the area. In Ogoniland food is now imported in an area once known as the food basket. Oil spills destroy bulk farmlands and crop. They also destroy aquatic life. When a situation like this occurs, economic activities are grossly reduced and good social life hampered. Damage to farm lands and aquatic life by oil spills cannot be overstated.

The efforts of the federal, state and local governments in Nigeria at ensuring sustainable development through numerous environmental legislations, fiscal incentives and grants to environmental ministries and agencies remain elusive, as Nigeria continues to experience complex environmental problems of atmospheric, noise and water pollution, oil pollution, climate change including flooding, coastal erosion and perennial oceanic surge and municipal solid waste management (Ajani, 2012). The reasons for such regulatory failure are many. Foremost among these reasons is the problem of designing best practices to secure effective and efficient enforcement and compliance with international and municipal environmental laws. A necessary conclusion that could be drawn from this analysis is that environmental benefits arising from existing legal and institutional frameworks are minimal and sub-optimal resulting in social and environmental welfare losses. The question that should engage the mind of many Nigerians is, why is environmental issues handled with little care? Why has the government made so much policies but little is done to effectuate it? Why is environmental pollution politicized in Nigeria? What changes in policy formulation and enforcement strategies are necessary to produce optimal environmental results? Providing answers to these puzzling questions is the main objective of this paper. The paper seeks to interrogate the politics of environmental pollution in Nigeria, and focuses on the emerging issues, trends and challenges of environmental pollution. Its general objectives are to examine the impact of policies on environmental sustainability on the Nigerian environment, investigate emerging environmental challenges in Nigeria such as flooding, erosion, bombing, gas flaring, Environmental Impact Assessment, among others.

BACKGROUND

Environmental pollution has been in existence since man began to live in settlements. In the earlier days of nomadic hunting communities, the tribal group moved on when food in their current location became depleted and the area around their camp became polluted or soiled. These nomads were a part of a balanced eco-system. As human societies developed, land became cultivated, livestock domesticated; and as permanent settlements became established, environmental pollution began to emerge. The problem became more serious as these permanent communities grew into cities. The increase in human population and consumption pattern also led to the increase in wastes generated, thereby creating environmental problems of collection and disposal. In response to waste disposal challenge, various societies developed waste collection and disposal systems that best suited their immediate environments.

Amokaye (2012) states, in traditional African societies, waste was dumped and burnt openly or centrally deposited in the bush where they later decomposed in hygienic manner or deposited in a flowing river. In developed societies, sewage systems were developed to collect and move the untreated sewage to the nearest river or sea where nature was left to deal with the problem. Apart from this river-borne pollution, however, the effect of man's activities tended to be local in nature. Energy sources for domestic heating were coal and wood; and, transportation power was provided by animal or the wind. None of these produced waste products at a rate greater than the ecosystem could absorb.

With the advent of science and technology, the nature, magnitude and impact of polluting activities began to expand. Science and technology made it possible for man to harness energy from burning fossil fuels to drive machines. They were developed at an ever accelerating pace, developing more and more products, including chemicals to meet the ever increasing and insatiable demands of the marketplace. The waste products from these new industries were discharged directly into the environment. As industries multiplied, so did the waste products of combustion and manufacturing processes, creating environmental problems of disposal of nuclear and other hazardous substances.

The scope of environmental pollution also increased to accommodate new ones such as atmospheric pollution, acidic rain, water and marine pollution, soil pollution, noise pollution, climate change and deforestation. The result being that humankind lives today in an environment where all life-supporting elements are polluted. The air we breathe in is no longer healthy. The water we drink is impure and decreases in volume and quality daily. The food we eat is contaminated. Humankind continues to witness persistent drought, low harvest, diseases and poverty as a result of climate change, deforestation, accumulation and disposition of nuclear and other hazardous substances. These ominous trends are symptoms of an unhealthy planet; a planet that can no longer cope with all the demands man is heaping upon it.

The direct and indirect effects of natural and anthropogenic perturbations are manifested in early death, diseases, physical deformities, genetic mutations and physiological malnutrition suffered by humankind. In developing countries where poverty, diseases and population explosions are prevalent, environmental issues become more confused and complex. The effects on the plants, animals and ecosystem are equally devastating and condemnable. Locally, the wild environmental facts confronting Nigeria as a nation are enormous. A few of Nigeria's glaring environmental problems include:

1. Excessive pressure on available resources, infrastructure and space due to unabated rural-urban migration in the past three decades; this stress has been reinforced by industrial and urban development that has caused a rising rate of pollution;
2. The high rate of soil degradation, sheet, gully and coastal erosion and flooding through non-judicious land use practices;
3. The depletion of natural forest resources through uncontrolled human activity, e.g, tree felling and over-grazing;
4. Unfettered bush burning and the risk of exterminating wildlife species as well as uncontrolled fishing and related activities which endanger the species of fish in Nigeria waters;
5. Pollution of surface and underground water systems through indiscriminate disposal of solid and liquid wastes;
6. destruction of valuable agricultural land through bad mining practices;
7. Permanent dangers posed by the encroachment of the desert on vast agricultural lands along northern borders; and
8. Oil pollution and related environmental consequences, particularly in the Niger Delta area of Nigeria (Aina 1992).

The efforts of the federal, state and local governments in Nigeria at ensuring sustainable development through numerous environmental legislation, fiscal incentives and grants to environmental ministries and agencies remain elusive as Nigeria continues to experience complex environmental problems of atmospheric, noise and water pollution, oil pollution, climatic change including flooding, coastal erosion and perennial oceanic surge and municipal solid waste management. The reasons for such regulatory failure are many.

Foremost among these reasons is the problem of designing best practices to secure effective and efficient enforcement and compliance with international and municipal environmental laws. A necessary conclusion that could be drawn from this analysis is that environmental benefits arising from existing legal and institutional frameworks are minimal and sub-optimal resulting in social and environmental welfare losses. The question that should puzzle the mind of policymakers is why this sub-optimal results? What can we do to maximize our environmental gains? What changes in policy formulation and enforcement strategies are necessary to produce optimal environmental results? The answers to these questions are also part of the focus of this paper.

MAIN FOCUS OF THE CHAPTER

Conceptualizing Environmental Issues and Pollution

Ashton-Jones (1998) sees environmental pollution as capable of derailing the psychological status of the citizens. Environmental pollution can occur from poor toilet facilities, smoking in the public, dumping of refuse around living houses and gas flaring, oil spillage to mention but a few.

Oil spillage is the most common and controversial of all the environmental impacts of oil exploration. Constitutional Rights Project (CRP, 1999) sees oil spills as uncontrolled releases of any product relating to oil production including crude oil, chemicals, or waste caused by equipment failure, operation mishaps, human error, or intentional damage to facilities. Similarly, Adewuyi, (2001) submits that oil spillage occurs during the drilling of oil wells and as a result of oil pipelines leakages and during the loading of oil into the tankers. Spills are potentially the most devastating on agricultural land and water resources.

In the same vein, UNDP (2006) reports that much of the environmental pollution in the oil-producing communities is the result of oil spillage due, essentially, to accidents based on human error and equipment failure. The report states that a total of 6,817 oil spills occurred between 1976 and 2001, with a loss of approximately three million barrels of oil. More than 70 percent was not recovered. Approximately 6 percent spilled on land, 25 percent on swamps and 69 percent in the offshore environment. Statistics from the Ministry of Petroleum Resources also indicate that between 1976 and 1996, a total of 4,836 incidents resulted in the spillage of at least 2,446,322 barrels (102.7 million U.S gallons) of which an estimated 1,896,930 barrels (79.7 million U.S gallons) were lost to the environment. The level damage done to the environment following oil spill is better imagined than experienced. Orubu, Odusola, and Ehwarieme, (2004) agree with this, when they reported that massive oil spills occurring in the riverine areas have done untold damage to the aquatic ecosystem, particularly in the mangrove swamp forest zone. Awobajo, (1981) also states that an average of one oil spill occurs every week and three oil spills are recorded each month in most of the oil-bearing areas of the Niger Delta, in which the proportion lost to the environment is quite disturbing. Thus, spills destroy the environment by poisoning the waters and destruction of vegetation and agricultural lands. Oftentimes oil and gas companies attribute the cause of oil spills to sabotage by the host community. However they (the oil companies cause oil spills out of negligence and carelessness, Sometimes, the cause may be due to the use of decrepit pipelines, some reportedly over 40 years old, which criss-cross oil communities. These pipelines are rusty, obsolete, and poorly maintained. Consequently, on-site oil leaks and ruptured pipelines are a serious problem in the Niger Delta.

Gas flaring is another major effect of oil exploitation on the environment and atmosphere; it generates air pollution and heat. The World Bank (1995) defines gas flaring as the deliberate burning of natural gas that is produced; it creates a ceaseless, high intensity flame. Natural gas is a by-product of oil extraction, which is removed from the earth crust along with the crude oil. Furthermore, The World Bank (1995) shows that gas flaring has been known to be the singular highest contributor to the problem of global warming or global climate change. In the same vein, Orubu (1999) adds that greenhouse gases such as methane and carbon dioxide emitted from gas flares contribute to global warming, which could lead to a rise in sea level, accelerate the problem of climatic change and harsh living conditions on earth if not checked. It also has negative effects on the immediate environment as it adversely affects plant growth, wildlife, and human beings. It has been estimated that the total emission of carbon dioxide (CO2) from gas flaring in Nigeria amounts to about 35 million tons per year and it is on record that Nigeria flares the highest amount of gas in the world (World Bank, 1995, 2000/2001).

The percentage of gas flared in Nigeria, which is about three times the OPEC average, is about 16 times the world average (Ajayi and Ikporukpo, 2005). Gas flaring close to residential areas increases the risk on life, forests, and waterways, thereby making living unbearable to human beings, terrestrial and aquatic animals. A study of gas flaring in South-Eastern Nigeria by Isiche and Sanford (1976) observed that at the global level, flaring in Nigeria contributes a measurable percentage of the world's total emissions of greenhouse gases. Specifically, due to the low efficiency of many of the flares, much of the gas is released as methane (which has a high warming potential) rather than carbon dioxide. At the local level, the low-lying Niger Delta is particularly vulnerable to the potential effects of sea level rising. Further, air and leaf temperatures increased up to eighty or one hundred meters from the stack, such that species compositions of vegetation are affected (Isiche and Sanford 1976).

The most noticeable effect of the flares is light pollution. Across the oil-bearing areas, the night sky is lit up by flares, which during the rainy season reflect brightly from the clouds. It is difficult to differentiate between night and day because of the polluted air and burning light. Communities close to flares complain that the light disturbs nocturnal animals, driving them away from the area. The economic and health cost of heat emissions from the flares and destruction of vegetation are part of the negative externalities of the wasteful flaring of oil-associated gas. Indeed, gas flaring has been identified as the major cause of respiratory infection among the Niger Delta people, as well as the cause of reduced growth potentials of farm crops (Egwaikhide & Aregbeyen, 1999). It may seem economical to the oil companies but gas flaring is environmentally destructive and the brunt of its effects on humans, environment and atmosphere is borne by the local communities. Ibeanu (2000) indicates that Nigeria exceeded the world average for natural gas flaring by seventy-two percent in 1991. In that year, while the world average for gas flared as a percentage of total production was four percent, Nigeria flared seventy-six percent of that total production. Onosode, (2003) opines that flaring of gas in Nigeria for about 50 years has also been estimated to constitute a waste of 12.5 million dollars annually, an amount that would have been earned if the oil-associated natural gas was captured, sold, or used as a local energy source. However, the lack of delivery and utilization of infrastructure for natural gas in Nigeria render it a useless by-product. Okonta and Douglas (2001) corroborated this and identified the reason for the high rate of gas flares in Nigeria as multinationals involved in oil exploration and production activities in Nigeria flared more gases in the course of their operation than in any other countries where they are involved in oil activities. They suggest that these oil companies found it economically expedient to flare non-associated gas on site rather than incur the expenses of putting in place facilities to re-inject the gas back into the wells or to collect it for commercial use.

The purpose of environmental impact assessment (EIA) is to provide decision makers and the public with a systematic, comprehensive and object assessment of environmental consequences of an action. This assessment is to show sufficient adverse environmental effects and to allay public fears over the consequences of such effects based on known methods of handling the effects.

The Dynamics of Environmental Pollution in Nigeria: The Case of Ogoniland

In the wake of the rapid dispersal of ideas which characterize the contemporary era of globalization, peoples across the globe are intoning and rallying around novel concepts, norms and standards of environmental protection as a social justice objective. The case of the Ogoni of South-southern Nigeria exemplifies this trend. This case captures a critical paradox of environmental protection which is that seemingly modest normative gestures do in fact publicize ecological problems and go some length in eliciting meaningful responses, even if often from marginal constituencies (Otu 2016).

The history of pollution from oil spills and fire in Ogoniland is very tragic. With this independent study, conducted at the request of the Federal Government of Nigeria, the United Nations Environment Programme (UNEP) reveals the nature and extent of oil contamination in Ogoniland. The assessment was unprecedented covering over a 14-month period, the UNEP team examined more than 200 locations, surveyed 122 kilometres of pipeline rights of way, reviewed more than 5,000 medical records and engaged over 23,000 people at local community meetings (Otu, 2016). Detailed soil and groundwater contamination investigations were conducted at 69 sites, which ranged in size from 1,300 square metres (Barabeedom-K.dere, Gokana local government area (LGA) to 79 hectares (Ajeokpori-Akpajo, Eleme LGA).

The Environmental Assessment of Ogoniland covers contaminated land, groundwater, surface water, sediment, vegetation, air pollution, public health, industry practices and institutional issues. The report reveals the best available understanding of what has happened and the implications for affected populations – and provides clear operational guidance as to how legacy can be addressed. United Nations Environmental Programme, UNEP recruited a team of international experts in disciplines such as contaminated land, water, forestry and public health, who worked under the guidance of senior UNEP managers. During aerial reconnaissance missions, UNEP experts observed oil pollution which was not readily visible from the ground, including artisanal refining sites. Information was provided by Ogoniland residents about oil contamination in their communities supplemented by official oil spill data supplied by the Nigerian Government and Shell Petroleum Development Company, SPDC. Following its initial investigations, UNEP identified 69 sites for detailed soil and groundwater investigations.

In addition, samples of community drinking water sediments from creeks, surface water, rainwater, fish and air were collected throughout Ogoniland and in several neighboring areas. Altogether more than 4,000 samples were analyzed including water drawn from 142 groundwater monitoring wells drilled specifically for the study and soil extracted from 780 boreholes.

According to the UNEP (2011) "The Ogoni people live with this pollution every day. An Ogoniland has high rainfall, any delay in clearing up an oil spill leads to oil being washed away, traversing farmland and almost always ending up into the creeks. When oil reaches the root zone, crops and other plants begin to experience stress and can die, and this is a routine observation in Ogoniland. At one site, Ejama-Ebubu, in Eleme local government area (LGA), the study found heavy contamination present 40 years after an oil spill occurred, despite repeated clean-up attempts. The assessment found that overlapping authorities and responsibilities between ministries and a lack of resources within key agencies has serious implication for environmental management on the ground, including enforcement. Remote sensing revealed the rapid proliferation in the past two years of artisanal refining, whereby crude oil is distilled in makeshift facilities. The study found that this illegal activity is endangering lives and causing pockets of environmental devastation in Ogoniland and neighboring areas. The report concludes that pollution of oil by petroleum hydrocarbon in Ogoniland is extensively in land areas, sediments and swampland. Most of the contamination is from crude oil although contamination by refined product was found at three locations. The assessment found there is no continuous day layer across Ogoniland exposing the groundwater in Ogoniland (and beyond) to hydrocarbons spilled on the surface.

The UNEP investigation found that the surface water throughout the creeks contains hydrocarbons. Floating layers of oil vary from thick black oil to sheers. The highest reading of dissolved hydrocarbon in the water column, of 7,420 pg/l was detected at Ataba-Otokroma, bordering the Gokana and Andoni LGAs. The Ogoni community is exposed to production hydrocarbons in outdoor air and drinking water, sometimes at elevated concentrations. They are also exposed through dermal contacts from contaminated soil, sediments and surface water. Since average life expectancy in Nigeria is less than 50 years, it is a fair assumption that most members of the current Ogoniland community have lived with chronic oil pollution throughout their lives. Of most immediate concern, community members at Nisisioken Ogale are drinking water from wells that is contaminated with benzene, a known carcinogen, at levels over 900 times above the World Health Organization (WHO) guideline (Otu 2016).

According to the UNEP report 2011, the environmental restoration of Ogoniland could prove to be the world's most wide-ranging and long term oil clean-up exercise ever undertaken if contaminated drinking water, land, creeks and important ecosystems such as mangroves are to be brought back to full, productive health.A major new independent scientific assessment, carried out by the UNEP, shows that pollution from over 50 years of oil operations in the region has penetrated further and deeper than many may have supposed.

The UNEP assessment found exercise that: Some areas, which appear unaffected at the surface, are in reality severely contaminated underground and action to protect human health and reduce the risks to affected communities should occur without delay.In at least 10 Ogoni communities where drinking water is contaminated with high levels of hydrocarbons, public health is seriously threatened. In one community, at Nisisioken Ogale, in western Ogoniland, families are drinking water from wells that is contaminated with benzene- a known carcinogen-at levels over 900 times above World Health Organization guidelines. The site is close to a Nigerian National Petroleum Company pipeline.UNEP scientists found an 8 cm layer of refined oil floating on the groundwater which serves the wells. This was reportedly linked to an oil spill which occurred more than six years ago.

While the report provides clear operational recommendations for addressing the widespread oil pollution across Ogoniland, UNEP recommends that the contamination in Nisisioken Ogale warrants emergency action ahead of all other remediation efforts. UNEP presented its report to the former President Dr.Goodluck Jonathan, in the Nigerian capital Abuja in 2012, yet no effort was made to implement the recommendation by the administration of Dr. Jonathan, despite that he is from the Niger Delta region. It was the current administration of President Mohammadu Buhari that flagged off the cleaning exercise. However, the communities still complain that nothing has been done since the flag off exercise.

The Politics of Environmental Pollution in Nigeria

A number of factors could be summarized here as the characteristics of the Nigerian environmental problems. These mostly include the consequences of rapid urbanization and industrialization which led to increased demand for variety of goods and services, resulting to construction of roads, houses and industrial buildings (which in one way or the other) has led to environmental problems. It is not uncommon to notice that many industries in the industrial areas of Lagos such as Apapa, Ilupeju, Ikeja and some other industrial areas in Nigeria, discharge their chemical wastes without the standard pre-treatment which could negate their effects in the environment.

As part of the characteristics of environmental problem, Cointreau, (1982) observed that, in most Nigerian cities, collection and disposal of refuse is usually more frequent and efficient in Government Reservation Areas (GRAs) and in other affluent neighborhoods than in poor neighborhoods. This observation cannot be far from the truth. The residents of this neighborhoods like those in Victoria Island, Lekki Pennisula, Ikoyi and so forth in Lagos

State of Nigeria for instance, are inhabited by very wealthy and with heavy weights political connections in the government and as such they use their influences to ensure that, their environments are kept as free as possible from refuse.

Most Nigerians who worked in petroleum industries and other similar hazardous industries are unaware of the environmental effects the chemicals that they are subjected to. Even when they are aware, they hardly have the thought of leaving these places because of fear of unemployment, while they are not really knowledgeable of the extent of the hazards they suffer due to their nature of illiteracy as well.

Olokesusi (1987) considers procedural environmental characteristics as the ways and means which environmental problems are tackled, planned and managed. We are again going to summarize the major lapses involved in the management of environmental problems.

The three levels of government in Nigeria lack an integrated approach to deal with environmental problems within it environment. Part of these lapses could be due to the difficulty in predicting (with some level of precision) the environmental impacts in different areas. Olokesusi (1987) once pointed out that, too many agencies are involved in environmental management. For instance, in Ibadan, the Oyo State capital, Solid Waste disposal is handled by State Waste Disposal Board, the Sanitary Inspection Division of the Ministry of Health, the Oyo State Environmental Task-Force, and not less than ten (10) certified waste disposal contractors as well as numerous uncertified contractors.

Obviously, considering this type of complex situation, it will be difficult to have an effective, integrated and efficient environmental management. At another level, the use of inappropriate technological measures also constitutes environmental problems. For instance, as Cointreau (1982) made observation in Onitsha, one of the metropolitan cities in Nigeria, that apart from the irregular collection of solid waste and refuse particularly from the poor neighborhoods, the trucks meant for evacuating these wastes are inadequate and epileptic in nature. Although, some efforts have been made in the passage of few legislations to control environmental pollution in all areas of life, but unfortunately many if not all the passed legislations are either not enforced or are poorly enforced.

In view of the Nigerian government's recognition of the damaging effects of the environment through floods, droughts, forest fires, technological accidents such as oil spills, industrial chemical effluents, dumping of toxic wastes and contamination of rivers, lakes, soil, air and other forms of pollutions of the environment. Consequently, the Nigerian government enacted various environmental policies between the 1930s and the 1990s. The table below shows the various enactments of the Nigerian government concerning its environment within the stated periods.

Table 1 shows that Nigerian government has been concerned with environmental regulations, formulating laws that could help secure the environment even before independence. So the problem is not really environmental legislation, but the implementation of these laws reflects political considerations.

Again, the dumping of toxic wastes in Koko in the former Bendel State in 1987, led the Nigerian government to promulgate the Harmful Wastes Decree which was meant to provide the legal framework for the effective control of the disposal of toxic hazardous waste into any environment within the confines of Nigeria. This was immediately followed by the creation of a regulatory body the Federal Environmental Protection Agency (FEPA) in 1988, charged with the overall responsibility of protecting and developing the Nigerian environment. States and local government councils were also encouraged to establish their own environmental regulatory bodies for the purpose of maintaining good environmental quality as it applied to their particular areas of jurisdiction.

FEPA Decree 58 of 1988 authorizes this agency among other things to establish and prescribe national guidelines, criteria and standards for water quality, air quality and atmospheric protection, noise levels, gaseous emissions and effluent limits etc, to monitor and control hazardous substance, supervise and enforce compliance. This Decree also gave the Agency broad enforcement powers, even without warrants, to gain entry, inspect, seize and arrest with stiff penalties of a fine or jail term on whosoever obstructs the enforcement in the discharge of their duties or makes false declarations of compliances (Gaiya, 2008).

Apart from the FEPA Decree of 1988, the Federal government through Decree 86 of 1992 promulgated another law on Environmental Impact Assessment (EIA) which aimed at protecting the Nigerian environment by making it compulsory for any project that may have an adverse effect on the environment. It sought to assess the likely or potential environmental impacts of proposed activities, including their direct or indirect cumulative, short-term and long-term effects on the environment (Nwilo & Badejo, 2008).

Table 1. Environmental policies enacted between 1930s and 1990s

S/N	Policy	Year of Enactment
1	The Forestry Ordinance	1937
2	The Eastern Region Forest Law	1955
3	The Oil Pipeline Act	1958
4	The Wild Animals Preservation Law (Western Region)	1959
5	The Forestry Ordinance with Amendment (Northern Nigeria)	1960
6	The Wild Animals Law (Northern Region)	1963
7	The Wild Animals Law (Eastern Region)	1965
8	The Forestry Amendment Edict (Western State)	1969
9	The Petroleum Drilling and Production Act	1969
10	The Sea Fisheries Decree	1971
11	The Sea Fisheries Regulation	1972
12	The Wild Animals Preservation (Law, Lagos State)	1972
13	The Forestry Amendment Edict (Western State)	1973
14	The Wild Animals Law Amendment Edict(North-Eastern State)	1975
15	The Wild Animals Law Amendment Edict (Kano State)	1978
16	Exclusive Economic Zone Decree	1978
17	The Kanji Lake National Park Decree	1979
18	The Endangered Species Decree	1985
19	National Conservation Strategy for Nigeria	1986
20	The Natural Resources Conservation Council Decree	1989
21	Federal Environmental Protection Decree	1989
22	National Parks Decree	1991

Source: Dibie (2000).

In addition, the EIA Decree had the responsibility of identifying the measures available to mitigate adverse environmental impacts of proposed activities and assessment of the identified measures. The activities of the EIA are meant to be policed by the Federal Environmental Protection Agencies and by the State environmental protection Agency respectively (ibid).

The point here is whether or not these extant laws have been implemented by the relevant agencies, and to point out some observable constraints which may hinder their being effectively functional. This observation should not be surprising as evidences abound that the Nigerian government is usually good at formulating public policies but grossly fall short when it comes to implementation of such policies.

First and foremost, we consider that it is relevant to examine the goals and principles on which EIA anchors its assessment. In addition, there are procedural frameworks which have been provided for EIA as follows:

The EIA processes start from the proposal to approval for implementation, resulting in the issuing of an Environmental Impact Statement (EIS) and then the certificate. It is pertinent to add here that, apart from the aforementioned provisions meant to guard against the pollution and degradation of the Nigerian environment, the Federal Republic of Nigeria 1999 Constitution - Section 20 clearly states that, "the state shall protect and improve the environment and safeguard the water, air and land forest and wild life of Nigeria" (Federal Republic of Nigeria Constitution 1999).

It is unfortunate to state that the paternalistic attitude of some of the Nigerian Judges towards matters relating to environmental hazards created by the companies have rendered the enforcement of environmental laws ineffective. Some members of the judiciary as noted by Ebeku (2003) have been reluctant to give orders compelling companies whose operations are damaging to the environment to ease the action complained of. Perhaps, these judges consider the potential loss of income and their investments at the expense of the environmental protection. In addition to this could be the fact that Nigeria economy depends largely on the sales of crude oil. Whichever is the case, such an action retards the implementation of environmental laws and thereby encourages relegating these laws to mere "paper tiger".

Between 1990 and 2004 (Osho, 2008) asserted that there have been several oil related cases filed in the Nigerian Courts by affected Nigerians ranging from pollution from oil exploration, loss of incomes, loss of properties, contamination of drinking water leading to water borne diseases etc. few cases need to be mentioned here; Shell Vs. Tiebo, Shell Vs. Isaiah, Seismograph Services Vs. Mark, Ogiale Vs. Shell, Shell Vs. Ambah.

In most of these cases and similar ones, the courts are said to have refrained from making definitive orders on how to remedy the situation of the oil spillage claims, loss of income from fishing and farming, pollution of drinking water and crops, damage to health as a result of waterborne diseases (Ibid). Instead of making orders to address the complaints in terms of damages due to the physical environment of these communities, they settled for compensation of the affected complainants. In this wise, the environmental laws that were meant to protect human beings and other living things are thrown to the dogs. However, hope is not completely lost as it is hoped in some circles that, the judiciary may in future begin to base cases of environmental problems brought to the courts not merely to award monetary compensations without addressing the preservation of a healthy environment (Okorodudu, 1998). This type of thinking could be regarded as a mere speculation of things that may likely or not likely to be. The multinational oil companies who are normally being complained against by oil communities on gas flaring more likely will always win cases against them on legal grounds as they have a lot of financial capacity to fight their cases instead of obeying environmental laws meant to sanitize the environment.

Given the attitude of the judiciary as discussed above, FEPA (now called Federal Ministry of Environment) has a challenge of translating the laudable provisions of the various environmental acts into effective management of the environment. Experience has shown so far that, infrastructural projects suffers from environmental impact assessment because, approval for such projects are often given before the Environmental Impact Assessment take place (if at all) it takes place (Anago, 2001).

The key defaulters in this exercise are the various levels of government: Federal, State and Local. These levels of government are said to be involved in routine approval of projects within the mandatory study list before any kind of Impact Assessment is made. The

Niger Delta Development Commission (NDDC) decided to dredge Ayetoro Canal prior to any EIA as required by law. In fact, it is stated that any EIA reports are actually "post mortem just to fulfill all righteousness" and fence off resistance and complaints from Nongovernmental Organizations (NGOs), and the affected communities (Ango, 2001).

Based on this type of attitude, it could safely and certainly be deduced that, EIA reports merely serve to satisfy the needed paper work requirement. To further point out that EIA reports may merely serve to "fulfill all righteousness", our source further referred us to a particular case where the Federal government awarded a contract for the dredging of Imo River to improve vehicular access to the nation, and aluminum smelting company at Ikot Abasi in Akwa Ibom State, but some of the affected coastal communities resisted the projects on the grounds that EIA draft report was unfavourable to them. Unknown to these communities, the final report of EIA got the approval of the Ministry of Environment and so the dredging project commenced despite the resistance of these communities. Worse of all, it is stated the Federal government refused to pay the communities the requested compensation on the grounds that the dredging exercise has the potential economic benefits to the communities (Ibid). Where then does justice, fairness and most importantly the ability of the EIA to be in control of the environment which it is statutorily supposed to be in charge? It simply demonstrates the impotency of EIA on Nigeria's environment

Emerging Issues and Challenges of Environmental Pollution in Nigeria

The concern for the environment has been repeatedly expressed over the years in a series of international summits and consensus like the 1997 Kyoto Protocol, Global Initiative for Gas Flaring Reduction and the 2007 Bali Declaration, etc. While for most part of the developed countries ofthe world, environmental accounting and reporting has developed voluntarily (Uwalomwa & Uadiale, 2011),however, this is not the same in developing countries (Azzone, Manzini & Noci, 1996).

Nigeria has not been an exception in paying lip- service to issue of environmental management as demonstrated by successive extension of gas flaring deadline many times namely 2004, 2007, 2008, 2010 and lately December 31st, 2012 as the terminal date for gas flaring by the legislature.

Nigeria has been signatory to some environmental issues internationally, some of which are:

- The 1997 Kyoto Protocol
- World Bank Global Gas Flaring Reduction (GGFR) Public-Private Partnership, 2002
- 2007 Bali Climate Declaration by Scientists

The following initiatives are some of the responses made by Nigeria to safeguard the environment against degradation:

- Associated Gas Re-injection Act Cap 26, LFN 1990 and its attendant regulations
- The Oil in Navigable Waters Act Cap 331, LFN 1990 and its attendant regulations
- The Oil Terminal Dues Act Cap 339, LFN 1990
- The Federal Environment Protection Agency Act Cap, 131 LFN 1990
- Statement of Accounting Standards (SAS) 14 and 17 which regulate both the upstream and downstream sectors of the petroleum industry
- Companies and Allied Matters Act, (CAMA) 1990
- Security and Exchange Commission (SEC) Code of Corporate Governance, 2008
- Nigerian Gas Master Plan, 2008 which is a guide for the commercial exploitations and
- management of Nigeria's gas sector aimed at growing the economy with gas
- Final Deadline on Gas Flaring: The Nigerian legislative arm has finally chosen 31st December, 2012 as deadline for gas flaring fully backed up by the law (AYOOLA 2011).

Despite these regulations the challenges of environmental pollution remain unabated, for instance, gas flaring has been a contentious issue in Nigeria right from the beginning of commercial exploitation of crude oil in the country. The gas that is flared in the oil fields of the Niger Delta is called associated gas because it comes out of the earth along with the target crude oil and is separated from the crude so as to make that commodity useful (Bassey 2009). Furthermore Bassey believes that the separated gas could be handled in a number of ways. First it could be harnessed for use as liquefied natural gas. Secondly it could be re-injected into the earth where it came from. Thirdly it could be vented or flared. In Nigeria over 50% of the gas associated with crude oil extracted is flared. This act has been considered reprehensible because of its impacts, its wastefulness and its continuous and routine manner.

The politics of environmental pollution has seen the dates for ending gas flares being shifted as it suits the corporations and the government. They are the players as well as the umpires and can freely shift the goal posts as they please (Bassey 2009). The latest shift of deadline was announced by former President UmaruYar'Adua at an International Gas Stakeholders Forum held in Abuja in November 2007, about a month to the end of a subsisting deadline. During his speech, he simply moved the deadline for gas flaring from January 2008 to December 2008 (The Guardian Tuesday January 15, 2008) despite the clamor of Nigerians and citizens of the world that gas flaring should be stopped at the close of 2007.

There were indeed a lot of noises late 2007 and early 2008 about how gas flaring would be ended in Nigeria and in the mix of official voices policy incoherence was very apparent (The Guardian Tuesday January 15, 2008). The government also set up of an ad hoc committee to halt gas flaring. This committee facilitated by the World Bank Global Gas Flaring Reduction Partnership (GGFR) turned out to be little more than hot air. None of the official moves made in this regard has gained the confidence of community actors who are justifiably fed up with mere rhetorics (Bassey 2009).

Government's staggering on the gas flare issue commenced in 1969. That was when the first major move was made by the Nigerian state to halt gas flaring in the country. At that time the General Yakubu Gowon ordered that corporations should set up infrastructure to utilize associated gas within five years of their commencement of operations. When the oil companies paid scant attention to this order the government then moved the goal post to 1979 but could not enforce this new deadline before it was overthrown in 1975. The Associated Gas Re-Injection Act Number 99 of 1979 required that oil corporations operating in Nigeria should produce detailed plans for gas utilization as well as guarantee zero flares by January 1, 1984. The only way by which they could continue flaring after that date would be by the express permission of the responsible Minister on a case-by-case basis (Bassey 2009).

Since these deadlines are never respected, government has resorted to shifting them according to the pleasure of the corporations through executive orders embedded in speeches/remarks and without any backing by law. It was in response to local and international pressure that the Federal Government once more pledged to halt gas flares in Nigeria and set January1, 2008 as the new date. This was not to be because by December 17, 2007 a fresh shift was announced with a new target date of December 31, 2008.

After that announcement there has been a lot of muddling that rendered it virtually impossible to know what target date the government is pursuing. This is why Nigerians believe that the Senate is on the right track when they embarked on the drafting of a Bill to end illegal gas flaring by December 31, 2008. An enactment of such a law is the only measure by which Nigerians can say that there is a modicum of the rule of law with regard to gas flaring in Nigeria. Otherwise, it is clear that the acclaimed rule of law claims of the government may well be nothing beyond political posturing.

Looking at the response of key players in the oil and gas sector to end gas flares reveals reluctance. Shell estimated that an additional investment of $1.85 billion (about N247.9 billion) would be required to complete some projects that would enable it achieve zero flare level by the end of 2009. A Shell boss Basil Omiyi made a statement in (2005, 5) that

The construction of these facilities will only be completed by the end of 2009, which means that gas flaring from the relevant flow stations will not be eliminated until the end of 2009. On the commissioning of these facilities we will cover 95 per cent of all associated gas produced.

Environmental Sustainability

ALL biological life depends on a wholesome and well functioning ecosystem and the earth has no limitless supply of resources which must be proportionately shared by all living things (Okueso, 2008). Human beings disobey the law of nature at their risk while human success results from co-operation with nature, fitting into the web of life (Akinsola,2006). In Nigeria, Environmental Management has remained consistently bad over the years and the problem has become a monster that is difficult to solve in spite of all efforts from all arms of government. Consequently, the environment continues to deteriorate while environmental related diseases aggravate poverty (Ivbijaro, 2007).

Also, the population has risen rapidly to about 140 million going by the latest National Census (2006) without commensurate provision of infrastructure and supporting services to cope with the rising population. The implication is poor environmental sanitation which is visible in form of increase waste generation, overutilization of limited existing facilities, inadequate supply of potable water, poor land use, conversion of residential accommodation to commercial use, increase in urban screen and shanties on the edge of the city as a result of acute rural- urban migration and a general decline in the quality of human life.

The goal 7 of millennium development is to ensure environmental sustainability and achieving sustainability means ensuring that current actions do not lead to future declines in human well being. However, apart from the fact that environmental sustainability is far from being achieved in Nigeria in the different spheres of life, it is clearly evident that our current activities can lead to future declines in human well being if care is not taken.

Nigeria environment is not yet sustainably developed because it has not achieved a high quality of life in that the activities of man degrade the environment. According to Okueso, and Adekoya, (2006), it can be sustainably developed if attention can be positively focused at addressing problems in the following areas of environmental sanitation.

1. Housing and urbanization
2. Adequate potable water supply
3. Management of urban drainage system
4. School sanitation
5. Noise pollution
6. Energy and Environmental Sanitation
7. Pest and vector control
8. Disposal of the Dead (Human and livestock)
9. Control of reared and stray animals
10. Hygiene Education and Promotion
11. Partnership
12. Improved Governance
13. Municipal Solid Waste Management
14. Medical Waste Management

15. Excreta and sewage management
16. Food sanitation
17. Market and Abattoir Sanitation

Flooding

Concerns for flooding has increased in recent times due to climate change, sea level rise, rapid population growth and urbanization, the level of awareness of flood risk, the limited efforts towards flood disaster risk reduction in many places and the exposure and vulnerabilities of large numbers of human population (Peduzzi et al. 2011, Raaijmakers et al. 2008).

In Nigeria, flooding and solutions to its impacts are critical issues (Obeta 2014). With history of devastating floods which affected millions of human populations and caused fiscal losses amounting to billions of US dollars, the importance of exploring more realistic flood risk mitigation measures for Nigeria should be paramount (OCHA 2012). Flooding in Nigeria are fluvial (resulting from rivers overtopping their natural and man-made defences), coastal (affecting mainly the coastal areas) and pluvial (flash, arriving unannounced following a heavy storm) in nature and have been a major cause of concern for rural areas and cities within the country (Bashir et al. 2012, Douglas et al. 2008). Whilst stake holders' efforts towards tackling the hazard have not yielded satisfactory results, they have been criticized as ad-hoc, poorly coordinated, non-generalizable and not well established (Obeta 2014). However, in the light of 'best practices' in flood risk reduction and 'lessons learned' from other countries' experiences of flooding, it can be argued that such stake holders' efforts are limited due to lack of quality data, which are needed to systematically tackle flooding, poor perception of flooding among the general populace, lack of funds and improved technology as well as poor political will power. The growing number of flood victims and the constrained sustainable development caused by flooding within the country suggest that much of what is known regarding flooding in the country is deficient on remedies. More critical is the subject-matter of Nigeria being one of the most populated countries of the world with population size estimated at over 170 million people (World Bank 2013).

Bombing

The increase in terrorist activities using explosive devices is also affecting the environment negatively. Not only is there loss of lives but destruction of assets, climate and soil. Nigeria has experienced several bombing in recent times; for example the bombing of oil installations around the Niger delta area, bombing of terminus market in Jos on 20th April, 2014, bombing of Nyanya, Abuja on 14th April and 1st may, 2014 etc. There has also been in recent times in Nigeria suicide bombings which involve the perpetrator's death and causing massive death and injuries. Examples of suicide bombing in Nigeria were the bombing of Church of Christ in Nations (COCIN) headquarters church Jos, the St. Finbarr's Catholic Church Rayfield, Jos, Living Faith church in Bauchi state etc. The effects of the bomb blast leads to damage of land especially for agricultural purposes and the release of various harmful and hazardous substances into the land, water and air greatly pollutes the environment. Large hectares of forests are being destroyed and agricultural land rendered unproductive especially in the northern part of Nigeria as a result of these terrorist activities. Wildlife is sometimes also destroyed when explosives are released and this can lead to the extinction of some rare species of animals.

Terrorist activities also cause indirect impact on the environment as people migrate from areas of conflict to other areas. People are displaced and there is massive migration into urban areas. Refugee camps are usually set up and more pressure is put on the forests and urban areas. As more people migrate into areas where infrastructure is inadequate for the few residents, available resources are over stretched and much pressure is exerted on the meagre resources available. There is usually no proper waste disposal system and toilets in refugee camps and boreholes are usually dug next to sewages this has led to the poor quality of water and consequently led to cholera and other health related challenges.

Environmental Impact Assessment

Globally, Environmental Impact Assessment (EIA) is recognized as a tool for achieving sustainable development. The main objective of the EIA is to ensure that potential environmental impacts are foreseen at the appropriate stage of project design and addressed before any decision is taken on the project. The EIA can be defined as a systematic process for identifying, predicting and evaluating potential impacts associated with a development project. The EIA process must proffer mitigation measures to avoid, reduce or minimize the negative impacts on the environment, public health and property. It must also enhance positive impacts. The mitigation measures entail identifying possible alternative site, project, process design, including that of not proceeding with the project.

EIA legislations and the required procedural guidelines for carrying out the EIA process became effective since the 1970s in developed countries. Nigeria took a giant leap when she promulgated her main EIA legislation (i.e. EIA Act No.86) in 1992. EIA is proclaimed in Principle 17 of 'Agenda 21' (Agenda for the 21st century) of the United Nations Conference on Environment and Development (UNCED), which was held on 3-14 of June, 1992, in Rio de Janeiro, Brazil. It states that: "Environmental Impact Assessment as a national instrument shall be undertaken for proposed activities that are likely to have a significant adverse impact on the environment and subject to a decision of a competent authority."

The Nigerian EIA Act No. 86 of 1992 makes the EIA mandatory for development projects likely to have adverse impacts on the environment prior to implementation. (Section 14(1) (a) & (b)). Section 14 (1) (d) further states that the Federal, State, Local Council or any of its agencies, prior to environmental assessment of the project in accordance with the EIA Act, shallnot "under the provisions of any law or enactment, issue a permit or license, grant an approval or take any other action for the purpose of enabling the project to be carried out in whole or in part."

EIA is mandatory for activities in all sectors of the economy. The activities in the oil and gas sector which require an EIA include:

1. Oil and gas fields development;
2. Construction of off-shore, on-shore, and overland pipe line;
3. Construction of oil and gas processing facilities;
4. Construction of oil refineries;
5. Waste treatment and disposal.

The regulatory body for the EIA administration in Nigeria is the Federal Ministry of Environment, Housing & Urban Development (FMEH & UD) (Formerly FEPA/FMENV). The regulatory body, with the approval of the President, Commander – in – Chief of the Armed Forces, is vested with the power to make regulations on EIA Procedure (Section 61 of the EIA Act).

Regrettably, the EIA Procedure of August 1994 devised and contrived by the regulatory body is deficient and defective. In most cases, the provisions in the EIA procedure step outside the ambits of the EIA Act.The faulty EIA procedure and vagaries of administrative fiats are being used to undermine the EIA Act. Hence, there is a legacy of environmental degradation and public health impairments arising from the projects whose approvals were purportedly obtained under the EIA Act. The failings of the regulator in this respect give rise to violent agitations in the oil- bearing communities, to assert their rights to healthy and unpolluted environment. The EIA practice in Nigeria (in its present form) is a showcase for corruption and infraction of the EIA Act. Also, the public access to information through the public registry is yet to be honoured in compliance since the commencement of the EIA Act in 1992.

SOLUTIONS AND RECOMMENDATIONS

This paper therefore recommends that;

- Sectoral policies, environmental laws and regulations are important, but cannot alone, be expected to deal with the problems of environment and development. Prices, markets and governmental economic policies also play a complementary role in shaping attitudes and behaviour towards the environment. Sustainable development entails that the exploitation of resources, production of goods and services, direction of investments, orientation of technological development and institutional change are all in harmony. This will ensure that the needs of the present generation are met without endangering the future generations.
- Since our national development policy objective is to achieve rapid economic growth and improvement in individual welfare on a sustainable basis, a range of enabling policies, economic instruments and incentives are required to propel the development process in the desired direction.
- Government should Integrate social, environmental and other costs of negative environmental externalities into economic activities, so that prices will appropriately reflect the true and total value of resources. This will help prevent cases of environmental degradation;
- Institutionalize "Polluter Pays Principle" so that the polluter bears the cost of environmental degradation or pollution; thus providing the positive incentives to limit degradation or pollution of the environment.
- Develop and implement a mechanism for charging emission fees and fines for all pollutants and effluents (based on quantity, quality and detrimental effects) thereby internalizing all costs and other negative externalities into the production process and output prices.
- The Ogoni cleanup should be facilitated and action expedited. Rather than mere flag off, government should ensure full implementation of the project, ditto to all areas that suffer environmental problems across the country.

FUTURE RESEARCH DIRECTIONS

It is expected that future research in this area of study should focus on the best possible ways to divest from mostly fossil fuel to other sources of renewable energy. This is basically because the authors discovered that the major cause of environmental pollution in Nigeria is mainly from the oil sector. Therefore, the earlier the government of the day focuses attention on the future without oil, the better. Even so, they should try and carry out more research on the areas of the maximum use of the wastes that constitute pollution hazards, and create value chains on their uses, and perhaps how to recycle them for good uses.

CONCLUSION

The age we live in is a challenging one and the range of earth life forms, ranging from aquatic to terrestrial species is exposed to un-simmering human pressure. From the forest to the oceans, environmental resources are being depleted with vigour. Attitudes of people towards the non-human world have become so predatory that we are beginning soon to live as endangered species when the vary basis of our survival is in jeopardy. It is therefore a great concern that nations begin as a matter of urgency to follow the path of ensuring long-term harmony in the economic and ecological practices for all generations. Sustainable environmental development is not just a concept but also a process and a means to an end. It demands an understanding of development imperatives and an attitude of mind to make it work to support human security and peace through the conservation of natural resources for sustenance of livelihoods and ultimately economic development of Nigeria.

From the authors' findings it could be asserted without fear of contradiction that we are still largely scratching on the surface in terms of finding practical solutions to the existing pollution challenges. This is because as the title of the chapter indicates, politics has eaten deep into the implementation of policies to curb environmental pollution in the country. Moreover, even when the government means well, the political will to get things done is not there, mainly as a result of variegated political interests. Also, the several fissiparous tendencies tearing from several fronts contribute to lack of policy concentration and commitment to implementing them. All these issues as discussed must be properly addressed in order to achieve maximum results.

REFERENCES

Adewuyi, A. O. (2001, November). The implications of crude oil exploitation and export on the environment and level of economic growth and development in Nigeria. Paper Presented at *Nigerian Economic Society Annual Conference.*

Aina, E. O. (1992) The Journey So far. In The Making of the Nigerian Environmental Policy (pp. 17-25). Ibadan: Ibadan University Press.

Ajani, J. (2012). Rage of nature: Flood ravages communities across Nigeria. *Vanguard.* Retrieved October 28, 2016, from http://www.vanguardngr.com/2012/10/rage-of-nature-flood-ravages-communities-across-nigeria/

Ajayi, D. D., & Ikporukpo, C. O. (2005). An Analysis of Nigeria's environmental vision. *Journal of Environmental Policy and Planning*, *7*(4), 341–365. doi:10.1080/15239080500441137

Akinsola, H. A. (2006). *A – Z of community health in medical, nursing and health education practice.* Ibadan: College Press Publishers Limited.

Amokaye, G.O. (2012). Environmental pollution and challenges of environmental governance in Nigeria. *British Journal of Arts and Social Sciences, 10*(1).

Anago, I. (2001). *Environmental assessment as a tool for sustainable development: The Nigerian experience.* Retrieved October28,2016, from,www.jepplaw.net/page

Ashtons J. (2003). Covering ethnic conflicts: An interview with Silvia Poggiol. *Politics, 8*(4), 34-47.

Awobajo, S. A. (1981, November). *An analysis of oil spill in Nigeria 1976-1980.* Paper presented at the seminar on the Petroleum Industry and Nigerian Environment organized by NNPC, Warri.

Aworawo, D. (1999). The impact of environmental degradation on the rural economy of The Niger Delta. In A. Osuntokun (Ed.), *Environmental problems of the Niger Delta.* Lagos: Friedrich Ebert Foundation.

Ayoola, T. J. (2011). Gas flaring and its implication for environmental accounting in Nigeria. *Journal of Sustainable Development, 4*(5).

Azzone, G., Manzini, R., & Noci, G. (1996). Evolutionary trends in environmental reporting. *Business Strategy and the Environment, 5*(4), 219–230. doi:10.1002/(SICI)1099-0836(199612)5:4<219::AID-BSE69>3.0.CO;2-K

Bashir, O., Oludare, H., Johnson, O., & Aloysius, B. (2012). Floods of fury in Nigerian cities. *Journal of Sustainable Development, 5*(7), 69–79. doi:10.5539/jsd.v5n7p69

Bassey, N. (2009, April). *Gas flaring: Assaulting communities, jeopardizing the world.* Paper presented at the National Environmental Consultation hosted by the Environmental Rights Action in conjunction with the Federal Ministry of Environment at Reiz Hotel, Abuja.

Cointreau, S. J. (1992). Environmental management of urban solid waste in developing countries: A project guide. Washington, DC: World Bank TC No. 5.

Constitutional Rights Project (CRP). (1999). *Land, oil and human rights in Nigeria's Delta.* Lagos: CRP.

Dibie, R. (2000) Understanding public policy in Nigeria: A twenty-first century approach. Lagos: Mbeyi and Associates (Nig) Ltd.

Douglas, I., Alam, K., Maghenda, M., Mcdonnell, Y., McLean, L., & Campbell, J. (2008). Unjust waters: Climate change, flooding and the urban poor in Africa. *Environment and Urbanization, 20*(1), 187–205. doi:10.1177/0956247808089156

Ebeku, K. (2003). Judicial attitudes to redress for oil related damages in Nigeria. *RECIEL, 12*(2), 199–208.

Egwaikhide, F. O., & Aregbeyen, O. (1999). Oil production externalities in the Niger Delta: Is a fiscal solution feasible? In *Fiscal federalism and Nigeria's economic development. Selected Papers presented at the Annual Conference of The Nigeria Economic Society.* Department of Economics, University of Ibadan.

Gaiya, B. (2008). *Environmental pollution in the Niger-Delta: An economic perspective.* Paper presented at International Conference on the Nigerian State, the Oil Industry and Niger- Delta at the Department of Political Science, Niger-Delta University, Bayelsa State.

Ibeanu, O. (2000). Oiling the friction: Environmental conflict management in the Niger Delta, Nigeria: An Environmental Change and Security Profile Report, issue No 6. Woodrow Wilson Centre.

Isiche, A. O., & Sanford, W. W. (1976). The effect of waste gas flares on the surrounding vegetation of South-Eastern Nigeria. *Journal of Applied Ecology, 13*.

Ivbijaro, F. A. (2007, March). *Sustainable environmental management.* Paper presented at the Sensitization Workshop on Keep Nigeria Clean by the National Orientation Agency (NOA), Ibadan.

Obeta, M. C. (2009). Extreme river flood events in Nigeria: A geographical perspective of Nigeria. *Journal of Geography and the Environment, 1*, 170–179.

Okonta, I., & Douglas, O. (2001). *Where cultures feast: Forty years of Shell in the Niger Delta.* Benin: ERA/FOEN.

Okorodudu, M.T. (1998). Law of environmental protection: Material and text. *RECIEL, 12*(2), 207.

Okueso, S. A. (2008). Improving environmental management for sustainable development. *Nigeria. International Journal of Environmental Science, 4*(1), 35–39.

Okueso, S. A., & Adekoya, A. F. (2006). *Chemical, health and environmental care for sustainable development in Nigerian.* Journal of Nigerian School Health Press.

Olokesusi, A. (1987). Characteristics of Environmental Problems in Nigeria and management prospects. *The Environmentalist, 1*(2), 44–57.

Onosode, G. (2003). *Environmental issues and the challenges of the Niger Delta; Perspectives from the Niger Delta. Environmental Survey Process.* Yaba: The CIBN Press Limited.

Orubu, C. O. (1999). *The exploitation of non-timber forest resources in the Niger Delta: Problems and perspectives. Technical paper, Niger Delta Environmental Survey.* Port-Harcourt.

Orubu, C. O., Odusola, A., & Ehwarieme, W. (2004). The Nigerian oil industry: Environmental diseconomies, management strategies and the need for community involvement. *Journal of Human Ecology (Delhi, India), 16*(3), 203–214.

Osho-Adejonwo, O. (2005). *The evolution of human rights approaches to environmental protection in Nigeria.* Retrieved March 12, 2010, from, www.iucnael.org/index.php?Optiocom_documans&task

Otu, E. C. (2016). Restoring environmental justice and affection in Ogoni Land. *International Affairs and Global Strategy, 4*(4), 94–108.

Peduzzi, P., Dao, H., Herold, C., & Mouton, F. (2009). Assessing global exposure and vulnerability towards natural hazards: The disaster risk index. *Natural Hazards and Earth System Sciences, 9*(4), 1149–1159. doi:10.5194/nhess-9-1149-2009

Raaijmakers, R., Krywkow, J., & van der Veen, A. (2008). Flood risk perceptions and spatial multicriteria analysis: An exploratory research for hazard mitigation. *Natural Hazards, 46*(3), 307–322. doi:10.1007/s11069-007-9189-z

The Guardian. (2008, January 15). Editorial: Gas flaring, end of a moving target? *The Guardian.*

Udoudo, A. (2006). *Newspaper coverage of Environmental Degradation in the Niger Delta* (Unpublished PhD Dissertation). University of Uyo.

Ugochukwu, O. (2003). *Delivering on our mandate in the journey so far.* Port Harcourt: NDDC.

UN Office for the Coordination of Humanitarian Affairs (OCHA). (2012). *Nigeria: Floods, Emergency Situation Report no 2.* Retrieved on October, 18, 2016, from, www.ochaonline.un.org/rowca

UNDP. (2006). *Niger Delta Human Development Report Garki.* Abuja: UN House.

UNEP. (2011). *United Nations Environmental Programme Report 2011.* New York: United Nations.

Uwalomwa, U., & Uadiale, O. M. (2011). Corporate social and environmental disclosure in Nigeria: Comparative study of the building materials and brewery industry. *International Journal of Business, 6*(2), 258–264.

Williams, A. (1992). Nigerian environment and national health policy. In A. Williams (Ed.), *Towards industrial pollution Abatement in Nigeria* (Vol. 1). Lagos: FEPA.

World Bank. (1995). *Defining an environmental development strategy for the Niger* (Vol. 2). World Bank.

World Bank. (2013). *The World Bank: Working for a world free of poverty, Population (Total).* Washington, DC: World Bank Group.

KEY TERMS AND DEFINITIONS

Air Pollution: The accumulation in the atmosphere of substances that in sufficient concentrations endanger human health or produce other measured effects on living matter and other materials. Air pollution are powered heat generation, the burning of solid wastes, industrial processes of air pollutants like carbon monoxide from incomplete combustion engine and others, such as hydrocarbons, nitrogen oxides, Sulphur dioxide and photochemical oxidants.

Environmental Pollution: Environmental pollution can occur from petroleum production activities.

Noise Pollution: A composite of sounds generated by human activities ranging from blasting stereo system to the roar of transport vehicles. The most readily measurable physiological effect of noise pollution is damage of hearing, which may be either temporary or permanent and may cause distraction of normal activities or general annoyance.

Chapter 17
Integrating Ecosystem Management and Environmental Media for Public Policy on Public Health and Safety

Augustine Nduka Eneanya
University of Lagos, Nigeria

ABSTRACT

Over the past three decades, the relationship between ecology and public policy has changed because of the increasing role of scientific uncertainty in environmental policy making. While earlier policy questions might have been solved simply by looking at the scientific technicalities of the issues, the increased role of scientific uncertainty in environmental policy making requires that we re-examine the methods used in decision-making. Previously, policymakers use scientific data to support their decision-making disciplinary boundaries are less useful because uncertain environmental policy problems span the natural sciences, engineering, economics, politics, and ethics. The chapter serves as a bridge integrating environmental ecosystem, media, and justice into policy for public health and safety. The chapter attempts to demonstrate the linkage between the environmental policy from a holistic perspective with the interaction of air, water, land, and human on public health and safety.

INTRODUCTION

Since the 1980s, ecosystem management has helped to draw science into environmental policy process. Over the years, the relationship between ecology and public policy has changed because of the increasing role of scientific uncertainty in environmental policy making (Ludwig, Mangel & Haddad, 2001). While earlier policy questions might have been solved simply by looking at the scientific technicalities of the issue, recent environmental problems are often plagued by scientific uncertainty, such as international climate change and policy negotiations. Ludwig, et al. (2001) argue that this increased role of scientific uncertainty in environmental policy making requires that we re-examine the methods used in

DOI: 10.4018/978-1-5225-3194-4.ch017

decision making. Within this context, strong disciplinary boundaries are less useful because uncertain environmental policy problems span the natural sciences, engineering, economics, politics, and ethics. This interdisciplinary approach to environmental policy problems has led public managers and Administrators to ecosystem and adaptive management. The multi-faceted nature of ecosystem management and adaptive management make them particularly relevant for our discussion of the linkages between the natural system, physical system, the social system, and environmental policy decisions.

The purpose of this chapter is to examine how the ecosystem management perspective integrates with the environmental media to develop environmental health policy within a holistic ecological view and citizen engagement. It also highlights key issues that are important for a public manager or Administrator to consider when making policy decisions about the interactions and impacts of the natural system, concretion system and social system.

BACKGROUND

There has been a relatively recent push in the environmental policy community to encourage ecosystem management approaches in the management of natural resources (Steel & Weber, 2001). However, the local citizens must land their support for the management process to be successful. Across the world, learning from citizens in government policy has continued to grow to include a broad range of policy areas and national context. As participation of citizens in public policy has become more common around the world, and as the range of agencies that seek public input has expanded, the methods for soliciting public input have also grown in number and complexity. Steel & Weber, (2001) observe that the way an ecosystem management effort is institutionalized and implemented can be directly related to the level of citizens' support it receives.

Apart from citizens' participation, activities frequently involve a range of actors from non-governmental subsystem as well as from different levels of government. For practitioners, the basic requirement of participation is often obvious – a local, state, or federal law requires a government agency or department to hold a leaving, announce a change in zoning, or form an advisory committee on a given issue.

Following this thread of stakeholders' engagement, natural resource agencies have adopted ecosystem management principles as a way to increase stakeholder involvement in resource management (Koontz & Bodire, 2008). Eco-system management calls for "management based on stakeholder collaboration, inter-agency cooperation, integration of scientific, social, and economic information; preservation of ecological processes, and adaptive management" (p. 60). Given that eco-system management has such an integrative and collaborative focus, adoption of this approach by governmental agencies often signals that the agency is prioritizing both stakeholders' involvement and holistic perspective for environmental management.

Another aspect of stakeholder collaboration that can be addressed through an eco-system management approach involves dealing with large-scale, trans-boundary environmental problems. Management of trans-boundary environmental issues (such as air pollution that drifts across city or state lines) can require collaboration among a variety of stakeholders as well as the different owners of private property (Thompson, Anderson, & Johnson, 2004). Though it is not always easy, as Anti-trust laws often hinder the collaborative process because they do not allow industries to participate fully in the collaboration and to share information.

However, Keough and Blahna, (2006) prepared a comprehensive review of the collaborative environmental management and ecosystem management literature to identify eight key principles that managers should implement for successful collaborative eco-system management. These eight principles include:

- Integrated and balanced goals
- Inclusive public involvement
- Stakeholder influence
- Consensus group approach
- Collaborative stewardship
- Monitoring and adaptive management
- Multi-disciplinary data, and
- Economic incentives (Keough & Blahna, 2006, p. 1373).

With respect to integrated goals, Keough & Blahna, (2006) argued that managers need to pay particular attention to ecosystem management goals as they relate to ecological, economic and social outcomes.

There are key ecological and social issues to be explored when discussing environmental media – such as air, water and land. Air is a natural resource that we include as a component of the natural and concretion systems. Air quality can have an impact on land use and land health; poor air quality can hinder the growth and survival of plants, animals, and humans that live on (and in) the land. Angolson et al. (2000) discussed a comprehensive overview of the techniques available to use land as an air quality control measure. They explain the role of land use in developing state implementation plans (SIPs) to reduce air pollution. For example, transportation, land use, and air quality are often explicitly linked in state implementation plans (SIPs) to limit the development of new transportation systems in regions that are in noncompliance with air quality standards.

Against this background, the chapter is divided into four sections. Section one examines the introduction, background, purpose, methodical approach and research questions. Section two analyzes and explains the conceptual model and the dynamics of the interactions of natural, concretion and social systems that addressed environmental issues related to air, water and land. Section three describes three major environmental media comprising of air, water and land and explain how these media can link social and ecological systems for environmental health policy. Section four examines the challenges, prospects and concludes the chapter by evaluating the impacts of the policy based on interaction between the air, water, land and human interaction with environment on public health issues.

CONCEPTUAL AND EXPLANATORY FRAMEWORK

In this context, system-based framework in figure 1 below has been designed to help researchers and decision makers understand the extent to which environmental policy decisions are inter-twined with the natural, concretion and social systems of the environment. In this framework, there are three different systems. (Natural, Concretion and Social systems) that interact with each other within the environment to generate environmental problems that culminate to policy demands in the system-based policy framework as demonstrated in Figure 1.

Figure 1. System-based policy framework

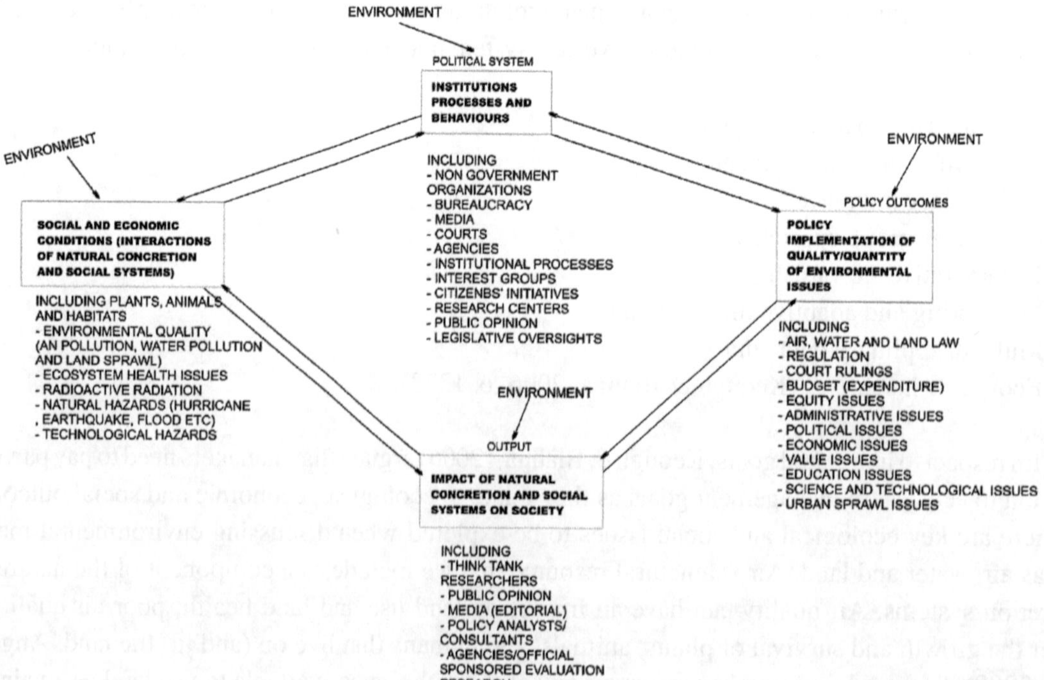

The system-based framework consists of (i) environmental issues or problem identifications (ii) Policy Agenda-setting (iii) Policy decision or non-decision (iv)Policy implementation of new program or change in an old public program (v) Policy or Program impact evaluation and (vi) Feedback (vii) Environmental health issues. The system-based framework helps us to explicate key environmental policy issues and relationships between them.

THE JUSTIFICATION FOR USING THIS SYSTEM MODEL AS EXPLANATORY FRAMEWORK

The justification for using this system-based explanatory framework is threefold. First, it captures key elements for environmental policy-making in a simplified structure that administrators, analysts, students, researchers and others can use to think through environmental policy questions. Second, it is designed to be general enough to be used across nations and even across regions of the world that have systematically different socio-cultural and governmental systems. Third, environmental policy administrators and analysts can use this model as a "checklist to help them think through all the effects of proposed policies" (Campbell & Corley, 2012, pp. 21-2). This would assist administrators and other decision makers focus on the importance of environmental health policy within the context of the federal, state, local government, foreign nations and international organizations. Its aim is to help current and future decision-makers, administrators and analyst focus on many and inter-related issues of environmental health policy with the goal of improving the quality of both policy and administrative decisions.

RESEARCH QUESTIONS

From the system-based policy framework, the following patterns of relationships or linkages are observed in figure 1 and they constitute the research questions:-

RQ 1: (Linkage A) What are the impacts of the interactions between natural system, concretion system, social system and environmental media on environmental policy making?
RQ 2: (Linkage B) To what extent does the interaction of air, water and land ecological resources generate policy decisions or non-decisions on environmental problems?
RQ 3: (Linkage C) What are the environmental health policy challenges and prospects in the interactions between air, water, and land ecological resources?

METHODOLOGY

This study adopts qualitative research method. This method is based on "array of interpretative technique which seeks to describe, translate and otherwise come to terms with the meaning, not the frequency of certain more or less naturally occurring phenomena in the social world" (Van Maanen, 1983, p.9). Data were collected from reviewed literature, Journal articles, textbooks, internet websites, government reports and Newspapers. Textual and secondary data analyses were adopted to explain or interpret important concepts of environmental political system and relationships between those sub-systems. As human dominate earth's ecosystem, it is justified that humans must be integrated into the interaction between air, water and land (Environmental media) for complete understanding of extant ecological systems (Grimm, et al, 2000). The study attempts to give an insight or illumination, which may guide action-oriented policy as environmental problem arises, and understand the parts of the process that need to be activated to bring a solution to fruition. The system-based framework is adopted in this chapter to explicate points and highlight how different areas of environmental health policy concerns fit into ecological system. This framework explains the patterns of relationships or linkages observed and constitute drivers to the study and answers to research questions as illustrated in analysis below:

ANALYSIS AND FINDINGS

RQ 1: (linkage A) what are the patterns of relationship between natural, concretion, social system and environmental media on environmental policy making?

Ecosystem Issues

The literature reviewed clearly demonstrates that the physical system of the environment is composed of the natural, and the concretion system. The social system consists of other governmental systems. The natural system is that part of the environment that is provided by nature. The Natural layer is the biogeophysical system that the city is placed upon and within. It includes climate, especially temperatures and precipitation. All these are affected by human activities within the ecosystem.

The concretion layer encompasses the physical parts of the environment that are created or have been created by humans. The concretion system represents the physical elements that people place upon and within the natural layer. They include buildings, and roads, dumps. These elements are affected by the natural layer.

The social system includes the social institutions that human create in this context, an "institution" is not a physical place, but a social organizations (Ostrom, 2005). Social system includes firms, nonprofit organizations, bureaucracy, state, local and federal institutions.

Both the natural and concretion system affect the social system and vice versa. All of the systems are part of "an interconnected system held together by flow of energy, water people, ideas and materials" (Gober, 2006, p. 9). Because of the complexity of each of these topics and other environmental issues, it is useful to have a conceptual framework in order to focus on specific factors important to the topic being discussed.

The concept of ecosystem refers to the existence of a biological community in some locale, and the physical and chemical factors that make up its non-living or abiotic environment. Example of ecosystems could "vary from composition to size and the relationship that exist amongst the components"(Abubakar, 2012). The ecosystem management in this context, therefore, refers to the integration of patterns of relationship between natural, concrete and social systems within holistic ecological perspective, as demonstrated in fig. 1.

Patterns of Relationship Between Natural, Concretion, and Social System in Development of Environmental Policy

Within the conceptual framework in fig. 1, the natural and concretion systems relate to the social system through the policy process in varying ways. The components of the natural systems for ecological air resource, include ambient air, naturally occurring particulate matter and naturally occurring chemicals. On the other and, the components of the concrete system for ecological air resource include human-generated particulate matter and human–generated chemicals. Different policy alternatives will likely have different positive and negative environmental impacts on the components of the natural system and the components of the concretion system. The natural and concretion systems relate to the social system through the policy process in varying ways. Different policy alternatives will likely have different positive and negative environmental impacts on the components of the natural system and the components of the concretion system. Even though we make policies within the social system, we can implement and enforce environmental policies within the natural, concretion or social systems – and usually policies affect some combination of all these. Therefore, it is imperative that policymakers and administrators think about the impacts of proposed environmental policies on each system. In short, policymakers must take a holistic view rather than a fragmented one (Campbell & Corley, 2012).

Environmental Media (Air, Water, and Land) Issues

RQ 2: (Linkage B) To what extent does the interaction of air, water and land ecological resources generate policy decisions or non-decisions on environmental problems?

Environmental Media Issues

Environmental media consists of air, water and land. To explore the key ecological and social issues for each, let us examine air, water and land resources.

- **Air Resource:** With respect to air as a resource, the natural system components of the model include the ambient air we breathe, naturally occurring particulate matter, and any naturally occurring chemicals that have an impact on air quality (either positively or negatively). The components of the concretion system include human generated chemicals and human-generated particulate matter that impact air quality. The relevant policy issues for air focus largely on air pollution, human-health issues and ecosystem health issues.
- **Water as a Natural Resource:** Regarding water as a natural resource, the components of the natural system include surface water (such as lakes and rivers), ground water, and existing aquatic habitats and ecosystems. On the other hand, the concretion system components of the model include the water delivery system, the wastewater system, and away artificial bodies of water. From a water perspective, the key issues that local policymakers have to address within an urban environment include issues of water availability, water pollution and human health issues. Water habitat health (suitability for non-human animals), though not yet a key issue, is becoming increasingly important as cities take a more comprehensive approach to sustainability.
- **Land Resource:** With respect to land resources, the general components of the natural system include soil, vegetation, animals, habitat and ecosystems and human. These components often overlap with each other (for example animals need habitat to survive) and across other components of the model (for example, humans are part of both the natural system and social system) with respect to land, the concretion system contains service infrastructure (for example, electrical, water, trash, transportation, other energy, and sewerage) buildings (including industrial, commercial residential) and transportation, vehicles (for example, cars, subway systems, buses, light rail systems and motor cycles).

The environmental issues that are relevant for policymaking with respect to land as a natural resource and development/ (also encompassing issues of sprawl) Brow fields (contaminated sites), soil erosion and ecosystem health issues.

One point that is apparent is that many of the policy-relevant issues for these three resources are related to either the quantity or quality of the natural resources: (1) how much of the resource is available for use and (2) the environmental quality of those resources. Therefore, "most of the policy alternatives that we explore for natural resources in an environment have to do with quantity and quality issues" (Campbell & Corley, 2012, p.132)

Interaction Between Three Environmental Media (Air, Water, and Land)

There is an explicit connection between the three environmental media of air, water and land. For example, air quality can have an impact on land use and land health; poor air quality can hinder the growth and survival of plants, animals, and humans that live on (and in) the land in urban areas. In fact, some urban regions have employed land-use restrictions as a way to control air quality. Ang-Olson et. al., (2000) examined a comprehensive overview of the techniques available to use land as an air quality control measure. In particular, they explain the role of land use in developing State Implementation Plans (SIPs) to reduce air pollution.

Another interaction between air, water and land occurs through the use of land and water as a carbon sink for carbon dioxide pollution in the air. Carbon dioxide (CO2), is a green -house gas that can be absorbed by water and land through two main mechanisms. The oceans can serve as a carbon-dioxide "sink" via mixing processes in the ocean that allow the water to absorb Co_2 from the atmosphere (McNeil, 2006). Plants and vegetation can serve as a carbon-dioxide sink through photosynthesis, which absorbs Co_2 and removes it from the air (Fluggae & Schilizzi, 2005; Sartori, et al., 2006; Sperrow, Eve & Panistian, 2003). This process of using water or vegetation to remove Co_2 from the air is called "carbon sequestration". Policy makers have an active interest in how this process can be used to reduce greenhouse gases and slow global warming. Natural scientists and social scientists alike have explored the practical application of carbon sequestration as a policy mechanism (Kennett, 2002, Kern et al., 2003; Wilson et al, 2007).

For environmental policy, the key ecological issues related to air quality, for example, are: the source of air pollutants, the type of air pollutants and the impacts of the pollutants on the environment and human health.

Challenges of Environment Issues on Human Health and Safety

RQ 3: (Linkage C) What are the environmental health policy challenges and prospects on the interactions between air, water, and land ecological resources?

This concept of environment is important because it represents the nexus between the physical and human components of the environment. It is not only of interest to geographers as a fertile theme of study, man-environment interactions are on lives and living things on the earth surface. Slight deviations of these intricate interactions from normal are known to spell doom for teeming organic populations in the earth's biospheric layers. The recurrent flooding being experienced in many parts of the globe has been linked to changes in the global climatic regime attributable to anthropo-genetic (human) factors.

Attempts by geographers to understand the relationship between man and his environment have produced three distinct concepts which explain the changing relationship between man and his environment. These changes are direct consequences of increasing man's scientific and technological capabilities to "force" his will on the environment.

The concept of environmental determinism captures the era when man's scientific and technologies achievements were so crude and limited that he could not but accept what nature (environment) offered him. This era marked the era of man's adaptation to whatever the environment offered. For instance, no farming activities could proceed in the arid area nor could man build on marshy terrain. Hence, settlements were confined to only dry relatively that terrain; routes avoided slugged and marshy terrain, including water bodies. A typical feature of the man – environment relationships was the maintenance of ecological equilibrium with the attendant lower risk of environmental disaster.

In relation to health, the environment is assumed to have a broader connotation and to refer to the totality of the external influences – natural and man-made which impinge on man and affect his well-being. It, thus, embraces life support systems and also the multiplicity of stimuli and hazards, direct and indirect, which man experiences. Life support systems are those environmental conditions which are necessary for human life such as oxygen, water, food and shelter.

According to Pat (2010), the environment influences human health in many ways – through exposures to physical, chemical and biological risk factors, and through related changes in behavior in response to those factors. According to the World Health Organization (WHO, 1981), 13 million deaths annually are due to preventable environmental causes. Mitigating environmental risk could save as many as four million lives as year in children alone, mostly in developing countries. The environment provides man with essential life support systems but it also presents him with a variety of hazards which may prejudice his health. If health is a "state of complete physical mental and social well-being and not the mere absence of disease of infirmity" (WHO, 1981), it represents a balanced relationship of the body and mind and complete adjustment to the total environment. Disease, on the other hand, is maladjustment or mal-adaptation in an environment, a reaction for the worse between man and hazards or adverse influences in his external environment.

Scholars have recognized that environmental issues have distinct international and regional dimensions. In addition to trans-boundary air pollution and water pollution, these manifest as global pandemics and global migration induced by climate change. And these issues tend to have noticeable second order effects. Global migration is creating political, security, and economic challenges for recipient countries. Global pandemics are shutting down travel, trade and tourism industries, thereby disrupting local economies.

The response of individual to these influences is conditioned by his genetic make-up or internal environment (Pyle, 1979). Thus, man is the product of the environment and at the same time the environment as the product of man. Man's health and well-being rest squarely on the environment, which can be divided into physical environment, biological environment, human environment and man's internal environment.

- **Physical Environment:** Even in the existence of man, the weather and climate have been postulated as influencing, favorably or unfavorably man's health. Solar radiation including cosmic rays, gamma rays, x-rays, ultra-violent rays, luminous rays and infra-red rays may induce skin burns, cancers, genetic mutation and other biological changes of exposed to for a longtime.

Water too provides a range of health hazards. Many diseases pathogens are essentially water born. Water impurities too may have adverse effect on health. The mineral content of rocks and soil may both be favorable or unfavorable in the sense that they may provide some trace elements essential for life or harmful for life.

- **Biological Environment:** Man's body provides a rich ground for parasites. Several parasites live in or within man permanently, without causing any structural change or functional disturbance. One of the intriguing features of the micro-organisms which attack man is their natural history and the ways they, the disease agents are transmitted from persons to person.

Whether it be causative organisms, intermediate host or vector, each element is the disease complex has its own specific environmental requirements. Each element, including man himself, is inescapable bound up with the geographical environment.

- **Human Environment:** The human or socio-cultural environment is essentially man-made and relates to the density, geographical distribution and mobility of population, to occupations, to socio-economic status, to housing to diets, to habits, customs and general life style.

- **Man's Internal Environment:** External environmental influences and hazards act on man, but the response is almost every case is thought to be conditioned by the genetic make-up of the individual. There appears to be a hereditary predisposition on the part of some individuals to certain diseases. On the other hand, some associations between the ABO blood groups have been proved beyond reasonable doubt (Howe, 1977).

Furthermore, there are ecological issues related to air quality. Air pollution, for example, is a challenge that threatens basic human welfare, damages natural and physical capital, and constrains economic growth. Exposure to air pollution increases a person's risk of illnesses, such as: lung cancer, stroke, heart disease and bronchitis. Those diseases and deaths from them also carry cost implications.

There are criteria of air pollutants that have negative health and environmental effects. Carbon monoxide (Co2) is one of them. A majority of the Carbon monoxide emissions come from mobile sources. Because Co2 bonds more readily to haemoglobin than oxygen does alone, these emissions can reduce the oxygen level in the heart and brain. They can also negatively impact the cardiovascular health of people with reduced heart capacity. From an environmental perspective, Co2 contributes to Cmog, which is particularly problematic in many urban areas. Carbon dioxide (C02) is the most prevalent long-lasting greenhouse gas. So much of it has built up in the atmosphere since fossil fuel burning began in earnest that those of us alive today are the first "homo sapiens" to breathe in air with carbon dioxide levels of 400 parts per million- up from 275 ppm before the industrial revolution. Scientists say the increase has been enough to raise global average temperatures by nearly 1oC from pre-industrial times, glaciers and pushing up sea levels. With coal so cheap and abundant, and energy so vital to development, another crop of fast-growing countries, including Vietnam and Indonesia, are driving "global coal renaissance". A coal power plant typically operates for at least 30 years, so countries racing to build them are locking in decades' worth of emissions. Gas plants are cleaner, but still produce a lot of Co2. In Sub-Saharan Africa and in Asian countries, according to a World Energy Outlook Special Report (IEA, 2016b), identified a link between energy, air pollution and health and urged the energy sector to contribute to the solution. The Agency reports that about 6.5 million deaths are attributed each year to poor air quality, making it the world's fourth largest threat to human health, behind high blood pressure, dietary risks and smoking. Key factors fuelling air pollution in Sub-Saharan Africa is poverty which has seen millions depend on biomass as energy source. About 2.7 billion depend on wood for cooking and kerosene for lightening which is blamed for around 3.5 million premature deaths each year (Uroko, 2016, p.36)

Sulphur dioxide (SO2) is another pollutant. The majority of Sulphur dioxide (SO2) emission come from electricity generation and fossil-fuel combustion. SO2 emissions can trigger breathing difficulties, cause respiratory illness, and significantly contribute to haze and reduced visibility. One of the most harmful aspects of SO2 for human health and the environment is the way it interacts with nitrogen oxides in the air and the resulting chemicals become part of precipitation. Acid rain can disrupt the sensitive balance of crops, ecosystems, water bodies and animal habitats, and it can damage parts of the concretion system as well. The use of fossil fuel is a contributory factor in air pollution. The industrial revolution brought about increased utilization of coal and oil and they have been catalysts for modern economies. Power plants, industrial facilities and automobile constitute major sources of outdoor pollution. According to International Energy Agency Report (2016), coal is responsible for about 60 per cent of global combustion-related sulphur dioxide emissions. Human health and the ozone layer often become casualties of a world heavily dependent on fossil-fuel. The impact of urban vehicle emissions is heightened by the fact that they are discharged not from the top of tall chimneys but directly into the street-level air that pedestrians breathe.

Nitrogen oxides (NOx) represent the third of the criteria air pollutants identified by US Environment Protection Agency (EPA). Major sources of NOx include mobile sources, electricity generated and fossil-fuel combustion. Nitrous oxide (which is a member of the NOx family) is a greenhouse gas and, therefore, may contribute to global warming. NOx pollution can lead to the deterioration of ecosystems, buildings, and water bodies, as well as plant and animal life. NOx particles also cause breathing and lung problems for human. Moreover, fuels used to power the global transportation system majorly diesel generate more than half the nitrogen oxides emitted globally. In addition, these chemicals can contribute to eutrophication in water bodies because of increased nitrogen loading. When NOx interacts with volatile organic compounds (Vocs) in the presence of sunlight, it forms ground-level ozone. Ground-level ozone can cause breathing problems, chest pain and a variety of lung problem for human. This pollutant can also significantly damage the environment by making plants more sensitive to disease, reducing crop yields and harming the ecosystems.

Lead is the last criteria of air pollutant. Since mobile sources represent one of the major sources of lead in the air, the EPA (USA), for example, required phase-out of lead in gasoline has profoundly reduced the amount of this pollutant in the air. Despite this effort, lead emissions from non-road (for example, construction) equipment and vehicles remain a problem. Additionally, humans can be exposed to lead in older homes through lead-based paint. Once it is ingested or inhaled, lead can compromise the immune system, the heart, the kidneys, the nervous system, and learning/developmental systems (especially in children). Since lead accumulates in the environment, negative impacts on plants, animals and ecosystems intensify over time.

Thus, premature deaths from air pollution are costing the global economy $5.1tn annually, or roughly twice the economic output of the United Kingdom, with more than half of that burden falling on China and other developing economies in Asia. According to the World Bank Report, (2016) and IHME Report (IHME, 2016), there "welfare cost" from the toll of indoor and outdoor air pollution have soared over the years. In 1990, the welfare cost was $2.6 trillion from air pollution globally. In East Asia, China alone contributed to $2.3 trillion in 2013. An estimated 5.5 million lives were lost to diseases associated with air pollution in 2013, the year global data are available (Donnan, September 09, 2016). Exposure to air pollution increases a person's risk of illnesses such as lung cancer, stroke, heart disease and bronchitis. Those diseases and deaths from them also carry a rising economic cost.

SOLUTIONS AND RECOMMENDATIONS OF IMPACT OF ENVIRONMENTAL ISSUES ON PUBLIC HEALTH AND SAFETY

The Paris Agreement on climate change, especially on how to phase out subsidies to fossil fuels, slash carbon emissions and build a low-carbon future. Carbon dioxide (C02) emissions are the driving force behind climate change. The impact of climate change on the environment has been attributed to rise of sea levels, high temperatures, changing landscape, increased risk of drought, fire and floods, stronger storm and heat-related sicknesses and so on.

Power generates contribute more to global warming than any other industry. More than 20 per cent of the world' electricity comes from natural gas plants and 41 per cent from coal power stations that pump out twice as much carbon dioxide.

Carbon dioxide (C02) is the most prevalent long-lasting greenhouse gas. So much of it has built up in the atmosphere since fossil fuel burning began in earnest that those of us alive today are the first "homo sapiens" to breathe in air with carbon dioxide levels of 400 parts per million- up from 275 ppm before the industrial revolution. Scientists say the increase has been enough to raise global average temperatures by nearly 1oC from pre-industrial times, glaciers and pushing up sea levels. With coal so cheap and abundant, and energy so vital to development, another crop of fast-growing countries, including Vietnam and Indonesia, are driving "global coal renaissance". A coal power plant typically operates for at least 30 years, so countries racing to build them are locking in decades' worth of emissions. Gas plants are cleaner, but still produce a lot of Co2. (Clark, 2015)

In Sub-Saharan Africa and in Asian countries, according to a World Energy Outlook Special Report (2016), identified a link between energy, air pollution and health and urged the energy sector to contribute to the solution. The Agency reports that about 6.5 million deaths are attributed each year to poor air quality, making it the world's fourth largest threat to human health, behind high blood pressure, dietary risks and smoking. Key factors fuelling air pollution in Sub-Saharan Africa is poverty which has seen millions depend on biomass as energy source. About 2.7 billion depend on wood for cooking and kerosene for lightening which is blamed for around 3.5 million premature deaths each year.

Solutions to climate change is a collective action of many countries emitting carbon dioxide (C02) into atmosphere to stop it. Ratification of Paris Agreement by countries would make it enter into force. Paris Agreement will enter into force after 55 countries, which contribute at least 55% of the world's total greenhouse gas emissions, enact their instruments of ratification, acceptance, approval or accession. Data on UNFF indicates that 61 countries have ratified the Paris Agreement accounting for 47.79% of global GHG emissions.

Countries agreed at UN climate talks five years ago that temperatures should not rise by more than 2oC, an aim the Paris climate agreement is supposed to deliver. Warming above that level raises the risk of severe pervasive and irreversible damage (IPCC, 2016).

To reduce the effect of climate change, scientist have advocated for geoengineering the environment as an accompanying strategy to emission control (Bussinessday 09 March, 2 016, p.3). Geoengineering is the deliberate and large-scale intervention in earth's climatic system with the aim of reducing global warming. Climate engineering has two categories of technologies-carbon dioxide removal and solar radiation management. Carbon dioxide removal addresses a cause of climate change by removing one of the greenhouse gases from the atmosphere. Solar radiation management attempts to offset effects of greenhouse gases by causing earth to absorb less solar radiation. However, the Inter-governmental Panel on Climate Change (IPCC) Report of 2007 concluded that geoengineering options for climate change "remained largely speculative and unproven". The costs, benefits, and risks of many geoengineering approaches to climate change are not well understood.

Researchers from University of California, Los Angelis (UCLA, and five other universities evaluated various approaches to climate change mitigation-focusing on reducing emissions, sequestering carbon through biological means on land and in the ocean, storing carbon dioxide in a liquefied form in underground geological formations and wells, and increasing the earth's cloud cover and solar reflection.

None came close to reducing emissions as much as limiting the amount of carbon being pumped into the atmosphere. Other proposed strategies, such as biochar production and geological carbon capture and storage, are rated somewhat lower, but deserve further research and development.

Scientists have also shown that the reasons for climate change current changes are said to be man-made and the biggest culprit by far is burning of fossil fuels. Government policies, it has been discovered are inadvertently contributing to carbon emissions. Subsidies like $4 billion annual fossil-fuel subsidy offered by the United States Treasury as tax breaks for drillers provides less incentives to decarbonize the world. Government in Africa, Middle East and Latin America who subsidize consumers of fossil fuels pose a greater threat to the environment. According to the International Energy Agency (IEA, 2016a), a global think tank Report of 2016, some 40 countries subsidize fossil-fuel consumption, creating a huge incentive to release carbon dioxide. Their governments spend about half-trillion dollars on subsidies, about four times more than global subsidies for renewables such as solar and wind energy.

Moreover, the "practical options" advanced by oil industry practitioners-essentially increasing energy efficiencies –will not fully address the challenge posed by fossil fuels. According to a National Oceanic Atmospheric Administration (NOAA), human activity adds roughly 7 billion metric tons of carbon to the environment every year, mostly from the burning of fossil fuels indicating that reducing carbon emission as strategy will not bring much respite (Businessday 09 March 2016, p.3).

In addition, the demand for environmental regulation is overwhelming a demand to prevent harm. However, it is difficult to predict precisely what activities will cause what harm. Often the first clue that something is harmful is provided by the discovery that harm has occurred. Scientific uncertainty makes tis "dilemma of preventative regulation" particularly acute with respect to regulation of toxic substances. Faced with is pervasive uncertainty, decision-makers have increasingly relied on a two-step process for making environmental policy decisions. "Risk assessment" is the process used to characterize the environmental effects of exposure to hazardous materials or situations. "Risk management" is the process of weighting policy alternatives to select the most appropriate regulatory action, if any, for responding to these risks. It is not surprising that there are different approaches to characterizing risks. There is just as wide a divergence of opinion augmented form of cost-benefit analysis to determine what activities are worth regulating and how stringently to regulate them. Others would regulate without regard to cost considerations, based either on the technologies available for controlling risks or on purely environmental or health-based goals (Percival & Leape, 2000).

Finally, of all the different risk factors for premature deaths, the air we breathe over which individuals have little control need to be addressed. Experts say investments in cleaner energy forms have become crucial. Fossil fuels dominate the production of electricity, accounting for more than 80 per cent of total power supply. In Sub-Saharan Africa, for example, "South Africa generates almost 60% of all the power generated in Sub-Saharan Africa, derives 94% of its power from coal. The country also accounts for around 25% of total oil consumption in Sub-Saharan Africa and Nigeria for more than 20%, meaning that the remaining 40%-plus countries collectively consume less oil that the Netherlands"(Businessday 27 September, 2016, p.36). The scope and availability of renewable energy resources in Africa is significant and can be deployed across grid-based, mini-grid and off-grid structures. Potentials in hydro, solar and wind remain Africa's surest path to curb air pollution. Government regulation can drive this initiative as well drive deep reforms. Kenya has passed a law that requires new buildings to be fitted with solar water heating systems. Regulations in Senegal is promoting a switch to liquefied petroleum gas (LPG) and Ghana and Cameron, have made strong commitments to increase the share of LPG for cooking and are developing related policy measures. Nigeria is ripe for such bold reforms. Rising energy consumption, rural-urban migration and poor energy access are problems that continue to drive air pollution. Governments in Sub-Saharan Africa who have made commitments towards sustainable energy now have to back words with action to curb the menace.

FUTURE RESEARCH DIRECTIONS

Future research should focus on climate change as it affects human health and safety. Carbon dioxide C02 emissions are the driving force behind climate change. Though, the Paris agreement is a step in the right direction, but the reality is it's weak and delay action.to the next decade. What is needed by future researchers is how to slash carbon dioxide (C02) emissions and build a low-carbon future in the interest of human health, justice and safety. Other areas of future research is: researching on the relationship between environmental justice as it concerns fossil-fuel removal and its benefits to both poor and wealthy countries.

CONCLUSION

As demonstrated in this chapter, it is important to keep in mind the connections between all the environmental media (air, water and land) components and their impacts on environmental health policy. There is need to consider the connections between air, water, land and human interactions in all environmental policies. Though, environmental policies often may be crafted in a fashion that treats pollution in one medium as separable from pollution in another, in most cases the systems are tightly coupled. For example, carbon dioxide (Co_2) in the air can be absorbed by the ocean. The oceans water is a "Carbon sink" from the perspective of the air, but the absorption of Co_2 is still a pollutant for the water; it affects the acidity of the water, directly changing its habitat characteristics (Williams, 2010) and also affecting its ability to dissolve and carry metals (Lenntech, 2009). Land pollution can directly affect the water table. Additionally, when land is polluted, flooding can spread the pollution widely across the land and this causes plants to die, then soil erosion may occur. When it comes to air, water and land, we should always be on the lookout for interaction effects.

Regulations are often governmental interventions to correct environmental challenges. Pollution is one type of environmental problem. While Federal, State and City-level governments are the key actors that enact and enforce environmental rules and laws, there are also increasing role for nongovernment actors. Sometimes, governments actively collaborate with other actors (Protoski & Prakash, 2004), and sometimes they step back and allow local communities or technical nongovernmental organizations (Lim & Prakash, 2015) to address environmental challenges.

However, the foot-print in environmental concerns extends beyond the traditional pollution problems. The new challenges are found in the areas of environment justice, for example, climate change (Bowen & Wells, 2002), public health, and the sustainable use of common pool resources. These include deforestation, depletion of fisheries, biodiversity loss, the conversion of rural arable land to urban settlements, lead in drinking water due to outdated infrastructure, water conflicts, and new global diseases. Because of the complexity of each of these topics, it is useful to have a model, or mental framework for thinking about environmental policy as it affects public health issues. Such a framework is designed to help researchers and policy makers understand the extent to which environmental policy decisions are inter-wined with the natural, built and social systems of the environment. Such a holistic system-based framework can provide a mental checklist to help reduce the likelihood of neglecting key elements of environmental issues.

From this holistic system-based framework, it is fairly easy to see how the natural and concretion layers of the environment affect each other and how the social and concretion layers affect each other. Floods, hurricanes, and tornadoes remind us that the natural system affects the social system. Ensconced in our own cultures, it can be more challenging to see how the social system affects the natural system, yet our cultural assumptions and habits definitely affect the natural system. These environmental issues or elements would inform the crafting of environmental policies as the eco-system management perspective integrates the policy perspective with a holistic ecological view and a citizen's focus on environmental health issues. To increase effectiveness of public safety management, environmental Administrators and Policymakers must think outside the box of a single-medium approach to environmental policy and focus instead holistically on linkages between air, water, and land.

REFERENCES

Abubakar, S. D. (2012). *Introduction to elements of physical Geography". The Basics of Social Sciences.* Lagos: Malthouse Press Ltd.

Ang-Olson, J., Fischer, M., & Dulla, R. (2000). Land use as an air quality control measure – Review of current practice and examination of policy options. *Energy, Air Quality, and Fuels, 1738*, 33–38.

Bowen, W. M., & Wells, M. V. (2002). The Politics and Reality of Environmental Justice: A History and Considerations for Public Administrators and Policy makers. *Public Administration Review, 62*(6), 688–698. doi:10.1111/1540-6210.00251

Campbell, H. E., & Corley, E. A. (2012). *Urban Environmental Policy Analysis.* New York: M.E. Sharpe.

Clark, P. (2015). *Carbon capture: miracle machine or white elephant?* Retrieved from https://www.ft.com/content/88c187b4-5619-11e5-a28b-50226830d644

Donnan, S. (2016, September 9). World economy chokes on $5tn cost of pollution as premature deaths soar. *The Businessday*, p. A1. Retrieved from http://www.businessdayonline.com

Fluggae, F., & Schilizzi, S. (2005). Greenhouse gas abatement policies and the value of carbon sinks: Do grazing and cropping systems have different destines? *Ecological Economics, 55*(4), 584–598. doi:10.1016/j.ecolecon.2004.12.033

FootprintE. (2011). Retrieved from: http://www.sustainablescale.org/conceptualframework/understandingscale

Gober, P. (2006). Metropolitan Phoenix: Place making and Community Building in the Desert. Philadelphia, PA: University of Pennsylvania Press. doi:10.9783/9780812205824

Grimm, N.B., Grove, J.M., Pickett, S.T.A & Redman, C.L. (2000). Integrated approaches to long-term studies of urban ecological systems. *Bioscience, 50*(7), 571-584.

Howe, G. M. (1977). *A Geography of Diseases.* London: Academic Press.

Institute for Health Metrics Evaluation (IHME). (2016). Rethinking Development and Health: Findings from the Global Burden of Disease Study. Seattle, WA: IHME.

Intergovernmental Panel on Climate Change (IPCC). (2007). Climate Change 2007: Synthesis Report. Contribution of Working Groups I, II and III to the Fourth Assessment Report of the Intergovernmental Panel on Climate Change. IPCC.

International Energy Agency. (2016b). *Energy and Air Pollution: World Energy Outlook-Special Report*. Retrieved from https://www.iea.org/publications/freepublications/publication/WorldEnergyOutlookSpecialReport2016EnergyandAirPollution.pdf

International Energy Agency (IEA). (2016a). *World Energy Outlook 2016*. Retrieved from https://www.docdroid.net/IOBt86G/world-energy-outlook-2016.pdf#page=4

Kennett, S. A. (2002). National Policies for biosphere greenhouse gas management: Issues and opportunities. *Environmental Management, 30*(5), 595–608. doi:10.1007/s00267-002-2716-5 PMID:12375081

Keough, H. L., & Blahna, D. J. (2006). Achieving integrative, collaborative ecosystem management. *Conservation Biology, 20*(5), 1373–1382. doi:10.1111/j.1523-1739.2006.00445.x PMID:17002755

Kerr, S., Liu, S. G., Pfaff, A. S. P., & Hughes, R. F. (2003). Carbon dynamics and land-use choices: Building a regional-scale multi-disciplinary model. *Journal of Environmental Management, 69*(1), 25–37. doi:10.1016/S0301-4797(03)00106-3 PMID:12927149

Koontz, T. M., & Bodine, J. (2008). Implementing ecosystem management in public agencies: Lessons from the US Bureau of Land Management and the Forest Service. *Conservation Biology, 22*(1), 60–69. doi:10.1111/j.1523-1739.2007.00860.x PMID:18254853

Lenntech. (2009). *FAQ air pollution: The question library on air-related issues*. Delft, The Netherlands: Lenntech Water Treatment & Purification Holding B.V. Retrieved February 11, 2011 from http://www.lenntech.com/faq-air-pollutionhtm

Lim, S., & Prakash, A. (2004). The Regulations and Innovation: The case of 15014001. *Public Administration Review, 74*(2), 233–244. doi:10.1111/puar.12189

Ludwig, D., Mangel, M., & Haddad, B. (2001). Ecology, Conservation, and Public Policy. *Annual Review of Ecology and Systematics, 32*(1), 481–517. doi:10.1146/annurev.ecolsys.32.081501.114116

McNeil, B. I. (2006). Oceanic implications for climate change policy. *Environmental Science & Policy, 9*(7-8), 7, 8, 595–606. doi:10.1016/j.envsci.2006.08.006

Ostrom, E. (2005). *Understanding Institutional Diversity*. Princeton, NJ: Princeton University Press.

Pat, R. (2010). *Health and Environment. Trance, Ferney Voltaire*. World Medical Association.

Protoski, M., & Prakash, A. (2004). The Regulation Dilemma: Cooperation and Conflict in Environmental Governance. *Public Administration Review, 64*(2), 152–163. doi:10.1111/j.1540-6210.2004.00357.x

Pyle, G. F. (1979). *Applied Medical Geography*. Washington, DC: V. H. Winston and Sons.

Sartori, F., Lai, R., Ebinger, M. H., & Parish, D. J. (2006). Potential soil carbon sequestration and Co_2 offset by dedicated energy crops in the USA. *Critical Reviews in Plant Sciences*, *25*(5), 441–472. doi:10.1080/07352680600961021

Sperow, M., Eve, M., & Paustian, K. (2003). Potential soil C sequestration on US agricultural soils. *Climatic Change*, *57*(3), 319–339. doi:10.1023/A:1022888832630

Steel, B. S., & Weber, E. (2001). Ecosystem management, decentralization, and public opinion. *Global Environmental Change*, *11*(2), 119–131. doi:10.1016/S0959-3780(00)00062-5

Sustainable Scale Project. (2003). Retrieved from http://www.sustainablescale.org/conceptualnderstandingscale/measuringscale/EcologicalFootprint.asp

Thompson, J. R., Anderson, M. D., & Johnson, K. N. (2004). Ecosystem management across ownerships: The potential for collision with antitrust laws. *Conservation Biology*, *18*(6), 1475–1481. doi:10.1111/j.1523-1739.2004.00266.x

Uroko, C. (2016, September 27). Nigeria joins world leaders on global efforts to reverse climate change impacts. *The Businessday*, p. 36. Retrieved from http://www.businessdayonline.com

Van Maanen, J. (1983). *Qualitative Method*. London: Sage.

Williams, P. L. (2010). Ice-free Arctic Ocean may not be of much use in soaking up carbon dioxide, a component of global warming according to new study. Athens, GA: Office of Public Affairs, University of Georgia.

Wilson, E. J., Friedmann, S. J. I., & Pollak, M. F. (2007). Research for deployment: Incorporating risk, regulation, and liability for carbon capture and sequestration. *Environmental Science & Technology*, *41*(17), 5955–5952. doi:10.1021/es062272t PMID:17937265

World Health Organization. (1981). Constitution of the World Health Organization. In H. E. Caplan Jr & J. McCartney (Eds.), *Concepts of Health and Disease – Interdisciplinary Perspectives*. Reading: Addison-Wesley.

KEY TERMS AND DEFINITIONS

Adaptive Management: Adaptive management generally describes an approach to managing ecosystems in a way that recognizes their complexity and dynamism. Consequently, the management of these systems should utilize flexible experiments that allow managers and scientists to react to changes in the ecological system by adapting their management and policy schemes. Adaptive management is a process that involves incremental management steps; these steps are regularly evaluated to determine of the outcomes are desirable at various time intervals.

Ecosystem: A system refers to a collection of independent parts or events that make up a whole. Each part has a specific function, even though the expression of their roles depends on the proper functioning of all the other parts. The whole system fails to function without input from outside the system which acts to produce some output. Ecosystems combine biological communities and the physical environments associated with them in which nutrients and energy move through in loops or cycles. In other words, ecosystem is the interacting system of a biological community and its environmental surroundings.

Ecosystem Management: Ecosystem management integrates scientific knowledge of ecological relationships within a complex socio-political and values framework toward the general goal of protecting native ecosystem integrity over the long term.

Environmental Justice: This connotes the extent of environmental injustice in the placement of "disamenities" (or environmental bads). It is widely accepted that disamenity placement –whether in land, air, or water-should not be affected by race or ethnicity. The ultimate goal of environmental policy is to improve society and policy makers should concern itself with justice irrespective of race or ethnicity.

Environmental Media (Air, Water, and Land) Interaction: This connotes three environmental media of air, water, and land. For example, air quality can have an impact on land use and land health; poor air quality can hinder the growth and survival of plants, animals, and humans that live on (and in) the land in urban areas. In fact, some urban regions have employed land-use restrictions as a way to control air quality.

Habitat: The place where a population (human, animal, plant, microorganism) lives and its surroundings, both living and nonliving.

Public Safety: This connotes public health issues within the environment that emerged because of the interaction between air, water, and human activities on the environment and their impacts on human and ecosystem health issues.

Compilation of References

Abah, J., Mashebe, P., & Denuga, D. D. (2015). Prospect of Integrating African Indigenous Knowledge Systems into the Teaching of Sciences in Africa. *American Journal of Educational Research*, *3*(6), 668–673.

Abdulrahman, A., & Zuwaira, H. R. (2016). Effects of Insecurity on the Internally Displaced Persons in Northern Nigeria: Prognosis and Diagnosis. Global Journal of Human Social Science: F Political Science, 16(1), 3-7.

Abiodun, O. E., Olaleye, J. B., Dokai, A. N., & Odunaiya, A. K. (2011). Land Use Change Analyses in Lagos State: From 1984-2005. FIG Working week 2011, Bridging the Gap between Cultures, Marrakech, Morocco.

Abiodun, B. J., Pal, J. S., Afiesimama, E. A., Gutowski, W. J., & Adedoyin, A. (2008). Simulation of West African monsoon using RegCM3 Part II: Impacts of deforestation and desertification. *Theoretical and Applied Climatology*, *93*(3-4), 245–261. doi:10.1007/s00704-007-0333-1

Abubakar, S. D. (2012). *Introduction to elements of physical Geography". The Basics of Social Sciences*. Lagos: Malthouse Press Ltd.

Acemoglu, D., & Robinson, J. A. (2012). *Why Nations Fail: The Origin of Power, Prosperity and Poverty*. Profile Books.

Achebe, C. (1983). *The trouble with Nigeria*. Enugu: Fourth Dimension Publishers.

Adamczyk, A. (2014). *Ukrainian immigration to Poland during the political crisis in Ukraine*. Poznań, Poland: Adam Mickiewicz University. doi:10.14746/ssp.2014.3.2

Adebowale, S. (2013). *Boko Haram: Security, Humanitarian Conditions Deteriorating UNHCR*. Retrieved from http// www.theeagleonline.com.ng/news/boko-haramsecurity-humanitarian-conditions-deterioratingunhrc/

Adedayo, V., Fasona, M., Olorunfemi, F., Elias, P., & Oloukoi, G. (2013). An Evaluation of Policy on Natural Resource Management and Its Impacts on Rural Livelihoods in Nigeria. *The Nigerian Journal of Business and Social Studies*, *7*(1), 49–71.

Adedayo, V., Fasona, M., Olorunfemi, F., Elias, P., & Oloukoi, G. (2013). An Evaluation of Policy on Natural Resource Management and Its Impacts on Rural Livelihoods in Nigeria. The Nigerian. *The Journal of Business and Social Studies*, *7*(1), 49–71.

Adejobi, S., & Akhakpe, I. (2012). Urbanization process and climate change: Lagos State, Nigeria Experience. *International Journal of Social Science and Humanities Review*, *3*(1).

Adelegan, J. A. (2005). The history of environmental policy and pollution of water sources in Nigeria (1960-2004). The Way Forward, 3.

Adelekan, I. (2011). Vulnerability assessment of an urban flood in Nigeria: Abeokuta flood 2007. *Natural Hazards*, *56*(1), 215–231. doi:10.1007/s11069-010-9564-z

Adeniyi, P. O. (2002). Guest Statement on "Change, Turbulence and Uncertainty in Nigeria". In Resource and Environmental Management (2nd ed.). Prentice Hall.

Adeniyi, P. O., Omojola, A. S., & Soneye, A. S. O. (1992) Mapping Landuse – Landcover Resources of Wamako Local Government Area, Sokoto State, Nigeria With SPOT Satellite Imaging System, *Proceedings of 35th Annual Conference of N.G.A.*

Adeola, F. (2000). Endangered community, enduring people: Toxic contamination, health, and adaptive responses in a local context. *Environment and Behavior, 32*(2), 209–249. doi:10.1177/00139160021972504

Aderogba, K. A, (2012b, October). Qualitative Study of recent floods and sustainanble growth and development of cities and towns in Nigeria. *International Journal of Basic and Applied Sciences*, 200-216.

Aderogba, K. A. (2012a). Global warming and challenges of flood in Lagos metropolis *Nigerian. Academic Research International, 3*(1), 448–468.

Aderogba, K. A. (2012a). Global warming and challenges of flood in Lagos Metropolis, Nigeria. *Academic Research International, 2*(1), 448–468.

Adewuyi, A. O. (2001, November). The implications of crude oil exploitation and export on the environment and level of economic growth and development in Nigeria. Paper Presented at *Nigerian Economic Society Annual Conference.*

Adilieje, C. (2011). Public policy and environmental management in Nigeria. *The Constitution: A Journal of Constitutional Development, 11*(1).

Afiesimama, E. A., Pal, J. S., Abiodun, B. J., Gutowski, W. J. Jr, & Adedoyin, A. (2006). Simulation of West African monsoon using the RegCM3.Part I: Model validation and interannual variability. *Theoretical and Applied Climatology, 86*(1-4), 23–37. doi:10.1007/s00704-005-0202-8

African Natural Resource Center-African Development Bank (ANRC-AfDB). (2016). *Review of land tenure policy, institutional and administrative systems of Botswana – Case study. African Development Bank.* Abidjan: Cote d'Ivore.

African Union Commission, Economic Commission of Africa, African Development Bank (AUC-ECA-AfDB) Consortium. (2011). Land policy in Africa: West Africa Regional Assessment. Addis Ababa: ECA Publications and Conference Management Section.

Agabi, C. E., & Wilson, J. P. (2005). Trauma, PTSD, and resilience: A review of literature. *Trauma, Violence & Abuse, 6*(3), 195–216. doi:10.1177/1524838005277438 PMID:16237155

Agbola, B. O., Ajayi, O., Taiwo, O. J., & Wahab, B. W. (2012). Anthropogenic causes and consequences. *International Journal of Disaster Risk Science, 3*(4), 207–217. doi:10.1007/s13753-012-0021-3

Agbonkhese, O., Agbonkhese, E. G., Aka, E. O., Joe-Abaya, J., Ocholi, M., & Adekunle, A. (2014). Flood Menace in Nigeria: Impacts, Remedial and Management Strategies. *Civil and Environmental Research, 6*(4), 32–40.

Agranoff, R. (2011). Collaborative Public Agencies in the Network Era. In *In the State of Public Administration: Issues, challenges and opportunities*. New York: M.E. Sharpe.

Agrawal, A. G., Yadama, R., & Bata, C. A. (2006). *Decentralization and Environmental Conservation: Gender effects from participation in joint forest management.* Washington, DC: US Agency for International Development Bureau for Africa.

Agrawal, A., & Gibson, C. C. (1999). Enchantment and Disenchantment: The Role of Community in Natural Resource Conservation. *World Development, 27*(4), 629–649. doi:10.1016/S0305-750X(98)00161-2

Aina, E. O. (1992) The Journey So far. In The Making of the Nigerian Environmental Policy (pp. 17-25). Ibadan: Ibadan University Press.

Airoldi, L., & Beck, M. W. (2007). Loss, Status and Trends for Coastal Marine Habitats of Europe. *Oceanography and Marine Biology - an Annual Review*, 387.

Ajani, J. (2012). Rage of nature: Flood ravages communities across Nigeria. *Vanguard.* Retrieved October 28, 2016, from http://www.vanguardngr.com/2012/10/rage-of-nature-flood-ravages-communities-across-nigeria/

Ajayi, D. D., & Ikporukpo, C. O. (2005). An Analysis of Nigeria's environmental vision. *Journal of Environmental Policy and Planning*, *7*(4), 341–365. doi:10.1080/15239080500441137

Ajugwo, A. (2013). Negative effects of gas flaring: The Nigerian experience. *J. Environ. Pollut. Hum. Health*, *1*, 6–8.

Akachuku, A. (2006). *Disappearing Forests, the consequences and the challenges of Sustainable development of Nigeria.* Proc. of Forestry Ass. of Nigeria.

Akali, K. (2007). The contribution of Environmental Right Action (ERA) in the struggle for ecological rights in the Niger-Delta. *The Constitution: A Journal of Constitutional Development*, *7*(1).

Ake, C. (Ed.). (1985). *Political Economy of Nigeria.* London: Longman.

Akhakpe, I. (2008). Food crisis, climate change and human development in Nigeria: Challenges and the Way Forward. *The Nigerian Journal of Energy and Environmental Economies*, *1*(2).

Akhakpe, I.B. (2016). Public policy and sustainable national development in Nigeria's Fourth Republic: challenges and possibilities. *International Journal of Management Perspectives, 13*(1).

Akinsola, H. A. (2006). *A – Z of community health in medical, nursing and health education practice.* Ibadan: College Press Publishers Limited.

Akintola, F. (1982). *Flooding problem at Ibadan.* Ibadan, Nigeria: Department of Geography, University of Ibadan.

Akowonjo, D. (2011). Ethno-religious Conflicts and National Security in Nigeria. In T. O. Ogundipe, A. Akinade, & M. Omoegun (Eds.), *Religious cum Ethnic Violence in Nigeria: Security and Socio-economic Perspectives.* Lagos: University of Lagos Institute of Security.

Akpalu, W., Arudt, C., & Matshe, I. (2015). Introduction to the Special Issue on Economics of Climate Change Impacts on Developing Countries: Selected studies on Sub-saharan Africa and South-East Asia. *Sustainability*, *7*(9), 12122–12126. doi:10.3390/su70912122

Akten, S., & Akten, M. (2014). Doğal Kaynak Yönetimi İçin Sürdürülebilir İndikatörler ve İlkeler, II. Ulusal Akdeniz Orman Ve Çevre Sempozyumu, 22-24 Ekim 2014, Isparta.

Alexander, D. (2000). *Confronting Catastrophe: New Perspectives on Natural Disasters.* New York: Oxford University Press.

Alkassim, B. (2013, December 17). NEMA: 16, 470 Nigerians in displaced persons camps. *Daily Trust Newspaper*, p. 10.

Allen, F. (2010). *Implementation of Oil-Related Environmental Policy in Nigeria: Government Inertia and Conflict in ...ger Delta* (Unpublished PhD Thesis). University of Kwazulu Netal.

...ii, S. (2016). Internal Displacement in Nigeria and the Case for Human Rights Protection of Displaced ...w. *Policy and Globalization*, *51*(2), 26–33.

Aloni, C., Daminabo, I., Alexander, B. C., & Bakpo, M. T. (2015). The Importance of Stakeholders Involvement in Environmental Impact Assessment. *Resources and Environment*, *5*(5), 146–151.

Alphonsa, A., & Ravi, G. (2016). Earthquake early warning system by IOT using Wireless sensor networks. *Proc. IEEE International Conference on Wireless Communications, Signal Processing and Networking (WiSPNET)*. doi:10.1109/WiSPNET.2016.7566327

Aluko, O. (2004). *Housing and Urban Development in Nigeria*. Kins Publications Series.

Aluko, O. (2011). Functionality of the Town Planning Authorities in Effecting Urban and Regional Planning Laws and Control in Nigeria: The Case of Lagos State. *African Research Review*, *5*(6), 156–171. doi:10.4314/afrrev.v5i6.14

Amadi, A. (2012). *Brief description of the communities threatened by coastal erosion and displacement*. Retrieved from https://ng.boell.org/sites/default/files/uploads/2012/12/eko_atlantic_project_-_opinion_papers.pdf

Amadi, L. (2013). Climate Change, Peasantry and Rural Food Production Decline in the Niger Delta Region: A case of the 2012 flood disaster. *J. Agric. Crop Res.*, *1*(6), 94-103.

Amadi, L., & Ogonor, M. (2015). Climate Change, Environmental Security and Displacement in Nigeria: Experience From the Niger Delta. *African Journal of Environmental Science and Technology*, *1*(9), 53–64.

American Psychological Association (2014). *The road to resilience*. Retrieved from apa.org/helpcenter/road-resilience.aspx

Amokaye, G.O. (2012). Environmental pollution and challenges of environmental governance in Nigeria. *British Journal of Arts and Social Sciences, 10*(1).

Amue, G. J., & Igwe, S. R. (2016). ICT resource management and sustainable development: An inquiry into the causes of Sustainable development. In A. Nkemnabe (Ed.), *Resource management for sustainable development. Awka: SCOA Heritage Nigeria Ltd.*

Amujiri, B. A. (2001). Urban Planning in Nigeria: Problems and Prospects. In *Issues in Urbanization and Urban Administration in Nigeria*. Enugu: Jamoe Enterprises.

Anago, I. (2001). *Environmental assessment as a tool for sustainable development: The Nigerian experience*. Retrieved October28,2016, from,www.jepplaw.net/page

Anderson, D. A. (2013). *Environmental Economics and Natural Resource Management*. Routledge.

Anfara, V. A., Brown, K. M., & Mangione, T. L. (2002). Qualitative analysis on stage: Making the research process more public. *Educational Researcher*, *31*(7), 28–38. doi:10.3102/0013189X031007028

Ang-Olson, J., Fischer, M., & Dulla, R. (2000). Land use as an air quality control measure – Review of current practice and examination of policy options. *Energy, Air Quality, and Fuels*, *1738*, 33–38.

Annan, K. (2005). *Towards Development, Security and Human Rights for All*. Available at www.ohchr.org/Documents/Publications/A.59.2005.Add.3.pdf

Anthoff, D., Nicholls, R. J., & Tol, R. S. J. (2001). *Global sea-level rise and equity weighting*. Working Paper FNU – 136.

Anthony, R. N., & Govindarajan, V. (2004). *Management Control Systems*. McGraw- Hill Inc.

Antonov, J. I., Levitus, S., & Boyer, T. P. (2005). Thermosteric sea level rise, 1955–2003. *Geophysical Research Letters*, *32*(12), L12602. doi:10.1029/2005GL023112

Anyanwu, C. N. (1999). *Introduction to Community Development*. Ibadan: Gabesther Educational Publishers.

Anyanwu, D., & Tanee, F. (2008). Tolerance of Cassava (Var. TMS 30572) to different concentrations of Post-planting cruse Oil Pollution. *Nigeria Journal of Botany*, *21*(1), 203–207.

Appiah-Opoku, S. (2007). Indigenous beliefs and environmental stewardship: A rural Ghana experience. *Journal of Cultural Geography*, *22*(2), 79–88. doi:10.1080/08873630709478212

Arat, Z. (1995). Çevre Yönetimi ve Planlama, *Yeni Türkiye Dergisi*, 5.

Arı, Y. & Soykan, A. (2006). Kazdaği Milli Parki'nda Kültürel Ekoloji ve Doğa Koruma. *Türk Coğrafya Dergisi*, (44), 11-32.

Aribigbe, A. (2007). *Urban Land Use Planning, Policies and Management in Sub-Saharan African Countries: Empirical Evidence From Akure, Nigeria*. Fourth Urban Research Symposium, 2007. Retrieved March 9th, 2017, sitesresources.worldbank.org/INTURBANDEVELOPMENT/

Armitage, D. (2005). Adaptive Capacity and Community-Based Natural Resource Management. *Environmental Management*, *35*(6), 703–715. doi:10.1007/s00267-004-0076-z PMID:15940398

Arndt, C., Asante, F., & Thurlow, J. (2015). Implications of Climate change for Ghana's Economy. *Sustainability*, *7*(6), 7214–7231. doi:10.3390/su7067214

Ashton-Jones, N. (19998). *The Human Ecology of the Niger Delta. An ERA Handbook*. Environmental Right Action (ERA).

Ashtons J. (2003). Covering ethnic conflicts: An interview with Silvia Poggiol. *Politics*, *8*(4), 34-47.

Asnani, P. U. (2006). *Solid Waste Management*. Retrieved from http://www.iitk.ac.in/3inetwork/html/reports/IIR2006/Solid_Waste.pdf

Atilola, O. (2010). *Land Administration Reform Nigeria: Issues and Prospects*. FIG Congress, 2010: Facing the Challenges-Building the Capacity, Sydney, Australia. Retrieved March 9th, 2017, fine.net/pub/fig2010/papers/invo4/invo4_atilola_4758

Atkinson, P. A., Martin, C. R., & Rankin, J. (2009). Resilience revisited. *Journal of Psychiatric and Mental Health Nursing*, *16*(2), 137–145. doi:10.1111/j.1365-2850.2008.01341.x PMID:19281544

Attoh, F. (2016). The Gender Dimensions of Terrorism in Nigeria: Instrumentalization of Borno Women as Gender Based Violence. Kogi Journal of Sociology, 1(3), 59-69.

Attri, S.D., & Tyagi, A. (2010). *Climate Profile of India*. New Delhi: Ministry of Earth Sciences. Retrieved from http://www.indiaenvironmentportal.org.in/files/climate_profile.pdf

Aubel, J., & Samba, K. (1996). Community participation lessons on sustainability for community health projects. *Journal of International Health Development*, *17*(1), 54–63.

Automative Research Association of India. (2007). *Draft report on "Emission Factor development for Indian Vehicles" as a part of Ambient Air Quality Monitoring and Emission Source Apportionment Studies*. Pune: Automative Research Association of India. Retrieved from http://www.cpcb.nic.in/Emission_Factors_Vehicles.pdf

Awobajo, S. A. (1981, November). *An analysis of oil spill in Nigeria 1976-1980*. Paper presented at the seminar on the Petroleum Industry and Nigerian Environment organized by NNPC, Warri.

Aworawo, D. (1999). The impact of environmental degradation on the rural economy of The Niger Delta. In A. Osuntokun (Ed.), *Environmental problems of the Niger Delta*. Lagos: Friedrich Ebert Foundation.

Aye, J. R. (2008). *Reforming the Africa public sector retrospect and prospects*. Dakar: Codesria Green Book.

Ayeni, Kapangaziwiri, Soneye, & Engelbrecht. (2015). Assessing the impact of global changes on the surface water resources of Southwestern Nigeria. *Hydrological Sciences Journal, 60*(11-12), 1956 – 1971.

Ayeni, A. (2010). Sustainable development through international cooperation in research and development. In L. Popoola & O. Olorunmisola (Eds.), *Contemporary issues in sustainable development: lessons for, and challenges to Nigeria.* Ibadan: University of Ibadan.

Ayeni, A. O., Cho, M. A., Mathieu, R., & Adegoke, J. O. (2016). The local experts' perception of environmental change and its impacts on surface water in Southwestern Nigeria. *Environmental Development, 17*, 33–47. doi:10.1016/j.envdev.2015.09.007

Ayeni, A. O., & Olorunfemi, F. B. (2014). Reflections on Environmental Security, Indigenous Knowledge and the Implications for Sustainable Development in Nigeria. *Journal of Research in National Development, 12*(1), 46–57.

Ayeni, A., Soneye, A., & Badru, F. (2014). Adaptation to Water Stress in Nigeria Derived Savanna Area: The Indigenous Knowledge and Socio-Cultural Nexus of Management and Humanitarian Services. *Journal of Management Policies and Practices, 15*(3), 78–87.

Ayoola, T. J. (2011). Gas flaring and its implication for environmental accounting in Nigeria. *Journal of Sustainable Development, 4*(5).

Azzone, G., Manzini, R., & Noci, G. (1996). Evolutionary trends in environmental reporting. *Business Strategy and the Environment, 5*(4), 219–230. doi:10.1002/(SICI)1099-0836(199612)5:4<219::AID-BSE69>3.0.CO;2-K

Baki, B. & Cengiz, E., (2002). Toplam Kalite Çevre Yönetimi. *Uludağ Üniversitesi, İktisadi ve İdari Bilimler Fakültesi Dergisi, 21*(1).

Balicki, J., & Chamarczuk, M. (2013). Wokol problematyki migracyjnej: Kultura przyjecia [Around migration policy: The culture of reception]. Uniwersytet Kardynała Stefana Wyszyńskiego, Międzyinstytutowy Zakład Badań nad Migracją.

Barnett, J. (1997). Environmental Security: Now What? In *Seminar proceedings*. Department of International Relations, Keele University.

Barrow, F. H., Armstrong, M. I., Vargo, A., & Boothroyd, R. A. (2007). Understanding the findings of resilience-related research for fostering the development of African American Adolescents. *Child Adolescent Clinics of North America, 16*(2), 393–413. doi:10.1016/j.chc.2006.12.004 PMID:17349515

Bartlett, A. A. (1997). *Environmental Sustainability*. Paper presented at the Annual Meeting of the American Association of Physics Teachers Meeting, Denver, CO. Retrieved from http://www.albartlett.org/articles/art1997aug16.html

Basedau, M. (2005). *Context Matters – Rethinking the Resource Curse in Sub-Saharan Africa.* German Overseas Institute (DÜI) Responsible Unit: Institute of African Affairs. Working Papers Global and Area Studies.

Bashir, O., Oludare, H., Johnson, O., & Aloysius, B. (2012). Floods of fury in Nigerian cities. *Journal of Sustainable Development, 5*(7), 69–79. doi:10.5539/jsd.v5n7p69

Bass, S., Dalal-Clayton, B., & Pretty, J. (1995). Participation in Strategies for Sustainable Development. *Environmental Planning Issues, 7*, 118. Retrieved from http://pubs.iied.org/pdfs/7754IIED.pdf

Bassey, N. (2009, April). *Gas flaring: Assaulting communities, jeopardizing the world.* Paper presented at the National Environmental Consultation hosted by the Environmental Rights Action in conjunction with the Federal Ministry of Environment at Reiz Hotel, Abuja.

Bates, P. D., & De Roo, A. P. J. (2000). A simple rasterbased model for flood inundation simulation. *Journal of Hydrology (Amsterdam), 236*(1-2), 54–77. doi:10.1016/S0022-1694(00)00278-X

Batwel, A. (2008). *Factors affecting community participation in primary education development projects in Makete District of Tanzania* (MA dissertation). Sokoine University of Agriculture, Morogoro, Tanzania.

Beck, J. T., & Bozeman, B. (2007). Public values: An Inventory. *Administration & Society, 39*(3), 354–381. doi:10.1177/0095399707300703

Behnken, J. A. (2013). *Natural Resource Management Knows No Bounds: A Case Study Of The Cache River Joint Venture Partnership* (Unpublished master dissertation). Southern Illinois University, Carbondale, IL.

Bekele, H. (2005). *Urbanization and Urban Sprawl.* Department of Infrastructure, Kungliga Tekniska Hogskolan.

Bell, M., & Davis, D. L. (2001). Reassessment of the lethal London Fog of 1952: Novel indicators of acute and chronic consequences of acute exposure to air pollution. *Environmental Health Perspectives, 109*(s3), 389–394. doi:10.1289/ehp.01109s3389 PMID:11427388

Benedyczak, J. (2014). *Polska granica w czasie rewolucji ukrainskiej* [Polish border during the Ukrainian revolution]. Warsaw, Poland: Fundacja im. Stefana Batorego. Retrieved from http://www.batory.org.pl/upload/files/Programy%20operacyjne/Otwarta%20Europa/Granica_W_czasie_kryzysu.pdf

Benkhelifa, I., Nouali-Taboudjemat, N., & Moussaoui, S. (2014, May). Disaster management projects using wireless sensor networks: An overview. In *Advanced Information Networking and Applications Workshops (WAINA), 2014 28th International Conference on* (pp. 605-610). IEEE. doi:10.1109/WAINA.2014.99

Berg, C. (2015). *Evaluating Collaborative Natural Resource Management in The Ace Basin National Estuarine Research Reserve* (Unpublished master dissertation). The University of Charleston.

Berkes, F., Colding, J., & Folke, C. (2000). Rediscovery of traditional ecological knowledge as adaptive management. *Ecological Applications, 10*(5), 1251–1262. doi:10.1890/1051-0761(2000)010[1251:ROTEKA]2.0.CO;2

Berkes, F., & Folke, C. (Eds.). (1998). *Linking Social and Ecological Systems. Management Practices and Social Mechanisms for Building Resilience.* Cambridge, UK: Cambridge University Press.

Bernstein, A. B., & Sweeney, M. H. (2012). Public Health Surveillance Data: Legal, Policy, Ethical, Regulatory, and Practical Issues. *Morbidity and Mortality Weekly Report, 61*(03), 30–34. PMID:22832995

Betts, A. (2009). *Forced migration and global politics.* Malden, MA: Wiley-Blackwell. doi:10.1002/9781444315868

Bevir, M. (2006). Democratic governance: Systems and radical perspective. *Public Administration Review, 66*(3), 426–436. doi:10.1111/j.1540-6210.2006.00599.x

Biersack, A. (2006). Reimagining political ecology: culture/power/history/nature. In A. Biersack & J. Greenberg (Eds.), *Reimagining political ecology* (pp. 3–40). Durham, NC: Duke University Press. doi:10.1215/9780822388142-001

Blaikie, P., & Brookfield, H. (1987). *Land Degradation and Society.* London: Methuen.

Bluestone, J. R. (2010). *Why the earth shakes: Pre-modern understandings and modern earthquake science* (PhD thesis). Faculty of the graduate school, University of Minnesota.

Boahen, A. (1985). *General History of Africa: Africa under Colonial Domination 1880-1935.* UNESCO.

Boateng, O. (2010). Carving Up Africa: 125 years of the Berlin Conference. *New African*, 492.

Bocharova, O., & Giza, T. (2015). Ludzie w sytuacji granicznej – cazus Donbasu [People in borderland situation- the case of Donbas]. In S. Jakonowska (Ed.), *Czlowiek w sytuacji granicznej. Filozoficzne, kulturowe I historyczne wymiary refleksji I jej implikacje pedagogiczne* (pp. 234–248). Radom, Poland: Wydawnictwo Naukowe Instytutu Technologii I Eksploatacji.

Bodungen, B., & Turner, R. K. (2001). Science and Integrated Coastal Management, An Introduction. In *Science and Integrated Coastal Management* (pp. 1–14). Berlin, Germany: Dahlem University Press.

Bogardi, J., & Brauch, H. (2005). Global Environmental Change: A Challenge for Human Security Defining and Conceptualising the Environment Dimension of Human Security. In A. Rechkemmer (Ed.), *UNEO-Towards an International Enviornmental Organization-Approaches to a Sustainable Reform of Global Environmental Governance* (pp. 85–109). Baden-Baden, Germany: Nomos.

Boin, A., & Hart, P. (2003). Public Leadership in times of crisis: Mission impossible? *Public Administration Review, 63*(5), 544–553. doi:10.1111/1540-6210.00318

Bonanno, G. A. (2004). Loss, trauma and human resilience: Have we underestimated the human capacity to thrive after extremely aversive events? *The American Psychologist, 59*(1), 20–28. doi:10.1037/0003-066X.59.1.20 PMID:14736317

Bonanno, G. A., Moscowitz, J. T., Papa, A., & Folkman, S. (2005). Resilience to loss in bereaved spouses, bereaved parents, and bereaved gay men. *Journal of Personality and Social Psychology, 88*(5), 827–843. doi:10.1037/0022-3514.88.5.827 PMID:15898878

Bonanno, G. A., Rennicke, C., & Dekel, S. (2005). Self enhancement among high-exposure survivors of the September 11th terrorist attack: Resilience or social maladjustment? *Journal of Personality and Social Psychology, 88*(6), 984–998. doi:10.1037/0022-3514.88.6.984 PMID:15982117

Bonanno, G., Galea, S., Bucciarelli, A., & Vlahov, D. (2007). What predicts psychological resilience after disaster: The roll of demographics, resources and life stress. *Journal of Consulting and Clinical Psychology, 75*(5), 671–682. doi:10.1037/0022-006X.75.5.671 PMID:17907849

Booth, K. (1991). Security and Emancipation. *Review of International Studies, 17*(: 4), 319.

Border Guard. (2017). *Headquarters of the Border Guard.* Retrieved from https://www.strazgraniczna.pl/

Borrini-Feyerabend, G. (1996). *Collaborative management of protected areas: Tailoring the approach to the context*. Gland.

Botes, L., & van Rensburg, D. (2000). Community participation in development: Nine plagues and twelve commandments. *Community Development Journal: An International Forum, 35*(1), 41–58. doi:10.1093/cdj/35.1.41

Botkin, D., & Keller, E. (1995). *Environmental Science*. John Wiley & Sons.

Bowen, W. M., & Wells, M. V. (2002). The Politics and Reality of Environmental Justice: A History and Considerations for Public Administrators and Policy makers. *Public Administration Review, 62*(6), 688–698. doi:10.1111/1540-6210.00251

Bozeman, B., & Moulton, S. (2011). Integrative Publiciness: A Framework for Public Management Strategy and Performance. *Journal of Public Administration Research and Theory, 21*(3), i363-i380.

Bozeman, B. (2007). *Public Values and Public interests: Counter-balancing Economics individualism.* Washington, DC: Georgetown University Press.

Brauch, H. (2005). *Environment and Human Security. Towards Freedom from Hazard Impacts, UNU-EHS, Intersection 2*. Bonn, Germany: UNU-EHS.

Brauch, H., Oswald, U., Grin, J., Mesjasz, C., & Kameri-Mbote, P et al. (Eds.). (2008). *Facing Global Environmental Change: Environmental, Human, Energy, Food, Health and Water security Concepts*. Berlin: Springer-Verlag.

Bray, M. (1996). *Decentralization of Education: Community financing*. Washington, DC: World Bank. doi:10.1596/0-8213-3724-6

Brian, T., & Laczko, B. (2014). *Fatal journeys: Tracking lives lost during migration* [Report]. Geneva, Switzerland: International Organization for Migration.

Bridgewater, P. B. (1993). Landscape Ecology, Geographic Informations Systems and Nature Conservation. In R. H. Young, D. R. Gren, & S. Cousins (Eds.), Landscape Ecology and GIS (pp. 23–36). Academic Press.

Bronfenbrenner, U. (1995). Developmental ecology through space and time: A future perspective. In P. Moen, G. H. Elder Jr, & K. Luscher (Eds.), *Examining lives in context: Perspectives on the ecology of human development* (pp. 619–647). Washington, DC: American Psychological Association. doi:10.1037/10176-018

Brooks, N., Nicholls, R. J., & Hall, J. (2006). *Sea Level Rise: Coastal Impacts and Responses*. Berlin: WBGU.

Brown, L., & Jacobson, J. (1986). *Our Demographically Divided World*. Worldwatch Paper 74. Washington, DC: Worldwatch Institute.

Brown, L. (1977). *Redefining National Security*. Worldwatth Inst.

Brown, N. (1989). Climate, ecology and international security. *Survival, 31*(6), 519–532. doi:10.1080/00396338908442497

Bruce, E. (2003). Nigeria: The Ogoni: Oil, Blood, and the Death of a Homeland. An Encyclopedia of Indigenous People and Environmental Issues, 1-11.

Bruun, P. (1988). The Bruun Rule of Erosion by Sea-Level Rise: A Discussion on Large Scale Two- and Three Dimensional Usages. *Journal of Coastal Research, 4*(4), 627–648.

Bruun, P. M. (1962). Article. *Journal of Waterways and Harbor Division, 88*(WW1), 117–130.

Bryant, R. (2011). How to manage the psychological effect of natural disasters. *The Conversation*. Retrieved from http://theconversation.com/profiles/richard-bryant-161

Bryant, R. (1998). Power, knowledge and political ecology in the third world: A review. *Progress in Physical Geography, 22*(1), 79–94. doi:10.1177/030913339802200104

Bryant, R., & Bailey, S. (1997). *Third world political ecology: an introduction*. London: Routledge.

Bryson, J., & Crosby, B. (1992). *Leadership for the Common Good*. San Francisco: Jossey – Bass.

Burke, C. S., Sims, A. E., Lazzara, E. H., & Salas, E. (2007). Trust in Leadership: A multi-level review and integration. *The Leadership Quarterly, 18*(6), 606–632. doi:10.1016/j.leaqua.2007.09.006

Burns, J. M. (1978). *Leadership*. New York: Harper & Row Publishers Inc.

Burton, I., Kates, R. W., & White, G. F. (1978). *The environment as hazard*. New York: Oxford University Press.

Butler, S. (2012). *Nigerian Land Markets and the Land Use Law of 1978*. World Resources Institute & Landesa.

Butler, D. (2012). Focusing events in the early twentieth century: A hurricane, two earthquakes, and a pandemic. In C. Rubin (Ed.), *Emergency Management: The American Experience 1900-2010* (pp. 13–50). Boca Raton, FL: CRC Press. doi:10.1201/b11887-3

Buzan, B. (1991). New Patterns of Global Security in the Twenty-First Century. *International Affairs, 67*(3), 431-451.

Buzan, B., Waever, O., & Wilde, J. (1997). *Security a New Framework for Analysis.* Lynne Rienner Publishers.

Bynander, F., Chmielewski, P., & Simons, G. (2008). *The politics of crisis management in transitional Poland from 1990-1999.* Vällingby, Sweden: Swedish National Defense College.

Çağlar, İ. (1998). Çevre Yönetimi. *Çorum Çevre Dergisi*, 5.

Cain, J. (2001). *Planning improvements in natural resources management; guidelines for using Bayesian networks to manage development projects.* Wallingford, UK: Institute of Hydrology.

Campbell, B., Sayer, J. A., Frost, P., Vermeulen, S., Ruiz-Pérez, M., Cunningham, A., & Prabhu, R. (2001). Assessing the performance of natural resource systems. *Conservation Ecology, 5*(2), 22. Retrieved from http://www.consecol.org/vol5/iss2/art22/

Campbell, H. E., & Corley, E. A. (2012). *Urban Environmental Policy Analysis.* New York: M.E. Sharpe.

Campese, J., Sunderland, T., Greiber, T., & Oviedo, G. (2009). *Rights-based approaches Exploring issues and opportunities for conservation.* IUCN-CIFOR. Retrieved from www.cifor.org/publications/pdf_files/Books/BSunderland0901.pdf

Canadian Centre for Occupational Health and Safety. (2017). *What makes chemicals poisonous?* Retrieved from http://www.ccohs.ca/oshanswers/chemicals/poisonou.htmal

Capra, F., & Luisi, P. L. (2014). *The systems view of life: A unifying vision* [Kindle]. New York, NY: Cambridge University Press. doi:10.1017/CBO9780511895555

Carney, D. (1998). *Sustainable Rural Livelihoods. What contribution can we make?* London: Department for International Development.

Carter, W. N. (1991). Disaster management: A disaster manager's handbook. In *Disaster management: A disaster manager's handbook.* ADB.

CBN. (2015). *Central Bank of Nigeria Draft Annual Report 2015.* Available at https://www.cbn.gov.ng/Documents/cbnannualreports.asp

Centre for Environmental Economics and Policy in Africa (CEEPA). (2006). Climate Change and African Agriculture. *The economic impacts of climate change on agriculture in Zambia.* Discussion Paper No. 27, CEEPA, University of Pretoria.

Centrum Badan Opinii Spolecznej (2014). Stosunek do protestow na Ukrainie (21).

CGT_NetRaising_8Earthquakeconclusion. (n.d.). Retrieved from https://www.rgs.org/.../CGT_NetRaising_8Earthquakeconclusion.pdf

Chacon, R. (2012). Conservation or Resource Maximization? Analysing Subsistence Hunting Among the Achuar (Shiwiar) of Ecuador. In R. Chacon & R. Mandoza (Eds.), *The Ethics of Anthropology and Amerindian Research: Reporting on Environmental Degradation and Warfare* (pp. 311–360). New York: Springer. doi:10.1007/978-1-4614-1065-2_13

Chalecki, E. (2002). *Environmental Security: A Case Study of Climate Change.* Pacific Institute for Studies in Development, Environment, and Security. Retrieved from www.pacinst.org

Challinor, A. J., Ewert, F., Arnold, S., Simelton, E., & Fraser, E. (2009). Crops and climate change: Progress, trends, and challenges in simulating impacts and informing adaptation. *Journal of Experimental Botany, 60*(10), 2775–2789. doi:10.1093/jxb/erp062 PMID:19289578

Chambers, R., & Conway, G. (1991). *Sustainable Rural Livelihoods: Practical Concepts for the 21st Century.* IDS discussion paper 296.

Chandan, J. S. (2003). *Management Theory and Practice.* New Delhi: Vikas Ltd.

Chapin, F. S. III, Randerson, J. T., McGuire, D. A., Foley, J. A., & Field, C. B. (2008). Changing feedbacks in the climate–biosphere system. *Frontiers in Ecology and the Environment, 6*(6), 313–320. doi:10.1890/080005

Chaplin, F. S., Zavaleta, E. S., Eviner, V. T., Naylor, R. L., Vitousek, P. M., Reynolds, H. L., & Diaz, S. et al. (2000). Consequences of changing biodiversity. *Nature, 405*(6783), 234–242. doi:10.1038/35012241 PMID:10821284

Chen, J., Díaz, M., Rubio, B., & Troya, J. M. (2013, June). RAISE: RAIlway infrastructure health monitoring using wireless sensor networks. In *International Conference on Sensor Systems and Software* (pp. 143-157). Springer International Publishing. doi:10.1007/978-3-319-04166-7_10

Chen, Y. L., Wang, N. C., & Lin, Y. C. (2012). Power-Efficient Gathering in Sensor Information System Architectures with a Phase-Based Coverage Algorithm in a Wireless Sensor Network. *Advanced Science Letters, 9*(1), 620–625. doi:10.1166/asl.2012.2600

Chigbu, U. E., & Kalashyan, V. (2015). Land-Use Planning and Public Administration in Bavaria, Germany: Towards A Public Administration Approach to Land-Use Planning. *Geomatics, Landmanagement and Landscape,* (1): 2015.

Chirenje, L. I. R. A. G., & Musamba, E. B. (2013). Local communities' participation in decision-making processes through planning and budgeting in African countries. *Chinese Journal of Population Resources and Environment, 11*(1), 10–16. doi:10.1080/10042857.2013.777198

Chisholm, V. (2015). *Preventative Conservation and Disaster Management Planning in Cultural Institutions* (Unpublished master dissertation). The State University of New Jersey.

Chmielewski, P., & Kuźmiński, L. (2006). *The biggest constructional catastrophe in Polish history: The collapsed building in Chorzów-Katowice.* Retrieved from https://dlc.dlib.indiana.edu/dlc/bitstream/handle/10535/4850/The%20biggest%20constructional%20catastrophe%20in%20polish%20history%20the%20collapsed%20building%20in%20chorzow%20katowice.pdf?sequence=1

Chmielewski, P. (2007). Errare humanum est, but sometimes costs are extremely high. The biggest constructional catastrophe in Polish history: The collapsed building in Chorzow-Katowice. *Polish Psychological Review, 3*(159), 321–339.

Christopherson, R. W. (1997). *Geosystems: An Introduction to Physical Geography* (3rd ed.). London: Prentice-Hall.

Chukwuma, Obinna, Richard, Ephraim, & Ikechukwu. (2012). *Biotechnological Tools for Environmental Sustainability: Prospects and Challenges for Environments in Nigeria—A Standard Biotechnology.* Research International. doi:10.1155/2012/450802

Church, J. A., Gregory, J. M., Huybrechts, P., Kuhn, M., Lambeck, C., Nhuan, M. T., & Woodworth, P. L. et al. (2001). Changes in sea level. In J. T. Houghton, Y. Ding, & D. J. Griggs et al. (Eds.), *Climate Change 2001: The Scientific Basis* (pp. 639–693). Cambridge, UK: Cambridge University Press.

Cianciara, D. (2015). Kryzys zwiazany z uchodzcami nie jest chwilowy [Refugee crisis is not temporary]. *Hygeia Public Health, 50,* 427-434. Retrieved from http://www.h-ph.pl/pdf/hyg-2015/hyg-2015-3-427.pdf

Cielecki, T. (2009). Koordynacja systemu bezpieczeństwa wewnętrznego w Polsce [Coordination of the system of internal security in Poland]. In, J. Supernat (Ed.), Między tradycją a przyszłością w nauce prawa administracyjnego [Between tradition and future in law administration education] (pp. 75-91). Wrocław, Poland: Wydawnictwo Uniwersytetu Wrocławskiego.

Clapper, J. R. (2016). *Statement for the record Worldwide Threat Assessment* [Report]. U.S. Intelligence Community Senate Armed Services Committee. Retrieved from http://www.armed-services.senate.gov/imo/media/doc/Clapper_02-09-16.pdf

Clark, P. (2015). *Carbon capture: miracle machine or white elephant?* Retrieved from https://www.ft.com/content/88c187b4-5619-11e5-a28b-50226830d644

Clark, R. E. (2003). Fostering the work motivation of individuals and teams. *Performance Improvement*, *42*(3), 21–29. doi:10.1002/pfi.4930420305

Claudio, S., Yih-Min, W., Aldo, Z., & Abd Hiroo, K. (2011). Earthquake early warning: Concepts, methods and physical grounds. *Soil Dynamics and Earthquake Engineering*, 106–118.

Coelho, C., & Veloso-Gomes, F. (2004). Crosshore Beach Profile Models – Application to Aveiro Coast. *Journal of Coastal Research, 39.*

Coelho, C., Silva, R., Veloso-Gomes, F., & Taveira Pinto, F. (2006). *Risk Analysis Approach for the Portuguese West Coas.* Conducted at Risk Analysis 2006 – Fifth International Conference on Computer Simulation in Risk Analysis and Hazard Mitigation, Malta.

Cohen, S. (2004). Social relationships and health. *The American Psychologist*, *59*(8), 676–684. doi:10.1037/0003-066X.59.8.676 PMID:15554821

Cointreau, S. J. (1992). Environmental management of urban solid waste in developing countries: A project guide. Washington, DC: World Bank TC No. 5.

Concern. (2005). *Approaches to Disaster Risk Reduction.* Retrieved from http://earthmind.org/files/risk/Concern_ApproachestoDRR-paper-final.pdf

Conroy, M. J., & Peterson, J. T. (2013). *Decision Making in Natural Resources Management: A Structured Adaptive Approach.* Wiley-Blackwell. doi:10.1002/9781118506196

Constitutional Rights Project (CRP). (1999). *Land, oil and human rights in Nigeria's Delta.* Lagos: CRP.

Cook, J., Nuccitelli, D., Green, S. A., Richardson, M., Winkler, B., Painting, R., & Skuce, A. et al. (2013, June). Quantifying the Consensus on Anthropogenic Global Warming in the Scientific Literature. *Environmental Research Letters*, *8*(2), 024024. doi:10.1088/1748-9326/8/2/024024

Copeland, J. (2008). *Emergency response: Unity of effort through a common operating picture* (Master's Thesis). Carlisle Barracks, PA: U.S. Army War College.

Cornwall, A. (2008). Unpacking 'Participation': Models, meanings and practices. *Community Development Journal: An International Forum*, *43*(3), 269–283. doi:10.1093/cdj/bsn010

Costa, C. (2007). Introduction to Environment Géotechnique – Fundamentals of Géotechnique, Faculty of Sciences and Technology, New University of Lisbon. *Civil Engineering (New York, N.Y.)*, 1–38.

Costanza, R., Hart, M., Posner, S., & Talberth, J. (2009). *Beyond GDP: The Need for New Measures of Progress.* Retrieved from https://www.bu.edu/pardee/files/documents/PP-004-GDP.pdf

Covey, S. M. R. (2009). Building Trust: How the best leaders do it. *Leadership Excellence*, *21*(1), 15.

Cranenbroeck, J., & Shlyakhova, M. (2014). *Geosensing the world in a new generation digital earth - The new role of the surveyors.* Academic Press.

Crooks, S. (2004). The effect of sea-level rise on coastal geomorphology. *British Ornithologists Union*, *1*, 18–20.

Crosby, B. C., & Kiedrowski, J. (2008). Integrative Leadership: observation from a University of Minnesota Seminar Series. *Integral Leadership Review, 8*(3), 1-14. Retrieved from www.integralleadershipreview.com/archives/2008-06/2008-06-article-crosby.html

Crosby, B. C., & Bryson, J. M. (2005). *Leadership for the Common Good: Tackling Public*. Jossey-Bass.

Cutter, S. L., Mitchell, J. T., & Scott, M. S. (2000). Revealing the vulnerability of people and places: A case study of Georgetown County, South Carolina. *Annals of the Association of American Geographers, 90*(4), 713–737. doi:10.1111/0004-5608.00219

Czuchnowski, W., & Wroński, P. (2014, March 21). Szef MSW: Jesteśmy przygotowani na eskalację konfliktu wokół Ukrainy [Chief of MSW: We are prepared for the escalation of conflict around Ukraine]. *Gazeta Wyborcza*. Retrieved from http://wyborcza.pl/1,75478,15659536,Szef_MSW__Jestesmy_przygotowani_na_eskalacje_konfliktu.html

da Silva, R. I., Almeida, V. D. D., Poersch, A. M., & Nogueira, J. M. S. (2010, April). Wireless sensor network for disaster management. In *Network Operations and Management Symposium (NOMS)*, 2010 *IEEE* (pp. 870-873). IEEE.

Daily, G. C., & Ehrlich, P. R. (1992). Population, Sustainability, and Earth's Carrying Capacity: A framework for estimating population sizes and lifestyles that could be sustained without undermining future generations. *Bioscience*. Retrieved from http://dieoff.org/page112.htm

Dalby, S. (2002). Security and Ecology in the Age of Globalization. *ECSP Report, 8*(95), 95-108.

Dalby, S. (2002). *Environental Security*. Minneapolis, MN: University of Minosota Press.

Dalby, S. (2013). Environmental dimensions of human security. In R. Floyd & R. Mathew (Eds.), *Environmental Security Approaches and Issues*. London: Routledge Taylor and Francis.

Danish Energy Agency. (2012). *Energy Policy in Denmark*. Danish Energy Agency. Retrieved from file:///C:/Users/mcaesar/Downloads/EnergyPolicyinDenmark.pdf

Davey, M., Eaker, D. G., & Walters, L. H. (2013). Resilience processes in adolescents: Personality profiles, self-worth and coping. *Journal of Adolescent Research, 18*(4), 347362.

Davis, P., & West, K. (2009). What Do Public Values mean for Public Action, Putting Public values in their plural place. *American Review of Public Administration, 39*(6), 602–618. doi:10.1177/0275074008328499

Davydov, D. M., Stewart, R., Ritchie, K., & Chaudieu, I. (2010). Resilience and mental health. *Clinical Psychology Review, 30*(5), 479–495. doi:10.1016/j.cpr.2010.03.003 PMID:20395025

Dawidczyk, A., Gryz, J., & Koziej, S. (2006). Zarządzanie strategiczne bezpieczeństwem. Teoria-Praktyka-Dydaktyka [Strategic security management. Theory-Practice-Pedagogy]. £ódź, Poland: Wydawnictwo Wyższej Szkoły Humanistyczno-Ekonomicznej w £odzi.

de Sherbinin, A., Carr, D., Cassels, S., & Jiang, L. (2007). *Population and Environment Annual Review of Environment and Resources*. Academic Press. doi:10.1146/annurev.energy.32.041306.100243 PMID:20011237

De Soto, H. (2000). *The Mystery of the Capital: Why capitalism Triumph in the West and Fails Everywhere Else*. London: Black Swan Books.

de Soyza, I. (2000). The resource curse: Are civil wars driven by rapacity or paucity? In M. Berdal & D. M. Malone (Eds.), *Greed and grievance: Economic agendas in civil wars* (pp. 113–135). Boulder, CO: Lynne Rienner.

Dearden, P., & Mitchel, B. (2012). *Environmental Change and Challenge* (4th ed.). Oxford University Press.

Dell, M., Jones, B., & Olken, B. (2012). Temparature shocks and Economic growth: Evidence from the last half century. *American Economic Journal. Macroeconomics, 4*(3), 66–95. doi:10.1257/mac.4.3.66

Denz-Penhey, H., & Mucdoch, J. C. (2008). Personal resilience: Serious diagnosis and Prognosis with unexpected quality outcomes. *Qualitative Health Research, 18*(3), 391–404. doi:10.1177/1049732307313431 PMID:18235162

Department of Environment. (1987). *Handling Geographic Information- Report of the Committee of Enquiry chaired by Lord Chorley*. London: HMSO.

DESA. (2008). *Achieving Sustainable Development and Promoting Development Cooperation*. United Nations. Retrieved from http://www.un.org/en/ecosoc/docs/pdfs/fina_08-45773.pdf

Deudney, D. (1990). The Case Against Linking Environmental Degradation and National Security. *Millennium, 19*(3), 465. doi:10.1177/03058298900190031001

Devasena, A., & Sowmya, B. (2015). Wireless sensor network in disaster management. *Indian Journal of Science and Technology, 8*(15). doi:10.17485/ijst/2015/v8i15/74191

DFID (Department for International Development). (1999). *Sustainable Livelihoods Guidance Sheets*. Retrieved from http://www.ennonline.net/dfidsustainableliving

Dibie, R. (2000) Understanding public policy in Nigeria: A twenty-first century approach. Lagos: Mbeyi and Associates (Nig) Ltd.

Dietrich, K. (2015). *When We Can't See the Enemy, Civilians Become the Enemy: Living through Nigeria's Six-Year Insurgency*. Centre for Civilians in Conflict.

Dinis, J. L., & Tavares, A. (2005). Geomorphological susceptibility of the Portuguese west coast to tsunamis. *Proceedings of the III Congress on Planning and Management of Coastal Areas of Portuguese Speaking Countries*. Hydric Resources Association.

Diukwu, I., Ijaiya, M., Mohammed, M., & Egila, A. (2013). The Effect of Gas Production, Utilization, and Flaring on the Economic Growth of Nigeria. *Nat. Resour, 4*, 341–348.

DMSG. (2001). *Disaster Management Support Group: "The Use of Earth Observing Satellites for Hazard Support Group." Final Report, NOAA*. Department of Commerce.

Doğanay, H., & Altaş, N. T. (2013). *Doğal Kaynaklar*. Ankara: Pegem Akademi.

Donnan, S. (2016, September 9). World economy chokes on $5tn cost of pollution as premature deaths soar. *The Businessday*, p. A1. Retrieved from http://www.businessdayonline.com

Douglas, I., Alam, K., Maghenda, M., Mcdonnell, Y., McLean, L., & Campbell, J. (2008). Unjust waters: Climate change, flooding and the urban poor in Africa. *Environment and Urbanization, 20*(1), 187–205. doi:10.1177/0956247808089156

Dow, K., & Dowing, T. E. (2006). *The Atlas of Climate Change: Mapping the World's Greatest Change*. Earthscan.

Downing, T. E., & Patwardhan, A. (2003). *Vulnerability Assessment for Climate Adaptation*. APF Technical Paper 3, United Nations Development Programme.

Drabek, T. (1991). *Emergency Management: Principles and Practice for Local Government*. Washington, DC: International City Management Association.

Duku, J. (2015, August 9). Boko-Haram. NEMA Receives 10,000 Cameroun Returnees. *The Nations Newspapers*.

Dumanski, J. (2006). *Soil Conservation and The International Environmental Conventions.* Paper presented at the 14th International Soil Conservation Organization Conference: Water Management and Soil Conservation in Semi-Arid Environments, Marrakech, Morocco.

Dunster, J., & Dunster, K. (1996). Dictionary of Natural Resource Management, the Comprehensive, Single-Source Guide to Natural Resource Management Terms. CAB International (International for Agriculture and Bioscience).

Durosaro, I., & Ajiboye, S. (2011). Problems and Coping Strategies of Internally Displaced Adolescents in Jos Metropolis, Nigeria. *International Journal of Humanities and Social Science, 1*(20), 256–262.

Durotoye, B. (1999). Human Occupation of Hazard Areas in Nigeria. In A. Oshuntokun (Ed.), *Environmental Problems of Nigeria*. Lagos: Friedrich Ebert Foundation.

Dworzecki, J. (2012). Crisis management system in Poland. *The Science for Population Protection, 2*, 1–15.

Dyson, M., Bergkamp, G., & Scanlon, J. (Eds.), *(n.d.). Flow. The Essentials of Environmental Flows*. IUCN.

Eagles, P. F. J., McCool, S. F., & Haynes, C. D. (2002). Sustainable Tourism in Protected Areas. World Commission on Protected Areas, Spain: Best Practice Protected Area Guidelines Series:8.

Easterling, W. E., Aggarwal, P. K., Batima, P., Brander, K. M., Erda, L., Howden, S. M., . . . Tubiello, F. N. (2007). Food, Fibre and Forest products. In Climate Change 2007: Impacts, Adaptation and Vulnerability. Contribution of working Group II to the Fourth Assessment report of the intergovernmental Panel on climate Change. Cambridge University Press.

Ebeku, K. (2003). Judicial attitudes to redress for oil related damages in Nigeria. *RECIEL, 12*(2), 199–208.

Eccleston, C. (2010). *Global Environmental Policy: Principles concepts and practice.* CRC Press Inc.

Echikwonye, R., & Beetseh, K. (2011). The role of public policy making and development in Nigeria. *Journal of Social Science and Public Policy, 3*.

Ecosystem Valuation. (2009). *Travel Cost Method: Ecosystem Valuation.* Retrieved Oct 12, 2008, from: http://www.ecosystemvaluation.org/travel_costs.htm

Edwards, M. (2009). Seeing Integral leadership through three important lenses: Developmental, ecological and governance. *Integral Leadership Review, 9*(1), 1–13.

Egwaikhide, F. O., & Aregbeyen, O. (1999). Oil production externalities in the Niger Delta: Is a fiscal solution feasible? In *Fiscal federalism and Nigeria's economic development. Selected Papers presented at the Annual Conference of The Nigeria Economic Society.* Department of Economics, University of Ibadan.

Eilam, D., Izhar, R., & Mort, J. (2011). Threat detection: Behavioral practices in animals and humans. *Neuroscience and Biobehavioral Reviews, 35*(4), 999–1006. doi:10.1016/j.neubiorev.2010.08.002 PMID:20727909

Ekpo, U. (2004). *The Niger Delta Oil and Politics*. Lagos: International Energy Communications Limited.

Emas, R. (2015). Successes and Shortcomings in the Implementation of National Sustainable Development Strategies: From the Greening of Governance to the Governance of Greening. Retrieved from http://digitalcommons.fiu.edu/cgi/viewcontent.cgi?article=3229&context=etd

Emedare, W. (2015). *Housing in Nigeria: Why Nigeria Desperately Need 17 million More Houses*. Retrieved 14th March, 2017, www.sapientvendors.com.ng

Emery, K. O., & Aubrey, D. G. (1991). *Sea levels, land levels and tide gauges*. New York: Springer-Verlag. doi:10.1007/978-1-4613-9101-2

Emordi, E., Osaghae, E., & Nwaokocha, M. (2008). *Niger Delta Crisis and the State's Response Up to 2007.* Paper presented at the 1st Annual Conference of the Society of Peace Studies and Practice, Institute for Peace and Conflict Resolution.

Eneanya, A. N. (2015). *Theory and Practice of Public Administration.* Lagos: University of Lagos Press.

EOLSS. (1996). *Encyclopedia of Life Support System, 1996 Workshop Reports.* EOLSS Publishers Company Ltd.

Eradiauwa, U. A. (1999). *The role of Traditional Institution in Nation-building in a democratic society.* Academic Press.

Eric, C. B. (2013). The Potential Impact of Climate Change on Soil Properties and Processes and Corresponding Influence on Food Security. *Agriculture, 3*(3), 398–417. doi:10.3390/agriculture3030398

Erickson, P. J. (2009). Conceptualizing Food Systems for global environmental Change research. *Global Environmental Change, 18*(1), 234–245. doi:10.1016/j.gloenvcha.2007.09.002

Erkal, T., & Değerliyurt, M. (2009). Türkiye'de Afet Yönetimi. *Doğu Coğrafya Dergisi, 22*, 147–164.

Eso, K. (2003). *Further Thoughts On Law and Jurisprudence.* Ibadan: Spectrum Books.

Etuonovbe, A. K. (2011). *The Devastating Effect of Flooding in Nigeria.* Academic Press.

European Comission. (2004). Living with coastal erosion in Europe: Sediment and Space for Sustainability, PART I - Major findings and Policy Recommendations of the EUROSION project. *Directorate General Environment European Commission., 1*, 1–57.

European Commission. (2002). *Analysis of Selected Concepts on Resource Management. A Study to Support the Development of a Thematic Community Strategy on the Sustainable Use of Resources.* Retrieved from http://ec.europa.eu/environment/archives/natres/pdf/cowlstudy.pdf

European Commission. (2014). *Shengen area.* Retrieved from http://ec.europa.eu/dgs/home-affairs/what-we-do/policies/borders-and-visas/schengen/index_en.htm

European Council of Foreign Relations. (2015). *Multilateral issues and crisis management.* Retrieved from http://www.ecfr.eu/scorecard/2015/issues

European Union. (2014). *A Common European Asylum System.* Publications Office of European Union. Belgium, Brussels.

Eweje, G. (2006). Environmental Cost and Responsibilities Resulting from Oil Exploration in Developing Countries: The Case of the Niger Delta of Nigeria. *Journal of Business Ethics, 69*(1), 27–56. doi:10.1007/s10551-006-9067-8

Ezeani, E. O. (2001). *Contending Theoretical Perspectives on Global Urbanization. In Issues in Urbanization and Urban Administration in Nigeria.* Enugu, Jamoe Enterprises.

Ezemonye, M. N., & Emeribe, C. N. (2014). Flooding andhousehold preparedness in Benin City, Nigeria. *Mediterranean Journal of Social Sciences, 5*, 547–553. doi:10.5901/mjss.2014.v5n1p547

Fairholm, M. R. (2004). Perspectives on the practice of Leadership. *Public Administration Review, 64*(5), 577–590. doi:10.1111/j.1540-6210.2004.00405.x

Fajonyomi, S. O. (2012). *When the state goes to sleep: of citizen and state relationship.* Lagos: LASU Inaugural Lecture Series.

Falkenmark, M., & Rockstrom, J. (2008). Building resilience to drought in desertification-prone savannas in sub-Saharan Africa: The water perspective. *Natural Resources Forum, 32*(2), 93–102. doi:10.1111/j.1477-8947.2008.00177.x

FAO. (2004). *The State of Food and Agriculture.* Rome: FAO.

FAO. (2006). *Global Forest Resources Assessment 2005: Progress Towards Sustainable Forest Management*. Rome: Forestry Department, Food and Agriculture Organization of the United Nations.

FAO. (2011). *Global Forest Resource Assessment (FRA 2010)*. Retrieved from FAO Yearbook-resources – Landuse: http://www.fao.org/economic/ess/ess-publications/ess-yearbook/ess-yearbook2010/yearbook2010-reources/en/

Farazmand, A. (2001). Crisis and Emergency Management. In A. Farazmand (Ed.), *Handbook of Crisis and Emergency Management*. New York: Marcel Dekker.

Fasihuddin, Q., Kolhare, N. R., & Daimi, S. S. (2012). Measurement and Analysis of Vibration and Seismic Signal. *Measurement*, *7*, 8.

Fasona, M., Fabusoro, E., Sodiya, C., Adedayo, V., Olorunfemi, F., Elias, P., & Oloukoi, G. et al. (2016). Some Dimensions of Farmers'-Pastoralists' Conflicts in the Nigerian Savanna. Journal of Global Initiatives –Policy, Pedagogy. *Perspectives*, *10*(2), 87–108.

Fasona, M., Oloukoi, G., Olorunfemi, F., Elias, P., & Adedayo, V. (2014). *Natural Resource Management and Livelihoods in the Nigerian Savanna*. Ibadan: Penthouse Publication.

Feagin, R. A., Wu, X. B., Smeins, F. E., Whisenant, S. G., & Grant, W. E. (2005). Individual versus community-level processes and pattern formation in a model of sand dune plant succession. *Ecological Modelling*, *183*(4), 435–449. doi:10.1016/j.ecolmodel.2004.09.002

Federal Environmental Protection Agency (FEPA). (1998). *Revised National Policy on the Environment*. Federal Republic of Nigeria.

Federal Government of Nigeria (FGN). (1999). The 1999 Constitution of the Federal Republic of Nigeria. Abuja: Government Press.

Federal Government of Nigeria (FGN). (2012). *Nigeria's path to sustainable development through green economy: country report to the Rio + 20 Summit Abuja*. Federal Ministry of Environment.

Federal Ministry of Environment. (2016). *National Policy on The Environment (Revised)*. Retrieved from http://environment.gov.ng/assets/downloads/revisednationalpolicyontheenvironmentfinaldraft.pdf

Federal Republic of Nigeria. (2004). *National Water Policy*. Federal Republic of Nigeria July 2004.

FEPA. (1991). National Environmental Protection (effluent limitation) Regulation (S.1.8). FEPA, FGPE, Lagos Nigeria.

Ferguson, E. (2013). Personality is of central concern to understand health: Towards a theoretical model for health psychology. *Health Psychology Review*, *7*(1), S32–S70. doi:10.1080/17437199.2010.547985 PMID:23772230

Fergus, S., & Zimmerman, M. A. (2005). Adolescent resilience: A framework for understanding healthy development in the face of risk. *Annual Review of Public Health*, *26*(1), 399–419. doi:10.1146/annurev.publhealth.26.021304.144357 PMID:15760295

Ferreira, B. R. (1996). The Use and Effectiveness of Community Policing in A Democracy. In *Policing in Central and Eastern Europe: Comparing Firsthand Knowledge with Experience from the West, College of Police and Security Studies*. Retrieved from https://www.ncjrs.gov/policing/use139.htm

Finazzi, F. (2016). The Earthquake Network Project: Toward a Crowdsourced Smartphone-Based Earthquake Early Warning System. *Bulletin of the Seismological Society of America*, *106*(3), 1088–1099. doi:10.1785/0120150354

Fiorino, D.J. (1990). Citizen participation and Environmental risk: A survey of institutional mechanisms. *Science, Technology and Human Value, 15*(2), 226-243.

Fischer, J., Kühnlenz, F., Ahrens, K., & Eveslage, I. (2009). Model-based Development of Self-organizing Earthquake Early Warning Systems. *Proceedings MATHMOD 2009*.

Flach, F. F. (1980). Psycho-biological resiience, psychotherapy and the creative process. *Comprehensive Psychiatry, 21*(6), 510–519. doi:10.1016/0010-440X(80)90054-1 PMID:7428355

Fleishman, E. A. (1973). Twenty – years of consideration and structure. In E. A. Fleishman & J. C. Hunt (Eds.), *Current Development in the study of Leadership*. Carbondale, IL: Southern Illinois University Press.

Fleming, K., Picozzi, M., Milkereit, C., Kuhnlenz, F., Lichtblau, B., Fischer, J., & Ozel, O. et al. (2009). The Self-organizing Seismic Early Warning Information Network (SOSEWIN). *Seismological Research Letters, 80*(5), 755–771. doi:10.1785/gssrl.80.5.755

Fletcher, D., & Sarker, S. (2012). A grounded theory of psychological resilience in Olympic champions. *Psychology of Sport and Exercise, 13*(5), 669–678. doi:10.1016/j.psychsport.2012.04.007

Floyd, R. (2008). The environmental security debate and its significance for climate change. *The International Spectator, 43*(3), 51–65. doi:10.1080/03932720802280602

Fluggae, F., & Schilizzi, S. (2005). Greenhouse gas abatement policies and the value of carbon sinks: Do grazing and cropping systems have different destines? *Ecological Economics, 55*(4), 584–598. doi:10.1016/j.ecolecon.2004.12.033

Flyod, R., & Mathew, R. (2013). *Environmental Security Approaches and issues*. London: Routledge.

FMARD. (2016). *The Agriculture Promotion Policy (2016–2020): Building on the Successes of the ATA, Closing Key Gaps. Federal Ministry of Agriculture & Rural Development (FMARD).* Policy and Strategy Document.

Folke, C., Hahn, T., Olsson, P., & Norberg, J. (2005). Adaptive governance of social-ecological systems. *Annual Review of Environment and Resources, 30*(1), 441–473. doi:10.1146/annurev.energy.30.050504.144511

Folorunsho & Awosika. (2001). *Flood Mitigation in Lagos, Nigeria, Through Wise Management of Solid waste.* Academic Press.

Food and Agriculture Organization. (2006). *World Agriculture: Towards 2030/2050.* Rome: Food and Agriculture Organization of the United Nations.

FootprintE. (2011). Retrieved from: http://www.sustainablescale.org/conceptualframework/understandingscale

Foster, J. B. (1999). Marx's Theory of the Metabolic Rift: Classical Foundations for Environmental Sociology. *American Journal of Sociology, 105*(2), 366–405. doi:10.1086/210315

Frankenberg, E., Sikoki, B., Sumantri, C., Suriastini, W., & Thomas, D. (2013). Education, vulnerability and resilience after a natural disaster. *Ecological Sociology, 18*(2), 16. Retrieved from http://www.ncbi.nlm.nib.gov/pmc/articles/PMC4144011/

Frohlich, C., & Brunt, M. (2013). Two-year survey of earthquakes and injection/production wells in the Eagle Ford Shale, Texas, prior to the Mw4. 8 20 October 2011 earthquake. *Earth and Planetary Science Letters, 379*, 56–63. doi:10.1016/j.epsl.2013.07.025

Frynas, G. (2000). *Oil in Nigeria: Conflicts and Litigations Between Oil Companies and Village Communities*. Hamburg, Germany: LIT Verlag Munster.

Füssel, H. M. (2007). Vulnerability: A generally applicable conceptual framework for climate change research. *Global Environmental Change, 17*(2), 155–167. doi:10.1016/j.gloenvcha.2006.05.002

Gaffey, C. (2016). Niger Delta Communities Sue Shell Over Oil Spills. *Newsweek*. Available at http://www.newsweek.com/niger-delta-communities-sue-shell-over-oil-spills 432300Accessed 10/7/2017

Gaiya, B. (2008). *Environmental pollution in the Niger-Delta: An economic perspective.* Paper presented at International Conference on the Nigerian State, the Oil Industry and Niger- Delta at the Department of Political Science, Niger-Delta University, Bayelsa State.

Gandile, Tessema, & Nake. (2017). Biodiversity conservation using the indigenous knowledge system: The priority agenda in the case of Zeyse, Zergula and Ganta communities in Gamo Gofa Zone (Southern Ethiopia). *International Journal of Biodeversity and Conservation*, *9*(6), 167–182.

Garmezy, N. (1971). Vulnerability research and the issue of primary prevention. *The American Journal of Orthopsychiatry*, *41*(1), 101–116. doi:10.1111/j.1939-0025.1971.tb01111.x PMID:5539483

Garmezy, N. (1991). Resilience and vulnerability to adverse developmental outcomes associated with poverty. *The American Behavioral Scientist*, *34*(4), 416–430. doi:10.1177/0002764291034004003

Gatrell, P. (2013). *The making of the modern refugee.* Oxford, UK: Oxford University Press. doi:10.1093/acprof:oso/9780199674169.001.0001

Geist, H. J., & Lambin, E. F. (2001). *What Drives Tropical Deforestation? A meta-analysis of proximate and underlying causes of deforestation based on subnational case study evidence. LUCC Report Series No. 4.* CIACO Louvain-la-Neuve.

Genç, F. N. (2007). Türkiye'de Doğal Afetler ve Doğal Afetlerde Risk Yönetimi. *Stratejik Araştırmalar Dergisi*, *9*(5), 201–226.

Geography Subject Guide, I. B. (2011). *The characteristics of killers: Earthquakes, typhoons, droughts and nuclear accidents*. Retrieved from http://geographylaunchpad.weebly.com/1 characteristics-of-hazards.html

Ghaus-Pasha, A. (2005). *Role of Civil Society Organizations in Governance.* 6th Global Forum on Reinventing Government Towards Participatory and Transparent Governance, Seoul, South Korea.

Gierszewski, J. (2011). Pozamilitarne ogniwa ochronne panstwa i ich miejsce w systemie obronnym [Non-military protection systems and their place in the defense system]. In M. Chrabkowski, C. Tatarczuk, & J. Tomaszewski (Eds.), *Bezpieczeństwo w administracji i biznesie we współczesnym świecie [Security in administration and business in modern world]* (Vol. 1, pp. 49–67). Gdynia, Poland: Wyższa Szkoła Administracji i Biznesu.

Gilman, E. (2005). *Assessing and Managing Coastal Ecosystem Response to Projected Relative Sea-Level Rise and Climate Change*. Honolulu, HI: Blue Ocean Institute.

Glantz, M. D., & Johnson, J. L. (1999). *Resilience and development: Positive life adaptations.* New York: Kluwer Academic/Plenum Publishers.

Gober, P. (2006). Metropolitan Phoenix: Place making and Community Building in the Desert. Philadelphia, PA: University of Pennsylvania Press. doi:10.9783/9780812205824

Godfray, J. C. J., Beddington, J., Cute, I., & Haddad, L. (2010). Food Security: The Challenge of Feeding 9 billion people. *Science*, *327*(5967), 812–818. doi:10.1126/science.1185383 PMID:20110467

Goltz, J. D., & Bourque, L. B. (2016). Earthquakes and Human Behavior: A Sociological Perspective. *International Journal of Disaster Risk Reduction*, *21*(2017), 251–265.

Gomes, P. (2009). *Impacte da subida do nível do mar sobre o turismo* (Master thesis). Faculty of Science and Technology of the New University of Lisbon.

Goodman, R., & West-Olatunji, C. A. (2008). Transgenerational trauma and resilience: Improving mental health counselling for survivors of Hurricane Katrina. *Journal of Mental Health Counseling, 30*(2), 121–136. doi:10.17744/mehc.30.2.q52260n242204r84

Gordon, E., & Johnson, J. L. (1994). Epilogue: Educational resilience: Challenges and prospects. In E. Gordon (Ed.), *Educational resilience in inner-city America: Challenges and prospects* (pp. 199–194). Hillsdale, NJ: Erlbaum.

Gornitz, V., & Kanciruk, P. (1989). Assessment of global coastal hazards from sea-level rise. *Proceedings of the 6th Symposium on Coastal and Ocean management.*

Government of Tripura. (2008). *City Development Plan-Agartala.* Retrieved from http://agartalacity.tripura.gov.in/PDF/Utilities/CityDevP.pdf

Grabbe, H. (2000). The sharp edges of Europe: Expanding Schengen eastwards. *International Affairs, 76*(3), 519–536. doi:10.1111/1468-2346.00149

Grafton, R. Q., Adadmowicz, W., Dupont, D., Nelson, H., Hill, R. J., & Renzetti, S. (2004). *Economics of the environment and natural resources.* Malden, MA: Blackwell Publishing. doi:10.1002/9780470755464

Graven, M. (2015). Between Law and History: Berlin Conference of 1884-1885 and the Logic of Free Trade, London. *Review of International Law, 3*(1), 31–59. doi:10.1093/lril/lrv002

Gray, R. H., Javad, M., Power, D. M., & Sinclair, C. D. (2001). Social and environmental disclosure and corporate characteristics: A research note and extension. *Journal of Business Finance & Accounting, 28*(3/4), 327–356. doi:10.1111/1468-5957.00376

Gregory, P. J., Johnson, S. N., Newton, A. C., & Ingram, I. S. I. (2009). Integrating Pests and Pathogens in to the Climate Change/Food Security Debate. *Journal of Experimental Botany, 60*(10), 2827–2838. doi:10.1093/jxb/erp080 PMID:19380424

Grimm, N.B., Grove, J.M., Pickett, S.T.A & Redman, C.L. (2000). Integrated approaches to long-term studies of urban ecological systems. *Bioscience, 50*(7), 571-584.

Gudzbeler, G., Nepelski, M., & Urban, A. (2011). Japanese support system of crisis management services. *International Security, 3*(1), 107–127.

Gül, A., & Özgüner, H. (2005). *Ülkemizdeki Korunan Doğal Alanlarda Yönetim Zonlarının Oluşturulması. Korunan Doğal Alanlar Sempozyumu.* Isparta: SDÜ Orman Fakültesi Poster Bildiri Kitabı.

Gunderson, L. H., & Holling, C. S. (Eds.). (2002). *Panarchy: Understanding Transformation in Human and Natural Systems.* Washington, DC: Island Press.

Güneş, G. (2011). Korunan Alanların Yönetiminde Yeni Bir Yaklaşım: Katılımcı Yönetim Planları. *Ekonomi Bilimleri Dergisi, 3*(1), 47–57.

Gyong, J. E. (2013). A social analysis of the transformation agenda of president Goodluck Ebele Jonathan. *European Scientific Journal, 8*(16).

Haase, T. W. (2014, October). *Administrative resilience in Lebanon.* Presentation at Sam Houston State University, Huntsville, TX.

Haddad, E. (2008). *The refugee in international society: between sovereigns.* Cambridge, UK: Cambridge University Press.

Haddow, G., Bullock, J., & Coppola, D. P. (2007). *Introduction to Emergency management.* Elsevier.

Hagmann, J., Chuma, E., Muriwira, K., Connoly, M., & Ficarelli, P. (2002). Success Factors in Integrated Natural Resource Management R & D: Lessons from Practice. *Conservation Ecology*, *5*(2), art29. doi:10.5751/ES-00298-050229

Happer, C., & Philo, G. (2013). The Role of the Media in the Construction of Public Belief and Social Change. *Journal of Social and Political Psychology*, *1*(1), 321–336. doi:10.5964/jspp.v1i1.96

Haque, M. (1999). The Fate of Sustainable Development under Neo-liberal Regimes in Developing Countries. *International Political Science Review*, *20*(2), 197–198. doi:10.1177/0192512199202005

Harris, J. M. (2000). *Basic Principles of Sustainable Development*. Global Development and Environment Institute, Tufts University. Retrieved from http://ase.tufts.edu/gdae/publications/working_papers/Sustainable%20Development.pdf

Harrison, M., & Coussens, C. (2007). *Global Environmental Health in the 21st Century: From Governmental Regulation to Corporate Social Responsibility*. Roundtable on Environmental Health Sciences, Research, and Medicine. Retrieved from https://www.ncbi.nlm.nih.gov/books/NBK53989/pdf/Bookshelf_NBK53989.pdf

Harvey, D. (1996). Justice, Nature and the Geography of Difference. Oxford, UK: Blackwell.

Harvey, D. (2005). *Brief History of Neoliberalism*. New York: Oxford University Press.

Heinman, M. (1997). Science by the people; Grass root environmental monitoring and the debate over scientific expertise. *Journal of Planning Education and Research*, *16*(4), 291–299. doi:10.1177/0739456X9701600405

Hennecke, W. G., & Cowell, P. J. (2000). GIS modelling of impacts of an accelerated rate of sea-level rise on coastal inlets and deeply embayed shorelines. *Environmental Geoscience*, *7*(3), 137–148. doi:10.1046/j.1526-0984.2000.73002.x

Hepcan, Ş. (1997). *Milli Parklarda Yönetim Zonlarının Belirlenmesi Amacıyla Manisa Spil Dağı Milli Parkı Örneğinde Bir Yöntem Araştırması* (Unpublished doctoral dissertation). Agean University, İzmir, Turkey.

History of the European Union. (2017) Retrieved from https://pl.wikipedia.org/wiki/Historia_Unii_Europejskiej

Hoffmann, W. A., & Jackson, R. B. (2000). Vegetation–Climate Feedbacks in the Conversion of Tropical Savannah to Grassland. *Journal of Climate*, *13*(9), 1593–1602. doi:10.1175/1520-0442(2000)013<1593:VCFITC>2.0.CO;2

Holgate, S. J., & Woodworth, P. L. (2004). Evidence for enhanced coastal sea level rise during the 1990s. *Geophysical Research Letters*, 31.

Homer-Dixon, T. (1991). On the Threshold: Environmental Changes as Causes of Acute Conflict. *International Security*, *16*(2), 76–116. doi:10.2307/2539061

Homer-Dixon, T. (1999). *Environment, Scarcity and Violence*. Princeton, NJ: Princeton University Press.

Hopf, S. M. (2010). *Risk and resilience in children coping with parental divorce*. Dartmouth Undergraduate Journal of Science.

Hora, E. A. (2014). Factors that affect Women Participation in Leadership and Decision-Making Position, *Asian Journal of Humanity. Art and Literature*, *1*(2), 97–118.

Houghton, J. T., Ding, Y., Griggs, D. J., Noguer, M., van der Linden, P. J., Dai, X., . . . Johnson, C. A. (2001). Climate Change 2001: The Scientific Basis: Contributions of Working Group I to the Third Assessment Report of the Intergovernmental Panel on Climate Change. Cambridge University Press.

Howe, G. M. (1977). *A Geography of Diseases*. London: Academic Press.

Howell, B. F. Jr. (1986). History of ideas on the cause of earthquakes. *Eos (Washington, D.C.)*, *67*(46), 1323–1326. doi:10.1029/EO067i046p01323

Hutton, D. (2012). Lessons Unlearnt: The (Human) Nature of Disaster Management. In Emergency Management (pp. 1-10). In Tech.

Iaccino, L. (2015). Nigeria's oil war: Who are the Niger Delta militants? *International Business Times.* Available at http://www.ibtimes.co.uk/nigerias-oil-war-who-are niger-delta militants-1520580.accessed12/21/2015

Ibeanu, O. (2000). Oiling the friction: Environmental conflict management in the Niger Delta, Nigeria: An Environmental Change and Security Profile Report, issue No 6. Woodrow Wilson Centre.

Ibitoye, M. O. (2006). *Geospatial data. A panacea for Flood Management.* Paper presented at the Nigerian Institution of Surveyors (NIS) Conference, Akure.

Igbaekemen, G. O. (2014). IMPACT of Leadership Style on Organisation Performance: A Strategic Literature. *Review Public Policy and Administration Research, 4*(9), 126–135.

Iglicka, K., & Ziołek-Skrzypczak, M. (2010, Sept.). *EU membership highlights Poland's migration challenges.* Migration Policy Institute. Retrieved from http://www.migrationpolicy.org/article/eu-membership-highlights-polands-migration-challenges

Ijaiya, H., & Joseph, O. T. (2014). Rethinking Environmental Law Enforcement in Nigeria. *Beijing Law Review, 5*(4), 306–321. doi:10.4236/blr.2014.54029

İlke, E.F., Uzun, O., Çetinkaya, G., Erduran, F., & Açıksöz, S. (2011). Doğal Kaynaklarin Kullaniminda Peyzaj Yönetimi, Koruma ve Planlamasi. *I. Konya Kent Sempozyumu*, 319-332.

Imasuen, E. (2015). Insurgency and humanitarian crises in Northern Nigeria: The case of Boko Haram. *African Journal of Political Science and International Relations, 9*(7), 284–296. doi:10.5897/AJPSIR2015.0789

INAG I.P. (2007). *Water quality of bathing areas.* Retrieved Oct 12, 2008, from: http://www.inag.pt/

INAG I.P. (2008). *Coastal Interventions.* Retrieved Oct 12, 2008, from: http://www.inag.pt/index.php?option=com_content&view=article&id=47&Itemid=90

Indian Oil Corporation Limited. (2005-2013). *Annual Sales Report of Fuel.* Mumbai: Indian Oil Corporation Limited. Retrieved from https://www.iocl.com/InvestorCenter/PastAnnualReport.aspx

Ingram, J. S. I. (2009). Food System concepts. In R. Rabbinge & A. Linneman (Eds.), *ESF/cost Forward Look on European food Systems in a Changing World.* Strasbourg: European Science Foundation.

Ingwe, R., Okoro, J., & Uim-Agbor, U. (2012). Neo-liberal policies and regimes in Post – independent Nigeria at the dawn of post neo-Liberalism (1960 – 2007): A Research Agenda. *Universal Journal of Management and Social Sciences, 2*(12).

Institute for Health Metrics Evaluation (IHME). (2016). Rethinking Development and Health: Findings from the Global Burden of Disease Study. Seattle, WA: IHME.

Intergovernmental Panel on Climate Change (IPCC). (2007). Climate Change 2007: Synthesis Report. Contribution of Working Groups I, II and III to the Fourth Assessment Report of the Intergovernmental Panel on Climate Change. IPCC.

Inter-governmental Panel on Climate Change (IPCC). (2007). *Summary for Policymakers. In climate Change 2007: The Physical Science Basis. Contribution of Working Group I to the Fourth Assessment report of the Intergovernmental panel on Climate Change.* Cambridge, UK: Cambridge University Press.

Internal Displacement Monitoring Centre. (2014). *Annual Report.* Geneva: IDMC.

Internal Displacement Monitoring Centre. (2015). *Report of the Displacement Tracking Matrix in Nigeria.* Geneva: IDMC.

International Centre for Investigative Reporting. (2017). *Report on IDPs Camps.* Abuja: ICIR.

International Energy Agency (IEA). (2016a). *World Energy Outlook 2016.* Retrieved from https://www.docdroid.net/IOBt86G/world-energy-outlook-2016.pdf#page=4

International Energy Agency. (2016b). *Energy and Air Pollution: World Energy Outlook-Special Report.* Retrieved from https://www.iea.org/publications/freepublications/publication/WorldEnergyOutlookSpecialReport2016EnergyandAir-Pollution.pdf

International Institute for Sustainable Development (IISD), International Union for Conservation of Nature and Natural Resources (IUCN) and Stockholm Environment Institute (SEI). (2003). *Livelihoods and Climate Change.* A Conceptual Framework Paper Prepared by the Task Force on Climate Change, Vulnerable Communities and Adaptation. Retrieved from https://www.iisd.org/pdf/2003/natres_livelihoods_cc.pdf

International Monetary Fund (IMF). (2015, December). Who are the biggest emitters of greenhouse gases? *IMF Research Bulletin, 16*(4), 4.

International research Institute (IRI). (2006). *A Gap Analysis for implementation of Global Climate Observing System Programme in Africa.* IRI Technical report No. 0601, International Research Institute for Climate and Society.

IOM. (2015). *Displacement Tracking Matrix (DTM) Round II Report in Nigeria.* IOM.

IPCC (2001). *Summary for policymakers climate change 2001: Mitigation.* A Report of Working Group III of the Intergovernmental Panel on Climate Change.

IPCC (Intergovernmental Panel on Climate Change). (2007). *Climate Change Impacts, Adaptation and Vulnerability. Contribution of Working group II to the Fourth assessment Report of the IPCC.* Geneva, Switzerland: IPCC.

IPCC. (2007). *Climate Change 2007: The Physical Science Basis.* Working Group I Contribution to the Fourth Assessment Report of the Intergovernmental Panel on Climate Change. Cambridge, UK: Cambridge University Press. Retrieved Oct 12, 2008, from http://ipcc-wg1.ucar.edu/wg1/wg1-report.html

IPCC. (2014). *Climate Change 2014: Impacts, Adaptation and Vulnerability.* IPCC Working Group II Contribution to AR5. Summary for Policymakers. Intergovernmental Panel for Climate Change.

IPCC. (2014). *Climate Change 2014: Livelihoods and Poverty. The Working Group II Contribution to the Intergovernmental Panel on Climate Change Fifth Assessment Report.* Cambridge, UK: Cambridge University Press.

Irikana, G. (2015). The Niger Delta Ecosystem and Oil production. An Analysis of Capitalist Incorporation of A Natural Economy. In Conflict and Development in the Niger Delta Sociological Perspectives. Enugu: ABIC Books.

IRP. (2010). *Guidance Note On Recovery: Livelihood.* International Recovery Platform, Kobe, Japan. Retrieved from http://www.recoveryplatform.org/

Irvin, R. A., & Stansbury, J. (1996). *Citizen Participation in Decision-Making: Is it Worth the Effort? 'CBEP (Community-Based Environmental Protection) is designed to maximize the use of scarce resources, encourage local support, and consider the economic well-being of communities'.* Environmental Protection Agency. Retrieved from https://smartnet.niua.org/sites/default/files/resources/citizen%20participation%20is%20it%20worth%20the%20effort.pdf

Isiche, A. O., & Sanford, W. W. (1976). The effect of waste gas flares on the surrounding vegetation of South-Eastern Nigeria. *Journal of Applied Ecology, 13.*

Isomaa, R., Väänänen, J.-M., Fröjd, S., Kaltiala-Heino, R., & Marttunen, M. (2013, August). is low? Low self-esteem as an indicator of internalizing psychopathology in adolescence. *Health Education & Behavior, 40*(4), 392–399. doi:10.1177/1090198112445481 PMID:22872582

IUCN. (1994). *Guidelines for Protected Area Management Categories.* Gland.

Ivancevich, M. J., Donnelly, J. H., & Gibson, L. J. (2003). *Management: Principles and Foundations.* AITBS Publishers.

Ivbijaro, F. A. (2007, March). *Sustainable environmental management.* Paper presented at the Sensitization Workshop on Keep Nigeria Clean by the National Orientation Agency (NOA), Ibadan.

Iyoha, F.E. (2008). *Local government and the elusive rural development in Nigeria: Constraints on local authorities capacity and the promise of decentralized participatory development.* Ekpoma: Ambrose Ali University. Inaugural Lecture Series.

Izac, A. M. N., & Sanchez, P. A. (2001). Towards a Natural Resource Management Paradigm for International Agriculture: The Example of Agroforestry Research. *Agricultural Systems, 69*(1-2), 5–25. doi:10.1016/S0308-521X(01)00015-4

Jacobs, L. (2014). The Contested Politics of Public Value. *Public Administration Review, 74*(4), 480–494. doi:10.1111/puar.12170

Jain, S. (2006). *An Empirical Economic Assessment of Impacts of Climate Change on Agriculture in Zambia.* Pretoria, South Africa: University of Pretoria, Centre for Environmental Economics and Policy in Africa (CEEPA), CEEPA discussion paper No. 27.

Jain, N. K. (2005). *Organisational Behaviour* (Vol. 1). Atlantic Publishers.

Jampolsky, J. A. (2014). *Integrated Resource Management Planning with The Ute Mountain Ute Tribe: Property, Place, and Governable Space* (Unpublished doctoral dissertation). University of Colorado.

Jazeerah, A. (2017). *Police Displaced Thousands in Nigeria's Otodo-Gbame.* Retrieved from www. aj jazeerah.com/news/2017/3/police-displaced-thousands-otodo-gbame-slum

Jike, V. (2004). Environmental Degradation, Social Disequilibrium, and the Dilemma of Sustainable Development in the Niger Delta of Nigeria. *Journal of Black Studies, 34*(5), 686–701. doi:10.1177/0021934703261934

Johnson, R. (2000). *GIS Technology for Disasters and Emergency Management.* An ESRI White Paper.

Jones, P. G., & Thornton, P. K. (2003). The potential impacts of climate change in tropical agriculture: The case of maize in Africa and Latin America in 2055. *Global Environmental Change, 13*, 51–59. doi:10.1016/S0959-3780(02)00090-0

Joseph, J. (1994). *The resilient child.* New York, NY: Insight Books.

Kahle, L. R., & Gurel-Atay, E. (Eds.). (2014). *Communicating Sustainability for the Green Economy.* New York: M.E. Sharpe.

Kaiser, R. B., Hogan, R., & Craig, S. B. (2008). Leadership and the fate of organizations. *The American Psychologist, 63*(2), 96–110. doi:10.1037/0003-066X.63.2.96 PMID:18284278

Kalafatic, C. (2007). *Indigenous peoples' sustainable livelihoods.* FAO Livelihood Support Programme. Retrieved from ftp://ftp.fao.org/docrep/fao/010/aj033e/aj033e02.pdf

Kalantary. (2010). Climate Change in Zambia: Impacts and Adaptation. *Global Majority e-Journal, 1*(2), 85-96.

Kaldor, M. (1999). *New and Old Wars: Organized Violence in a Global Era.* Cambridge, UK: Polity Press.

Kalita, H. K., & Kar, A. (2009). Wireless sensor network security analysis. *International Journal of Next-Generation Networks*, *1*(1), 1–10.

Kangiwa, H. (2012). *National Policy in Internally Displaced Persons in Nigeria. Report of the Federal Commissioner.* National Commission for Refugees.

Kaplan, R. (1994, February). The Coming Anarchy:How security crime, overpopulation, tribalism and disease are rapidly destroying the social fabric of our planet. The Atlantic Monthly, 49.

Kapucu, N., & Van Wart, M. (2006). The evolving role of the public sector in managing catastrophic disasters: Lessons learned. *Administration & Society*, *38*(3), 1–30. doi:10.1177/0095399706289718

Kapucu, N., & Van Wart, M. (2008). Making matter worse: Anatomy of Leadership failures in catastrophic events. *Administration & Society*, *80*(7), 711–740. doi:10.1177/0095399708323143

Karamidehkordi, E. (2012). Sustainable Natural Resource Management, a Global Challenge of This Century. In Sustainable Natural Resources Management (pp. 105-114). In Tech.

Keengwe, J., Mbae, J. C., & Onchwari, G. (2016). Handbook of Research on Global Issues in Next-Generation Teacher Education. Information science Reference.

Keleş, R., & Hamamci, C. (2002). *Çevrebilim*. Ankara: İmge Kitabevi.

Kennedy, V. S., Twilley, R., Klepas, J., Cowan, J., Jr., & Hare, S. (2002). Coastal and Marine Ecosystems & Global Climate Change: Potential Effects on U.S. Resources. Report for the Pew Center on Global Climate Change.

Kennedy, P. (1992). *Preparing for the 21st century*. New York: Random House.

Kennett, S. A. (2002). National Policies for biosphere greenhouse gas management: Issues and opportunities. *Environmental Management*, *30*(5), 595–608. doi:10.1007/s00267-002-2716-5 PMID:12375081

Keough, H. L., & Blahna, D. J. (2006). Achieving integrative, collaborative ecosystem management. *Conservation Biology*, *20*(5), 1373–1382. doi:10.1111/j.1523-1739.2006.00445.x PMID:17002755

Kerr, S., Liu, S. G., Pfaff, A. S. P., & Hughes, R. F. (2003). Carbon dynamics and land-use choices: Building a regional-scale multi-disciplinary model. *Journal of Environmental Management*, *69*(1), 25–37. doi:10.1016/S0301-4797(03)00106-3 PMID:12927149

Kervankıran, İ., & Eryılmaz, A. G. (2015). Milli Parkların Sürdürülebilir Kullanımı ve Yönetim Planı Önerisi: Isparta İli Örneği. *SDU Faculty of Arts and Sciences Journal of Social Sciences*, *34*, 173–190.

Kettl, D. (2006). Managing boundaries in American administration: The collaboration imperative. *Public Administration Review*, *66*(s1), 10–19. doi:10.1111/j.1540-6210.2006.00662.x

Kibert, C. J., Thiele, L., Peterson, A., & Monroe, M. (2012). *The Ethics of Sustainability*. Retrieved from http://rio20.net/wp-content/uploads/2012/01/Ethics-of-Sustainability-Textbook.pdf

Kilewo, E. G., & Frumence, G. (2015). Factors that hinder community participation in developing and implementing comprehensive council health plans in Manyoni District, Tanzania. *Global Health Action*, *8*, 26461. Retrieved from https://www.ncbi.nlm.nih.gov/pmc/articles/PMC4452651/pdf/GHA-8-26461.pdf

Klare, M. (1996, November). Redefining Security: The New Global Schisms. *Current History*.

Klein, R. J. T., Nicholls, R. J., & Thomalla, F. (2003). resilience to natural hazards: How useful is this concept. *Environmental Hazards*, *5*(1), 35–45. doi:10.1016/j.hazards.2004.02.001

Klepner, T. (2014). Wspolpraca transgraniczna panstw grupy Wyszehradzkiej na rzecz zapewnienia bezpieczenstwa granic. *Poliarchia, 2*, 99-122.

Komolafe, A. A., Adegboyega, S. A. A., Anifowose, A. Y. B., Akinluyi, F. O., & Awoniran, D. R. (2014). Air pollution and climate change in Lagos, Nigeria: Needs for proactive approaches to risk management and adaptation. *American Journal of Environmental Sciences, 10*(4), 412–423. doi:10.3844/ajessp.2014.412.423

Kong, Q., Kwony, Y. W., Schreierz, L., Allen, S., Allen, R., & Strauss, J. (2015, July). Smartphone-based networks for earthquake detection. In *Innovations for Community Services (I4CS), 2015 15th International Conference on* (pp. 1-8). IEEE. doi:10.1109/I4CS.2015.7294490

Kong, Q., Allen, R. M., Schreier, L., & Kwon, Y. W. (2016). MyShake: A smartphone seismic network for earthquake early warning and beyond. *Science Advances, 2*(2), e1501055. doi:10.1126/sciadv.1501055 PMID:26933682

Koontz, T. M., & Bodine, J. (2008). Implementing ecosystem management in public agencies: Lessons from the US Bureau of Land Management and the Forest Service. *Conservation Biology, 22*(1), 60–69. doi:10.1111/j.1523-1739.2007.00860.x PMID:18254853

Koponen, J. (1993). The Partition of Africa: A Scramble for a Mirage? *Nordic Journal of African Studies, 2*(1), 117–135.

Korvenoja, T. (1993). The Environmental Problems and Politics of Power Review on the African Elite. *Nordic Journal of African Studies, 2*(1), 140–153.

Kour, H., & Sharma, A. K. (2010). Hybrid energy efficient distributed protocol for heterogeneous wireless sensor network. *International Journal of Computers and Applications, 4*(6), 1–5. doi:10.5120/828-1173

Krajewski-Siuda, K., & Romaniuk, P. (2006). *Can pigeons change the emergency system in Poland? European Journal of Public Health*. doi:10.1093/eurpub/ckl065446

Krech, S. (2005). Reflections on Conservation, Sustainability, and Environmentalism in Indigenous North America. *American Anthropologist, 107*(1), 78–86.

Krokfors, C. (1995). Povery, Environmental Stress and Culture as Factors in African Migrations. In J. Baker & T. Akin Aina (Eds.), *The Migration Experience in Africa* (pp. 54–64). Nordiska Afrikainstitutet.

Küçükali, U. F. (2005). *Havza Tanımı ve Doğal Kaynak Yönetimi* (Unpublished master dissertation). Yıldız Teknik Üniversitesi, İstanbul, Turkey.

Kufoniyi, O. (2009). *Enabling environment for geospatial information technology uptake in Africa*. UNECA Committee on Development Information, Science and Technology, Addis Ababa, Ethiopia.

Kumari, N., Patel, N., Anand, S., & Bhattacharya, P. P. (2013). Designing low power wireless sensor networks: a brief survey. International Journal of Advanced Research in Electrical, Electronics and Instrumentation Engineering, 2(9).

Kundzewicz, Z. W., Mata, L. J., Arnell, N. W., Doll, P., Kabat, P., Jimenez, B., . . . Oki, T. Z-Sen and I.A. Shiklognov (2007). Fresh water Resources and their Management. In Climate Change 2007; Impacts, Adaptation and Vulnerability. Contribution of Working group II to the 4th Assessment report of the Intergovernmental Panel on climate Change. Cambridge University Press.

Kurdoğlu, O. (2007). Dünyada Doğayi Koruma Hareketinin Tarihsel Gelişimi ve Güncel Boyutu. *Artvin Çoruh Üniversitesi Orman Fakültesi Dergisi, 8*(1), 59–76.

Kusek, J. Z., & Rist, R. C. (2004). *Ten Steps to a Results-Based Monitoring and Evaluation System*. The International Bank for Reconstruction and Development / The World Bank. doi:10.1596/0-8213-5823-5

Kusky, T. M. (2003). *Geological hazards: A sourcebook.* Westport, CT: Greenwood Press.

Kuvan, Y. (1991). Avrupa topluluğu tarafından yapılan korunan alan sınıflandırması ve ülkemizdeki korunan alanların bir değerlendirilmesi. *Istanbul Üniversitesi Orman Fakültesi Dergisi, 41*(1-2), 133–137.

Ladan, M. T. (2012). Impact of insecurity in the north on internally displaced people and migration flows between Nigeria and neighbouring countries. Paper presentation made at the Forum of European Union Working Group in Nigeria and Development, Abuja.

Laird & Frank, N. (1993). Participatiory analysis, democracy, and technological decision making. *Science, Technology and Human Values, 18*(3), 341-361.

Lambie, E., Wilson, T. M., Johnston, D. M., Jensen, S., Brogt, E., Doyle, E. E., & Helton, W. S. (2016). Human behaviour during and immediately following earthquake shaking: Developing a methodological approach for analysing video footage. *Natural Hazards, 80*(1), 249–283. doi:10.1007/s11069-015-1967-4

Lanicci, J. M., Ramsay, J. D., & Murray, E. H. (2017). Re-conceptualizing environmental security as resilience: Strategic planning for human and national security. *Journal of Human Security and Resilience, 1*, 1–37. Retrieved from http://www.thinkhumansecurity.org/sft965/re-conceptualizingenvironmentalsecurityasresilience-strategicplanningforhumanandnationalsecurity.pdf

Le Mehaute, B., Hwang, L. S., & Van Dorn, W. (1971, September). Methods for improving tsunami warning system. In Engineering in the Ocean Environment, IEEE 1971 Conference on (pp. 329-331). IEEE.

Leggett, J., Pepper, W. J., & Swart, R. J. (1992). *Emissions Scenarios for the IPCC: an Update, Climate Change 1992: The Supplementary Report to The IPCC Scientific Assessment.* Cambridge University Press.

Lennon, J. (2006). Cultural Heritage Management. In M. Lackwood, G. L. Worboys, & A. Kothari (Eds.), *Managing Protecting Areas: A Global Guide* (pp. 448–473). London: Earthscan.

Lenntech. (2009). *FAQ air pollution: The question library on air-related issues.* Delft, The Netherlands: Lenntech Water Treatment & Purification Holding B.V. Retrieved February 11, 2011 from http://www.lenntech.com/faq-air-pollutionhtm

Lewis, F., Johnston, R., & Hamdard, M. (2014). *Tools and Approaches for Integrated Natural Resource Management.* European Commission / FP7 Collaborative project.

Lim, S., & Prakash, A. (2004). The Regulations and Innovation: The case of 15014001. *Public Administration Review, 74*(2), 233–244. doi:10.1111/puar.12189

Liu, G., Tan, R., Zhou, R., Xing, G., Song, W. Z., & Lees, J. M. (2013, April). Volcanic earthquake timing using wireless sensor networks. In *Proceedings of the 12th international conference on Information processing in sensor networks* (pp. 91-102). ACM. doi:10.1145/2461381.2461396

Liverman, D. M., & Kapadia, K. (2010). Food Skystems and Global environment: An Overview. In J. I. Ingram, P. J. Erickson, & D. Liverman (Eds.), *Food Security and Global Environmental change.* London: Earthscare.

Lobell, D. B., & Bruke, M. B. (2008). Why are Agricultural Impacts of Climate Change so uncertain? The importance of Temperature Relative to Precipitation. *Environmental Research Letters, 3*(3), 034007. doi:10.1088/1748-9326/3/3/034007

López-Marrero, T., & Tschakert, P. (2011). From theory to practice: Building more resilient communities in flood-prone area. Environment and Urbanization. *International Institute for Environment and Development, 23*(1), 229–249.

Low, P. S. (Ed.). (2005). *Climate Change and Africa.* New York, NY: Oxford University Press.

Ludwig, D., Mangel, M., & Haddad, B. (2001). Ecology, Conservation, and Public Policy. *Annual Review of Ecology and Systematics*, *32*(1), 481–517. doi:10.1146/annurev.ecolsys.32.081501.114116

Luthans, F., & Youssef, C. M. (2007). Emerging Positive organizational behavior. *Journal of Management*, *33*(3), 321–349. doi:10.1177/0149206307300814

Luthar, S. S., Cicchetti, D., & Becker, B. (2000). The construct of resilience: A critical evaluation and guidelines for future work. *Child Development*, *71*(3), 543–562. doi:10.1111/1467-8624.00164 PMID:10953923

Luthar, S. S., Sawyer, J. A., & Brown, P. J. (2006). Conceptual Issues in Studies of resilience: Past, present and future research. *Annals of the New York Academy of Sciences*, *1094*(1), 105–115. doi:10.1196/annals.1376.009 PMID:17347344

Maak, T., & Pless, W. M. (Eds.). (2006). *Responsible Leadership: A Rational Approach*. London: Routledge.

Mabogunje, A. (2009). *Land Reform in Nigeria: Progress, Problems & Prospects*. Retrieved March 10th, 2017, siteresources.worldbank.org/EXTARD/resources/Mabogunje

Mabogunje, A. L. (2009). *Land Reform in Nigeria: Progress, Problems & Prospects*. Retrieved March 11th, 2017, siteresources.worldbank.org/.../mabogunje

Machado, G. (2007). *Morpho-sedimentary analysis of the beach, and internal continental shelf of the coastline of the Jurubatiba National Park* (PhD dissertation). Federal University of Rio de Janeiro, Rio de Janeiro, Brazil.

Machida, M., Irwin, B., & Feltz, D. (2013). Reliance in competitive athletes with spinal cord injury: The role of sports participation. *Qualitative Health Research*, *23*(8), 1054–1065. doi:10.1177/1049732313493673 PMID:23771633

Maduabum, C. P. (2014), What has public administration got to do with it? Lagos: Inaugural lecture, National Open University of Nigeria.

Majchrzak-Czarnecka, M., Rafalik, N., & Zasadzińska-Baraniewska. (2015, October 8). Masowy napływ cudzoziemców – wyzwanie humanitarne I polityczne. 5-9. Rzadowe Centrum Bezpieczeństwa (RCB). Centrum Operacyjno-analityczne, Warsaw, Poland.

Mamdani, M. (2004). *Citizen and Subject: Contemporary Africa and the Legacy of Late Colonialism*. Fountain Publishers.

Mann, S. C. (2011). *Creating Resilient, Sustainable Local Governments: Merging Human Resource Management and Emergency Management* (Unpublished doctoral dissertation). Mississippi State University.

Mansbridge, J. (Ed.). (1990). *Beyond Self interest*. Chicago: University of Chicago Press.

Mardis, R. M. (2015). *Crisis, Strategy, And Response: Investigating The State of Strategic Leadership in Emergency Management* (Unpublished doctoral dissertation). Capella University.

Marriott, C., Hamilton-Giachritis, C., & Harrop, C. (2014). Factors promoting resilience following childhood sexual abuse: A structural, narrative review of the literature. *Child Abuse Review*, *23*(1), 17–34. doi:10.1002/car.2258

Maslow, A. (1954). *Motivation and Personality*. New York: Harper.

Masten, A. S. (2013). Risk and resilience in development. In P. D. Zelado (Ed.), Oxford Handbook of Developmental Psychology: Vol. 2. Self and others (pp. 579-607). New York, NY: Oxford University Press.

Masten, A. S., Cutuli, J. J., Herbers, J. E., & Reed, J. M.-G. (2007). Resilience in development. In The Oxford Hand Book of Positive Psychology. Oxford, UK: Oxford University Press.

Masten, A. S., & Obradovic, J. (2006). Competence and resilience in development. *Annals of the New York Academy of Sciences*, *1094*(1), 13–27. doi:10.1196/annals.1376.003 PMID:17347338

Masten, A. S., & Reed, M. J. (2002). Resilience in development. In C. R. Snyder & S. S. Lopez (Eds.), *Handbook of positive psychology* (pp. 74–78). New York: Oxford University Press.

Matczak, P., & Abgarowicz, G. (2013). *Analysis of civil security systems in Europe. Country study: Poland* [Report]. Poznań, Poland: University of Adam Mickiewicz. Retrieved from http://www.anvil-project.net/wp-content/.../01/Poland_v1.0.pdf

Mathews, J. (1989). Redefining Security. *Foreign Affairs*, *68*(2), 2. doi:10.2307/20043906 PMID:12343986

Mbow, C. (2010). *Africa's risky gamble*. IGBP.

McCandless, E., & Karbo, T. (2011). *Peace, Conflict, and Development in Africa*. University for Peace. Retrieved from http://www.upeace.org/pdf%5CREADER_webpages.pdf

McCormick J. (2001). *Environmental policy in the European Union*. The European Series. Palgrave.

McCrimmon, M. (2007). Reframing Leadership for a modern age. *Integral Leadership Review, 7*(2), 1-6.

McEntire, D. A. (2004). *The Status of Emergency Management Theory: Issues, Barriers, and Recommendations for Improved Scholarship*. Paper Presented at the *FEMA Higher Education Conference*, Emmitsburg, MD.

McFadden, L., & Green, C. (2007). Defining 'vulnerability': Conflicts, complexities and implications for Coastal Zone Management. *Journal of Coastal Research*, 120–124.

McGuire, M. (2006). Collaborative Public Management: Assessing what we know and how we know it. *Public Administration Review*, *66*(s1), 33–43. doi:10.1111/j.1540-6210.2006.00664.x

McLuskey, D. (2001). *Extension practice in creating a sustainable agriculture*. Australasia-Pacific Extension Network International Conference 3-5 October 2001, Toowoomba, Australia.

McNeil, B. I. (2006). Oceanic implications for climate change policy. *Environmental Science & Policy*, *9*(7-8), 7, 8, 595–606. doi:10.1016/j.envsci.2006.08.006

Meadows, D. (1972). *The Limits to Growth: A report for the Club of Rome's project on the predicament of mankind, Part 1 Club of Rome*. Potomac Associates.

Mehran, M. G., & Kiani, A. (2014). A Study on the Relationship Between Commercial Land Use and Increase of Urban Travels (Case study: City Center of Urmia). *Trends in Life Sciences: An International Review Journal*, *3*(4), 21–27.

Meier, B. M., Pardue, C., & London, L. (2012). Implementing community participation through legislative reform: A study of the policy framework for community participation in the Western Cape province of South Africa, International. *Health and Human Rights*, *12*(15), 1–14.

Meiners, R., & Yandle, B. (1998). *The common law: How it protects the environment*. PERC. Retrieved from http://www.perc.org/articles/common-law-0 Retrieved 12/3/2017

Mendis-Millard, S., & Reed, M. G. (2007). Understanding community capacity using adaptive and reflexive research practices: Lessons from two Canadian biosphere reserves. *Society & Natural Resources*, *20*(6), 543–559. doi:10.1080/08941920601171915

Mensah, A. M., & Castro, L. C. (2004). *Sustainable Resource Use & Sustainable Development: A Contradiction?!* Zentrum fur Entwicklungsfourschung (ZEF), Center for Development Research University of Bonn. Retrieved from http://www.zef.de/fileadmin/downloads/forum/docprog/Termpapers/2004_3b_Mensah_Castro.pdf

Messner, F., & Meyer, V. (2006). Flood Damage, Vulnerability and Risk Perception-Challenges for Flood Damage Research. In J. Schanze, E. Zeman, & J. Marsalek (Eds.), *Flood Risk Management: Hazards, Vulnerability and Mitigation Measures* (pp. 149–167). Springer Netherlands. doi:10.1007/978-1-4020-4598-1_13

Migotto, M., Davis, B., Carletto, G., & Kathleen, B. (2006). *Measuring Food Security using repondents' perception of food Consumption Adequacy.* Research paper No. 2006/88, UNV WIDEr.

Millennium Ecosystem Assessment (MEA). (2005). *Millennium Ecosystem Assessment*. Washington, DC: Island Press. Available at www.millenniumassessment.org/en/index.aspx

Ministry of Environment and Forests. (2012). *India's Second National Communication to the United Framework Convention on Climate Change*. Government of India. Retrieved from http://pib.nic.in/newsite/PrintRelease.aspx?relid=82211

Ministry of Finance and National Planning (MFNP). (2009). *Annual Progress Report of the Fifth National Development Plan*. Lusaka, Zambia: MFNP.

Mische, P. (1989, February 18). Arctic Expedition Finds Chemical Threat to Ozone. *New York Times*.

Mitchell, B. (1989). *Geography and Resource Analysis*. Longman Group.

Mmon, P., & Arokoyu, S. (2010). Mangrove forest depletion, biodiversity loss and traditional resources management practices in the Niger Delta, Nigeria. *Research Journal of Applied Sciences, Engineering and Technology*, 2, 28–34.

Mogborukor, J. (2014). The Impact of Oil Exploration and Exploitation on Water Quality and VegetalResources in a Rain Forest Ecosystem of Nigeria. *Mediterranean Journal of Social Sciences, 5*(27).

Mohammed, T. M. (2013). National framework for the protection of internally displaced persons (IDPs) in Nigeria. Presentation made at a workshop for Judges and Kadis on refugee law, organized by The National Judicial Institute, Abuja.

Mooney, E. (2005). The Concept of Internal Displacement and the Case for Internally Displaced Persons as a Category of Concern. *Refugee Survey Quarterly*, *24*(3), 9–26. doi:10.1093/rsq/hdi049

Moore, M. H. (1995). Creating Public value: Strategic Management in Government. Cambridge, MA: Harvard University Press.

Morag, N. (2011). *Comparative Homeland Security: Global lessons* [Kindle]. Hoboken, NJ: John Wiley & Sons, Inc. doi:10.1002/9781118048252

Morel, B., & Igor, L. (Eds.). (2006). *Environmental Security and Environmental Management: The Role of Risk Assessment*. Springer. doi:10.1007/1-4020-3893-3

Morse, R. S., Buss, T. F., & Kinghorn, C. M. (Eds.). (2007). *Transforming Public Leadership for the 21st Century*. Armonk, NY: M.E. Sharpe.

Mountjoy, N. J. (2014). *Community-Based Natural Resource Management: Group Capacity, Resource Management Planning and Assessing Success* (Unpublished doctoral dissertation). Southern Illinois University.

Mozambique, D. N. G. A. (2005). *Vulnerability assessment for climate change and adaptation strategies, Ministry for the coordination of environmental action*. National Directorate of Environmental Management.

Muehe, D. (2004). Definition of boundaries and typologies of the border under morphodynamic and evolutionary aspects. In *Ministry of environment and planning, budget and management. Subsidies for a management project* (pp. 13–32). Brasília: MMA and MPO.

Muhanna, E. (2006). Sustainable Tourism Development and Environmental Management for Developing Countries. *Problems and Perspectives in Management*, *4*(2), 14–30.

Mukuna, T. E., & Shisanya, C. A. (2015). *Milestones in Green Transition and Climate Compatible Development in Eastern and Southern Africa*. Organisation for Social Science Research in Eastern and Southern Africa (OSSREA). Retrieved from http://www.ossrea.net/publications/images/introduction-ccd.pdf

Musa, S. D. (2005). Sustainable Grassroots Development: A major challenge of Globalization. In The Humanities and Globalization: the Africa perspective. Aboki Publishers.

Myer, R. A., James, R. K., & Moulton, P. (2010). *Or a Tornado or Earthquake Drill. This is Not a Fire Drill: Disaster Intervention and Prevention on College Campuses*. Hoboken, NJ: John Wiley& Sons. doi:10.1002/9781118269909

Myers, N. (1989). Environment and Security. *Foreign Policy*, *74*(74), 23–41. doi:10.2307/1148850

Naanen, B., & Tolani, P. (2014). *Private Gain Public Disaster Social Context of Illegal oil Bunkering and Artisanal Refining in the Niger Delta*. NIDEREF.

Naiken, L. (2003). FAO Methodology for Estimating the Prevalence of Undernourishment, Measures and Assessment of food Deprivation and Under nutrition. In *Proceedings of the International Scientific Symposium*. Rome: FAO.

Nath, V. (2001). *White Collar Invasion: Developed country policies leading to environmental degradation in South, Inlaks Fellow*. London School of Economics. Retrieved from www.vikasnath.org

National Ocean Service. (2016). *What is coastal resilience?* Retrieved from http;//oceanservice.noaa.gov/ecosystems/resilience/welcome.htm

National Planning Commission. (2016). *Nigeria's National Budget*. Abuja: NPC.

NCFRMI. (2009). *No End to Internal Displacement. Geneva: International Displacement Monitoring Centre (IDMC)*. Nigeria: Global Project.

Nehal, A. F., Majila, B. S., & Kala, C. P. (2004). Indigenous Knowledge Systems and Sustainable Management of Natural Resources in a High-Altitude Society in Kumaun Himalaya, India. *Journal of Human Ecology (Delhi, India)*, *16*(1), 33–42.

NEMA. (2017). *Internally displaced persons*. Retrieved June 24, 2017 from www.nema.gov.ng/internallydisplacedpersons

New Delhi Intergovernmental Panel on Climate Change (IPCC). (2007). Fourth Assessment Report of the Intergovernmental Panel on Climate Change. *Intergovernmental Panel on Climate Change*. Retrieved from https://www.ipcc.ch/pdf/assessment-report/ar4/syr/ar4_syr_full_report.pdf

Newell, T., Reeher, G., & Ronayne, P. (Eds.). (2008). Relationships that make Government work. Washington, DC: CQ Press.

Nhantumbo, I., Norfolk, S., & Pereira, J. (2003). *Community Based Natural Resources Management in Mozambique: A Theoretical or Practical Strategy for Local Sustainable Development? The Case Study of Derre Forest Reserve*. Sustainable Livelihoods in Southern Africa Research Paper 10, Institute of Development Studies.

Nicholls, R. J. (2003). *Case Study on Sea-Level Rise Impacts*. Working Party on Global and Structural Policies, OECD Workshop on the Benefits of Climate Policy: Improving Information for Policy Makers. Ref: ENV/EPOC/GSP(2003)9/FINAL. OECD.

Nicholls, R. J., & Small, C. (2002). Improved Estimates of Coastal Population and Exposure to Hazards Released. *EOS Transactions*, *83*(2), 301–305. doi:10.1029/2002EO000216

Nicholson, S. (2000). Land surface processes and Sahel climate. *Reviews of Geophysics*, *38*(1), 117–339. doi:10.1029/1999RG900014

Nigerian Environmental Study/Action Team (NEST). (2004). Available at www.climatenetwork.org/profile/.../nigerian-environmental-studyaction-team-nes

Nigerian Monitor. (n.d.). *List of Nigerian States at risk of experiencing massive earthquakes before 2020*. Retrieved from www.nigerianmonitor.com/nigerian-states-risk-experiencing-massivarthquake-2020-see-full-list/

Nittel, S. (2009). A survey of geosensor networks: Advances in dynamic environmental monitoring. *Sensors (Basel)*, *9*(7), 5664–5678. doi:10.3390/s90705664 PMID:22346721

Noordwick, M., Tomich, T. P., & Verbist, B. (2001). Negotiation Support Models for Integrated Natural Resource Management in Tropical Forest Margins. *Conservation Ecology*, *5*(2).

Nour, A. M. (2011). Challenges and Advantages of Community Participation as an Approach for Sustainable Urban Development in Egypt. *Journal of Sustainable Development*, *4*(1), 79–91.

Nriagu, J., Udofa, E., Ekong, I., & Ebuk, G. (2016). Health Risks Associated with Oil Pollution in the Niger Delta. *NigeriaInternational Journal of Environmental Research and Public Health*, *13*(3), 346. doi:10.3390/ijerph13030346 PMID:27007391

Nwanna, C. (2004). *Rural-Urban Migration and Population Problems in Nigerian Cities. In Industrialization, Urbanization and Development in Nigeria 1950-1999*. Lagos: Concept Publications.

O'Flynn, J. (2007). From New Public Management to Public value: A paradigmatic change and managerial implications. *Australian Journal of Public Administration*, *66*(3), 353–366. doi:10.1111/j.1467-8500.2007.00545.x

O'Leary, R. C., General, R. C., & Bingham, L. B. (2006). Intrdocution to the symposium on collaborative public management. *Public Administration Review*, *66*(s1), 6–9. doi:10.1111/j.1540-6210.2006.00661.x

O'Shea, P. G., Foti, R. J., Havenstein, N. M. A., & Bycio, P. (2009). Are the best leaders both transformational and transactional? A pattern analysis. *Leadership*, *5*(2), 237–259. doi:10.1177/1742715009102937

Oakley, P. (1989). *Community Involvement in Health Development: An examination of critical issues*. Geneva, Switzerland: World Health Organization.

Oakley, P. (1991). *Projects with People: The Practice of Participation in Rural Development*. Geneva: International Labour Office.

Obeta, M. C. (2009). Extreme river flood events in Nigeria: A geographical perspective of Nigeria. *Journal of Geography and the Environment*, *1*, 170–179.

Obi, C. (2001). *The Changing Forms of Identity Politics in Nigeria under Economic Adjustment: The Case of the Oil Minorities Movement of the Niger Delta*. Research Report Nordiska Afrikainstitutet No.11.9.

Obi, C. (2000). Globalised Images of Environmental Security in Africa. *Review of African Political Economy*, *27*(83), 47–62. doi:10.1080/03056240008704432

Obikaeze, V. C., & Onuoha, C. B. (2016). The Nigerian State and the Management of Internally Displaced Persons 2012-2016. *African Journal of Politics and Society*, 4-21. Retrieved from www.academix.ng

Odusote, A. (2016). Nigeria: The Matrix between Fragility of Livelihoods and Conflicts. Journal of Global Initiatives –Policy, Pedagogy. *Perspectives*, *10*(2), 9–30.

Oduwole, T. A., & Fadeyi, A. D. (2013). Issues of refugees and displaced persons in Nigeria. *Journal of Sociological Research.*, *4*(1), 1–18.

OECD. (2001). *Policies to Enhance Sustainable Development*. Organization for Economic Co-Operation and Development (OECD). Retrieved from https://www.oecd.org/greengrowth/1869800.pdf

OECD. (2007). *Instrument Mixes for environmental policy*. Retrieved from http://www.oecd.org/department

OECD. (2008). *Promoting Sustainable Consumption Good Practices in OECD Countries*. OECD Publications. Retrieved from https://www.oecd.org/greengrowth/40317373.pdf

OECD. (2013). *Local Strategies for Youth Employment: Learning from Practice*. Retrieved from http://www.oecd.org/employment/leed/Local%20Strategies%20for%20Youth%20Employment%20FINAL%20FINAL.pdf

OECD/IEA. (2010). *IEA Energy Statistics*. Available online: http://www.iea.org/statist/index.htm

Office of Technology Assessment. (1995). *U.S. Congress, Office of Technology Assessment, Environmental Policy Tools: A User's Guide. U.S. Congress, Office of Technology Assessment, OTA-ENV-634*. Washington, DC: U.S. Government Printing Office, September.

Ogata, S., & Sen, A. (2003). *Human Security Now: Commission on Human Security*. Final Report. New York: United Nations.

Ogunleye-Adetona, C. I., & Oladeinde, C. (2013). The role of community self-help projects in rural development of Kwara state, Nigeria. *International Journal of Development and Sustainability*, *2*(1), 28–45.

Okoko, K. (2002). The Politics of Resource Extraction, Distribution and Sustainable Development. Nigerian Journal of Oil and Politics, 2(1), 118-128.

Okonta, I., & Douglas, O. (2001). *Where cultures feast: Forty years of Shell in the Niger Delta*. Benin: ERA/FOEN.

Okorodudu, M.T. (1998). Law of environmental protection: Material and text. *RECIEL, 12*(2), 207.

Okorodudu-Fubura, M. T. (2013). *Environmental disasters in Nigeria: Policy and management.* Paper presented at the National Conference on the Environment (House of Representative Committee on the Environment), Transcorp Hilton Hotel, Abuja.

Okueso, S. A. (2008). Improving environmental management for sustainable development. *Nigeria. International Journal of Environmental Science*, *4*(1), 35–39.

Okueso, S. A., & Adekoya, A. F. (2006). *Chemical, health and environmental care for sustainable development in Nigerian*. Journal of Nigerian School Health Press.

Olagunju, O. (2006). *Management of Internal Displacement in Nigeria* (Unpublished Thesis). Brandeis University.

Olaniran, A.O. (2002). *Review of the state of agricultural enterprise in Nigeria*. Being the 2nd LASU CDDS lecture series held at the Nigerian Institute of International Affairs, Lagos, Nigeria.

Olaopa, T. (2016). *The state of public administration and sustainable development in Africa: Towards a bold and functional public administration*. Guest lecture at the 2nd International Conference, Department of Public Administration LASU.

Olawoye, J. E. (2010). Sustainable rural development: Challenges, opportunities and lessons from Nigeria. In L. Popoola & O. Olorunmisola (Eds.), Contemporary Issues in Sustainable Development: Lessons for, and Challenges to Nigeria. Ibadan: University, of Ibadan.

Olojede, I., Fajonyomi, B., & Akhakpe, I. (2013). The Ecology – security nexus: oil, environmental pollution, community unrests and national security in Nigeria. In A. Ikelegbe (Ed.), *Oil, Enviornment and Resource Conflicts in Nigeria*. Zurich: UT VERLAG Gmb IT and Co.

Olokesusi, A. (1987). Characteristics of Environmental Problems in Nigeria and management prospects. *The Environmentalist*, *1*(2), 44–57.

Olujimi, J. (2009). Evolving a Planning Strategy for Managing Urban Sprawl in Nigeria. *Journal of Human Ecology (Delhi, India)*, *25*(3), 201–208.

Olujimi, J., & Gbadamosi, K. (2007). Urbanization: Innovation Policy and Sustainable Development. *Proceedings of International Conference Organised By Covenant University*, 246-252.

Oluyede, P. A. (1985). Development in Land Law and Law of Conveyancing. In *The Challenge of the Nigerian Nation: An Examination of its Legal Development 1960-1985*. Nigerian Legal Institute of Advanced Legal Studies.

Omotosho, J. B., & Abiodun, J. (2007). A numerical study of moisture build-up and rainfall over West Africa. *Meteorological Applications*, *14*(3), 209–225. doi:10.1002/met.11

Omran, A., & Read, A. D. (2008). "Waste Not, Want Not" A Study of Household Attitude toward Recycling of Solid Wastes. *Environmental Engineering and Management Journal*, *7*, 1–8.

Önen, Y. (2002). *Kentsel Mekan, Çevre, Çoğulculuk ve İnsan Haklarına Genel Bir Yaklaşım, İnsan, Çevre, Kent*. İstanbul: Demokrasi Kitaplığı.

OneWorld. (2010). *Climate Change in Zambia*. London, UK: One World, Briefing Guide. Available at http://UK.Oneworld.net/guides/Zambia/climate-change

Onibokun, A. G., & Faniran, A. (1995). "Urbanization and Urban Social Problems in Nigeria." In Urban Research in Nigeria. Ibadan: IFRA, CASSAD.

Onosode, G. (2003). *Environmental issues and the challenges of the Niger Delta; Perspectives from the Niger Delta. Environmental Survey Process*. Yaba: The CIBN Press Limited.

Orubu, C. O. (1999). *The exploitation of non-timber forest resources in the Niger Delta: Problems and perspectives. Technical paper, Niger Delta Environmental Survey*. Port-Harcourt.

Orubu, C. O., Odusola, A., & Ehwarieme, W. (2004). The Nigerian oil industry: Environmental diseconomies, management strategies and the need for community involvement. *Journal of Human Ecology (Delhi, India)*, *16*(3), 203–214.

Osagioduwa, E., & Toluwanimi, O. O. (2016). Management of Internally Displaced Persons in Africa: Comparing Nigeria and Cameroon. Africa Research Review, 10(1), 193-210.

Oshio, E. (1990). The Indigenous Land Tenure System and Nationalization of Land in Nigeria. *Boston College Third World Law Journal*, *10*(1), 42–43.

Osho-Adejonwo, O. (2005). *The evolution of human rights approaches to environmental protection in Nigeria*. Retrieved March 12, 2010, from, www.iucnael.org/index.php?Optiocom_documans&task

Ostrom, E. (2005). *Understanding Institutional Diversity*. Princeton, NJ: Princeton University Press.

Osuide, S. (1990). Environmental pollution in Nigeria. *Habitat International*, *14*(1), 5–15. doi:10.1016/0197-3975(90)90014-R

Otubu, T. (2010). Land Reforms and the Future of Land Use Act in Nigeria. *Nigerian Current Law Review, 2007-2010*, 126–143.

Otu, E. C. (2016). Restoring environmental justice and affection in Ogoni Land. *International Affairs and Global Strategy*, *4*(4), 94–108.

Owei, E., Obinna, P. N., & Akarolo, C. I. (2008). Land Market Distortions in Nigerian Cities and Urban Sprawl. *44th ISOCARP Congress.*

Oyebande, L. (1998). *The Water Cycle: Courier of Health and Wealth, How Sustainable is it? Inaugural Lecture Series.* University of Lagos Press.

Oyeshola, O. P. (2008). *Sustainable development: issues and challenges for Nigeria*. Ibadan: Daily Graphics Nigeria Ltd.

Oyesiku, O. K. (2010). *New Cities in Urban and Regional Planning*. Lagos: Longman.

Ozer, E. J., Best, S. R., Lipsey, T. L., & Weiss, D. S. (2003). Predictors of posttraumatic stress disorder and symptoms in adults: A meta-analysis. *Psychological Bulletin*, *129*(1), 52–71. doi:10.1037/0033-2909.129.1.52 PMID:12555794

Palabiyik, H. (1998). Çevre Sorunu Olarak Kentsel Katı Atıklar (Çöpler) ve Entegre Katı Atık Yönetimi. *Türk İdare Dergisi*, 420.

Paris, R. (2001). Human Security: Paradigm Shift or Hot Air? *International Security*, *2*(26), 87–102.

Pasteur, K. (2011). *From Vulnerability to Resilience (V2R)*. Warwickshire, UK: Practical Action. doi:10.3362/9781780440583

Patil, H. A., & Patil, V. K. (2015, July). Survey paper on multiple input multiple output (mimo) system analysis and simulations. Journal of Emerging Technologies and Innovative Research, 2(7).

Pat, R. (2010). *Health and Environment. Trance, Ferney Voltaire*. World Medical Association.

Paul, S. (1987). *Community Participation in Development Projects: The World Bank Experience*. World Bank Discussion papers. Retrieved from http://www.ircwash.org/sites/default/files/Paul-1987-Community.pdf

Paul, B. D., & Demarest, W. J. (1984). Citizen participation over planned: The case of health project in the Guatemala community of San Pedro Laguna. *Social Science & Medicine*, *19*(3), 185–192. doi:10.1016/0277-9536(84)90209-0 PMID:6484608

Pęcillo, M. (2016). The concept of resilience in OSH management: A review of approaches. *International Journal of Occupational Safety and Ergonomics*, *22*(2), 291–300. doi:10.1080/10803548.2015.1126142 PMID:26652938

Peduzzi, P., Dao, H., Herold, C., & Mouton, F. (2009). Assessing global exposure and vulnerability towards natural hazards: The disaster risk index. *Natural Hazards and Earth System Sciences*, *9*(4), 1149–1159. doi:10.5194/nhess-9-1149-2009

Peet, E., & Watts, W. (1996). Liberation Ecology: Development, Sustainability and Environment in an Age of Market Triumphalism. In R. Peet & M. Watts (Eds.), *Liberation Ecologies: Environment, Development, Social movements*. London: Routledge. doi:10.4324/9780203286784

Peres, J., Moreira-Almeida, A., Nasello, A., & Koinig, H. (2007). Spirituality and resilience in trauma victim. *Journal of Religion and Health*, *46*(3), 343–350. doi:10.1007/s10943-006-9103-0

Peters, C. B. (2016). *Theory and Information for Natural Resource Management* (Unpublished doctoral dissertation). University of California.

Petery, B. G., & Pierre, J. (1998). Government without Governance?:Rethinking. *Public Administration Research and Theory*, *8*(2), 223–243. doi:10.1093/oxfordjournals.jpart.a024379

Pfefferbaum, R. L., Pfefferbaum, B., van Horn, R. L., Klomp, R. W., Norris, F. H., & Reissman, D. B. (2013). The communities advancing resilience toolkit (CART): An intervention to build community resilience to disasters. *Journal of Public Health Management and Practice*, *19*(30), 250–258. doi:10.1097/PHH.0b013e318268aed8 PMID:23524306

Pielke, R. A. Sr, Adegoke, J., Beltran-Przekurat, A., Hiemstra, C. A., Lin, J., Nair, U. S., & Nobis, E. et al. (2007). An overview of regional land-use and land-cover impacts on rainfall. *Tellus*, *59B*(3), 587–601. doi:10.1111/j.1600-0889.2007.00251.x

Pigou, A. (1920). *The Economics of welfare*. London: Macmillan and Company.

Pirages, D. (1991). Social Evolution and Ecological Security. *Bulletin of Peace Proposals*, *22*(3), 329–334. doi:10.1177/096701069102200310

Poland Ukraine Locator. (2008). Retrieved from https://commons.wikimedia.org/wiki/File:Poland_Ukraine_Locator.png

Polska Agencja Prasowa (Polish Press Agency). (2014, August 21). UDSC: w 2014 r. już 1200 wniosków obywateli Ukrainy o status uchodźcy [Bureau for Foreigners: In 2014 already 1200 asylum petitions by Ukrainian citizens]. *Gazeta Wyborcza*. Retrieved from http://wyborcza.pl/1,91446,16508207,UDSC__w_2014_r__juz_1200_wnioskow_obywateli_Ukrainy.h

Prasad, M. S. R., & Khasim, S. (2013). Structural Monitoring Of Buildings Using Wireless Sensor Networks. *International Journal of Modern Engineering Research*, *3*(3).

Preet, V. D. (2006). *Natural Hazard and Disaster Management*. Delhi: Shakarpur.

Premium Times. (2015). *Nigeria sets new record, now has Africa's highest number of displaced persons*. Retrieved from http//premiumtimesng.com/news

Pretty, J. (1998). Supportive policies and practices for scaling up sustainable agriculture. In N. G. Röling & M. A. E. Wagemakers (Eds.), *Facilitating sustainable agriculture: participatory learning and adaptive management in times of environmental uncertainty*. Cambridge, UK: Cambridge University Press.

Prizzia, R. (2005). Community involvement in protecting the environment: The role of Restoration Advisory (RABS). *The Innovation Journal*, *10*(1).

Protection Sector Working Group. (2015). *IDPs Protection Strategy*. Abuja: PSWG.

Protoski, M., & Prakash, A. (2004). The Regulation Dilemma: Cooperation and Conflict in Environmental Governance. *Public Administration Review*, *64*(2), 152–163. doi:10.1111/j.1540-6210.2004.00357.x

Prusty, A. R., Sethi, S., & Nayak, A. K. (2015, December). Analysis of energy complexity under mobility for cluster routing in wireless ad hoc sensor networks. In *Microwave, Optical and Communication Engineering (ICMOCE), 2015 International Conference on* (pp. 314-317). IEEE. doi:10.1109/ICMOCE.2015.7489755

Prusty, A. R., Sethi, S., & Nayak, A. K. (2016b). Energy Aware Optimized Routing Protocols for Wireless Ad Hoc Sensor Network. Handbook of Research on Advanced Wireless Sensor Network Applications, Protocols, and Architectures, 156. DOI:10.4018/978-1-5225-0486-3.ch006

Prusty, A. R. (2012). The Network and Security Analysis for Wireless Sensor Network: A Survey. *International Journal of Computer Science and Information Technologies*, *3*(3), 4028–4037.

Prusty, A. R., Sethi, S., & Nayak, A. K. (2016a). A hybrid multi-hop mobility assisted heterogeneous energy efficient cluster routing protocol for Wireless Ad hoc Sensor Networks. *Journal of High Speed Networks*, *22*(4), 265–280. doi:10.3233/JHS-160548

Pryor, P. (2012). *HaSPA (Health and Safety Professional Alliance). The core body of knowledge for generalist OHS Professionals. Tullamarine*. VIC Safety Institute of Australia.

Pyle, G. F. (1979). *Applied Medical Geography*. Washington, DC: V. H. Winston and Sons.

Quarantelli, E. L. (1993). Community crises: An exploratory comparison of the characteristics and consequences of disasters and riots. *Journal of Contingencies and Crisis Management*, *1*(2), 67–78. doi:10.1111/j.1468-5973.1993.tb00009.x

Raaijmakers, R., Krywkow, J., & van der Veen, A. (2008). Flood risk perceptions and spatial multicriteria analysis: An exploratory research for hazard mitigation. *Natural Hazards*, *46*(3), 307–322. doi:10.1007/s11069-007-9189-z

Rameesh, V. M., Rajan, P., & Pullarkat, D. (2014). Development of a resilient wireless sensor network for real-time outdoor applications. In *Proceeding on International Symposium on Technology Management and Emerging Technologies (ISTMET)*. IEEE. Doi:10.1109/ISTMET.2014.6936476

Ramesh, V. M. (2014). Design, Development, And Deployment of a Wireless Sensor Network for Detection of Landslides. *Ad Hoc Networks, 13*(2014), 2-18.

Ramsay, J., & O'Sullivan, T. (2013). There's a Pattern Here: The Case to Integrate Environmental Security into Homeland Security Strategy. *Homeland Security Affairs, 9*, Article 6. Retrieved from https://www.hsaj.org/articles/246

Randall, H. (2015-2016). All hazards Response starts at the local level. *PA Times Winter*, *2*(1), 22-3.

Rashed, T., & Weeks, J. (2003). Assessing vulnerability to earthquake hazards through spatial multicriteria analysis of urban areas. *International Journal of Geographical Information Science*, *17*(6), 547–576. doi:10.1080/1365881031000114071

Raymond, P. (1974). Community Ideology. In *An essay in applied social philosophy*. London: Routledge and Kegan Paul Ltd.

Reivich, K. J., Seligman, M. E. P., & McBride, S. (2011). Master resilience training in the U.S. Army. *The American Psychologist*, *66*(1), 25–34. doi:10.1037/a0021897 PMID:21219045

Rennie, J. K., & Singh, N. (1996). *Participatory Research for Sustainable Livelihoods*. Winnipeg: IISD, Small study for Directorate B Environmental Instruments Directorate General Environment European Commission (B4-3040/2000/258081/MAR/B2). Retrieved from http://ec.europa.eu/environment/enveco/waste/pdf/zeeuw.pdf

Reporters, S. (2015). *Maroko Evictees Mark 25 Years of July ""90 Massacre Over Lekki" "*. Retrieved 14th March, 2017, http:// secure.saharareporters.com/2015/11/11/maroko-evictees-25-years

Reuters. (2012). *Nigeria floods kill 137, displace thousands*. Retrieved from https://en.wikipedia.org/wiki/2012_Nigeria_floods

RIANovosti. (2014, September 24). Over 300,000 Ukraine Refuges seek Asylum in Russia: Federal Migration Service. *RIANovosti*. Retrieved from http://en.ria.ru/russia/20140924/193243399/Over-300000-Ukrainians-Seek-Asylum-in-Russia-Federal-Migration.html

Richardson, D. (2003). *Community-based solid waste management systems in Hanoi, Vietnam* (MSc Thesis).

Richardson, G. E. (2002). The metatheory of resiience and resiliency. *Journal of Clinical Psychology*, *58*(3), 307–321. doi:10.1002/jclp.10020 PMID:11836712

Riche, B. (2007). *Climate Change Vulnerability Assessment in Zambia*. Geneva, Switzerland: The World Conservation Union.

Ries, C. P., Jenkins, J., & Wise, O. (2009). *Improving the Energy Performance of Buildings: Learning from the European Union and Australia. Technical Report.* Rand Corporation.

Rim-Rukeh, A., Irerhievwie, G., & Agbozu, I. E. (2013). Traditional beliefs and conservation of natural resources: Evidences from selected communities in Delta State, Nigeria. *International Journal of Biodeversity and Conservation, 5*(7), 426–432.

Roach, B., Harris, J. M., & Codur, A. (2015). *Microeconomics and the Environment*. Medford, MA: Global Development and Environment Institute, Tufts University. Retrieved from http://www.ase.tufts.edu/gdae/education_materials/modules/Microeconomics_and_the_Environment.pdf

Robbins, P. (2004). What is political ecology? In P. Robbins (Ed.), *Political ecology: A critical introduction* (pp. 5–16). Malden, MA: Blackwell Publishing.

Roberts, A. (1995). Civic Discovery as a Rhetorical Strategy. *Journal of Policy Analysis and Management Policy Analysis and Management, 14*(2), 291–307. doi:10.2307/3325154

Roe, D., Nelson, F., & Sandbrook, C. (Eds.). (2009). *Community management of natural resources in Africa: Impacts, experiences and future directions, Natural Resource Issues No. 18.* London, UK: International Institute for Environment and Development.

Roggers, K. (1997). Article. *Ecological Security and Multinational Corporations, 2*, 7–17.

Rohlmann, M. (1992*). Integrated Natural Resources Management: A Question of Property Institutions?* (Unpublished master thesis). York University, Canada.

Rushefsky, M. E. (2002). Public policy in the united states at the down of the twenty-first century network. M. E. Sharpe, Inc.

Ruther, M. (1985). Resilience in the face of adversity: Protective factors and resilience to psychiatric disorders. *The British Journal of Psychiatry, 147*(6), 598–611. doi:10.1192/bjp.147.6.598 PMID:3830321

Ruther, M. (1987). Psychosocial resilience and protective mechanism. *The American Journal of Orthopsychiatry, 57*(3), 316–331. doi:10.1111/j.1939-0025.1987.tb03541.x PMID:3303954

Rutther, M. (1979). Protective factors in children's response to stress and disadvantage. In M. W. Kent & J. E. Rolf (Eds.), Primary prevention of psychopathology: vol. 3. Social competence in children (pp. 49-74). Hanover, NH: University Press of New England.

Rzadowe Centrum Bezpieczenstwa (Government Security Center). (2014). *Działania resortu spraw wewnętrznych w związku z sytuacją na Ukrainie* [Activities of the interior affairs sector related to the situation in Ukraine]. Author. Retrieved from rcb.gov.pl/wp-content/uploads/biuletyn/6.pd

Saita, J., Sato, T., & Nakamura, Y. (2008). *What is the useful application of the earthquake early warning system?* In The 14th World Conference on Earthquake Engineering, Beijing, China.

Samuel, C. (2014). *The Current Status of Hazard Mitigation in Local Emergency Management: An Examination of Roles, Challenges, and Success Indicators* (Unpublished doctoral dissertation). University of North Texas.

Sandal, A. (2001). Sekizinci Beş Yıllık Plana Çevre Yönetimi Açısından Bakış. *Çevre ve İnsan Dergisi*, 48.

Sanderson, N. (2004). *Application of Collaborative Resources Management to Developing a Coastal Resources Management Plan for Olango Island, Philippines* (Unpublished master dissertation). University of Calgary.

Sartori, F., Lai, R., Ebinger, M. H., & Parish, D. J. (2006). Potential soil carbon sequestration and Co_2 offset by dedicated energy crops in the USA. *Critical Reviews in Plant Sciences, 25*(5), 441–472. doi:10.1080/07352680600961021

Sasnal, P., & Uzelac, A. (2014). Unaided refugees make ISIS stronger: Europe and the worst humanitarian crisis of the 21st century. *Polish Institute of International Affairs, 102*(19), 1-8. Retrieved from https://www.pism.pl/files/?id_plik=18847

Sayer, J. A., & Campbell, B. (2001). Research to Integrate Productivity Enhancement, Environmental Protection, and Human Development. *Conservation Ecology, 5*(2).

Sayer, J. A., & Campbell, B. M. (2003). Research to Integrate Productivity Enhancement, Environmental Protection and Human Development. In J. A. Sayer & B. M. Campbell (Eds.), *Integrated Natural Resources Management, Linking Productivity, the Environment and Development* (pp. 1–14). Cromwell Press. doi:10.1079/9780851997315.0001

Schallenberg, M., Hall, C. J., & Burns, C. W. (2003). Consequences of climate-induced salinity increases on zooplankton abundance and diversity in coastal lakes. *Marine Ecology Progress Series, 251*, 181–189. doi:10.3354/meps251181

Schernewski, G., & Löser, N. (2004). *The modern tourist's perception of the beach: Is the sandy beach a place of conflict between tourism and biodiversity?* Managing the Baltic Sea. Coastline Reports.

Scherr & McNeely. (2008). McNeely Biodiversity conservation and agricultural sustainability: towards a new paradigm of 'ecoagriculture' landscapes. *Phil. Transactions of the Royal Society. B, 363*, 477–494

Schnytzer, A., & Schnytzer, Y. (2011). *Animal modeling of earthquakes and prediction markets* (No. 2011-20). Working Papers, Bar-Ilan University, Department of Economics.

Schriesheim, C. A., Castro, S. L., Zhou, X., & Dechurch, L. A. (2006). An investigation of path-goal and transformational leadership theory predictions at the individual level of analysis. *The Leadership Quarterly, 17*(1), 21–38. doi:10.1016/j.leaqua.2005.10.008

Schulze, E. D., & Mooney, H. A. (1993). *Biodiversity and ecosystem function*. Berlin, Germany: Springer. doi:10.1007/978-3-642-58001-7

Sehlke, G. (2016). *A Conceptual Framework for Developing and Implementing Integrated Water Resources Management (Iwrm) Programs In The United States* (Unpublished doctoral dissertation). University of Idaho.

Shaeffer, S. (1998). *Partnerships and participation in basic education: A series of training modules and case study for educational planners and managers*. Paris: UNESCO, International Institute of Educational Planning.

Shafritz, J. M., Russell, E. W., & Borick, C. P. (2011). *Introducing Public Administration* (7th ed.). New York: Longman.

Shamroukh, R., & Awad, F. (2009, March). Detection of surviving humans in destructed environments using a simulated autonomous robot. In *Mechatronics and its Applications, 2009. ISMA'09. 6th International Symposium on* (pp. 1-6). IEEE. doi:10.1109/ISMA.2009.5164820

Sharad, R. P., & Thorat, P. R. (2013). Live Human Detecting Robot for Earthquake Rescue Operation. *International Journal on Recent and Innovation Trends in Computing and Communication, 1*(11).

Shastri, C. M., Bhat, D. M., Nagaraja, B. C., Murali, K. S., & Ravindranath, N. H. (2002). Tree species diversity in a village ecosystem in Uttara Kannada district in Western Ghats, Karnataka. *Scientific Correspondence, 82*, 1080–1084.

Shepard, A., & And Bowler, C. (1997). Beyond the requirement improving public participation in EIA. *Journal of Environmental Planning and Management, 40*(6), 725–738. doi:10.1080/09640569711877

Sherwood, H., Smith, H., Davies, L., & Grant, H. (2014, June 1). Europe faces 'colossal humanitarian catastrophe' of [illegible]gees dying at sea. *The Guardian*. Retrieved from http://www.theguardian.com/world/2014/jun/02/europe-refugee-[illegible]rica-processing-centres

[illegible] *alive: Women, ecology, and development*. London: Zed Books.

Sholanke, O. (1990). Is the Grant of Governor's Consent under the Nigerian Land Use Act automatic? *Journal of African Law*, *34*(1), 42–52. doi:10.1017/S0021855300008184

Silva, S. A. F. (2012). *Management of Beaches of Costa de Caparica. Load capacity, wave value, security and information* (Master thesis). Faculty of Science and Technology of the New University of Lisbon.

Silva, C. G. (2004). *Coastal Sedimentation Environments and Morphodynamic Processes Acting on the Coast Line. In Introduction to Marine Geology* (pp. 175–218). Rio de Janeiro: Intersciences.

Simon, D. (2007). Global change and urban risk: The challenge for African cities. *The Constitution. A Journal of Constitution Development*, *7*(1).

Singh, R., & Nagla, K. S. (2012). Proposed model of an earthquake detector by using UBG-04LX-F01 laser rangefinder. *International Journal of Advanced Research in Computer Science and Software Engineering*, *2*(5).

Skusiewicz, K. (2016). The "Ukrainian" crisis as a chance for the certain sectors of Polish economy. In Przedsiębiorczosc i zarzadzanie: Bezpieczenstwo i zarzadzanie kryzysowe – nowe wyzwania bezpieczenstwa narodowego. Lodz, Poland: Wydawnictwo Spolecznej Akademii Nauk.

Small, C., & Nicholls, R. J. (2003). A Global Analysis of Human Settlement in Coastal Zones. *Journal of Coastal Research*, *19*(3), 584–599.

Smith, D. (2009, July 6). Devastation in Zambia as Climate Change Brings Early flooding. *The Guardian*.

Smith, R. H. (2004). Focusing on Public Value: Something New and Something Old. *Australian Journal of Public Administration*, *63*(4), 68–79. doi:10.1111/j.1467-8500.2004.00403.x

Śnierzyński, M. (2009). Struktura bezpieczenstwa narodowego w Polsce [The structure of national security in Poland]. In, M. Włodarczyk & A. Marjański (Eds.), *Bezpieczeństwo i zarządzanie kryzysowe – aktualne wyzwania* [Security and crisis management: Current challenges] (pp. 57-68). Stołeczna Wyższa Szkoła Przedsiębiorczości i Zarządzania w £odzi, £ódz, Poland. Retrieved from http://www.swspiz.pl/wydawnictwo/pliki/2009/X-6.pdf

Solomon, S., Qin, D., Manning, M., Alley, R. B., Berntsen, T., Bindoff, N. L., & Wratt, D. et al. (2007). Technical Summary. In S. Solomon, D. Quin, M. Manning, Z. Chen, M. Marquis, K. B. Averyt, & H. L. Miller et al. (Eds.), *Climate Change 2007: The Physical Science Basis. Contribution of Working Group I to the Fourth Assessment Report of the Intergovernmental Panel on Climate Change*. Cambridge, UK: Cambridge University Press.

Sonwa, D. J., Walker, S., Nasi, R., & Kaanninen, M. (2011). Potential synergies of the main current forestry efforts and climate change Mitigation in Central Africa. *Sustainability Science*, *6*(1), 59–67. doi:10.1007/s11625-010-0119-8

Spaargaren, G. (2000). Ecological modernization theory and domestic consumption. *Journal of Environmental Policy & Planning*, *2*(4), 323-335.

Speake, S., & Gismondi, M. (2005). Water: A Human Right. In Consuming Sustainability Critical Social Analysis of Ecological Change. Fernwood Publishing.

Sperow, M., Eve, M., & Paustian, K. (2003). Potential soil C sequestration on US agricultural soils. *Climatic Change*, *57*(3), 319–339. doi:10.1023/A:1022888832630

Spyke, N. P. (1999). Public Participation in Environmental Decision making in the new millennium structuring new spheres of public influence. *Boston College Environmental Affairs Law Review. Boston College. Law School*, *26*(2), 263–313.

Steel, B. S., & Weber, E. (2001). Ecosystem management, decentralization, and public opinion. *Global Environmental Change*, *11*(2), 119–131. doi:10.1016/S0959-3780(00)00062-5

Stohlgren, T. J., Chase, T. N., Pielke Snr, R. A., Kittel, T. G. F., & Baron, J. S. (1998). Evidence that local land use practices influence regional climate, vegetation, and stream flow patterns in adjacent natural areas. *Global Change Biology*, *4*(5), 495–504. doi:10.1046/j.1365-2486.1998.t01-1-00182.x

Stoker, G. (2006). Public Value Management: A New Narrative for Networked Governance. *American Review of Public Administration*, *36*(1), 51–57. doi:10.1177/0275074005282583

Sun, X. (2015). Selected Enforcement of Land Regulations: Why Large-Scale Violators Succeed. *China Journal (Canberra, A.C.T.)*, *74*, 66–90. doi:10.1086/681938

Sustainable Scale Project. (2003). Retrieved from http://www.sustainablescale.org/conceptualnderstandingscale/measuringscale/EcologicalFootprint.asp

Szarpak, L. (2009). Aspekt prawny zarzadzania kryzysowego w Polsce (Legal aspect of crisis management in Poland). In M. Chrabkowsi, C. Tatarczuk, & J. Tomaszewski (Eds.), *Bezpieczeństwo w administracji i biznesie we współczesnym świecie [Security in administration and business in modern world]* (Vol. 1, pp. 145–155). Gdynia, Poland: Wyższa Szkola Administracji i Biznesu.

Szpak, A. (2016). Kryzys migracyjny w Europie a bezpieczenstwo – podloze spoleczno-ekonomiczne, zagrozenia, pomoc rozwojowa [Migration crisis in Europe and security – socio-economic background, threats, and development aid]. *Sprawy Miedzynarodowe, 1,* 111-132. Retrieved from https://www.ceeol.com/search/article-detail?id=425293

Taubenbock, H., Post, J., Roth, A., Zosseder, K., Strunz, G., & Dech, S. (2008). A conceptual vulnerability and risk framework as outline to identify capabilities of remote sensing. *Natural Hazards and Earth System Sciences*, *8*(3), 409–420. doi:10.5194/nhess-8-409-2008

Tawari-Fufeyin, P., Megbuwe, P., & Adams, O. G. (2015). Some aspects of a historic flooding in Nigeria and its effects on some Niger-Delta Communities. *American Journal of Water Resources*, *3*(1), 7–16.

The Conversation. (n.d.). Retrieved from theconversation.com/black-americans-may-be-more-resilient-to-stress-than-the-white-americans

The Guardian. (2008, January 15). Editorial: Gas flaring, end of a moving target? *The Guardian.*

The UNEP - DTIE International Environmental Technology Centre. (2016). *Participation: Enhancing Local Programmes.* Author.

Thisday. (2016). *Land-Grabbing Prohibition Law: All Eyes On Lagos State Government.* Retrieved March 14th, 2017, http://www. thisdaylive.com/index/2016/08/02

Thisday. (2017). *Feeding Fat.* Retrieved June 18, 2017 from www.thisday.com/index.php/feeding-fat

Thomas, R. (2002). *What is Integrated Natural Resource Management. Integrated Natural Resource Management: Putting Practice into Action.* Paper presented at the meeting of Fourth INRM Task Force, Aleppo, Syria.

Thomas, L. (2006). *Türkiye Korunan Alan Yönetiminde IUCN Kategori Sistemi. Çevre ve Orman Bakanlığı.* Doğa Koruma ve Milli Parklar Genel Müdürlüğü. Biyolojik Çeşitlilik ve Doğal Kaynak Yönetimi Projesi.

Thompson, A. M., & Perry, J. L. (2006). Collaboration Processes: Inside the black box. *Public Administration Review*, *66*(s1), 20–32. doi:10.1111/j.1540-6210.2006.00663.x

Thompson, J. R., Anderson, M. D., & Johnson, K. N. (2004). Ecosystem management across ownerships: The potential for collision with antitrust laws. *Conservation Biology*, *18*(6), 1475–1481. doi:10.1111/j.1523-1739.2004.00266.x

Thomson, T. M. (1998). *Management by Objectives* (2nd ed.). Jossey-Bass/Pfeiffer. Retrieved from http://home.snu.edu/~jsmith/library/body/v20.pdf

Thornton, P. K. (2010). Livestock Production: Recent Trends, Future Prospects. *Philosophical Transactions of the Royal Society of London. Series B, Biological Sciences, 365*(1554), 2853–2867. doi:10.1098/rstb.2010.0134 PMID:20713389

Thornton, P. K., Jones, P. G., Alagarswamy, G., & Andresen, J. (2009). Spatial Variation of Crop Yield Responses to Climate Change in East Africa. *Global Environmental Change, 19*(1), 54–65. doi:10.1016/j.gloenvcha.2008.08.005

Tichy, N. M., & Bennis, W. G. (2007). *Judgment: How Winning Leaders make Great calls*. New York: Penguin.

Touval, S. (1966). Treaties, Borders and the Partition of Africa. *Journal of African History, 7*(2), 279–293. doi:10.1017/S0021853700006320

Transport Research Wing, Ministry of Road Transport and Highways. (2011-12). *Road Transport Yearbook*. Government of India.

Treasure, D. C., Lemyre, P. N., Kuezka, K. K., & Standage, M. (2007). Motivation in elite-level sport. In Intrinsic motivation and self-determination in exercise and sports (pp. 153-165). Champaign, IL: Human Kinetics.

Tributsch, H. (1982). *When the Snakes Awake: Animals and Earthquake Prediction*. Cambridge, MA: MIT Press.

Trottier, T. C., Van Wart, M., & Wang, X. (2008). Examining the nature and significance of leadership in government organizations. *Public Administration Review, 68*(2), 319–333. doi:10.1111/j.1540-6210.2007.00865.x

Tugade, M. M., & Fredrickson, B. L. (2004). Resilient individuals use positive emotions to bounce back from negative emotional experiences. *Journal of Personality and Social Psychology, 86*(2), 320–333. doi:10.1037/0022-3514.86.2.320 PMID:14769087

Türköz, O. İ. (1997). Çevre Yönetimi ve Metotları. *Standart Ekonomik ve Teknik Dergi*, 421.

Turner, N. J., Ignace, M. B., & Ignace, R. (2000). Traditional Ecological Knowledge and Wisdom of Aboriginal Peoples in British Columbia. *Ecological Applications, 10*(5), 1275–1287. doi:10.1890/1051-0761(2000)010[1275:TEKAWO]2.0.CO;2

Ubani, E., & Onyejekwe, I. (2013). Environmental impact analyses of gas flaring in the Niger delta region of Nigeria. *American Journal of Scientific and Industrial Research, 4*(2), 246–252. doi:10.5251/ajsir.2013.4.2.246.252

Udoudo, A. (2006). *Newspaper coverage of Environmental Degradation in the Niger Delta* (Unpublished PhD Dissertation). University of Uyo.

Ugochukwu, O. (2003). *Delivering on our mandate in the journey so far*. Port Harcourt: NDDC.

Ulleberg, I. (2009). *The role and impact of NGOs in capacity development: From replacing the state to reinvigorating education*. International Institute for Educational Planning. Retrieved from http://unesdoc.unesco.org/images/0018/001869/186980e.pdf

Ullman, R. (1983). Redefining Security. International Security, 31, 129.

UN High Commissioner for Refugees (UNHCR). (2014). *Ukraine: Overview of population displacement*. Retrieved from http://www.refworld.org/docid/542a9cc04.html

UN Office for the Coordination of Humanitarian Affairs (OCHA). (2012). *Nigeria: Floods, Emergency Situation Report no 2*. Retrieved on October, 18, 2016, from, www.ochaonline.un.org/rowca

UN/ISDR. (2002). *Living with risk: A global review of disaster reduction initiatives. Preliminary version prepared as an inter-agency effort co-ordinated by the*. Geneva, Switzerland: ISDR Secretariat.

UNCHA. (2012). *Floods Situation Report* (Vol. 2). UN Office for the Coordination of Humanitarian Affairs.

UNDP. (2006). *Niger Delta Human Development Report Garki*. Abuja: UN House.

UNDP. (2006). *Niger Delta Human Development Report*. Oxford Press.

UNDP-UNEP. (2009). *Mainstreaming Poverty-Environment Linkages into Development Planning: A Handbook for Practitioners*. Available online at www.unpei.org

UNEP (United Nations Environment Programme). (2011). Environmental Assessment of Ogoniland. UNEP Report.

UNEP. (2002). *Global Environmental Outlook 3 – Past, present and future perspectives. Synthesis Report. UNEP, Earthscan United Nations Environment Programme*. Cambridge, UK: UNEPWCMC.

UNEP. (2011). *United Nations Environmental Programme Report 2011*. New York: United Nations.

UNHCR. (2016). *Nigeria Humanitarian Update on the North East*. UNHCR.

UNISDR. (2009). *Terminology on disaster risk reduction. United Nations international strategy for disaster risk reduction*. Retrieved from http://www.nat-hazards-earth-syst-sci.net/13/2707/2013/nhess-13-2707-2013.pdf

UNISDR. (2009a). *Risk and Power in a Changing Climate: Invest today for a safer tomorrow. United Nations International Strategy for Disaster Reduction*. UNIDR.

UNISDR. (2013a). *Global Assessment report on Disaster Risk Reduction: From shared risk to shared value: the business case for disaster risk reduction*. United Nations Offices.

United Nations Development Programme (UNDP). (2004). Reducing Disaster Risk – A challenge for development – A Global Report. New York: UNDP.

United Nations Development Programme (UNDP). (2010). *Climate Change: the Zambian Story*. New York: UNDP News Report.

United Nations Development Programme. (1994). *Human Development Report: Redefining Security*. Oxford, UK: Human Dimensions.

United Nations Framework Convention on Climate Change. (1994). *UNFCCC Report*. Author.

United Nations High Commission on Refugees. (2014, September 2). *Number of displaced inside Ukraine more than doubles since early August to 260,000*. United Nations High Commission on Refugees. Retrieved from http://www.unhcr.org/540590ae9.html

United Nations Human Settlements Programme. (2010). *State of the World's Cities, 2010/2011*. Nairobi, Kenya: UN-Habitat.

United Nations Humanitarian Commissioner for Refugees (UNHCR). (2015). *Internally Displaced Persons Matrix in the North East*. Nigeria: UNHCR.

United Nations. (1991). Mitigating Natural Disasters: Phenomena, Effects, and Options: a Manual for Policy Makers and Planners. New York: UN.

United Nations. (2006). *The Millennium Development Goals Report*. Available online: http://unstats.un.org/unsd/mdg/Resources/Static/Products/Progress2006/MDGReport2006 .pdf

United Nations. (2015). *Getting started with sustainable development goals: a guide for stakeholders*. New York: United Nations.

United Nations. (2015). *Strengthening resilience to natural disasters crucial, speakers stress as second committee takes up sustainable development*. Seventieth Session, 13th& 14th Meetings. Retrieved from http://www.un.org/press/en/2015/gaef3427.doc.htm

United Nations. (2016). *The sustainable development goals report*. New York: United Nations.

United Nations. (2017). *Nigeria-North-East: Humanitarian Emergency Report No.12, 31st May, 2017*. Author.

UNU-IHDP & UNEP. (2012). *Inclusive Wealth Report 2012.Measuring progress toward Sustainability*. Cambridge, UK: Cambridge University Press.

ur Rahman, M., Rahman, S., Mansoor, S., Deep, V., & Aashkaar, M. (2016). Implementation of ICT and Wireless Sensor Networks for Earthquake Alert and Disaster Management in Earthquake Prone Areas. *Procedia Computer Science, 85*, 92-99.

Uroko, C. (2016, September 27). Nigeria joins world leaders on global efforts to reverse climate change impacts. *The Businessday*, p. 36. Retrieved from http://www.businessdayonline.com

Uwalomwa, U., & Uadiale, O. M. (2011). Corporate social and environmental disclosure in Nigeria: Comparative study of the building materials and brewery industry. *International Journal of Business*, *6*(2), 258–264.

Valiela, I. (2006). *Global Coastal Change. Science Singapore*. Blackwell.

van Breugel, L. (2013). *Community-based tourism: Local participation and perceived impacts - A comparative study between two communities in Thailand Research Master Social and Cultural Science*. Faculty of Social Sciences Radboud University Nijmegen.

Van Maanen, J. (1983). *Qualitative Method*. London: Sage.

Vanclay, Esteves, Aucamp, & Franks. (2015). *Social Impact Assessment: Guidance for Assessing and Managing the Social Impacts of Projects*. International Association for Impact Assessment. Retrieved from https://www.iaia.org/uploads/pdf/SIA_Guidance_Document_IAIA.pdf

Vecchio, R. P., Justin, J. E., & Peace, C. L. (2008). The utility of transactional and transformational leadership for predicting performance and satisfaction with a path-goal theory framework. *Journal of Occupational and Organizational Psychology*, *81*(1), 71–82. doi:10.1348/096317907X202482

Veloso-Gomes, F. (2006). Coastal Erosion and Spatial Planning in Costa de Caparica. *Journal of Integrated Coastal Management*, *7*, 83–95.

Vigoda-Gadot, E., & Mizrahi, S. (2008). Public Sector Management and the democratic ethos: A 5 – year study of key relationship in Israel. *Journal of Public Administration: Research and Theory*, *18*(1), 79–108.

Villagran de Leon, J. C. (2006). Vulnerability – A conceptual and methodological review. Studies Of the University: Research Council Education, 4.

Villarica, H. (2011). *Maslow: A New and Improved Recipe for Happiness*. Retrieved June 21, 2017 from www.thecable.ng

Voumik, L. C. (2014). A green Economy in the Context of Sustainable Development and Poverty Eradication: What are the Implications for Bangladesh? *Journal of Economics and Sustainable Development*, *5*(3), 119–131.

Wang, D., & Ni, Y. (2012). Wireless sensor networks for earthquake early warning systems of railway lines. In *Proceedings of the 1st International Workshop on High-Speed and Intercity Railways* (pp. 417-426). Springer Berlin Heidelberg. doi:10.1007/978-3-642-27963-8_38

Wangu, G., Nyariki, D., & Sakwa, M. (2015). The Role of Indigenous Knowledge in Socio-Economic Development. *International Journal of Science and Research*, *4*(5), 32–37.

Wang, X., & Van Wart, M. (2007). When Public participation in administration leads to trust: An Empirical Assessment of manager's perceptions. *Public Administration Review*, *67*(2), 265–278. doi:10.1111/j.1540-6210.2007.00712.x

Wart, M. V. (2011). Changing Dynamics of Administrative leadership. In *In the State of Public Administration: Issues, Challenges and Opportunities*. New York: M. E. Sharpe.

Wasilwa, C. (2015). *Effect of Community Participation on Sustainability of Community Based Devt. Projects in Kenya*. Retrieved from https://www.linkedin.com/pulse/effect-community-participation-sustainability-based-calebwasilwa Retrieved 5/3/2017

Watson, P. J., Friedman, M. J., Gibson, L. E., Ruzek, J. I., Norris, F. H., & Richie, E. C. (2003). Early intervention for trauma-related problems. In R. J. Uesano & A. E. Norwood (Eds.), *Trauma and disaster: response and management*. Washington, DC: American Psychiatric Publishing.

WCED. (1987). *Global Policies for Sustainable Agriculture, a Report of the Advisory Panel on Food Security, Agriculture, Forestry and Environment to the World Commission on Environment and Development*. London: Zed Books.

Wehrmann, B. (2011). *Land use planning: concept, tools and applications*. Deutsche Gesellschaft für Internationale Zusammenarbeit (GIZ) GmbH, Division Agriculture, Fisheries and Food Sector Project Land Policy and Land Management, Eschborn/Germany.

Wehrmann, B. (2011). *Land use planning: concept, tools and applications. Deutsche Gesellschaft für Internationale Zusammenarbeit (GIZ)*. GmbH, Division Agriculture, Fisheries and Food Sector Project Land Policy and Land Management.

Welch, J., Rimes, H., & Bozeman, B. (2014). Public Value Mapping. In Valuing Public value. Minneapolis, MN: University of Minnesota.

Welp, M., & Stoll-Kleemann, S. (2006). Integrative Theory of Reflexive Dialogues. In S. Stoll-Kleemann & M. Welp (Eds.), *Stakeholder Dialogues in Natural Resources Management* (pp. 43–78). New York: Springer. doi:10.1007/978-3-540-36917-2_3

Westing, A. (Ed.). (1986a). *Global Resources and International Conflict: Environmental Factors in Strategic Policy and Action*. Oxford, UK: Oxford University Press.

White, G. F. (1964). *Choice of Adjustment to Floods*. Research paper. Department of Geography, University of Chicago.

White, P., Pelling, M., Sen, K., Seddon, D., & Russell, S. (2005). *Disaster risk reduction. A Development Concern*. DFID.

Wild, A. (2003). *Soils, Land and Food: Managing the Land During the Twenty -First Century*. Cambridge, UK: Cambridge University Press. doi:10.1017/CBO9780511815577

Williams, M. A. (2016). *A Whole Community Approach to Emergency Management for the United States Virgin Islands* (Unpublished doctoral dissertation). Walden University.

Williams, P. L. (2010). Ice-free Arctic Ocean may not be of much use in soaking up carbon dioxide, a component of global warming according to new study. Athens, GA: Office of Public Affairs, University of Georgia.

Williams, A. (1992). Nigerian environment and national health policy. In A. Williams (Ed.), *Towards industrial pollution Abatement in Nigeria* (Vol. 1). Lagos: FEPA.

Williams, L., & Shearer, H. (2011). Appraising Public Value: Past, Present and Future. *Public Administration Review, 89*(4), 1367–1384. doi:10.1111/j.1467-9299.2011.01942.x

Wilson, E. J., Friedmann, S. J. I., & Pollak, M. F. (2007). Research for deployment: Incorporating risk, regulation, and liability for carbon capture and sequestration. *Environmental Science & Technology, 41*(17), 5955–5952. doi:10.1021/es062272t PMID:17937265

Wilson, J. O. (1974). *Political Organizations*. Princeton, NJ: Princeton University Press.

World Bank. (1986). *Poverty and Hunger: Issues and Options for Food Security in Developing countries*. World Bank Publications.

World Bank. (1995). *Defining an environmental development strategy for the Niger* (Vol. 2). World Bank.

World Bank. (2005). *Agriculture Investment Sourcebook*. Washington, DC: World Bank.

World Bank. (2011). *Gender and climate Change: 3 things you should know*. Washington, DC: World Bank Group.

World Bank. (2013). *The World Bank: Working for a world free of poverty, Population (Total)*. Washington, DC: World Bank Group.

World Bank. (2017). *Doing Business 2017*. Retrieved March 14th, 2017,www.doingbusiness.org/media/WBG/Doing-Business/Documents/AnnualReport

World Commission on Environment and Development (WCED). (1987). *Our Common Future*. Oxford Press.

World Health Organization. (1981). Constitution of the World Health Organization. In H. E. Caplan Jr & J. McCartney (Eds.), *Concepts of Health and Disease – Interdisciplinary Perspectives*. Reading: Addison-Wesley.

World Population Review. (2017). *Lagos Population, 2017*. Retrieved March 11th, 2017, worldpopulationreview.com/world-cities/lagos/population.

Wyatt, R.J. (2016). *Encouraging a Neighborhood contingency for America's First Responders*. Retrieved in http//patimes.org/encouraging.neighborhood_contingencypolicy-america

Yavaş, H. (2001). Doğal Afet Yönetimi ve Yerel Gündem 21 Çalışmaları Kapsamında İzmir'de Deprem Riski. *Dokuz Eylül Üniversitesi Sosyal Bilimler Enstitüsü Dergisi, 3*(3), 118–138.

Yılmaz, E. (2002). Integrated Natural Resource Management and Its Workshop. *Eastern Mediterranean Forestry Research Institute Journal of DOA, 8*, 1-30.

Yılmaz, A., Bozkurt, Y., & Taşkın, E. (2005). Doğal Kaynakların Korunmasında Çevre Yönetiminin Etkinliği. *Dumlupınar Üniversitesi Sosyal Bilimler Dergisi, 13*(1), 15–30.

Youssef, M., Yousif, A., El-Sheimy, N., & Noureldin, A. (2007, April). A novel Earthquake warning system based on virtual MIMO-wireless sensor networks. In *Electrical and Computer Engineering, 2007. CCECE 2007. Canadian Conference on* (pp. 932-935). IEEE. doi:10.1109/CCECE.2007.238

Zapałowski, A. (2016). *Gra wokół ochrony południowo-wschodniej granicy Polski* [Play around the defense of the south-eastern border of Poland]. European Center of Geopolitical Analysis. Retrieved from http://www.geopolityka.org/analizy/andrzej-zapalowski-gra-wokol-ochrony-poludniowo-wschodniej-granicy-polski

Zdanowicz, M. (2009). *Polska w Shengen* [Poland in Shengen]. Białystok, Poland: Temida 2.

Zdziarski, E. L. (2006). *Crisis in the context of higher education. Crisis management: Responding from the heart.* Academic Press.

Zeeuw, A. de (2000). *Resource Management: Do we need public policy?* Retrieved from http://ec.europa.eu/environment/enveco/waste/pdf/zeeuw.pdf

Zennaro, M., Pehrson, B., & Bagula, A. (2008, October). Wireless Sensor Networks: a great opportunity for researchers in Developing Countries. *Proceedings of WCITD2008 Conference*, 67.

Zenter, H. (1964). The State and the Community. *Sociology and Social Research*, *48*, 414.

Zielinski, J. (2015). *Interpelacja poselska w sprawie zabezpieczenia polskich granic przez Straż Graniczną oraz braku wyposażenia funkcjonariuszy Straży Granicznej we właściwe elementy uzbrojenia* [Appeal about securing Polish borders by Border Guard and lack of proper arming of BG]. Directed to Polish Senate, Warsaw, Poland. Retrieved from http://ciechanowiec.info/downloads_2/IT%20227.76%20Straz%20Graniczna.pdf

Zimmerman, M. A., & Brenner, A. B. (2010). Resilience in adolescence: Overcoming neighbourhood disadvantages. In J. Reich, A. J. Zautra, & J. S. Hall (Eds.), *Handbook of Adult Resilience* (pp. 283–308). New York, NY: Guildford Press.

About the Contributors

Augustine Nduka Eneanya (Ph.D.) is a Senior Lecturer in the Department of Political Science, University of Lagos, Yaba, Lagos, Nigeria. He has worked in Public Sector for 16 years and private sector for 12 years and advanced to the post of General Manager. He also worked as the Managing Consultant of a Human Resources Consulting Firm in Lagos, Nigeria. He is a member of many professional associations, among which are America Society for Public Administration (ASPA), Chartered Institute of Personnel Management (CIPM), Nigerian Institute of Management (NIM) and Nigerian Institute of Management Consultants. He is the author of several books, and published scholarly papers in local and international journals. He was awarded "Distinguished Scholarship Award" by Zambia Society of Public Administration, Lusaka, Zambia on December 6, 2016.

* * *

Waziri B. Adisa lectures criminology, theory and urban governance in the Department of Sociology, Faculty of Social Sciences, University of Lagos, Akoka, Lagos.

Ighodalo Akhakpe is an Associate Professor in the Department of Public Administration, Lagos State University, LASU. He obtained his B.Sc, M.Sc and Ph.d. degrees from the Department of Political Science, University of Ibadan, Nigeria. He has published many books including: Leading Issues in Political Thought (1999); Management and Administration of Public Enterprises in (2005); Bureaucracy and Good Governance in Nigeria (2014); The Dynamics of Public Sector Management: A Reader (Eds). Dr Akhakpe has published extensively in national and international journals on Public Policy and Development Studies.

Henry Alapiki is a Professor of Political science with specialization in Development studies, Political Economy and Comparative Politics. Professor Alapiki has published widely in National and International journals and served as a resource person and lead researcher for notable international organizations on contemporary Development challenges in Nigeria. He was Dean, Faculty of Social Sciences, University of PortHarcourt and currently Executive Director of University of PortHarcourt Foundation.

Luke Amadi is a Senior Research Fellow at the Educational Support and Development Initiatives for the Less Privileged(ESDIL).He has several publications in scholarly journals.

Franca Attoh, PhD, is a Senior Lecturer in the Dept. of Sociology, University of Lagos, Nigeria. She is a sociologist with specialization in Criminology and Gender Studies. She has interests also in political sociology and development studies. She has contributed over twenty publications as chapters in books and learned journals. In addition, she has carried out numerous researches in the area of trafficking in persons for the International Organization on Migration (IOM) and the National Agency for the Prohibition of Trafficking in Persons and other related matters (NAPTIP). Dr. Attoh was involved in several research works for Fredrich Ebert Stiftung and Independent National Electoral Commission (INEC) especially prior to the elections of 2011. Dr. won a Tertiary Education Trust Fund (TETFUND) in 2015 to carry out a research on the Spatial Distribution of Crime in the University of Lagos. Dr. Franca Attoh, was picked by the University of Texas in Austin to represent Africa on a Global Research on Christian Marriages. In addition, she was the country representative for WageIndicator Foundation in Amsterdam Netherlands for a research on cost of living in Lagos, Nigeria. She consults for the Nigeria Institute of Security (NIS), the Nigeria Police Force (NPF) and the ECOWAS Early Warning Directorate. Dr. Attoh has participated in numerous conferences both locally and internationally. She is a fellow of the Institute of Security Nigeria. And currently the Sub-Dean Faculty of Social Sciences, University of Lagos, Nigeria.

Amidu Owolabi Ayeni research focus is on "Water resources, Climate change studies, and Geo-information sciences". He has attended various scientific training institutes/centers, including ICTP, Trieste – Italy (2009) and Council for Scientific and Industrial Research (CSIR), Pretoria, South Africa in 2011/2012. Presently, is a Senior Lecturer at the University of Lagos, Nigeria. He has published several articles in reputable journals and member of different scientific bodies including African Association of Remote Sensing of the Environment (AARSE) and International Association of Hydrological Science (IAHS).

Magdalena Denham is an Assistant Professor at the College of Criminal Justice, Department of Security Studies, Sam Houston State University. Her research interests include emergency management, community resilience, and comparative aspects in Homeland Security Studies. She has published in Oxford Bibliographies, Journal of School Violence, Law Enforcement Executive Forum, International Journal of Qualitative Research. She is an engaged scholar and supports service-learning in pedagogy and andragogy.

Austin Eneanya graduated in Systems Engineering of University of Lagos with B.Sc (Hons) and currently doing MBA Program in Department of Business Administration, University of Lagos, Yaba, Lagos, Nigeria. Eneanya is proficient in project management, graphic design and is a Consultant of Small and Medium Enterprises in Nigeria. He has organized series of training ranging from learning and development on Human Resources, small business feasibility studies and entrepreneurial development.

Mayowa Fasona is a Senior Lecturer in Environmental change and natural resource management.

Pedro Gomes holds a Msc in Environmental Engineering from the Faculty of Sciences and Technology - New University of Lisbon. Currently working in the Department of Economic Statistics - Enviroment and Agriculture at Statistics Portugal where the main tasks consist in collecting information internally or by external sources in order to prepare and analyse statistics on environment and economics applied to companies in order to provide information with National and International responsibilities. His work is highlighted in the contributions for publications developed in Statistics Portugal regarding Environmental Statistics and Environmental Management and Protection in Industrial Enterprises. He is an element of the Working Group "2030 Agenda for Sustainable Development Goals". He is the author of four chapters for the SPRINGER book "BEACH MANAGEMENT TOOLS: Concepts, Methodologies and Case Studies" edited by Camilo M. Botero, Omar D. Cervantes and Charles W. Finkl.

Francisco Sacramento Gutierres (Male) is a researcher in the Big Data Analytics group at Eurecat. Mr Gutierres holds a PhD in Physical Geography by the Institute of Geography and Spatial Planning of the University of Lisbon (IGOT-UL), MSc in Conservation Biology, Post-graduation in GIS Applied to Sciences of the Earth by the Faculty of Sciences of University of Lisbon and a degree in Biophysical Engineering – Environmental Planning and Management by University of Évora. Before joining Eurecat he participated in different European and National projects such as LUCAS, a pilot study on the provision of harmonized land use/cover statistics by means of data integration through Geographical Information Technology (Synergies between LUCAS and the national systems); MYGEOSS initiative, an European expert research contract for Innovative applications in the environmental and social domains promoted by the European Commission (EC), and EUREGIO NL-Germany – monitoring impacts of biodiversity conservation policies in a highly fragmented landscape across international borders. He has been involved in Geographic Information Systems (GIS), Remote Sensing (RS), GIS Cloud computing, Predictive modelling, statistics, and monitoring and modelling of habitats of the Natura 2000 Network (EU Habitats Directive) and assessment of Land Use and Land Cover (LULC) changes in his previous positions in Institute of Geography and Spatial Planning of the University of Lisbon (IGOT-UL), National Statistics Institute(NSI)/Eurostat, and Faculty of Geo-Information Science and Earth Observation (ITC) – University of Twente. Francisco Gutierres has contributed to over 50 publications (including 6 book chapters), 1 patent application (Geoapp ʽMohana - MOnitoring of HAbitats at the level of a NAtura 2000 siteʼ), and serves as Teaching Assistant and Ad hoc reviewer of 6 international scientific journals.

Urooj Afshan Jabeen holds Ph.D in Economics and working as Assistant Professor of Economics at Government Degree and Postgraduate College for Women, Nalgonda, India. She is the author of research papers in Economics published in international journals.

Geetanjali Kaushik is armed with a Masters from TERI and a BSc from University of Delhi (Silver and Golden Jubilee Awardee). Dr. Geetanjali Kaushik completed her PhD in Environmental Sciences, from the prestigious Indian Institute of Technology Delhi with 45 papers in International Journals of repute. She has participated in International Conferences held in Germany (won first prize) and USA (received American Society of Nutrition travel grant). Subsequent to her PhD she has also worked with UC Berkley on an air quality monitoring project for over 40 locations across Delhi. She has also edited 4 International books. Her taught assignments include stints as Assistant Professor with the University of Petroleum and Energy Studies, Dehradun and Amity University, Noida. In 2012 Dr. Kaushik received full scholarship from School of Management, Swansea University UK to undertake MBA. She topped the MBA class and has worked in UK and Turkey in various positions. She returned to India in 2015 and worked as Associate Professor in Environmental Management before joining Clean Air Asia. In Clean Air Asia as an Air Quality Specialist she has worked on creating Air Quality Profiles for 30 Indian cities with the application of the ADB's Clean Air Scorecard Tool.

Bolajoko Ibiyemi Malomo (Ph.D) is a Personality and Social Psychologist and lectures in the Department of Psychology, University of Lagos, Nigeria. Her area of interest is Behaviour Modification and this is reflected in her various research studies on Road User Behaviour and Public Health. She is a member of The Nigerian Psychological Association (NPA), Consortium of Adolescent Road Safety (CADROSA) and Association of Psychological Science (APS). She has attended and presented papers in several local and international conferences and also published scholarly papers in local and international journals.

Ankita Mohanty is pursuing B.Tech in Computer Science and Engineering at Indira Gandhi Institute of Technology (IGIT), Sarang, Odisha, INDIA. She has completed her (10+2) in science from St. Joseph's Girls' High School and Stewart Science College respectively with special paper in Computer Science and Information Technology. She believes in importance of cultural heritage. Along with excel in studies, she loves writing articles, poems, short stories. She received Jr. Scholarship in Odishi Dance from CCRT (Centre for Cultural Research and Training). She is graduated in Odishi and Kathak dance and a trained singer. She has co-authored an international journal publication in 2015. She has attended technical seminars and workshops at national levels. She has worked as a student organizer and volunteer in many technical events at both college and UG level. She has successfully executed her technical assignments as student intern with different reputed organizations. She is a student member with CSI at CSI IGIT student chapter.

G. S. Mmaduabuchi Okeke is a Senior Lecturer in the Department of Political Science, University of Lagos, Nigeria. He specializes in International Relations and African Government and Politics. He has traveled widely and published in many international and local journals, with many chapter contributions in books. He is currently working on a co-edited book, Nigerian Politics: Issues and Perspectives.

Alok Ranjan Prusty, presently working as an Assistant Professor in School of Computing Science & Engineering at prestigious Vellore Institute of Technology, VIT Bhopal University, (M.P.), India. Previously, he was working as Assistant Professor at Indira Gandhi Institute of Technology(IGIT), Sarang (An Autonomous Institute of Government of Odisha). Dr. Prusty has received both M.Tech and B.Tech degree in Computer Science & Engineering under Biju Patnaik University of Technology, Odisha. He has received his Ph.D. in Computer Science & Engineering from Siksha 'O' Anusandhan University, India. He was involved in SERB, DST Govt. of India research project during his Ph.D. research. He has excelled his research activities with research papers published in reputed Web of Science and Scopus indexed International journals, IEEE Conferences at IIT and other institutes of national repute. He has contributed research chapters in reputed international book volumes. He has authored international book "Sustainable Green Business Concepts and Practices for Startups". He has attended, organized and presented research papers in several international and national conferences, seminars, workshops and FDPs. He has guided number of students of Masters and Bachelors in their thesis/ projects / research assignments and Seminars. Dr. Prusty is a reviewer in distinguished research forums including IEEE Transactions, IGI Global, SAGE, Hindawi, Bentham, etc. and invited speaker and member to various scientific events. He is an Editor of SIGIE Magazine. His area of interest includes Sensor Networks, IoT, Green IT, Energy Complexity, Routing, etc.He is Life Member with CSI, SIG-IE, AIE(India), etc. Apart from these, Dr. Prusty has also bagged many awards and recognition at state and central level.

Ali Syed holds Ph.D in Economics from Osmania University, India. At present he is working as Associate Professor of Economics at Mulungushi University in Zambia. Prof. Ali was the Executive Dean of the School of Social Sciences from 2009 to 2013 in this University. He has more than 30 years of teaching and research experience in Economics and published many research papers in reputed journals of international level. He also served as United Nations Economist in the Republic of Kyrghyzstan during 1996-97, taught and trained the civil servants of the Ministry of Finance.

Scott Vautrain holds a Master's degree in Homeland Security Studies. Currently, Mr. Vautrain is a Security professional for the largest healthcare system in the larger Houston Metropolitan area. He has conducted graduate research on various homeland and national security topics as well as failed states.

Index

www.ingramcontent.com/pod-product-compliance
Lightning Source LLC
LaVergne TN
LVHW062115300926
849161LV00010B/2466
* 9 7 8 1 5 2 2 5 3 1 9 4 4 *